S0-BIY-808

THE GULF AND THE WEST

THE GULF AND THE WEST

Strategic Relations and Military Realities

Anthony H. Cordesman

Westview Press • Boulder

Mansell Publishing Limited • London

Published in 1988 in the United States of America by Westview Press, Inc.,
5500 Central Avenue, Boulder, Colorado 80301

Published in 1988 in Great Britain by Mansell Publishing Limited,
A Cassell Imprint, Artillery House, Artillery Row, London SW1P 1RT, England

Library of Congress Cataloging-in-Publication Data
Cordesman, Anthony H.
The Gulf and the West : strategic relations and military realities / Anthony H. Cordesman.
p. cm.
Bibliography: p.
Includes index.
ISBN 0-8133-0768-6
1. Persian Gulf—Strategic aspects. 2. Persian Gulf Region—Defenses. 3. Iraqi-Iranian Conflict, 1980– . 4. United States—Military relations—Persian Gulf Region. 5. Europe—Military relations—Persian Gulf Region. 6. Persian Gulf Region—Military relations—United States. 7. Persian Gulf Region—Military relations—Europe. I. Title.
UA832.C67 1988 88-10752
355'.033053—dc19 CIP

British Library Cataloguing in Publication Data
Cordesman, Anthony H.
The Gulf and the West : strategic relations and military realities.
1. Western bloc countries. International relations with Persian Gulf countries.
2. Persian Gulf countries. International relations with western bloc countries
I. Title
303.4'82'1713'053
ISBN 0-7201-1990-1

Printed and bound in the United States of America

The paper used in this publication meets the requirements of the American National Standard for Permanence of Paper for Printed Library Materials Z39.48-1984.

10 9 8 7 6 5 4 3 2 1

To friends whom I cannot thank by name—
in and outside government in the Middle East
and Gulf states and in the Department of Defense
and U.S. intelligence community

Contents

Tables and Maps

Tables

Maps

Preface

The last two years (mid-1986 to mid-1988) have transformed the security problems in the Southern Gulf from theoretical threats to the West's main source of imported oil to a major military confrontation between the West and Iran. At this writing, there is no way to predict how this confrontation will end, any more than it is possible to predict the end of the Iran-Iraq War. It is clear, however, that virtually any outcome will leave a strong anti-Western regime in Iran, and an Iran that will remain a threat to all its neighbors.

It is equally clear that the problems in securing the safety of the friendly Southern Gulf states will grow, rather than diminish. The West needs to find solutions that can build up its military capabilities and needs to develop an effective power projection capability to deal with the threat from Iran and other radical states and with Soviet ambitions in the region.

The focus of this book is the problems the West and the Gulf states face in developing the kind of strategic relations that can deal with these threats and with the opportunities for cooperation. It examines the military capabilities of various potential threats, the capabilities of Saudi Arabia and other friendly Gulf states, and the capabilities of Western power projection forces. It examines the factors that had blocked the development of effective strategic relations in recent years, and how U.S. intervention in the Gulf has affected this situation. Finally, it examines the steps both the West and the GCC states will have to take to create the proper kind of strategic relations and military capabilities in the future.

This book draws heavily on the author's previous writings on the Gulf and the Middle East. Some portions are adapted from an earlier work, *Western Strategic Interests in Saudi Arabia,* which was published by Croom Helm in London, but which saw only limited circulation due to the problems of various mergers in the British and American publishing world.

For those who are interested in the broader history of military developments in the region, this book supplements *The Gulf and the*

Search for Strategic Stability, Westview, 1984. A detailed account of the Iran-Iraq War can be found in *The Iran-Iraq War, 1984–1987*, Jane's, 1987. A detailed analysis of the Arab-Israeli military balance can be found in *The Arab-Israeli Military Balance and the Art of Operations*, University Press of America/AEI, 1986.

The reader should also be aware that this book is based on a wide range of interviews during the author's trips to the Gulf area and in the U.S. and Europe. At the same time, it draws heavily on computer data bases and news services. These sources cannot be referenced in depth, either because of the political sensitivity of the interviews or because computerized data bases are constantly in flux and do not lend themselves to precise references. In order to ease the problems other researchers may find in referencing source materials, alternative media sources are footnoted as references wherever possible.

Anthony H. Cordesman

Map 1 The Gulf

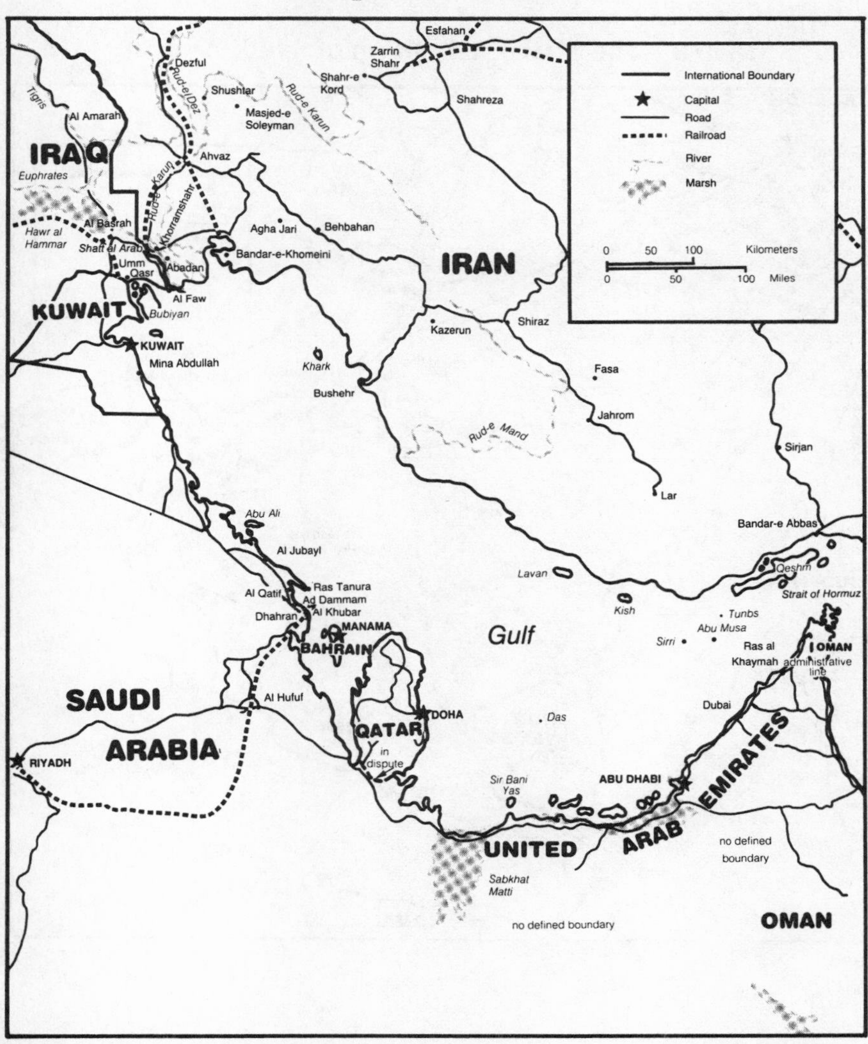

Adapted from CIA 504822 (546740) 7-81

Map 2 The Strategic Position of Saudi Arabia

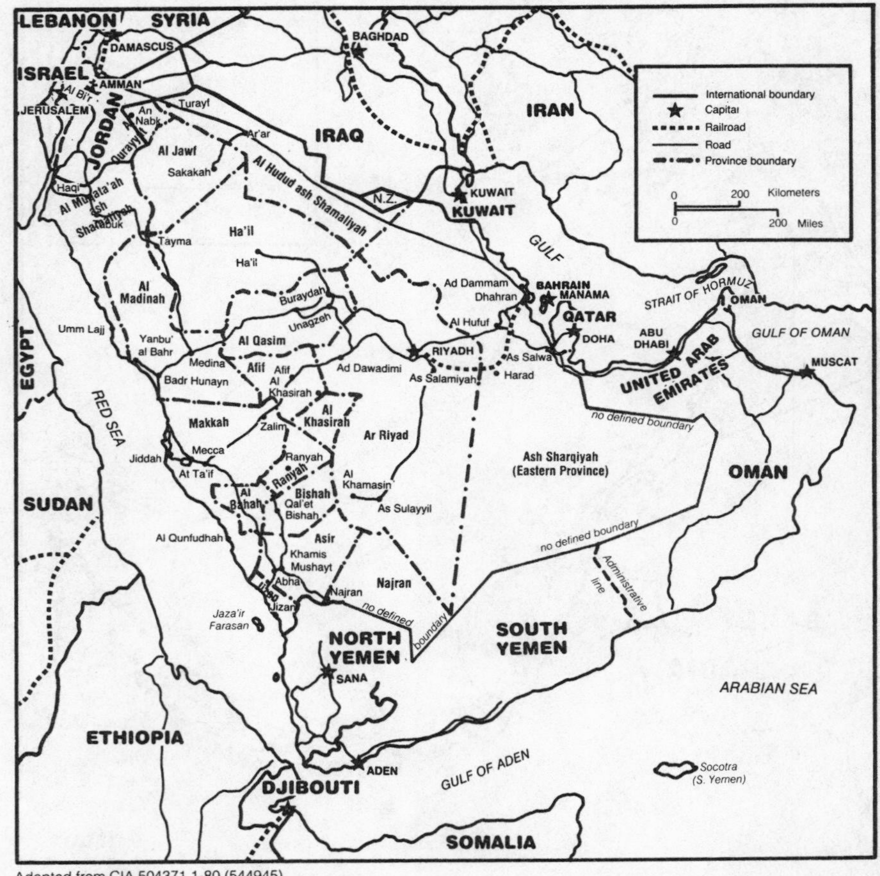

Adapted from CIA 504371 1-80 (544945)

Map 3 United Arab Emirates

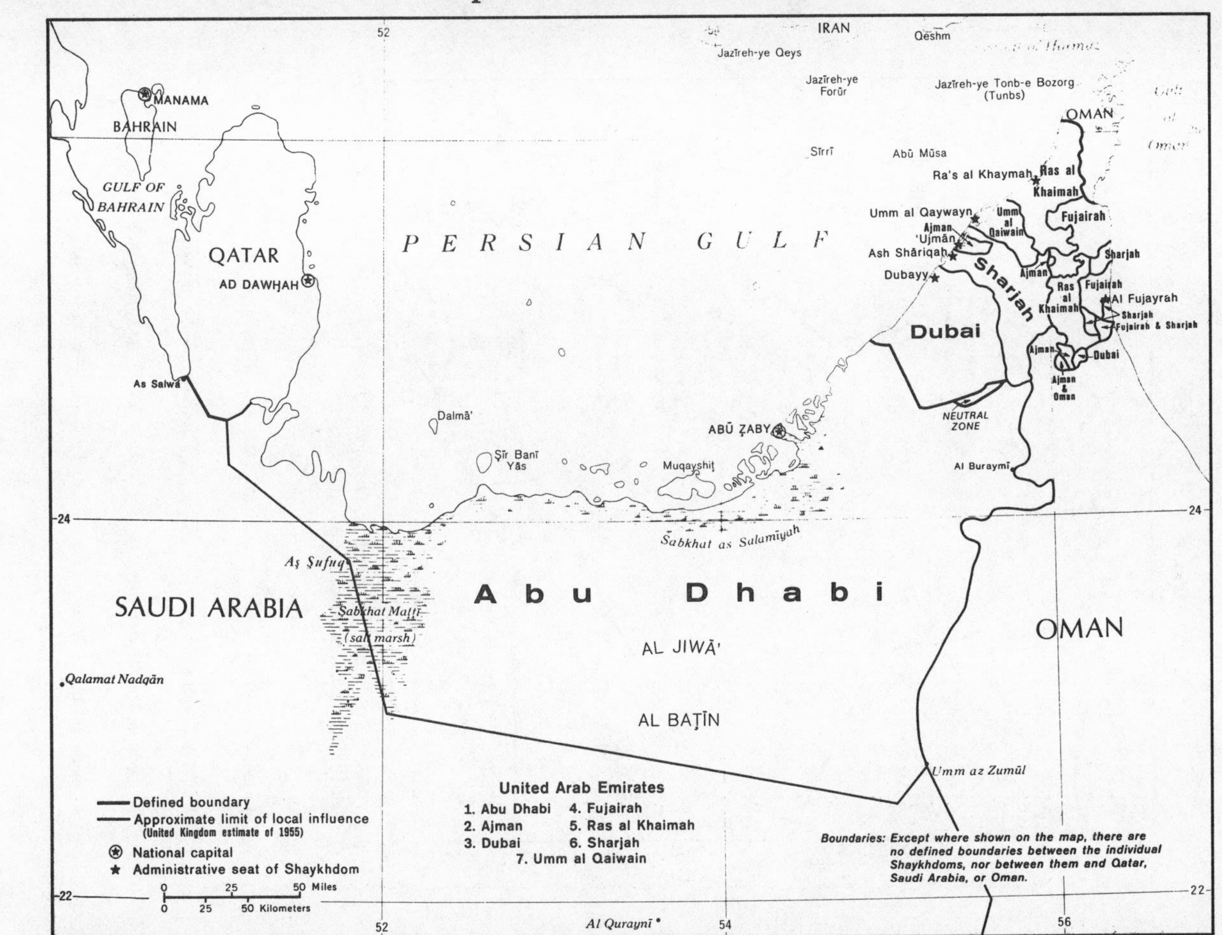

Map 4 Iraq

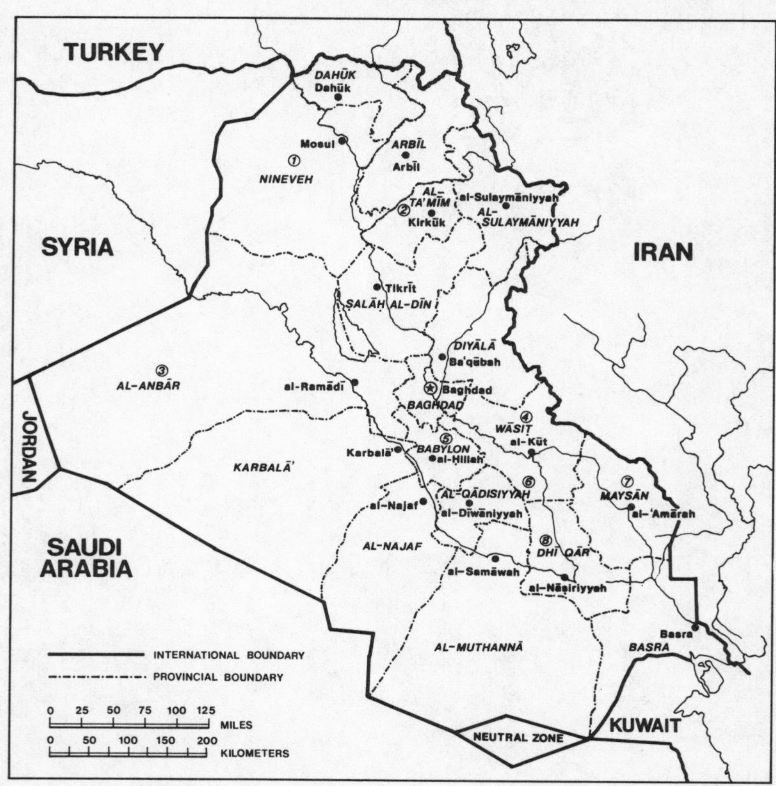

Source: Phebe Marr, *The Modern History of Iraq* (Boulder, Colo.: Westview Press, 1985), p. 30. Reprinted by permission.

① formerly Mosul

② formerly Kirkūk

③ formerly al-Ramādī and al-Dulaym

④ formerly al-Kūt

⑤ formerly al-Ḥillah

⑥ formerly al-Dīwāniyyah

⑦ formerly al-ʿAmārah

⑧ formerly al-Nāṣiriyyah and al-Muntafiq

Map 5 Iran

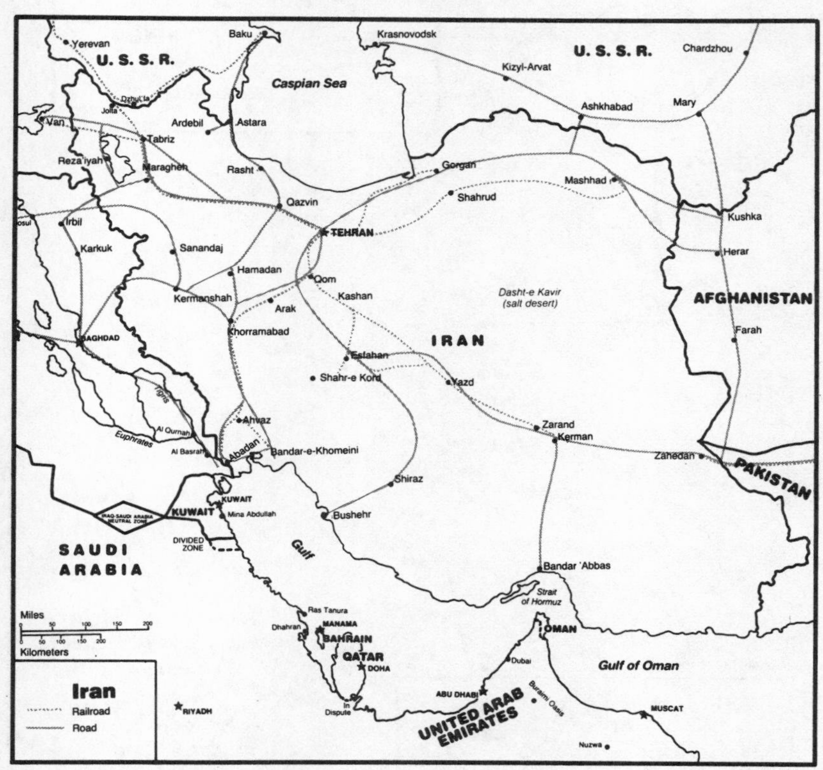

SOURCE: CIA 504190 (544499) 7-79

Map 6 The Red Sea

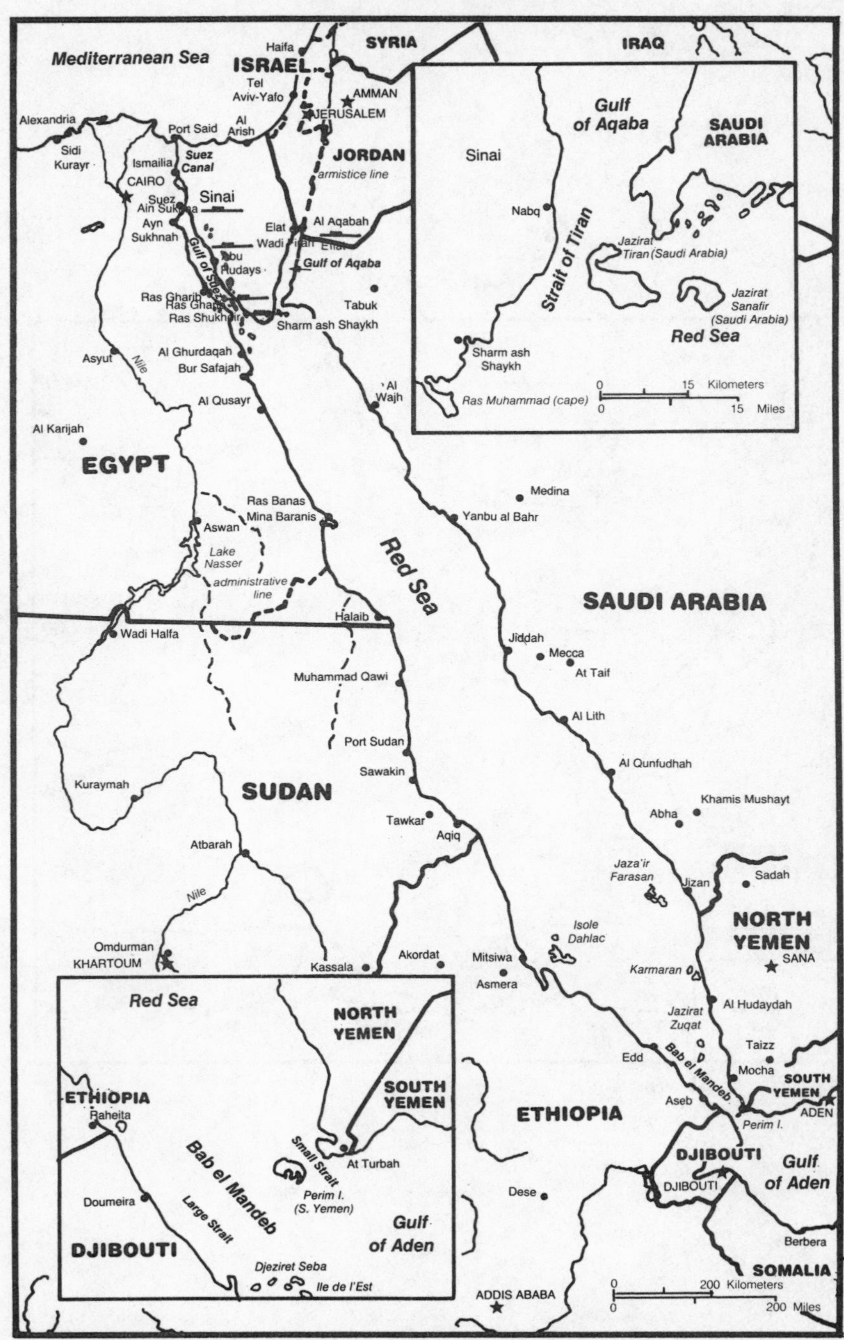

Adapted from Defense Mapping Service 504306 9-79 (541978)

Map 7 The Strategic Position of Bahrain

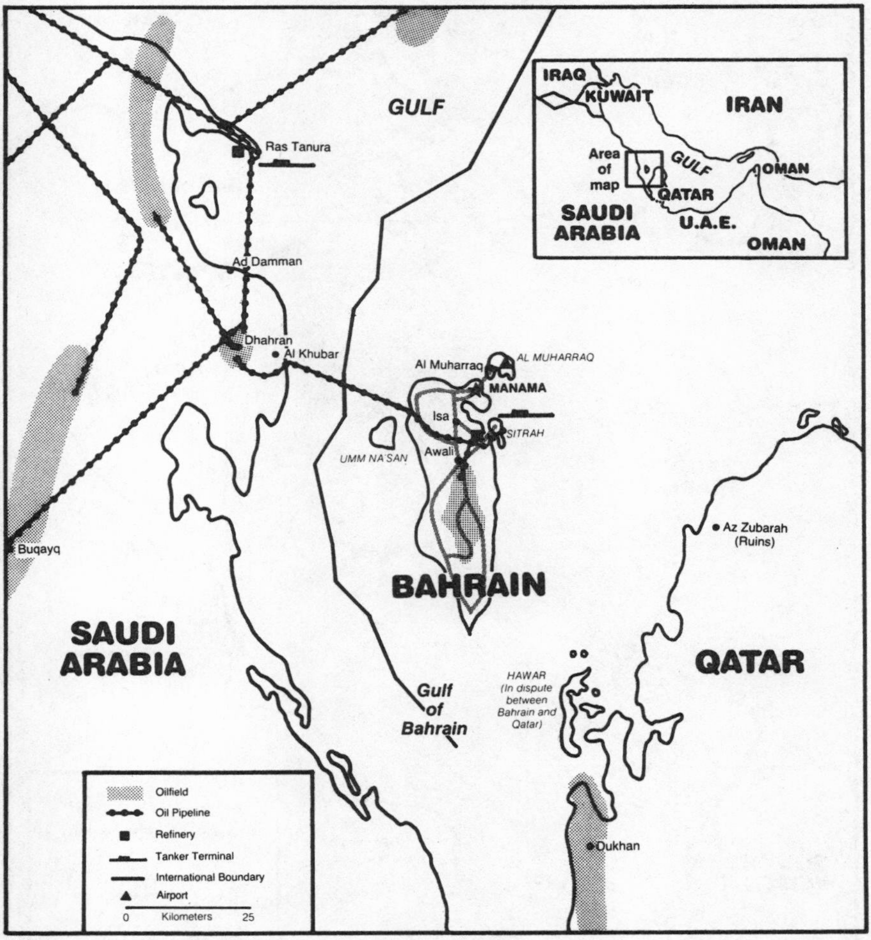

Adapted from CIA 626961 4-80

Map 8 Oman

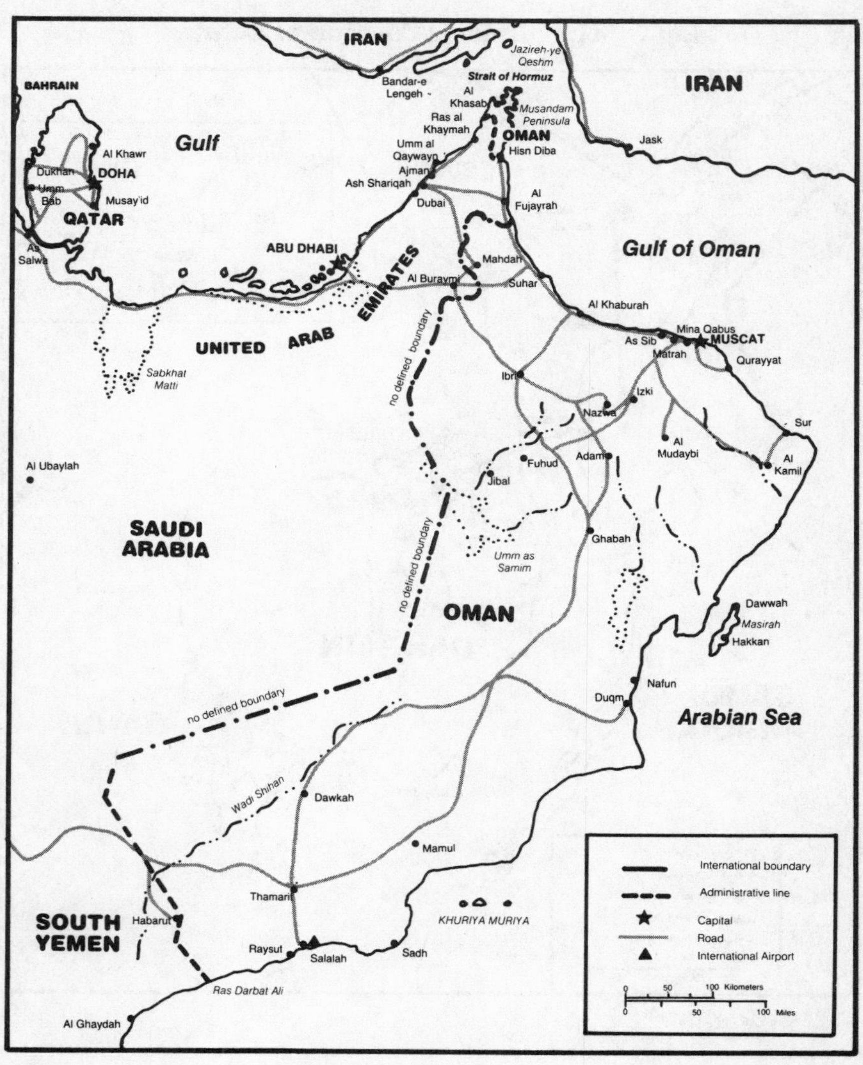
IRAN
Jazireh-ye Qeshm
Strait of Hormuz
Bandar-e Lengeh
IRAN
BAHRAIN
Al Khasab
Musandam Peninsula
Ras al Khaymah
OMAN
Jask
Gulf
Umm al Qaywayn
Hisn Diba
Al Khawr
Dukhan
DOHA
Ajman
Umm Bab
Ash Shariqah
Musay'id
Al Fujayrah
QATAR
Dubai
As Salwa
ABU DHABI
Gulf of Oman
EMIRATES
Mahdah
Al Buraymi
Suhar
Al Khaburah
Mina Qabus
As Sib
MUSCAT
UNITED ARAB
Matrah
Qurayyat
Sabkhat Matti
no defined boundary
Ibri
Izki
Nazwa
Sur
Al Mudaybi
Adam
Fuhud
Al Kamil
Al Ubaylah
Jibal
SAUDI ARABIA
Ghabah
Umm as Samim
no defined boundary
OMAN
Dawwah
Masirah
Hakkan
Nafun
no defined boundary
Duqm
Arabian Sea
Wadi Shihan
Dawkah
Mamul
Thamarit
KHURIYA MURIYA
SOUTH YEMEN
Habarut
Raysut
Salalah
Sadh
Ras Darbat Ali
Al Ghaydah
International boundary
Administrative line
Capital
Road
International Airport
0 50 100 Kilometers
0 50 100 Miles

Map 9 The Strait of Hormuz

Adapted from CIA 625450 7-79

1
Overview

Ever since the Arab oil embargo of 1974, the West has sought to find alternatives to its dependence on imported oil. The West, however, has had little overall success. It has made progress in conservation and in reducing its need for energy as a percentage of GNP, but it has fallen far short of its goals in finding new sources of energy. Developing new energy sources like coal and nuclear power has proved to present far more problems than the West estimated when it began its search for alternative fuels. Synthetic fuels, and wind and geothermal energy, have proved technically difficult and too costly for commercial scale production.

While a major global recession and financial crisis in the Third World have created a temporary "oil glut", the West's long-term dependence on oil imports is now projected to be even worse than any experts projected in the early 1970s. Further, nearly 71% of the surplus production capacity that makes up the current "oil glut" is in the Gulf, with 36% of that capacity in Saudi Arabia.

More than half the world's proven oil reserves are located in the Gulf region, and on the territory of eight nations: Iran, Iraq, Kuwait, Saudi Arabia, Bahrain, Qatar, the United Arab Emirates, and Oman. In spite of more than a decade of intense exploration in other parts of the world, the Gulf is still the region with the highest potential for major new discoveries and is the only area in the world where the discovery of new reserves is out-pacing oil production.

During 1987, the free world relied on over 12 million barrels of oil a day from the Gulf states. The U.S. imported 8% of its oil from the Gulf, Western Europe imported 25%, and Japan imported 60%. U.S. reliance on the Gulf seems certain to increase. U.S. domestic oil production is dropping steadily, and total U.S. imports rose to roughly 35% of U.S. consumption, and were higher than before the 1973–1974 oil embargo. The U.S. Department of Energy estimated that nearly 25% of all U.S. oil would come from the Gulf by the late 1990s.

The West has done a little better in creating the military forces and regional security arrangements necessary to ensure the security of its leading source of oil imports than it has in its efforts to find substitutes for these imports. The Shah's fall and the Iranian hostage crisis deprived the West of its only regional military pillar in 1979. Since that time, the Iran-Iraq War has created a steadily increasing risk that the Gulf could be dominated by a radical anti-Western power.

Nevertheless, Western military capabilities remain limited. The U.S. naval build-up in the Gulf in 1987 has shown the U.S. can deploy massive amounts of sea power, but it has also revealed important limits in the naval power that the U.S. and other Western states can project. The U.S. can also project large amounts of air power and significant amounts of land power. It has made major improvements in its capability to project these kinds of power since the fall of the Shah, but its Central Command (USCENTCOM) forces have limited land force strength and major shortfalls in strategic lift, and its land-based aircraft are dependent on friendly forward bases for effective military action.

European power projection capabilities have declined steadily for the last two decades. Western Europe can still deploy military forces to the Gulf, but these forces are now extremely limited. Britain no longer has full attack carriers in its navy, and France is the only European nation that can deploy even one light attack carrier to the region. No European power can deploy enough land and air power to deal with anything but the most low-level threat. In any other contingency, Europe's military role in the Gulf must be limited to providing the specialized forces the U.S. lacks and providing support for the arms and technology it has transferred to regional states.

Fortunately for the West, it does not have to rely solely on its own forces to secure its interests in the Gulf. Western interests largely coincide with those of the Southern Gulf states. The oil exporting states do not export to earn luxuries; they export to survive. Even the most radical and hostile regimes will normally sell to the highest bidder. Under most conditions, the West will not need to maintain large amounts of military power in the Gulf.

The Iran-Iraq War has demonstrated, however, that the Gulf can be threatened by domination by an anti-Western regime that would constantly exploit its influence and control over the other Gulf nations at serious political and economic cost to the West. The need that all the Gulf states now have to sell their oil to virtually any buyer will also decline steadily as world demand rises in the 1990s, and as the oil reserves outside the region decline. Even temporary reductions or sudden surges in oil prices can have a massive effect on the world's economy, and the West will badly need friendly exporting states in the event of any global war or crisis.

The West cannot afford to ignore the broader risks affecting the region. The threat from the Soviet bloc is now somewhat theoretical, and more political and economic than military. Similarly, the threat from radical Red Sea states, and from radical states like Syria, is muted by their own internal divisions and weakness. It will take years, however, to create a secure mix of regional defense capabilities, Western power projection capabilities, and the overall level of deterrence to ensure that these threats will remain under control in the future. Above all, it will take close strategic relations between the West and the Southern Gulf states.

The events of 1986 and 1987 have made all too clear that there is nothing theoretical about the need for sound Western strategic relations with the Southern Gulf states. The West must be ready for the fact that the present "oil glut" will end by the mid-1990s. It must be ready to face growing competition for oil from the Soviet bloc. It must also be ready for the fact that it will have depleted much of its own oil reserves, and that most Asian and African oil exporting nations will be cutting production as their oil reserves are depleted. The security problems caused by the Iran-Iraq War may yet be only the first phase of a steadily gathering storm.

THE PIVOTAL ROLE OF THE SOUTHERN GULF STATES

The West is being thrust towards a strategic partnership with the Southern Gulf states for several reasons. Iran is in the midst of a hostile religious revolution, and is deeply involved in a war with Iraq whose outcome is still uncertain. No one can predict whether Iran will win or lose the conflict or how much regional influence it will gain from the war. No one can predict whether Iran will tilt back towards the West, tilt toward the USSR, or become a chronic source of regional instability. The only thing that seems certain is that Iran will not become a support to Western security, and that it may well be a serious threat.

Iraq has become steadily more moderate since the late 1970s. It is, however, engaged in a brutal and uncertain war with Iran. It is unclear that Iraq's present regime will survive, or that Iraq will preserve its moderate course once it has come to grips with the cost of the Iran-Iraq War and with the problem of financing its war debts and economic recovery. While it now seems most likely that Iraq will remain relatively moderate, and concentrate on economic development and its trade ties with the West, the West can scarcely count on such a future. Further, there are no serious prospects for any direct military or strategic alliance between Iraq and the West that would give the West significant contingency capabilities in the Gulf area.

This leaves the West dependent on strategic relations with these Southern Gulf countries: Saudi Arabia, Bahrain, Qatar, the United Arab Emirates, and Oman. All have moderate regimes, all have shown a good ability to cope with the past tensions and problems in the region, and all have strong trade ties with the West. Further, all five states are now members of the Gulf Cooperation Council (GCC), which is making slow but steady progress towards creating a collective regional security structure.

The five Southern Gulf states, however, are by no means equal. Four of the five states are so small that they cannot hope to create a reasonable deterrent to deal with the military threat from the Northern Gulf states or from outside the region, and several are deeply divided. Only Saudi Arabia has the combination of population, wealth, stability, and geography necessary to build up a reasonable national deterrent, and to give the Gulf Cooperation Council real meaning.

Sheer geography also makes Saudi Arabia an essential security partner in any Western effort to create stable security relations with the Gulf states. Its borders and coastline dominate the Southern Gulf, Arabian Peninsula, and the northern coast of the Red Sea. While the West has contingency bases in states like Egypt, Oman, and Turkey that can be of great value, the West must have access to Saudi Arabia to deal with any serious Iranian or Soviet threat to the region.

STRATEGIC RELATIONS WITH SAUDI ARABIA

In theory, it should be relatively easy for the West to create a sound strategic relationship with the Southern Gulf states. While Saudi Arabia and the West have different cultures and political systems, they share common strategic interests in the Gulf area and throughout the Near East. While they differ in terms of their trade and oil prices, they share a common commitment to private enterprise and to maintaining a stable balance of world trade.

Saudi relations with the West are a good case in point. Saudi Arabia often differs with given Western states in terms of specific policy issues and tactics, but these differences are rarely significant enough to prevent close cooperation. Even the differences over the Arab-Israeli peace issue are largely ones of timing and tactics. While Saudi Arabia has differed with the U.S. over the specific approach that should be taken to achieving an Arab-Israeli peace settlement, so have most of America's NATO allies.

For the last decade, Saudi Arabia has consistently supported an Arab effort to reach a peace settlement with Israel based on Palestinian rights and the return of the occupied territories. Saudi Arabia has backed the West in key policy thrusts like its peace-keeping effort in Lebanon in 1982. Saudi Arabia has also joined the West in its support of the Afghan freedom fighters, in efforts to stabilize world oil supplies and prices, and in efforts to check the expansion of radical and Soviet influence in the Near East and Southwest Asia.

More recently, Saudi Arabia has strongly supported the U.S. and its West European allies in their efforts to secure freedom of navigation through the Gulf and to end the Iran-Iraq War. It has combined with Jordan to bring Egypt fully back into the Arab world, and it has strongly resisted the spread of Khomeini's brand of anti-Western fundamentalism.

If there is any major difference between Saudi Arabia and the West, it has been in Saudi Arabia's support of Syria, and in Saudi funding of Syria's military build-up. Such Saudi support of Syria has, however, been more the result of Saudi weakness, and need for outside military support, than the result of any political agreement with Syria's political aims and military ambitions. The Saudis have provided such aid as the price of eliminating a Syrian threat to Saudi Arabia's military and political security. It is important to note that Saudi Arabia has been Jordan's chief source of financial support within the Arab world in spite of the fact that Jordan has directly opposed Syrian efforts to block an Arab-Israeli peace agreement.

It is also notable that Saudi Arabia's strategic cooperation with the West has expanded steadily since the Shah's fall and the beginning of the Iran-Iraq War. Saudi Arabia has openly turned to the West for arms and military advice and support, and has quietly turned to the West for security in the event of a major military threat or attack by Iran or South Yemen.

THE WEST AND THE PROBLEM OF U.S. DOMESTIC POLITICS

Western strategic relations with the Southern Gulf states are constantly threatened, however, by factors which have little to do with Western strategic interests, and which have led to a split between the U.S. and Europe in how they deal with the region. The pressures of U.S. domestic politics have constantly threatened to block any effective relationship between the Gulf states and the one Western state with the power projection forces to provide significant "over-the-horizon" reinforcements.

The U.S. has been unable to combine its strategic relations with Israel with equally effective relations with friendly Arab states. Largely artificial fears regarding Israel's security have led to bitter and divisive Congressional and media debates over every recent U.S. arms sale to a Southern Gulf state. While each debate has touched upon the strategic and technical merits of the particular sale in question, the main focus of these debates has been the threat that U.S. arms sales to Bahrain, Saudi Arabia, Kuwait, and the UAE could pose to Israel, and the fear that close U.S. military relations with friendly Gulf states could somehow undermine U.S. support of Israel.

By the mid-1980s, these debates had virtually paralyzed the Reagan Administration's inability to get Congressional support for virtually every major U.S. effort to help modernize the forces of the Southern Gulf states. Saudi Arabia and most of the other Southern Gulf states responded by turning to other Western nations like France and Britain. The most dramatic of these shifts was the Kingdom's decision in late 1985 to turn to Britain for the sale of 72 Tornado aircraft, Hawk trainers, and the other equipment necessary to modernize its air force. There is little doubt that this situation would have become far worse because of covert U.S. arms sales to Iran if the Iranian attacks on the shipping moving through the Gulf had not forced the U.S. and Southern Gulf states to cooperate.

The recent crisis in the Gulf over the Iran-Iraq War has created another opportunity for the U.S. to create close military relations with Saudi Arabia and the other Southern Gulf states. It has also shown that neither the U.S. nor Europe has to choose between relations with the Southern Gulf states and Israel. Iran currently forces the Southern Gulf states to give priority to their own security and to realize they need American support. The issue is whether the U.S. can capitalize on this opportunity, or will be paralyzed by its own domestic politics.

THE COST OF "TAKING SIDES"

America is scarcely the root cause of most of the problems in the Gulf. The Iran-Iraq War is strictly the result of the ambitions and mistakes of the leadership of those two states. The divisions within the Southern Gulf states are self-inflicted, as are most of their mistakes in trying to create the kind of military forces that can protect them. Europe has often chosen to abdicate its responsibilities in the Gulf while performing an almost totally negative role as America's chief critic.The Soviet Union has chosen short-term opportunism over the search to bring peace and stability to the region. The continuing struggle between Israel and its Arab neighbors, which poisons the politics and stability of the entire region, must be blamed on both sides alike.

As the following chapters show, however, the U.S. has made many serious mistakes in dealing with the region. It has failed to clearly define its strategic objectives, and it has failed to pursue them. It has bypassed its own national security and policy planning systems, involved itself in fiascoes like covert arms sales to Iran, and then rushed into a massive convoy protection effort in the Gulf.

The most permeating problem affecting U.S. policy, however, has been the impact of domestic politics in blocking effective strategic relations with the Gulf states. The U.S. tendency to "take sides" between Israel and the Southern Gulf states has divided the U.S. from friendly states like Bahrain and Oman. At the same time, it has made it far harder for the U.S. to reach any arrangements in the region that improve its contingency basing capabilities and its ability to project military forces.

This taking sides has failed to serve Israel's interests as well as those of the U.S. The U.S. has found that it does not have any monopoly on the kind of military technology pouring into the region. Bahrain, Kuwait, Oman, Qatar, Saudi Arabia, and the UAE can obtain any arms they want from Western Europe or the Soviet bloc. Recent efforts to block U.S. arms sales to the Southern Gulf states have also seriously eroded the U.S. capability to restrain friendly Arab states from supporting Israel's enemies, and U.S. influence in shaping a peace process that can benefit both Israel and its neighbors.

This has been illustrated by the impact of the recent debates within the U.S. Congress over arms sales to Saudi Arabia. Where close ties between the Saudi and U.S. air forces created the kind of political and military relations that reinforced Saudi Arabia's already strong reluctance to become involved in any military engagement with Israel, the undermining of U.S. and Saudi military relations created growing Saudi hostility towards Israel. It also has led to the creation of British-equipped Saudi air units, which theoretically can attack Israel without regard to U.S. support and future replacement. Further, it ended in giving Saudi Arabia advanced attack capability much more quickly than if the Kingdom had bought aircraft from the U.S.

In contrast, if the U.S. had met Saudi Arabia's arms request, the new and upgraded F-15s would have been based on U.S.-supported facilities in the Gulf and lower Red Sea, rather than on bases near Israel. The conversion of the existing Saudi F-15C/Ds to the improved "MSIP" version to be used in the USAF would have taken place in separate batches of 3 aircraft each. They would not have begun in less than 46 months (6 months contracting time and 40 months lead time) after a letter of offer (LOA) was signed and could not have been completed in less than three years. Even assuming that a letter of offer had been signed on 1 January 1985, the conversions would not have been completed until 1992.

The Saudi purchases of 40 more F-15C/D MSIP fighters would have involved the same lead times for contracting and production. The earliest deliveries could not have taken place before early 1989 and could not be completed before 1990. Even if full operational training had begun in the summer of 1990, it could not have been completed before 1991–1992.

The British deliveries of the 48 Tornado IDS attack aircraft will be completed in 1988, and the remaining 24 F.3 air superiority fighters will be delivered by 1990. U.S. deliveries would have had a much slower pace. It would have been at least 1992 before the Saudi Air Force would have been fully able to operate its new force of 102 upgraded F-15C/Ds. By then, many of Saudi Arabia's F-5s would be twenty-one years old (many are already thirteen years old), and at least half would be converted to the trainer and light support role. Saudi Arabia would also have had to withdraw all its existing 23 Lightning and 37 BAC-167 aircraft from service.

Equally important, any follow-on upgrading of the Saudi F-15s to give them full dual capability in the air-to-ground role would not have taken place before the early 1990s. The Kingdom would have been limited to upgrading from the F-5E/F bomb racks that the Saudi Air Force has modified for use on its F-15 to the use of multiple ejection racks (MER) like the MER-200, tangential fuel tanks which carry air-to-ground munitions, and the ability to deliver Maverick.

Instead, Saudi Arabia is now acquiring Tornado fighters which are primarily designed for the attack mission and which will replace its F-5s much more quickly. Given Saudi Arabia's basing and support structure, it is almost certain to eventually base at least some of its initial deliveries of European fighters at its facilities in Tabuk—the only Saudi main base which does not have major U.S. support facilities and the only Saudi main operating base near Israel.

These same patterns have emerged in less public efforts to bloc U.S. arms sales to the other Southern Gulf states. Bahrain, Kuwait and Oman have been forced into complex political struggles virtually every time they have sought to make major arms purchases from the U.S. In those cases where they have gotten U.S. arms, the end result has often been unnecessary bitterness, which has undercut the political value of the arms transfers. In those cases where they are denied the arms, they have simply turned to Europe or even to communist countries. Qatar and the UAE have often turned away from the option of buying U.S. weapons systems because of these political problems in dealing with the U.S. They always, however, have been able to obtain similar weapons systems and technologies from Europe.

THE IMPACT OF WESTERN MILITARY RELATIONS WITH THE SOUTHERN GULF STATES ON ISRAEL'S SECURITY

For all the U.S. domestic political posturing and debates over the "greater Arab" threat to Israel, the U.S. can develop and maintain close strategic relations with Saudi Arabia and the other GCC states without creating any serious incremental threat to Israel. This does not mean Western arms sales can be made to Saudi Arabia and its neighbors without *any* risk to Israel. There is no doubt, for example, that Britain's Tornado sale will significantly improve the theoretical capability of the Saudi Air Force to attack Israel.

While Israel and its supporters sometimes lump every Arab state together in the threat that Israel must face, the Camp David agreement and the Iran-Iraq War have left Israel with only one real military opponent. This opponent is Syria, and Syria is unlikely to obtain significant military forces from any moderate Arab state in a future conflict against Israel.

None of the smaller Gulf states can pose any threat to Israel. Oman is relatively friendly to Israel, the UAE's forces are divided, and Bahrain, Kuwait, and Qatar have such small forces that the limited portion of their combat units they could actually deploy would have an insignificant impact on the balance.

All the smaller Gulf states face local threats that "pin down" their forces. Their combat units are critically dependent on foreign technicians and are trained and equipped to fight near fixed bases, rather than organized with the combat and service support necessary to deploy to foreign countries. They are not standardized in training, equipment, command and control, communications, or IFF with any of the front-line Arab states. They are too far away from Israel to support any forces they deploy, and even if they sent such forces they would simply force a front-line Arab state to devote the necessary support facilities at the cost of cutting the fighting capability of its own forces.

Saudi Arabia does have larger forces, and has bases near Jordan, but it too does not present a meaningful military threat to Israel. The Kingdom faces serious threats on its Gulf and Red Sea coasts and borders to which it must give all its attention. It at best can build a limited regional military deterrent and enough defensive capability to deal with local low- and medium-level threats. Even then, it will be hard pressed to use the improvements in its air force to offset the continuing weaknesses in its ground and naval forces. It will encounter even more problems in trying to offset the weaknesses in the small military forces of the other members of the Gulf Cooperation Council.

Saudi ground forces have no meaningful capability to engage Israeli forces, and they will not acquire this capability in the foreseeable future. The Saudi Air Force cannot obtain delivery of, and absorb, the advanced combat aircraft it has ordered until the early 1990s. By that time, improvements to the Israeli Air Force will have out-paced the improvements in the Saudi Air Force. This Israeli technical edge is ensured by U.S. ability to transfer newer and far more lethal technologies to Israel.

This does not mean the U.S. should sell the Southern Gulf states everything they ask for. The U.S. cannot choose between Israel and the Southern Gulf states on the grounds of strategic expediency. The U.S. is Israel's only major ally, and the only state that can and will aid it in the face of another war or major crisis. Israel's military forces cannot retain their strength without U.S. grant aid and access to U.S. military equipment and technology.

The U.S., however, should not try to protect one friend at the cost of keeping another weak. The U.S. has the ability to provide Israel with a large enough aid program to preserve its security, and with the kind of technology transfer that will allow it to keep a decisive qualitative edge over true confrontation states like Syria.

The key to balancing the U.S. commitment to Israel with Western strategic interests in the Gulf is to strengthen the U.S. security assistance program to Israel, rather than to weaken U.S. strategic relations with the Southern Gulf states. The U.S. needs to change many aspects of its policy in the Gulf, but the key step is to break out of the trap inherent in "taking sides" and to provide the flow of grant aid and technology transfer that will ensure that the risks Israel faces do not increase in the future.

THE IMPACT OF WESTERN POWER PROJECTION CAPABILITIES ON THE GULF

All of these issues would be far less important if arms sales were the only issue. The problem for the West is that recent trends in U.S. relations with Saudi Arabia and the other GCC states have undermined a strategic relationship which is critical to the West.

Equally important, if the U.S. does not help Western Europe meet the military needs of the Southern Gulf states, it may well cripple both their individual defense capabilities and efforts to build up collective security through the Gulf Cooperation Council. This seems likely to have drastic consequences for both the region and the West — given the Soviet presence in the People's Democratic Republic of Yemen (PDRY) and Ethiopia, the massive Soviet arms transfer effort in the Yemeni Arab Republic (YAR), the uncertainties surrounding the Iran-Iraq War and the Iranian revolution, and the instability of key Red Sea states like the Sudan. The West also cannot afford to ignore the long-term implications of a Soviet victory in Afghanistan or the growing economic problems the Soviet bloc is experiencing because of its lack of a secure future source of oil.

The U.S. also cannot afford to ignore the severe limitations in the power projection capabilities of its Western allies. Only three NATO European states can still project any significant amount of power in the Gulf: Britain, France, and Turkey.

Britain can still deploy several RAF squadrons, and a significant number of frigates, escorts, submarines, and mine warfare ships. Britain cannot, however, deploy a full brigade to the region. It can only deploy one light carrier at a time, with a maximum of eight to twelve short-range Harrier VSTOL fighters. Britain is rapidly losing amphibious ship strength, and will steadily reduce its power projection capabilities during the next decade.

France can deploy several air force squadrons, and as many as two brigades if it could obtain U.S. air and sealift support. France can also deploy a small carrier task force with one carrier equipped with 30–40 fixed-wing aircraft. This force, however, compares with U.S. carrier task forces in which each carrier displaces more than 70,000 tons and has 86 combat aircraft. Further, because the U.S. has a serious shortfall of strategic lift, any French force deployed to the Gulf would have to come at the direct expense of U.S. deployment of forces with better equipment.

Turkey can play an important role in its border areas near Iraq and Iran. It has considerable land and air forces in Eastern Turkey, but these forces are not equipped for operations far from Turkey's borders. Turkey also would have to strip its eastern defenses against the USSR.

The practical reality is that the United States is the only Western state which retains significant power projection forces that can help defend the Gulf. The U.S., however, has no full-time bases in the region. Its only contingency facilities in the Gulf are in Oman, and these are suitable only for a defense of the Straits of Hormuz, and not the Gulf as a whole. The closest facility at which the U.S. can station aircraft and troops in peacetime is Diego Garcia, which is nearly as far away from Kuwait City as Dublin, Ireland.

The U.S. has previously compensated for this weakness by establishing the kind of military relations with Oman that give it staging facilities in the Indian Ocean and lower Gulf. These facilities have already been of great value in helping U.S. naval forces to try to secure freedom of navigation in the Gulf. The U.S. has also established the kind of relations with Bahrain which allow it to use Bahrain's air and naval facilities for a wide variety of staging and support facilities.

Further, the U.S. has carried out the kind of arms sales and advisory efforts in Saudi Arabia that could give the U.S. suitable basing facilities in the Kingdom in a major crisis. These relations created Saudi air and air defense forces that are standardized and interoperable with the U.S. forces in USCENTCOM. This has made U.S. military relations with Saudi Arabia a critical part of U.S. efforts to create an effective "over-the-horizon" deterrent in the Gulf.

The Saudi Air Force operates U.S.-made F-15C/D and F-5E/F fighters, and many Saudi squadrons now approach U.S. Air Force proficiency and qualification levels. Saudi Arabia has equally modern naval facilities and ground bases. These now use only moderate amounts of major combat equipment, which is standard in USCENTCOM forces, but they have extensive stocks of parts and munitions, and service and support equipment which can be used by U.S. units. The Saudi base at Hafr al-Batin (which is located in the critical border area near Kuwait and Iraq) will also have two mechanized brigades equipped with U.S. armor.

Unlike Oman, Saudi Arabia and Bahrain have never formally agreed to provide the U.S. with contingency bases, but they have often requested U.S. reinforcements. Saudi and Bahraini officials and officers have conducted extensive talks with the commander of USCENTCOM and other senior U.S. officials. Like Oman, Saudi Arabia and Bahrain have played a strong role in forging the Gulf Cooperation Council, and during the last year, all three states have shown they will cooperate with U.S. forces in the face of any common threat to the Gulf or Red Sea area.

In fact, the problems the U.S. has encountered in deploying naval forces in the Gulf to defend its convoys to Kuwait have strongly indicated that USCENTCOM cannot function in any major contingency in the Gulf area without local cooperation and access to bases and facilities in the Gulf. While Oman, Turkey, and Egypt provide useful contingency facilities on the periphery of the Gulf and lower Red Sea, they cannot make up for the range and reinforcement problems

USCENTCOM would face in defending the critical oil facilities in the upper and central Gulf.

This Western dependence on U.S. power projection capabilities is the reason why the Gulf cannot be defended by an arrangement whereby Europe maintains close relations with friendly Arab states and the U.S. serves as the military sponsor and protector of Israel. Europe can meet the Arab world's needs for military equipment and advice, but it cannot provide the power projection forces needed to defend the Gulf. Only the U.S. can now deploy the sea, air, and land power necessary to check a major radical threat to a key Arab state and to ensure that the USSR cannot put military or political pressure on the region.

Nations like Britain and France can—and do—play a critical role in stabilizing the Gulf and creating the kind of strategic relations with the Gulf states the West so badly needs. Nevertheless, the U.S. needs the kind of military relations with Saudi Arabia and the other Southern Gulf states which would give these states the confidence that they can rely on the West in an emergency. As its intervention in the Gulf in 1987 showed all too clearly, the U.S. needs access to their facilities and their support in refueling, logistics, and resupply. It needs the maximum possible interdependence in terms of weapons, ammunition, and spare parts to ease the strain of projecting American power half-way around the world.

ESTABLISHING THE PROPER BALANCE IN WESTERN STRATEGIC RELATIONS

It will not be easy for the West to establish the proper balance of strategic relations with the Southern Gulf states. Nothing the West does can eliminate the risks in the region. The key to minimizing these risks, however, is to establish a stable balance in its approach to the region. It is also important to understand that the basic weaknesses in the West's approach to the region are weaknesses in U.S. policy and actions.

This is not simply a lack of balance in the U.S. approach to Israel and friendly Arab states. The U.S. cannot afford another series of policy adventures like covert arms sales to Iran or rushing into reflagging Kuwait tankers as compensation. It cannot hope for either regional pillars or permanent bases and a large military presence. It must define its policies and power projection capabilities in the region so that its policy is predictable and consistent, and clearly defines its long-term objective as one of creating a mix of regional political structures and military capabilities that do as much to ensure stability as possible.

In the short run, the U.S. will have to play out its current intervention in the Gulf in a way that will push Iran towards a cease-fire while minimizing any lasting increase in Iranian hostility towards the West, demonstrate Western resolve and military capability, improve relations with Iraq, and reassure the Southern Gulf states that they can place their trust in the West.

In the long run, the U.S. must establish the kind of stable military relations with the Southern Gulf states that will allow them to demonstrate that their "linkage" to the West will provide them with the security they need. This linkage does not require a "blank check" in terms of arms transfers or U.S. politico-military support, but it does require the U.S. to be able to join its European allies in providing the GCC states with enough arms and technology to ensure that they can develop a reasonable level of military capability and cooperate with the U.S. and other Western states in regional security efforts.

2
Western Strategic Interests in the Southern Gulf

In order to understand the West's policy priorities in the Gulf, it is necessary to understand its strategic interests, not only as they exist today but as they are likely to change in the future. The West has four main strategic interests in the Gulf:

- To ensure access to Gulf oil at a reasonable price, and without fear of political, economic, or military blackmail or intimidation.
- To maintain a stable and productive balance of trade.
- To ensure that Gulf holdings of hard currency are invested in the West or in forms of external aid that help develop and stabilize friendly nations in the Third World.
- To maintain the kind of politico-military relations with Gulf states that encourage their support of Western political positions and the peaceful resolution of the Arab-Israeli conflict.

The key Western strategic interest is access to oil. It not only is the most critical strategic interest; it is the one which shapes the Gulf's impact on trade, investment, aid, and world politics. Oil, however, is a remarkably complex issue. It both divides Western interests from those of the Gulf states and unites them. Its importance varies sharply over time, and is subject to unpredictable periods of glut and panic. It gives small Gulf states immense importance, but it does nothing to keep them from being weak.

OIL REVENUES AND OIL EXPORTS

As noted, the free world relied on over 12 million barrels of oil a day from the Gulf states during 1987. U.S. domestic oil production dropped by 4.8% in 1987 alone. America's 1987 output was only 14% of the output in 1970, the nation's peak production year. U.S. imports were up 5% in 1987, and one out of every five barrels of imports came from the Gulf.

This flow of oil creates a strong interdependence between the economies of the Western industrialized states and those of the Gulf. The West and the Southern Gulf states will always be "competitors" in the sense that the West will seek the lowest possible oil price and the lowest possible level of dependence on imports, while the Southern Gulf states will seek to maximize its oil revenues and market share. Both will continue to compete over issues like price, volume of production, increases in Gulf production of oil "products" versus increases in Western production, protection of its refineries, and a host of other issues. No arrangements between the West and the Southern Gulf states can eliminate the normal frictions of international trade—nor should they. The free market is ultimately the best way of minimizing energy costs and stabilizing the balance of trade.

Nevertheless, the West and the Southern Gulf states share important common strategic interests in regard to oil. These include avoiding artificially high or low oil prices, and avoiding unstable swings in Western and Gulf trade and the world economy. This common interest has led Saudi Arabia, Kuwait, and most of the other Southern Gulf states to cooperate with the West in recycling "petrodollars", and in trying to smooth out the effects of the kind of "boom and bust" cycle reflected in Table 2.1, which could repeat itself in the mid- to late 1990s as the world shifts from the "oil glut" of 1986 and 1987 to a new hardening of world energy prices.

A good example of such cooperation between the West and the Southern Gulf states took place in the period following the Shah's fall and at the beginning of the Iran-Iraq War. During 1980 and 1981, Saudi Arabia increased its oil production to 10 million barrels per day (MMBD) to reduce the pressure of near panic demand for oil and did so at the request of the United States and Europe. Saudi Arabia obviously benefitted from this increase in terms of revenues, but it also made a major effort to reduce the impact of the resulting oil crisis. When oil prices rose to $35 per barrel, Saudi Arabia had to dispose of excess oil income equal to roughly $20 billion a year. The Kingdom cooperated closely with key European states and the U.S. Treasury in "recycling" these oil revenues and in finding patterns of investment that would not disturb Western economies. Saudi Arabia also cooperated with Western governments in efforts to stabilize the oil market, to aid oil importing developing states, and to create a stable pattern of investment in the West.

This cooperation continued even though the rise in oil prices during the late 1970s and early 1980s triggered a global recession and mass conservation, and then decreased demand and forced oil prices down to well under $20 a barrel. The resulting "oil glut" reduced the demand for

Table 2.1: OPEC Revenues and Oil Exports: 1973–1986

Country	Oil Revenues ($Billions Per Year)				Oil Production (MMBD)			
	73	*80*	*84*	*86*	*73*	*80*	*84*	*86*
Saudi Arabia	4.3	102.2	43.7	28.5	7.4	10.0	4.7	4.6
Kuwait	1.8	17.9	10.8	9.2	3.0	1.4	1.0	1.4
Iraq	1.8	26.1	10.4	9.9	2.0	2.6	1.2	1.7
Iran	4.4	13.5	16.7	10.4	5.9	1.5	2.2	1.9
Libya	2.2	22.6	10.9	6.2	2.2	1.8	1.1	0.9
Qatar	0.3	5.4	4.4	1.9	0.3	0.3	0.3	0.3
UAE	0.9	19.5	13.0	10.5	1.5	1.7	1.2	1.7
Algeria	1.0	12.5	9.7	3.9	1.1	1.1	1.0	0.6
Ecuador	0.2	1.4	1.6	1.8	0.2	0.2	0.3	0.3
Gabon	0.2	1.8	1.4	1.1	0.1	0.2	0.2	0.2
Nigeria	2.1	25.6	12.4	9.9	2.0	2.1	1.4	1.5
Indonesia	0.2	12.9	11.2	8.5	1.3	1.6	1.4	1.3
Venezuela	3.0	17.6	13.7	9.7	3.5	2.2	1.9	1.7
Other	0.6	8.0	5.7	4.4	0.9	0.8	0.8	0.7
OPEC Total	22.5	278.8	159.4	112.0	30.8	26.9	17.9	17.5
Mexico	–	–	–	–	0.6	2.2	3.0	2.6

Sources: The data for 1973–1984 are actual data based on reporting in the *Petroleum Economist*, July, 1985; Worldwatch Institute, *Worldwatch Paper 66*, *Petroleum Institute Weekly*, and BP *Statistical Review of Energy*. The data for 1986 are preliminary estimates based on the trends for the first half of 1986 as reported in William L. Randol and Ellen Macready, *Petroleum Monitor*, First Boston Corporation, New York, May, 1986.

Saudi oil to monthly average levels as low as 1 MMBD in August, 1985, and cut Saudi oil revenues from a peak of $102 billion a year in 1980, to $28 billion in 1985, and as low as $22 billion in 1986. The Kingdom's projected income in 1987 was only marginally higher. Although oil prices increased in dollar terms, the dollar dropped in value relative to other currencies, and nearly 1 MMBD of Saudi Arabia's production of roughly 4 MMBD was tied up in low-income-producing barter deals.[1]

Even so, Saudi Arabia has cooperated with the West on payment scheduling and investment flows to minimize the impact of the shift in revenues during a time when it has faced a budget deficit of nearly $20 billion a year.[2] Further, it is Saudi Arabia which led the break-up of OPEC efforts to maintain prices by cutting world oil production. Saudi Arabia instead chose a strategy of increasing production that ensured a massive drop in prices. It did so, at least in part, to keep prices at levels that would ensure high long-term Western demand.[3]

WESTERN DEPENDENCE ON GULF OIL

The West and the Southern Gulf states also share a common interest in securing every Gulf state's ability to freely export oil. Wide swings take place in the West's dependence on Gulf oil from year to year, depending upon the health of the world's economy. Table 2.2 shows both the details of current export patterns and these swings in the imports of key OECD countries from the Gulf area, during a period ranging from peak demand to a record low. While no one can hope to accurately predict the future, these patterns should provide a reasonably accurate picture of the future dependence of individual Western countries on oil imports. Table 2.2 also shows just how low Western dependence on the Gulf states can be, and includes data for the record low in Gulf exports in the third quarter of 1985.

These swings take place in the West's oil demand, however, largely because of global economic conditions and they are extremely dependent on the ability of states outside the Gulf area to export large amounts of oil. These non-Gulf oil exporting states have only a limited portion of the world's oil reserves, and Table 2.2 shows that they will drop sharply in total production and export capacity over the next ten years.

Saudi Arabia is obviously the most important Gulf state in this regard. Table 2.3 shows Saudi Arabia has proven oil reserves of over 170 billion barrels. This is nearly 25% of the world's proven oil reserves and 28% of the free world's reserves. Even these percentages are artificially low because Saudi Arabia has had little incentive to prove more reserves over the last five years. The current "oil glut" exists almost solely because of Gulf surplus capacity. Some 71% of surplus capacity came from the Gulf in 1987, with 36% of this capacity in Saudi Arabia alone.

Saudi reserves are probably at least 30% higher than the proven reserves shown in Table 2.3, and the same is probably true of many other Gulf states. Abu Dhabi, for example, may have three times the reserves officially estimated by the U.S., and Iraq may have twice as many. While some nations like Iran have recently made exaggerated claims, more reputable sources estimate that world oil reserves may be as high as 887.348 billion barrels, rather than the 697.450 billion barrels reflected in official U.S. government estimates. This is a 27% rise, but it is important to note that 162.8 billion barrels, or 85% of this 189.9 billion

Table 2.2: Recent Patterns in Western Dependence on Gulf Oil, Part One—Western Dependence on Oil Shipped Through the Straits of Hormuz in 1986

Western Dependence on Gulf Oil Versus Total Consumption

Importer	Through Straits of Hormuz MBD	Through Straits of Hormuz Percent	From Other Sources MBD	From Other Sources Percent	Total MBD
Japan	2,226	52	2,022	48	4,248
Italy	705	35	1,308	65	2,013
France	471	24	1,474	76	1,945
Netherlands	438	24	1,393	76	1,831
Spain	210	20	849	80	1,059
FRG	215	9	2,292	91	2,497
U.S.	781	5	15,734	95	16,515
U.K.	102	3	3,556	97	3,658
Canada	55	2	2,196	98	2,251
Other	743	14	4,606	86	5,349
Total	5,946	14	35,420	86	41,336

Average Western Consumption of Oil in 1986 Passing Through the Straits of Hormuz from Selected Gulf Countries (MBD)

Importer	Saudi Arabia	UAE	Iran	Kuwait	Qatar	Bahrain	Total MBD	Total Percent
Japan	797	831	214	180	168	36	2,226	37
U.S.	635	44	19	68	13	2	781	13
Italy	364	50	135	151	4	1	705	12
France	321	34	65	22	28	1	471	8
Netherlands	216	14	81	120	6	1	438	7
FRG	151	4	41	19	–	–	215	4
Spain	114	9	66	20	1	–	55	1
U.K.	73	–	10	13	6	–	102	2
Canada	25	–	30	–	–	–	55	1
Other	409	69	111	134	15	5	743	12
Total								
–MBD	3,105	1,055	772	727	241	46	5,946	100
–Percent	52	18	13	12	4	1	–	100

Note: Figures show oil moving through the Straits, and not total oil exports as are shown in Part Two of Table 2.2.

Source: Adapted from Alexia I. Suma, "Basic Statistics on Persian Gulf Energy: Resources and Shipments", Washington, The Middle East Institute-Sultan Qaboos Center, December, 1987, and *National Security Policy Implications of the United States Operations in the Persian Gulf*, Report of the Defense Policy Panel and the Investigations Subcommittee on Armed Services, House of Representatives, July, 1987, p. 15.

(continued)

Table 2.2: Recent Patterns in Western Dependence on Gulf Oil, Part Two

OECD Imports by Country: 1982–1985 (In Millions of Barrels Per Day) a

	1982		1983		1984		1985	
	Total	Gulf	Total	Gulf	Total	Gulf	Total	Gulf
Total OECD	17.9	8.1	16.6	6.6	17.0	6.5	15.9	5.6
U.S.	4.3	.7	4.3	.4	4.7	.5	4.3	.3
Canada	–.1	.7	–.3	.1	–.3	–	–.4	–
Main Europe	8.7	4.4	7.9	3.2	7.7	3.0	7.4	2.6
France	1.8	.9	1.7	.6	1.7	.5	1.7	.5
Germany	2.1	.5	2.1	.3	2.1	.2	2.2	.2
Italy	1.8	.9	1.6	.8	1.6	.7	1.6	.6
Netherlands	.6	.3	.5	.3	.5	.3	.5	.3
Spain	1.0	.5	1.0	.4	.9	.4	.8	.3
European Exporters								
Norway	–.3	–	–.5	–	–.5	–	–.6	–
U.K.	–.6	.3	–.9	.2	–.8	.1	–1.0	.1
Smaller Europe								
Austria	.2	.1	.2	–	.2	–	.2	–
Belgium	.4	.2	.4	.1	.4	.1	.4	.1
Denmark	.2	–	.2	–	.2	–	.2	–
Ireland	.1	–	.1	–	.1	–	.1	–
Finland	.2	–	.2	–	.2	–	.2	–
Greece	.2	.2	.2	.1	.2	.1	.2	.1
Portugal	.2	.1	.2	.1	.2	.1	.2	.1
Sweden	.4	.1	.4	–	.3	–	.3	–
Switzerland	.2	–	.3	–	.2	–	.3	–
Turkey	.3	.2	.3	.2	.3	.3	.3	.3
Japan	4.5	2.7	4.4	2.6	4.5	2.8	4.3	2.6
Other OECD	.5	.3	.3	.2	.3	.2	.3	.2

Lowest Recent Imports from Given Gulf Countries (3rd Quarter 1985)

	Total	Bahrain	Iran	Iraq	Kuwait	UAE	Qatar	Saudi Arabia[b]
Total OECD	5.14	.03	1.12	.87	.52	1.06	.31	1.2
Total Europe	2.49	.0	.82	.74	.32	.11	.03	.47
U.K.	.10	.0	.01	.07	.01	.01	.0	.01
France	.47	–	.13	.17	.02	.05	.03	.07
FRG	.167	.0	.09	.01	.02	.01	–	.04
Italy	.50	.0	.16	.11	.13	.02	.0	.74
Other	1.26	–	.43	.39	.14	.03	.01	.271
U.S.	.21	–	.06	.08	.01	.03	.0	.04
Canada	.02	–	.02	–	–	–	–	–
Japan	2.21	.03	.18	.05	.15	.89	.27	.65
Other	.22	.01	.08	.01	.04	.03	.01	.05

Source: OECD computer data base, *International Petroleum Statistics Report*, July 25, 1986.

[a] Total includes oil imports regardless of source. Gulf includes all imports from the Gulf states.

[b] U.S. imports from Saudi Arabia fluctuated from 0.04 to 0.55 MMBD during 1982–1985, Japanese imports fluctuated from 0.70 to 1.54 MMBD, OECD Europe fluctuated from 0.47 to 2.40 MMBD, France fluctuated from 0.7 to 0.56 MMBD, West Germany fluctuated from 0.04 to 0.35 MMBD, and Italy fluctuated from .07 to .39 MMBD.

Table 2.3: World Oil Reserves, Part One

Region and Country	Estimated Proved Reserves: Billions of Barrels	Percent of World Total
Gulf[a]	396.18	56.7
Bahrain	.17	.02
Neutral Zone	(5.4)	.8
Iran	48.5	6.9
Iraq	44.5 (65.0)[b]	6.4
Kuwait[c]	92.7	13.3
Oman	3.5	.5
Qatar	3.35	.5
UAE	32.49	4.6
Abu Dhabi	(30.5)	–
Dubai	(1.44)	–
Ras al Khaimah	(0.1)	–
Sharjah	(.45)	–
Saudi Arabia[d]	171.7	24.6
Other Middle East	2.2	.3
Israel	.75	.1
Syria	1.45	.2
Total Middle East	398.38	57.0

a. Breakdown by individual countries includes only major exporting or reserve holding countries

b. The current official estimate is 44.1 billion, which has not been revised because of the Iran-Iraq War, but most U.S. officials now estimate Iraqi proved reserves at 65 billion or more.

c. Kuwait's reserves are probably in excess of 100 billion and Saudi Arabia's are near 200 billion. The reserves for Kuwait and Saudi Arabia include half of the Neutral Zone.

d. Neither Kuwait or Saudi Arabia provides up to date estimates of proven reserves.

Source: Adapted from *Oil and Gas Journal,* December 1984 and December 1985. The U.S. Department of Energy shows a slightly higher percentage of total reserves in the Middle East. See DOE/EIA-0219(84), pp. 79–81, and *Annual Energy Review, 1986*, Washington, GPO, 1986.

(continued)

Table 2.3: World Oil Reserves, Part Two

Region and Country	Estimated Proved Reserves: Billions of Barrels	Estimated Proved Reserves: Percent of World Total
Africa [a]	55.54	7.9
Algeria	9.0	1.3
Angola	1.8	.3
Egypt	3.2	.5
Libya	21.1	3.0
Nigeria	16.65	2.4
Western Hemisphere [a]	117.69	16.8
U.S.	27.3	3.9
Mexico	48.6	7.0
Canada	7.075	1.0
Venezuela	25.845	3.7
Western Europe [a]	24.425	3.5
Britain	13.59	1.9
Norway	8.3	1.2
Asia-Pacific [a]	18.5299	2.7
Australia	1.5	.2
Brunei	1.4	.2
India	3.0	.42
Indonesia	8.65	1.2
Malaysia	3.5	.5
Total Non-Communist	614.567	88.0
Communist	84.1	12.0
USSR	63.0	9.0
China	19.1	2.7
Other	2.0	.2
TOTAL WORLD	698.667	100.0

[a] Totals include entire region, not just countries shown.

barrel rise, would come from the Gulf. In any case, Saudi Arabia can continue to prove more reserves than it consumes and exports for at least the next half-decade and may well be able to do so for the next decade.[4]

This mix of Saudi production capability and reserves will be of steadily greater importance as time goes by. The expansion of alternative fuel supplies—such as nuclear power and synthetic fuels—and of new discoveries has fallen far below the levels predicted in the 1970s and early 1980s. Even the most conservative estimates of oil demand indicate that oil will provide at least 30% of the energy for OECD countries through the end of the 1990s, and the Department of Energy's projections in Table 2.4 show that these percentages will probably be far higher.[5]

After a decade of intense Western effort to find alternative energy supplies and oil outside the Middle East, the U.S. Geological Survey (USGS) shows virtually no increase in either the proven oil reserves or the estimated undiscovered oil reserves outside the Gulf. Further, the ratio of reserves to consumption had increased in the Gulf, and decreased virtually everywhere else. A 1983 study by the USGS concluded that

> Demonstrated reserves of crude oil has declined over the past 10 years...discoveries have lagged over the same period....Rates of discovery have continued to decline over the past 20 years even though exploratory activity has increased....[6]

A 1985 report by the USGS on efforts to discover new oil reserves outside the Gulf is even more discouraging.

> Clearly discoveries are on a downward trend from a high in the 1950s of some 35 billion barrels per year to a present day total of 10–15 billion barrels of new oil per year. Production of about 20 billion barrels per day has now out-paced discovery by a factor of two. The reality is...that the Middle East increasingly will monopolize world petroleum supplies...most of the world's conventional oil resources lie within the narrow confines of the Middle East and so does the production capacity. The economies of the western world rest on the daily production from the Middle East and indeed even the amount of oil transiting the Straits of Hormuz daily, some 7 to 8 million barrels is about two times the surplus producing capacity found outside the Middle East.[7]

The USGS cut its estimate of U.S. offshore oil reserves by 55% in 1985. Europe has no prospects of further major discoveries and North Sea oil production will decline steadily in the 1990s. The USSR is experiencing major problems in exploiting its proven reserves. Expected major

Table 2.4: Free-World Oil Consumption and Production, 1985–1995

Region and Country	Oil Consumption in MMBD			Percentage of Total National Energy Use	
	1985	1990	1995	1985	1995
U.S.	16	16–17	16.7–18	42	39–41
Canada	1.5	1.6–1.7	1.6–1.8	31	29–31
Japan	4.3	4.5–4.7	4.4–4.8	56	49–51
Europe	11.7	12.2–12.9	12.0–13.1	45	39–42
Other OECD	–	–	–	50	42–44
Total OECD	–	–	–	44	40–42
Other	9.6	10.0–10.6	10.3–11.0	–	–
OPEC	3.4	3.7	4.2	–	–
Free World	46.4	47.9–50.4	49.1–53	–	–

Region and Country	Oil Production in MMBD			Percentage of Total Free World Output	
	1985	1990	1995	1985	1995
U.S.	11.2	9.2–10.1	7.6–8.9	24.5	15.5–16.7
Canada	1.8	1.5–1.8	1.4–2.0	3.9	2.9–3.8
Europe	4.3	3.7–4.5	3.2–3.7	9.4	6.6–6.9
Other	9.4	9.5–10.6	9.5–11.7	20.6	19.5–22.0
Total Non–OPEC	26.7	24.0–27.0	21.7–26.3	58.4	44.5–49.3
OPEC	17.2	20.1–25.4	22.0–30.0	37.6	45.0–56.3
Communist Exports	1.8	1.0–1.6	0.5–1.6	3.9	1.0–3.0
Free World	45.7	48.0–50.9	48.8–53.3	100.0 *	100.0 *

*Totals may not add due to rounding.

Adapted by the author from the statistical data base for John S. Herrington, *Energy Security, A Report to the President of the United States*, Washington, Department of Energy, March, 1987.

discoveries in the PRC have not occurred, and increases in Mexican consumption are now expected to consume the increase in production.

In spite of the fact major oil exploration activities began in 30 new countries between 1979 and 1987, the total increase in all world reserves due to new discoveries during that period was less than 1% and most occurred in the Gulf area. While some sources have increased oil reserve estimates due to changes in methodology, it is important to note that these increases have actually cut the percentage of world oil reserves in

Asia and the Pacific, Western Europe, Africa, and Communist states. Virtually all the rise has been in the Gulf, with limited rises in the rest of the Middle East and a 22% rise in the Western Hemisphere that is almost solely the product of claims by Venezula.[8] Further, several key OPEC countries outside the Gulf are now rapidly consuming their reserves: these include Algeria, Ecuador, Gabon, Indonesia, and Nigeria. None are likely to be oil exporters by the mid-1990s.[9]

The West has an equal interest, however, in ensuring that the five smaller conservative Gulf states remain independent and can freely export their oil. Kuwait, Qatar, Bahrain, the UAE, and Oman now have a total of 130 billion barrels worth of proved reserves. As Table 2.3 shows, these five smaller Gulf states have a total of 19% of the world's proven oil reserves and 21% of the free world's oil reserves. In fact, these five small Gulf states have 10% more proven oil reserves than the entire Western Hemisphere, including Canada, Mexico, Venezuela, and the United States.[10]

Equally important, all the the Southern Gulf states combined have a total maximum sustainable oil production capacity of 15.8 million barrels a day. This compares with a total Gulf production capacity of about 20–25 million barrels per day, and is equivalent to roughly 50% of all OPEC production capacity. Saudi Arabia alone can produce up to 10.2 million barrels per day, or more than 40% of all the oil production capacity of all the Gulf states.[11]

The total oil import needs of the West and the free world will fluctuate with political and economic conditions, but today's "oil glut" is almost certain to disappear by the mid- to late 1990s, as the result of world economic recovery and economic growth. Table 2.4 shows the Department of Energy's projections of free world oil demand from 1985 to 1995, based on the price trends, economic trends, and efforts to find alternative energy supplies that existed in early 1987. It is clear that both Western and free world oil consumption will increase through the year 2000.[12]

A Department of Energy study conducted in 1987 indicated that the U.S. alone could double its imports over the next ten years, accounting for 50% or more of all U.S. oil consumption. With over 70% of the world's surplus oil production capacity in the Gulf, clearly the U.S. will sharply increase its dependence on Gulf oil.[13]

The "oil glut" could also disappear overnight if the Southern Gulf states could not export freely, and the Iran-Iraq War made a major cut in Gulf exports a constant possibility. Recent reports by the U.S. Secretary of Energy indicate that even with today's supply and demand conditions, a cutoff of Gulf oil could cause world oil exports to fall as much as 3 to 4 million barrels below world demand.

Such cuts would have an immediate impact on Europe and Japan, which get much of their oil from the Middle East, but they would also affect the U.S. The U.S. now gets only about 3% of its oil from the Gulf, but this low percentage of U.S. imports disguises the true nature of U.S. vulnerability. The U.S. now buys its oil from nations closer to the U.S.—largely because of lower transportation costs. In the event of a reduction in Gulf oil exports, however, the U.S. would have to compete with other importing nations on the world market. Even if it used its Strategic Petroleum Reserves, the U.S. might still face another massive oil price crisis.[14]

By 1995, the impact of any interruption could be much worse. Even before the 1986 drop in oil prices, the U.S. Department of Energy projected that Western and other importing nations' demand for OPEC oil would recover from about 17 million barrels per day in early 1985, to 24–26 million barrels in 1990, as alternative sources like North Sea oil are depleted. The same projections indicated that world demand would rise to the point where real prices would begin to rise in real terms by 1992–95, and slowly increase to levels of at least $36 a barrel in 1985 dollars by 2000, and $55 by 2010. The most likely rise was projected to be $57 a barrel by 2000, and $84 by 2010. The study also projected that if the West continued to develop alternative energy supplies like nuclear power, synthetic fuels, and coal at the rate these supplies increased in the early and mid-1980s, oil prices could rise to $80 a barrel by 2000 and $110 by 2010.[15]

The rapid drop in oil prices in 1985 through 1987 has increased these risks. It has forced producers to permanently close many low-production and high-cost stripper wells, and domestic production in the U.S. dropped by as much as 1 MMBD. It has also triggered a major race among the oil producing nations to produce as much oil as possible to try to compensate for the nearly 50% drop in oil prices between mid-1985 and mid-1986.

This price drop has been good news for the West in the short run, but bad news in the long run. Oil demand has risen while domestic production of oil and all other energy resources has fallen. This has meant that Western alternative energy and conservation projects have been severely curtailed and are now falling even further behind past goals.[16] It has also meant that exporting nations outside the Gulf are depleting their reserves 30–50% more quickly than past projections had indicated.[17]

As a result, the U.S. Department of Energy has developed new projections which indicate that oil prices will rise from a low of $14 a barrel in 1986 to between $15 and $23 a barrel by 1990, and between $22 and $28 a barrel in 1995. Even under these conditions, oil's share of total free world energy consumption is expected to drop only from 46% in 1986 to 42–43% in 1995. Further, total free world consumption of oil is expected to increase from 46 MMBD to 49–53 MMBD.[18]

Under these conditions, any sustained cut in the flow of oil from the Gulf could be expected to trigger a massive recession, or full-scale depression, in the OECD countries. It could also be expected to trigger a crisis in the less developed countries (LDCs) that could lead to large-scale famine or the collapse of their economies. The LDCs more than doubled their demand for OPEC oil between 1970 and 1984, and even after the price shocks of 1979–1908 their demand grew at 2.2% per year. It will continue to grow at a rate of 1.1–1.6%, even under the most favorable foreseeable conditions.[19] If the control of Gulf oil should fall into radical or pro-Soviet hands, it would give a hostile power major leverage over every aspect of free world economic activity.

THE IMPACT OF WESTERN AND GULF TRADE RELATIONS

All of the Gulf states—even Iran—are major trading partners with the West. The Southern Gulf states, however, are particularly heavy trading partners with the OECD states. They not only import substantial amounts of goods; they generally have a large enough trade surplus to make substantial overseas investments, and virtually all of these investments occur in the West.

This trade has been of major importance in spite of the "oil glut". In 1986, the Gulf nations and their immediate neighbors experienced a $9.5 billion trade deficit. Neverthless, they had a total import/export flow of some $168 billion. The trade between the U.S. and the nations in the region was worth over $16 billion, and the U.S. had a $3.6 billion trade advantage.[20]

Saudi Arabia is a particularly important trading partner, and plays a significant role in shaping world monetary policy and foreign aid activities. The patterns of Saudi trade have fluctuated sharply in recent years because of the rapid shifts in oil prices and in the Saudi market share, and accurate recent data are not readily available. In 1984, however, Italy, France, West Germany, Britain, Holland, and the other West European states all had a favorable trade balance with Saudi Arabia. Italy exported about $2.2 billion worth of goods and imported about $1.75 billion. France exported about $2.1 billion and imported $2 billion. West Germany exported about $2 billion and imported $0.8 billion. Britain exported about $1.9 billion and imported $0.7 billion. The Netherlands exported about $0.7 billion and imported

$0.4 billion. The rest of Europe exported about $1.1 billion and imported $1.0 billion.[21]

In 1985, the European Economic Community (EEC) exported about $10 billion worth of goods to Saudi Arabia in spite of the fact that the Kingdom's imports had fallen 15%, oil production had hit a ten year low, and the annual average production level had decreased to well below its 4.353 million barrel a day OPEC quota. The volume of trade continued to decline in 1986 and 1987, and is likely to do so in 1987 and 1988, but it should exceed $8 billion annually and favor Europe until oil prices begin to rise again in the early 1990s.[22]

Ironically, Saudi trade has been especially high with the U.S., although the U.S. has recently imported comparatively little Saudi oil.[23] There are now about 271 U.S.-Saudi joint ventures in Saudi Arabia worth about $3.7 billion. U.S. contractors won about 25% of all Saudi government contracts in 1983, worth about $4 billion. More broadly, U.S. exports to Saudi Arabia grew at an average annual rate of 35% before the recent decline in oil prices and the rise in the value of the dollar. Even today, Saudi Arabia is the sixth largest importer of U.S. goods, and U.S. exports take up roughly 20% of the Saudi market.

Since the early 1980s, the balance of U.S.-Saudi trade has followed the same pattern as European-Saudi trade, and has shifted decisively in favor of the U.S.. The U.S. imported about $73 billion worth of oil from Saudi Arabia between 1974 and 1984, and exported $48 billion worth of goods and services to Saudi Arabia. In 1982, however, Saudi Arabia imported $39 billion worth of goods and exported $50 billion. In 1983, Saudi Arabia imported $7.9 billion worth of goods from the U.S. and exported $3.8 billion. This left Saudi Arabia with a trade deficit of $4.1 billion.

The U.S. Department of Commerce estimates that Saudi Arabia imported $39.5 billion worth of goods in 1984, and exported $51.7 billion. It exported $4.7 billion worth of goods to the U.S. and imported $7 billion. This created a $2.3 billion trade advantage for the U.S. in a year when the high value of the dollar was otherwise creating a massive U.S. trade deficit. The Southern Gulf states as a whole account for over 60% of all U.S. exports to the Arab countries in the Middle East and Africa, and U.S. trade with Saudi Arabia accounts for the bulk of U.S. trade with the Southern Gulf state countries.[24]

The recent cuts in oil prices have sharply reduced the size of the West's trade surplus with Saudi Arabia, and any dollar estimates have been complicated by an increasing volume of barter deals. The Saudi cumulative current account deficit exceeded $67 billion, however, between 1983 and 1987, and barter deals are unlikely to alter the basic patterns in the flow of trade established in the mid-1980s, however, until oil prices again rise in real terms in the early to mid-1990s.

As in Saudi Arabia, Kuwait's trade patterns have fluctuated sharply in recent years, as have its major trading partners. Its foreign trade balance declined from a surplus of $2.6 billion in 1982 to $904 million in 1984, but its surplus has since risen to up to $6 billion per year. It typically, however, gets about 60% of its more than $6 billion in merchandise imports and $4 billion in services from the major Western states. For example, in 1984, Kuwait got roughly 20% of its imports from Japan, 10% from the U.S., 10% from France, 10% from the FRG, 7% from Italy, and 6% from the U.K.[25]

The trade statistics of the UAE are very complex because they are reported both for the federation and by individual sheikdom. Nevertheless, the UAE has shifted from a major trade surplus in the early 1980s to a deficit of $2 billion in 1987. It also has done most of its trade with the West. About 25% of its recent imports have come from Japan, 13–15% from the U.S., 11–14% from the U.K., 8–10% from the FRG, 5–9% from Italy, 6–8% from France, and 3–4% from the Netherlands.

Qatar has exhibited the same patterns in total balance of trade as the other Southern Gulf states, and has dropped from a $1.8 billion surplus in 1980 to one of only $0.2 billion in 1987. Japan now accounts for 20% of all Qatari imports, the U.K. for 15–17%, France for 5–9%, the U.S. for 9–10%, the FRG for 6–8%, Italy for around 5%, and the Netherlands for around 3%. It is interesting to note that the U.S. has accounted for around 10% of all Qatari imports during the last five years, although it imports less than 1% of Qatar's oil exports.

Bahrain has shifted from a trade surplus of $0.3 billion in the early 1980s to one of $0.1 billion in 1987. It too gets nearly 60% of its annual imports from the West—around 20% from Japan, 12–20% from the U.S., 14–15% from the U.K., 6–8% from the FRG, and 7–12% from Italy. Oman has been more careful to avoid trade deficits, although its balance on current account dropped from a surplus of nearly $1 billion in 1980 to only about $100 million in the mid-1980s and probably now is in deficit. Oman gets about 20% of its annual imports from Japan, 14–16% from the U.K., 4–8% from the FRG, 6–8% from the U.S., 3% from France, 3–5% from the Netherlands, and 2% from Italy.

These patterns are likely to continue for most of the 1980s, but they will then shift back towards the pattern of Gulf trade surpluses which existed in the early 1980s. This will further increase the importance of sound Western strategic relations with the Southern Gulf states.[26]

THE IMPACT OF SOUTHERN GULF CAPITAL INVESTMENT AND AID

There is no question that the drop in oil revenues has sharply affected the overseas investments of the Southern Gulf states. Saudi Arabia, for example, has seen its oil revenues drop from $119 billion in 1981 to roughly $22 billion in 1986 and 1987. It has had a budget deficit since 1982. It had a $24 billion deficit on current account in 1984, and may have had a $30 billion deficit in 1985. Saudi Arabia had at least a $20 billion deficit in 1986. It devalued the riyal against the U.S. dollar in June, 1986, and had to sell holdings in U.S. Treasury bills to keep its liquid international reserves of $25–27 billion.[27] Saudi Arabia's budget deficit in 1987 was probably over 40 billion riyals.[28]

The Saudi riyal has been devalued by 12% over the last three years, from 3.75 to the dollar to 3.35. Its FY1988 budget is projected to be 141.2 billion riyals ($37.7 billion), down some 17% from the 1987 level of 170 billion riyals. Oil revenues were unofficially projected at 65.2 billion riyals. The 1988 deficit was projected to be 35.9 billion riyals ($9.57 billion). Saudi Arabia sought to reduce the deficit through utility surcharges, 12–20% import duties, hospital charges, airline fare increases, foreign borrowing in the form of some $8 billion in bonds, and the imposition of taxes on foreigners up to 30%.[29]

The practical impact on Saudi Arabia has been to cause a limited net decline in GDP, and to force it into a series of budget cuts. Its 1988 budget will be 141 billion riyals ($38 billion), or 17% less than in 1987. Saudi Arabia has cut back on many areas of public expenditures, and has had to raise taxes and duties and issue bonds.

Saudi Arabia has seen its foreign reserves drop sharply over the last six years, from a height of around $150 billion. It still has over $50–55 billion in liquid foreign reserves, however, and $30 billion in fixed investments in the West. Many of these investments were invested in the U.S. and in U.S. government securities. This investment ties much of Saudi Arabia's non-oil income to the performance of Western economies. In fact, these ties are part of a broad link between the investment policies of the Southern Gulf states and the West.[30]

The Southern Gulf governments invested about $9 billion per year in the U.S. alone during 1980–1984, and private investment totaled billions more. Although U.S. Treasury figures are uncertain, Southern Gulf governmental investment in the U.S. peaked at $82 billion in early 1983, with $46 billion in U.S. government securities. Other sources report that Kuwait alone had over $95 billion invested in Europe and the United States, and the UAE had $55–$60 billion.

While this investment declined sharply during 1984–1988, as the result of the fall in oil revenues, and is likely to decline further until the early 1990s, it will probably recover by the mid–1990s. It represents a major source of foreign investment in the West, and is far larger than all other Middle Eastern and African investment combined.[31]

Gulf foreign aid has also been of continuing value to the West. For example, Saudi Arabia contributed some $48 billion in official foreign aid between 1970 and 1984, or roughly 5.8% of its GNP. This compared with an average of 0.3% of GNP for the industrialized countries, and neither the U.S. nor any other OECD nation currently gives significantly more than 1% of its GNP in foreign aid. Saudi Arabia provided $3.3 billion in foreign aid in 1984, or 3.2% of its GNP, in spite of the fall in its oil revenues and the fact that the OPEC countries as a whole cut their aid by nearly $1 billion. This meant Saudi Arabia remained the fourth largest source of economic aid to developing countries after the U.S. ($8.7 billion), Japan ($4.3 billion), and France ($3.8 billion). Saudi aid has been reduced since 1984, but the trends are difficult to estimate. The Kingdom has provided substantial aid to Iraq, the Sudan, the Afghan Mujahideen, and other recipients where the aid transfers are not publicly reported.

Saudi aid has gone to 70 different countries: 25 in Asia, 38 in Africa, and 7 in the Middle East, although over 80% has gone to other Arab and Islamic states.[32] With the exception of aid to Syria, virtually all Saudi aid has gone to recipient nations where such aid has been of strategic value to the West. Saudi Arabia has raised its IMF quota from 90 million SDR in 1974 to $3.2 billion SDR in 1984. This is ten times the rate of increase of the other six nations that sit on the IMF's Executive Board of Directors.[33]

Saudi Arabia has also been a primary source of Special Drawing Rights, or the equivalent of balance of payments loans, for the IMF's various aid funds. For example, Saudi Arabia has provided 2.3 billion SDRs—about 1/3 of the total—to the IMF's Special Oil Facility. Over 50 countries have used the facility—including 6 Western industrialized states. It has provided about 2 billion SDRs to the Supplementary Financing Facility, 8 billion SDRs to the IMF's Enlarged Access Program, and 1.5 billion SDRs to the General Agreement to Borrow. This had the net effect of providing over $10 billion worth of additional aid. Virtually all of this IMF aid went to nations where it directly served U.S. strategic interests as well as those of Saudi Arabia.[34]

Equally detailed data are not available on the other Southern Gulf states. Kuwait, however, has acted as a major aid donor, and the UAE has made substantial contributions to some of the other Arab states. Bahrain, Oman, and Qatar have never been significant sources of aid. It is clear, however, that Saudi Arabia and Kuwait are by far the major sources of both investment and aid. Further, while the volume of such investment and aid will drop significantly until oil prices recover, Saudi and Kuwaiti aid will still play an important role in the Arab world, the Horn, North Africa, and Afghanistan. Their aid is also almost certain to increase again in the 1990s, as oil revenues rise along with world oil prices.

THE BROADER ROLE OF THE SOUTHERN GULF STATES IN REGIONAL AFFAIRS

The political objectives of the Southern Gulf states differ from those of the West in several important ways—most notably in regard to the Palestinian movement and Israel. At the same time, the politics and culture of the Southern Gulf states do have a core of common interest with the West.

Saudi Arabia is the center of the conservative and moderate Islamic forces that offer the Arab world the option of modernization without radicalization. This position has generally led the Saudi government to take a pro-Western and anti-communist stand, and to systematically oppose Soviet efforts to expand its influence in the Gulf area and the rest of the Arab world. Saudi Arabia has consistently supported a policy of political and religious moderation in the Arab world, and the Kingdom has been one of the forces that has consistently attempted to move the Palestinian movement away from a posture of terrorism and armed struggle.

Saudi Arabia provided Oman with substantial aid in ending the radical and Soviet-backed Dhofar rebellion. It helped prevent a Nasserite and radical takeover in North Yemen. Its efforts to displace the USSR from South Yemen came close to success in 1978, helped push South Yemen into a peace with Oman in 1982, and are still the primary hope the West has of reducing Soviet influence in the PDRY. Though only partially successful, Saudi aid to North Yemen has been the main factor checking the growth of Soviet influence in that state.

Saudi Arabia has informally cooperated with the U.S. in aid to Jordan, Pakistan, the Sudan, and Somalia. In 1984 and 1985, it provided $550 million in aid to the Afghan freedom fighters, and exactly matched the U.S. aid program. It continued to provide substantial aid to them through the announcement the USSR would withdraw beginning May 15, 1988.[35]

The Kingdom played an important role in the 1981 negotiations of Special Ambassador Philip Habib, who was then attempting to remove all foreign forces from Lebanon. The Saudis worked actively and publicly to try to reach a solution in Lebanon after Israel's 1982 invasion, and for the September, 1983, cease-fire in the Shuf mountains. They helped secure Syrian agreement to several Lebanese attempts at peace negotiations.

The Kingdom has provided aid to friendly Sub-Saharan African states, and has made increases in its aid to states like Somalia and the Sudan, at the request of the U.S. Saudi Arabia has consistently encouraged Iraq and Syria to reduce their dependence on the USSR and is the only major pro-Western state with significant influence in both countries.

The other Southern Gulf states have played a more passive role. Kuwait has tended to shape its political role to protect it from the more radical Arab states. Its regional role has generally been one of accommodation, and one of supporting nations like Iraq and Syria, and movements like the PLO, whenever it could do so without any risk to its own internal political stability. Qatar has followed a similar pattern, often taking a slightly more neutral and anti-Western line. Bahrain has been cautiously pro-Western, shifting more towards an Arab centrist position whenever its ties to the U.S. have proved threatening. Oman has been more independent, has been much more openly pro-Western, and has been the only Southern Gulf state which could afford to ignore much of the Arab world's hostility to Israel. Even Oman, however, has had to show considerable caution whenever a crisis has arisen in the Arab-Israeli conflict or in Arab hostility towards the U.S.

The practical problem for the West is that all of the smaller Southern Gulf states are vulnerable to political pressure from the more radical and hardline Arab states. None have any real reason to be anti-Western, and all have tended to support the West in those areas where they can do so without risk to their internal security. All, however, must bow to the political currents in the Arab and Third World.

As for the net impact of such patterns on the West, the practical results have been twofold. The first has been to limit overt military ties to the West—a pattern whose causes and results will be discussed in far more detail in later chapters of this book. The second has been to support Syria. Kuwaiti and Saudi aid to Syria has been an important factor in the Syrian arms build-up. Kuwait and Saudi Arabia have, however, generally used their oil wealth to moderate Syrian actions in Lebanon and to try to work out a solution to the continuing civil war. Kuwaiti and Saudi diplomacy has helped reduce the risk of conflicts between Syria and Jordan, and the tensions between Jordan and the Palestinians that grew out of the conflict of September 1970. In 1985, for example, Saudi Arabia and Kuwait tacitly supported King Hussein's efforts to persuade the PLO to reject armed struggle, supported the easing of tensions between Syria and Jordan, and supported efforts to bring Syria and Iraq together to help end the Iran-Iraq War.

VULNERABILITY AND THE ARAB-ISRAELI PEACE ISSUE

The differences between the West and the Southern Gulf states over the Arab-Israeli peace issue have stemmed largely from two causes: first, from their vulnerability and need to maintain friendly relations with radical Arab states and movements, and second, from serious differences between their views and those of the U.S. over the proper tactics for reaching a peace agreement and preserving Palestinian rights.

Saudi Arabia has been the only Southern Gulf state to take an active role in the Arab-Israeli peace process, and it has steadily refused to react to U.S. efforts to persuade it to take a more aggressive and open position in moving towards direct negotiations with Israel. Like the other Southern Gulf states, the Kingdom has been extremely cautious about linking itself to U.S. peace initiatives. This would make it a target for Arab radicals and hostile states like Iran, would encourage radical and fundamentalist challenges to its legitimacy as the guardian of Islam's holy places, and would risk alienating Syria and creating a new political threat on its western border. Saudi diplomats are also deeply concerned with taking risks that end in failure. The history of recent peace efforts is one that makes them afraid they would simply increase Saudi vulnerability without producing any benefit.

Any stable pattern of Western strategic relations with the Southern Gulf states must recognize both that they are Arab states and are committed to Arab causes, and that their foreign policy is often the policy of weakness. The diplomacy of the Southern Gulf states has long been tailored to their vulnerability. They tried to compensate for their limited population and relatively small armed forces by maintaining friendly relations with radical political movements and states. They have constantly had to balance the conflicting demands of their ties to the West, their role in the Arab world, and, increasingly, their need to preserve their Islamic legitimacy.

As a result, the Southern Gulf states have either simply passively backed the "confrontation states" or have adopted a quiet "behind the scenes" style of diplomacy that avoids open confrontation.

Saudi Arabia is the key actor in this process, and makes the most useful case study. It has often used economic aid to build ties with states like Syria to minimize any political threat to the Kingdom. While the Saudi regime has long been privately hostile to Qaddafi, it has never ceased to try to maintain correct relations, and has often sent Qaddafi pro forma messages of support as a means of reducing his hostility.[36]

For all its differences with the U.S. over American support of Israel, Saudi Arabia played an important role in persuading President Sadat to make his initial break with the USSR and in giving Egypt the financial and military support it needed to reach the initial Sinai accords. In 1975, Saudi Arabia helped persuade Syria to accept Egypt's second disengagement agreement with Israel by trading Saudi aid to Syria and recognition of its right to head an Arab Deterrent Force in Lebanon for Syrian support of the Egyptian position and a cease-fire between Syria and the Lebanese Christians.

Saudi Arabia opposed the Camp David accords on the grounds that they made no provision either for Palestinian sovereignty on the West Bank and Gaza, or for the creation of a peace settlement that would support Jordanian, Syrian, and Lebanese stability. Saudi Arabia had little choice other than to reject the Camp David accords. Like Jordan, it could not accept the vague language regarding the settlement of the issues of the West Bank and Palestinian rights.

This Saudi opposition was not only a matter of principle; it reacted to the realpolitik of the region. Any other position would have severely threatened the Kingdom's internal stability and popular support for the Saudi government. Unlike Egypt—which could afford to make such concessions because its primary interests lay in reducing its defense burden, expanding its economic ties with the West, and obtaining the return of the Sinai—Saudi Arabia could not turn a blind eye to these aspects of the Camp David accords.

Even if Saudi Arabia had been willing to ignore the West Bank issue, in spite of 30 years of commitment to the Palestinian cause, it could not sacrifice its legitimacy as the custodian of the Islamic holy places and weaken the basic underpinning of the Saudi regime. Saudi religious legitimacy is not only a factor that increases Saudi Arabia's strategic importance to the West; it is a key factor in the Kingdom's regional and internal security. Saudi support of Camp David would have cost it the support of many of its citizens, many of its foreign workers, and most of the smaller Gulf states.

This Saudi commitment to the Palestinian cause did not, however, preclude a willingness to recognize Israel.[37] As early as May 11, 1977, Crown Prince Fahd talked of "complete, permanent peace" and a normalization of relations with Israel, stating that "all Arabs, including the Palestinians," were ready to negotiate a Middle East settlement with Israel if Israel recognized the full rights of the

Palestinian people. Prince Fahd asked President Carter to urge the Israelis to keep an open mind on a settlement that would be "just and lasting" and expressed "his strong hope that Israel would be reassured about the inclinations of his country toward the protection of Israel's security".

While upheaval in the Arab world following the Camp David accords led Saudi Arabia to step back from the peace issue, Prince Fahd reiterated these ideas once political tensions eased. In August 1981 he presented a formal eight-point peace plan that closely followed the UN resolutions supported by the United States. The Fahd plan called for

- Israeli withdrawal from all territory occupied in the 1967 war;
- Recognition of the Palestinian refugees' right to return to their homeland, or to compensation if they did not wish to return;
- Establishment of an independent Palestinian state in the West Bank and Gaza Strip, with East Jerusalem, which had been annexed by Israel, as its capital;
- A U.S. trusteeship of the occupied territories that would last only for a transitional period of a few months;
- Removal of all Israeli settlements from the occupied territories;
- Guarantees of freedom of worship in the Holy Land for all religions;
- Guarantees of the right of all states in the area to live in peace; and
- Guarantees of such an agreement by the United Nations or "some of its members," presumably the United States and the Soviet Union.

The first five points in the Saudi plan, and possibly the last three, were unacceptable to the Begin government in the precise form presented by Prince Fahd. Nevertheless, advancing the Fahd plan constituted tacit recognition of Israel's right to exist. Further, Saudi Arabia spent nearly two years quietly advancing these proposals to the Arab world, and made them the cornerstone of its diplomacy at the Arab summit at Fez, Morocco, in 1982.

Saudi Arabia has consistently pursued its own peace strategy ever since. It took a leading role in trying to bring peace to Lebanon after the Israeli invasion of June, 1982, and worked closely with the U.S. when it was trying to reach a joint arrangement with Israel and Syria. It supported the agreement that the U.S. worked out between Israel and Lebanon in the spring of 1983 and the U.S. efforts to persuade Syria to accept some modus vivendi with Israel after the agreement failed to bring peace to Lebanon. It has since made major diplomatic efforts to limit the conflict in Lebanon.

Saudi Arabia has responded favorably to most European peace initiatives, and initially supported President Reagan's September 1982 peace proposal. It backed the call for peace negotiations at the subsequent Arab summit meeting at Fez. Saudi Arabia still disagreed with some aspects of the U.S. position towards a peace settlement with Israel. It could not formally link itself to King Hussein's new peace initiative in 1985 because of the risk of alienating Syria and triggering new radical pressure from Syria, and from radical Palestinian elements in the Gulf. It did, however, strongly back the holding of the Palestinian National Council meeting in Amman, which was the critical step in making this peace initiative possible. Further, there are strong indications that Saudi Arabia has quietly begun informal direct contacts with Israel.

More generally, the Southern Gulf states severed formal diplomatic ties with Egypt after the Camp David accords, but most never reduced trade or their numbers of Egyptian workers. They tacitly supported the reintegration of Egypt into the Arab world during the mid-1980s, and steadily increased their low-level contacts. The Saudis and Kuwaitis quietly aided Egypt in gaining readmission to the Islamic conference, and publicly supported Jordan when it resumed relations with Egypt.[38]

Most of the Southern Gulf states joined Jordan in taking the lead in seeking the readmission of Egypt to the Arab League at its summit meeting in November, 1987. All resumed full diplomatic relations with Egypt in November, 1987. They also have supported UN resolutions that tacitly recognize Israel's right to exist and the need for Arab peace negotiations with Israel. Their objections to UN Resolution 242 have dealt with its treatment of the Palestinians solely as a "refugee problem" and not with its call for peace for all nations in the area. Their main differences with the West now stem largely from the political requirement to pay careful attention to the views of Syria and various Palestinian factions, and to the timing and nature of U.S. recognition of the PLO. While they are often forced to adopt a harder line of Arab political rhetoric, all at least tacitly accept the fact that a peace cannot be achieved with arms, and that any settlement must eventually recognize Israel's right to exist and its need for secure borders.

MOVING TOWARD A STRATEGIC PARTNERSHIP

Given this background, it should be clear that the basic strategic relationship between the West and the Southern Gulf countries is one of mutual dependence. The West is already heavily dependent on Gulf oil in spite of the present "oil glut", and will grow steadily more dependent until it can create new energy sources. Every major current projection of world energy balances now indicates that new energy sources like fusion will not be available until well after the year 2025. The Southern Gulf countries are the only part of the world with sufficient reserves and surplus production capacity to meet the West's near- and mid-term needs.

At the same time, the Southern Gulf states are dependent on the West's ability to pay for its oil imports and on Western exports. While they may have Arab and Islamic cultures, their economies are tied to the West and not the Arab world. Their societies have also become structurally dependent on the West for a wide spectrum of goods and services, ranging from food and most consumer goods to most development activity. When the Gulf states earn a cash surplus from their oil exports, they also must turn to the West for productive investment. One can spend money on the Third World; one cannot invest in it.

As long as the Southern Gulf states are free to export oil on the basis of a free market, both the West and Southern Gulf will share the same strategic interests. This will be even more true in the future as other exporting nations deplete their reserves. The competition for world oil exports is almost certain to increase with time. As long as the Southern Gulf states are free to export, however, the West should be able to make the transition to new energy resources without paralyzing rises in energy costs, critical energy interruptions, or being vulnerable to military or political attacks on their sources of oil imports.

Neither the West nor the Gulf can afford to ignore these realities because of temporary shifts in the oil market or in the Arab-Israeli conflict. This is as true of the U.S. as of any other Western state. The need for close strategic relations can be deferred, but in the long run, it cannot be ignored. This need will not diminish with neglect, and the only alternative is to increase the risk that far less friendly governments will come to control the economic life blood of the West.

NOTES

1. *Economist*, October 17–23, 1987, pp. 81–82.

2. For a good summary discussion of the trends involved, see the 1984 and 1985 annual reports of the Saudi Arabian Monetary Agency; David Fairlamb, "Why the Saudis Are Switching Investments", *Dun's Business Month*, May, 1985; pp. 56–60; "Saudi Arabia, Foreign Economic Trends and Their Implications for the United States", Department of Commerce, FET-85–79, September, 1985;

and the "The Gulf Cooperation Countries: A Survey", *Economist*, February, 1986, pp. 29–48, and October 17–23, 1987, pp. 81–82.

While the statistics involved are uncertain, Saudi Arabia seems to have averaged about 2.2 MMBD in 1985, with monthly production falling as low as 1.6 MMBD. The net Saudi revenue from crude oil and refined product in 1985/86 may have fallen as low as SR 90 billion ($24.61 billion). This forced Saudi Arabia to cut its budget target from SR 200 billion to SR 130 million. Saudi Arabia responded by raising its production level and triggering a major oil price war in late 1985. As a result, Saudi oil production rose to about 4.5 MMBD.

Saudi Arabia then set a 1986/87 budget target of SR 170 billion. The basic trends in the Saudi budget during the period between oil scarcity and oil glut are shown below. Defense amounted to roughly 25% of the Saudi budget, with manpower, operations, and maintenance costs rising to a level of $10 billion annually in 1986/87. It should be noted, however, that these projections predate the drop in the spot market to less than $15 a barrel and that further Saudi budget cuts are likely.

Saudi Arabian Public Expenditure: 1982/83 to 1986/87
(In Current Billions of SR with 1SR=$3.6507

	82/83	83/84	84/85	85/86	86/87
Expenditure	244.9	230.2	214.8	150/180	170/180
Revenue	246.2	206.4	169.6	130.0	150.0
Balance	+1.3	–23.8	–45.2	–20 to –50	–20 to –50

Source: SAMA, *MEED*, March 1, 1986, p. 24; and *Wall Street Journal*, March 22, 1985, p. 28. Data for 85/86 and 86/87 are estimates.

3. For typical reporting on the Saudi role in OPEC's 1986 price war see the *Economist*, February 8, 1986, p. 61 and April 26, 1986, p. 70.

4. These data are highly controversial, particularly because they ignore cost of recovery, which is far lower in the free-flowing wells of most of the Southern Gulf than in the Northern Gulf or the rest of the world. The USGS has working estimates of world oil reserves which are closer to 795 billion barrels than the figures shown in Table 2.3. The higher figures referenced above come from the *Oil & Gas Journal*.

5. The details of oil exploration in Saudi Arabia are summarized in the ARAMCO *Yearbook* and in *Facts and Figures*. For a good discussion of overall world reserves, see the *International Energy Annual* of the U.S. Energy Information Administration, Washington, D.C., DOE/EIA-02(84), pp. 79–81, and "Worldwide Report", *The Oil & Gas Journal*, December 31, 1984, p. 71. This latter report shows total world proved oil reserves as totaling 699 billion barrels, with 84 billion in Communist countries (63 billion in the USSR and 2 billion in Eastern Europe), 118 billion in the Western Hemisphere (27 billion in the U.S.), 55.5

billion barrels in Africa, 19 billion in Asia, 24 billion in Western Europe, and 398 billion in the entire Middle East.

6. Charles D. Masters, David H. Root, and William D. Dietzman, "Distribution and Quantitative Assessment of World Crude-Oil Reserves and Resources", USGS, Washington, unpublished, 1983.

7. Charles D. Masters, "World Petroleum Resources—A Perspective", USGS Open File Report 85-248, 1985. It is interesting to note that for all the talk of major discoveries, no major oil reserves have been discovered outside the Gulf in the last half decade, and such promising areas as China and Mexico have not proved to ease the world oil supply situation. The PRC now has only 19 billion barrels of proved reserves—well below the reserves required for its own demand levels in the year 2000—and Mexico has 49 billion.

8. James Tanner, "World Oil Reserves Rose 27% in Year as Producer Nations Boosted Estimates," *Wall Street Journal*, February 9, 1988, p. 42.

9. See Charles D. Masters, "World Petroleum Resources—A Perspective", USGS Open File Report 85-248, 1985, and Christopher Flavin, *World Oil: Coping With the Dangers of Success*, Worldwatch Paper 66, Washington D.C., Worldwatch Institute, 1985.

10. These estimates are confirmed by the U.S. Department of Energy. The DOE estimates that Kuwait has 92.7 billion in reserves, Bahrain has 170 million, Oman has 3.5 billion, Qatar has 3.4 billion, and the UAE has 32.5 billion (Abu Dhabi 30.5 billion, Dubai 1.44 billion, Ras al Khaimah 100 million, and Sharjah 450 million).

11. The estimation of both reserves and production capacity is now very complex because the data on Iran and Iraq have not been updated since 1980. The figures quoted here also count only production that can be sustained over a prolonged period. Installed capacity is significantly higher. See Anthony H. Cordesman, *The Gulf and the Search for Strategic Stability*, Westview, Boulder, 1984, pp. 16–18.

12. The reader should be aware that a major controversy exists over oil statistics, although not over the basic distribution of oil reserves discussed in the text. Sources like the *Oil & Gas Journal* add some 27% to previous estimates of reserves in February, 1988. These increased world oil reserves from 697.450 billion barrels to 887.348 billion. Virtually all of this rise, however, took place in the Middle East, where reserves rose from 401.879 billion to 564.680 billion. The Gulf states led this rise. Abu Dhabi alone tripled its estimate of reserves from 25.26 to 92.2 billion barrels. The other main source of increase was Venezula, which drove the rise in Western Hemisphere reserves from 120.165 to 146.417 billion barrels. Some of these claims, however, ignore economic factors. Iran also doubled its reserves to 92.85 billion barrels, but indicated that they "came from the highest authority—in the name of God."

Any rise in reserves was also offset by data from the International Energy Agency, which indicated that consumption was much higher than had previously been believed. It estimated that the consumption estimated understated actual world consumption by some 1.2 MMBD. The rise was driven by rising demand in developing countries, which rose from 13% of world demand in 1973 to more than 20% in 1986. *Wall Street Journal,* January 11, 1998, p. 2, February 9, 1988, p. 42.

13. Statement by General George B. Crist, *Status of the United States Central Command,* Defense Subcommittee of the House Appropriations Committee, February 22, 1988, p. 11.

14. See John S. Herrington, *Energy Security, A Report to the President of the United States,* Washington, Department of Energy, March, 1987. For other good projections of overall energy trends, and the impact of cuts in the flow of Gulf oil, see Energy Information Administration, *Impacts of World Oil Market Shocks on the U.S. Economy,* DOE/EIA-0411, July, 1983; Department of Energy, *Energy Projections to the Year 2000,* DOE/PE-0029/2, October, 1983, pp. 1–16 and Chapter II; and Secretary of Energy, *Annual Report to the Congress,* DOE-S-0010(84), September, 1984, pp. 1–12. For a more recent and more pessimistic projection, see CONOCO, *World Energy Outlook Through 2000,* April, 1985. Also see Richard B. Schmitt, "U.S. Dependence on Oil, Gas Imports May Grow", *Wall Street Journal,* April 23, 1985. For the details of European dependence on Gulf oil see OECD/IEA, *Oil and Gas Statistics, 1985,* No. 4, Paris, 1986. For an analysis of U.S. ability to substitute for any prolonged reductions in the flow of oil imports see the Office of Technology Assessment, *U.S. Vulnerability to an Oil Import Curtailment: The Oil Replacement Capability,* GPO 052-003-00963-3, 1984.

15. DOE, *World Energy Outlook Through 2000,* April, 1985, p. 5 and Chapter 5. The estimate of U.S. reserves has been cut significantly since this report was issued, and the "oil glut" is having the side effect of leading to overproduction or shutting in of marginal wells much sooner than had been previously estimated. Current projections for North Sea oil show a steady and sharp decline in production after 1986-87, depending on oil prices. See the *Wall Street Journal,* March 17, 1986, p. 2; *Washington Post,* March 9, 1986, p. K; and *New York Times,* March 18, 1986, p. D-7.

16. See John S. Herrington, *Energy Security, A Report to the President of the United States,* Washington, Department of Energy, March, 1987, pp. 223–228.

17. For a good summary of the impact on oil production and oil exporting states, see "Oil Turns Manic Depressive", *Economist,* February 15, 1986, pp. 61–62; Louis Uchitelle, "Oil Imports a Problem Again", *New York Times,* November 19, 1987, p. D–1; and "Energy: Another Deficit", *Economist,* November 14, 1987, pp. 35–36.

18. See John S. Herrington, *Energy Security, A Report to the President of the United States,* Washington, Department of Energy, March, 1987, pp. A-2 to A20.

19. See John S. Herrington, *Energy Security, A Report to the President of the United States,* Washington, Department of Energy, March, 1987, pp. A-2 to A-7. Also see Energy Information Administration, *Impacts of World Oil Market Shocks on the U.S. Economy,* DOE/EIA-0411, July, 1983; Department of Energy, *Energy Projections to the Year 2000,* DOE/PE-0029/2, October, 1983, pp. 1–16 and Chapter II; and Secretary of Energy, *Annual Report to the Congress,* DOE-S-0010(84), September, 1984, pp. 1–12.

20. Statement by General George B. Crist, *Status of the United States Central*

Command, Defense Subcommittee of the House Appropriations Committee, February 22, 1988, p. 11.

21. *Economist,* January 18, 1986, p. 59, and April 26, 1986, pp. 70–71. For detailed historical breakouts, see *Saudi Arabia, Foreign Trade Statistics, 1984 AD.*

22. See the Economist Intelligence Unit, *Regional Review: The Middle East and North Africa, 1986,* London, Economic Publications, 1986, pp. 210–214; and *Economist,* January 18, 1986, p. 59.

23. According to the Department of Energy publication, *Petroleum Supply Monthly,* Saudi Arabia provided the U.S. with less than 450,000 barrels per day of crude oil and product during 1984–1985 out of average monthly import levels ranging from 3.9 to 6.2 million barrels per day. Indonesia, Nigeria, Venezuela, Canada, and Mexico all provided equal or greater imports.

24. The previous statistics are taken from the U.S. Department of Commerce, "Saudi Arabia", *Foreign Economic Trends and Their Implications for the U.S.,* FET 84–80, July, 1984. For other recent trend data see the summary in Appendix A of *Saudi Arabia, A Country Study,* Department of the Army, DA Pam 550–51, Washington, D.C., 1985, pp. 32–322.

25. The trade statistics for the smaller Gulf states come from various annual editions of the Economist Intelligence Unit, *EIU Regional Review: The Middle East and North Africa,* London, Economist, 1985, 1986, and 1987.

26. The drop in oil prices affected Saudi Arabia the most before it led the cut in oil prices in 1986. Its average production dropped from 10 million to as low as 2.5 million barrels per day, and its current account swung from a surplus of almost $40 billion in 1981 to a deficit of nearly $25 billion in CY1985. This was equivalent to 30% of Saudi Arabia's GNP.

In the first quarter of 1986, however, the Kingdom ended its role as OPEC swing producer and increased oil output to 4.5 MMBD. This triggered a massive drop in oil prices, and the fall in income per barrel offset the increase total production. This forced the Kingdom to defer issuing its 1986/87 budget. This makes it almost impossible to estimate future trade patterns.

Similar problems affect all the other GCC and Gulf oil states except Kuwait. Kuwait's $85 billion in savings nearly equal those of Saudi Arabia, and it had a current account surplus of $5 billion in 1985. Kuwait may just be able to maintain a current account surplus in 1986, although it predicts $4 billion less in government revenues in 1986 than in 1985. *Economist,* April 26, 1986, pp. 71–72.

27. EIU, Annual Regional Review: *Middle East and North Africa, 1986,* p. 210.

28. *Economist,* October 17–23, 1987, pp. 81–82, and January 16, 1988, pp. 59–60.

29. *Wall Street Journal,* December 31, 1987, p. 4, January 5, 1988, p. 21, January 6, 1988, p. 12, January 7, 1988, p. 16, January 12, 1988, p. 2; *Washington Post,* December 31, 1987, p. E-3, January 12, 1988, p. C-3; *Economist,* January 16–22, 1988, p. 59; *New York Times,* January 6, 1988, p. A-1; *Chicago Tribune,* January 27, 1988, pp. 3–7.

30. For a typical estimate, see "New Saudi Budget Skims the Fat," *The Middle East,* February, 1987, pp. 19–21, and *Economist,* October 17–23, 1987, pp. 81–82. The most negative recent estimate of Saudi wealth is by Eilyahu Kanovsky, an economist at Bar-Ilan University in Israel. He estimates that the Saudi budget

surplus has dropped from $36 billion in 1980–81 to a $17 billion deficit in 1985–86. He estimates that Saudi monetary reserves have dropped from a peak of $150 billion to about $70 billion because $30 billion of the $100 billion reported to the IMF consists of uncollectable loans to states like Iraq. He also estimates Saudi defense expenditures for 1986–87 as being below $16 billion, but his estimate ignores oil barter and exchange agreements. See Hobart Rowen, "Reassessing Saudi Arabia's Economic Viability", *Washington Post,* July 20, 1986.

31. The figures used here are based on unofficial estimates provided by sources in the U.S. Treasury. Also see James Tanner, "Oil Glut May Cause Price Slide by Spring", *Wall Street Journal,* November 25, 1987, p. 6. David Fairlamb, op. cit., pp. 56–60; Osama Faquih, "Similarities in Economic Outlook Between the U.S. and Saudi Arabia", February 22, 1985; and Yusuf Nimatallah, "Arab Banking and Investment in the U.S.", IMF, February 22, 1985; and "The Economic and Fiscal Strategy of Saudi Arabia", Center for Strategic and International Studies, Georgetown University, Middle East Conference, March 20–21, 1985.

32. For detailed breakdowns of the distribution of Saudi aid, see *The Kingdom of Saudi Arabia: Relief Efforts,* Ministry of Finance and National Economy, Saudi Arabia, 1985, and *Annual Report of the Saudi Fund for Development, 1984–1985,* Saudi Arabia, 1985.

33. It is almost impossible to estimate the flow of Saudi aid to Syria with any precision. U.S. government officials feel the largest component of Saudi aid to Syria is an annual $529 million payment made under the terms of the 1978 Baghdad summit. An Israeli source, Shemuel Meir, estimates total OAPEC aid to Syria and Jordan as follows:

	1979	*1980*	*1981*	*1982*	*1983*	*% Change 1981–1983*
Jordan	1,008	1,212	1,172	948	691	–41
Syria	1,651	1,484	1,792	1,376	1,245	–30.5

See S. Meir, *Strategic Implications of the New Oil Reality,* Westview Press, Boulder, 1986, p. 55.

34. Ibid.

35. *Washington Post,* June 20, 1986, p. A-30.

36. The U.S. government stated that it had no evidence of Saudi aid to Libya in any aspect of its military development or terrorist activities in its background briefing papers for the Congress in justifying its 1986 arms sales to Saudi Arabia. See *Saudi Arms Sale: Questions and Answers,* February 24, 1986, p. 56.

37. The U.S. government officially rejected Israeli charges that the Saudis had provided significant arms to the PLO in its background papers defending its 1986 arms sales to Saudi Arabia. The only evidence of such transfers was an unattached shipping label Israel claimed had been found on a Palestinian arms cache near Damour. This label proved to be from a 1977 shipment to Saudi Arabia of .22 caliber long rifle and .30 caliber linked ball and tracer ammunition. No such ammunition has been found in captured PLO stores, and Israel

refused to reply to repeated U.S. requests for further ammunition and has never claimed to have found such ammunition in Palestinian hands. As for charges that Saudi Arabia provided M-16 rifles to the PLO, the U.S. never sold M-16 rifles to Saudi Arabia, and the serial numbers on arms captured by Israel turned out to be from M-16s sold to the Lebanese Army.

38. See U.S. Department of Defense, *Saudi Arms Sale Questions and Answers*, February 24, 1986, informal briefing paper for the U.S. Congress, p. X.17.

3

The Military Build-Up in the Gulf and the Role of the USSR

The West's strategic dependence on Gulf oil would be far less important in military terms if the the Southern Gulf states were less vulnerable. Unfortunately, the most powerful of the Gulf states is in the throes of a violently anti-Western revolution. Iraq's friendly relations with the West and the Southern Gulf states are too recent and too threatened by Iran to make Iraq's future posture predictable. At the same time, the Gulf is threatened by Soviet strategic ambitions and the developments in the more radical states on the fringes of the region.

All of the Southern Gulf states have small populations and severe limits on their ability to create modern and effective military forces. While Saudi Arabia is by far the largest of these states, it can play a significant strategic role in the Gulf, Red Sea, and Near East only if it cooperates with its neighbors and has the strength to deter or defend itself against possible threats. The Gulf and Red Sea areas are not only one of the world's most tempting strategic targets; they are one of the world's most unstable regions. Current threats, however, are only part of the problem. The Southern Gulf states cannot tailor their military forces or arms purchases against today's threat. They will not be able to make any major new equipment purchases fully operational and effective until the early to mid-1990s, and they must then be able to employ their "first line" weapons systems for a useful life of at least ten years. They are not buying or updating the forces to meet today's threat: They must buy a mix of modern equipment which will have a credible "life cycle" for at least the period of 1991–2001, and ideally for a life cycle of 1991–2015.

No one can predict the exact rate at which each potentially hostile nation will build up its military forces and technology, or the extent to which currently friendly or neutral states will shift their attitudes. No one can predict just how much the Soviet bloc will need oil, or the extent of Soviet ambitions. At the same time, some aspects of the future threats to the Southern Gulf states, the Gulf, and the free world's sources of imported oil are clear:

- A regional arms build-up is occurring throughout the region. The size of current and potential hostile forces and the weapons strength of current and potential threats are increasing far more quickly than Saudi Arabia and the smaller Gulf states can hope to increase their forces;
- There is a need to extend Saudi Arabia's limited deterrence and defense capabilities to cover five smaller, and often unstable, conservative Gulf states;
- There is a growing Soviet capability to intervene directly in the region which will take a far more tangible form if the USSR can fully suppress the Afghan freedom fighters;
- A major Soviet arms transfer and military advisory effort virtually surrounds the Kingdom;
- The threat from the Iran-Iraq War is almost certain to lead to continuing political tension and military build-ups through the end of this century;
- There is a threat of radical hostility and growing Soviet penetration in the Yemens;
- There is a threat of radical pressure from Ethiopia and possibly the Sudan;
- A massive military build-up in Syria may ultimately become as much a threat to Saudi Arabia and Jordan as to Israel;
- Inevitable shifts in the technology of the weapons deployed in the region will give hostile states technology roughly equivalent to the E-3A, Tornado B, and F-16A/Band F-15C/D by the mid-1990s.

BROAD PATTERNS IN THE REGIONAL ARMS RACE

The sheer rate of military build-up in the region presents a major challenge to the Southern Gulf states. They must compete with neighboring nations with far larger manpower pools, and seek stability in a region characterized by an arms race that has now lasted for two decades.

Tables 3.1 and 3.2 illustrate this point. These tables provide a rough estimate of the force trends in the region through 1990—the last date at which even rough regional estimates are possible. At the same time, they illustrate the rate of total force build-up in recent years, and the competition between major arms exporters and importers.

There is no rigid correlation between expanding military forces and arms imports and a conflict, or between the growth of neighboring military forces and the threat to the Southern Gulf states.

Nevertheless, their wealth and strategic position make them an obvious target for any state or combination of states that feels it can bring sufficient pressure to bear.

This "logic of arms" has long shaped regional perceptions of the strategic threats in the region—just as the U.S. must base the size of its forces against the potential threat posed by the USSR, and not on its current forces or political intentions (and seems certain to do so in the future). The Southern Gulf states must also take full account of the volatile nature of many neighboring regimes.

Egypt, for example, was once the strongest Arab radical state. It is now the strongest friendly and moderate Arab state in the Middle East. Libya has shifted from a conservative kingdom to a radical threat to every moderate and conservative Arab state. Syria has grown from a minor military power, whose internal political turmoil had little impact on the Gulf, to a military power whose forces numerically exceed those of Israel and whose current political stability depends on the life of one man and rule by a small religious minority.

Lebanon's stability has been transformed into a constant state of civil war, and Lebanon is now the scene of an emerging Shi'ite radicalism that has joined with Iran in becoming a growing threat to every moderate Arab state. The Sudan seems to be drifting into chaos, and the Yemens and Ethiopia have been a scene of political turmoil and military build-up for more than a decade.

While the detailed statistics in Tables 3.1 and 3.2 may be confusing to anyone other than an area expert, the broad trends are clear. Even if one makes the assumption that the current fall in oil revenues will continue to reduce the past rate of arms imports—and that Iran will only be beginning to rebuild the military strength it had under the Shah by 1990—the land and air threat to the Southern Gulf states will increase far more quickly than the Southern Gulf states can hope to increase their forces. This is not simply a matter of tanks and aircraft; it is a matter of increases in the capability of nations like Iran, Syria, and Iraq to project land and air forces outside their own territory:

- Syria is gradually becoming a major regional power as well as an Arab-Israeli confrontation state. It will increase its tank and aircraft strength while vastly increasing force quality. It will acquire the ability to strike deep into Iraq, Jordan, and Western Saudi Arabia, and to threaten the road and air routes to the Gulf from Jordan and Turkey.
- Iran will rebuild its forces. It will increase its present tank strength by at least 50%, and its air strength by well over 100% and possibly over 200%. It will strengthen its navy and improve its battle

Table 3.1: Major Weapons in Middle Eastern Forces Directly or Indirectly Affecting the Military Balance in the Gulf

Country	Main Battle Tanks						Combat Aircraft					
	73	79	82	84	88	92	73	79	82	84	88	92
Iran	920	1735	1110	1000	1000	1500	159	447	90	95	60	200
Iraq	990	1800	2300	4820	4500	4800	224	339	330	580	500+	530
Sub-Total	1910	3535	3410	5820	5500	6300	383	786	420	675	560+	730
Bahrain	0	0	0	0	60	90	0	0	0	0	12	18
Kuwait	100	280	240	240	260	300	34	50	49	49	80	95
Oman	0	0	18	18	39	60	12	35	37	52	53	70
Qatar	0	12	24	24	24	40	4	4	9	11	23	36
Saudi Arabia	85	350	450	450	550	700	70	178	191	203	226	239
UAE	0	0	118	118	136	160	12	52	52	43	65	72
Total GCC	185	642	850	850	1069	1350	132	319	338	358	459	530
North Yemen	30	232	714	664	683	700	28	11	75	76	73	85
South Yemen	50	260	470	450	470	550	20	109	114	103	68	120
Total Gulf	2175	4669	5444	7784	7722	9000	563	1225	947	1212	1160	1465
Egypt	1880	1600	2100	1750	2,250	1950	620	563	429	504	441	430
Jordan	420	500	569	750	986	900	52	73	94	103	109	136
Israel	1700	3050	3600	3600	3900	4000	488	576	634	555	586	530
Lebanon	60	0	0	142	90	120	18	16	8	3	7	21
Syria	1170	2600	3990	4100	4000	4200	326	389	450	503	478	490
Sub-Total	5230	7750	10259	10342	11226	11170	1504	1617	1615	1668	1621	1607

Algeria	400	500	630	700	910	1250	206	260	306	330	346	370
Libya	221	2000	2900	2800	2100	2300	44	201	555	535	544	550
Morocco	120	140	135	120	110	160	48	72	97	106	117	130
Tunisia	0	0	14	14	68	90	12	14	8	8	31	52
Sub-total	741	2640	3679	3634	3188	3800	310	547	966	979	1038	1102
TOTAL NEAR EAST	8146	15059	19382	21760	22136	23970	2377	3389	3528	3859	3819	4174
Djibouti	–	–	–	0	0	30	–	–	–	0	8	15
Ethiopia	50	624	790	1020	750	900	37	100	113	160	138	150
Sudan	130	150	190	73	175	180	50	36	30	34	43	57
Somalia	150	80	140	240	303	300	100	25	55	64	71	72
Turkey	1400	3500	3550	3532	3700	3700	288	303	402	458	412	420
TOTAL OTHER	1730	4354	4670	4865	4928	5110	475	464	600	716	672	714

Note: Numbers are generally adapted from IISS, *Military Balance*, JCSS, *Middle East Military Balance*, and SIPRI, *Year Book* for the appropriate year. All estimates for 1994 are made by the author.

Table 3.2: Near-Term Annual Trends in Arms Imports Impacting on the Gulf and Near East
(In Current $ millions)

	72	74	76	78	80	82	84	86	88	90	92	94
Gulf												
Iran	525	1,000	2,000	2,200	400	1,500	2,200	1,800	1,750	2,200	2,300	2,400
Iraq	140	625	1,000	1,600	1,600	4,600	7,700	4,500	4,800	4,500	4,500	4,600
Iran-Iraq Total	665	1,625	3,000	3,800	2,000	5,600	9,900	6,300	6,550	6,700	6,800	7,000
Saudi Arabia	100	340	440	1,300	1,800	2,600	2,600	2,400	2,400	2,700	2,900	2,900
Kuwait	5	10	80	300	40	110	390	500	550	600	640	690
Bahrain	–	–	–	–	80	5	40	80	80	70	70	75
Qatar	–	–	5	20	90	250	200	170	190	230	240	260
UAE	10	50	100	50	170	40	190	190	190	180	170	170
Oman	5	10	10	270	100	130	310	240	240	230	230	250
GCC Total	120	410	635	1,940	2,280	3,160	3,730	3,580	3,650	4,010	4,250	4,345
Gulf Total	780	2,035	3,635	5,740	4,280	8,760	13,630	9,880	10,200	10,710	11,050	11,345
Red Sea												
Sudan	20	30	50	120	100	170	110	80	120	140	140	140
Ethiopia	10	10	50	1,100	575	300	575	480	520	540	470	470
Somalia	20	90	100	240	190	70	70	80	100	100	110	120
North Yemen	10	10	20	90	550	240	100	220	240	260	280	290
South Yemen	20	40	40	140	240	50	90	90	110	120	130	140
Sub-Total	80	180	260	1,690	1,655	830	945	950	1,090	1,160	1,130	1,160

Levant												
Israel	300	950	975	900	825	950	675	1,200	1,400	1,400	1,550	1,600
Syria	280	825	625	900	2,700	2,300	1,500	1,200	1,500	1,800	2,000	2,400
Jordan	30	70	140	170	260	1,000	210	300	400	350	350	360
Lebanon	20	10	10	20	40	50	240	200	210	240	240	260
Sub-Total	630	1,855	1,750	1,990	3,825	4,300	2,625	2,900	3,510	3,790	4,140	4,620
North Africa												
Mauritania	–	–	20	30	–	10	20	15	20	25	30	30
Morocco	–	20	210	440	350	260	190	230	250	280	300	310
Algeria	10	20	320	725	525	1,300	525	700	650	650	720	750
Libya	160	330	1,000	2,000	2,200	2,900	1,800	1,400	1,800	2,000	2,300	2,300
Chad	–	–	10	5	1	3	40	75	75	60	60	60
Tunisia	10	10	10	35	140	60	140	280	300	310	320	330
Egypt	550	230	150	400	550	2,100	1,500	1,500	1,500	1,700	1,700	1,700
Sub-Total	730	610	1,720	3,639	3,766	6,633	4,215	4,200	4,595	5,025	5,430	5,480
Other												
Turkey	150	150	320	220	290	420	480	480	500	530	550	580
India	210	190	490	290	700	1,400	800	2,000	1,900	2,100	2,300	2,400
Pakistan	110	100	190	210	380	440	550	580	540	580	600	630
Afghanistan	20	80	50	90	10	160	400	200	230	230	250	250
Sub-Total	490	520	1,050	810	1,380	2,220	2,230	3,260	3,170	3,440	3,700	3,860
Total Region	2,710	5,200	8,415	13,869	14,906	22,743	23,645	21,190	22,565	24,125	25,451	26,465

Source: Author's estimates based on computer data provided by the U.S. Arms Control and Disarmament Agency. Data are historical through 1986. Estimates are provided from 1988 on. Figures represent current dollar value of actual deliveries.

readiness. By the early 1990s, its mix of new European and Communist bloc–made equipment is likely to roughly equal the quality of the equipment in NATO and Warsaw Pact forces, and Iran's forces will be supported by greatly improved C^3I and electronic warfare equipment.
- The Yemens will expand their forces but, more important, will improve in force quality. They will pose a significantly greater threat to Oman and Saudi Arabia.
- Libya will continue to expand its capability to arm and finance radical threats to the Gulf states.
- Ethiopia will begin to emerge as a significant regional military power, with the ability to fly and fight its own aircraft, and threaten traffic through the Red Sea as well as the Asir and the southeastern regions of Saudi Arabia.

The air build-up reflected in Table 4.1 will be particularly important. The bulk of the Soviet-made aircraft now in the Gulf, Levant, and Red Sea areas will be replaced during 1988–1995, and the USSR has already begun to transfer extremely advanced fighters like the MiG-29 to the Third World. Given the probable pattern in Soviet arms exports, the existing fighters in nations like the PDRY and Ethiopia will be replaced with fighter types with roughly three times the range, and five times the attack payload and lethality, of the existing fighters. The Soviet export versions of the MiG-21 and MiG-23 now in Iraq and Syria will be replaced by fighters with twice their range and three times their current attack payload and lethality.

By the early 1990s, the Southern Gulf states will confront Soviet fighter types with air to air combat capabilities roughly equivalent to the Saudi Air Force's F-15 C/Ds and Tornados, and probably forces with Soviet-made AWACS and electronic warfare aircraft. The threat the Southern Gulf states must plan for goes far beyond the issue of the growth in threat numbers. It includes the growth in threat capabilities and aircraft quality.

These arms trends reinforce the need for an effective Western arms transfer policy towards the Southern Gulf states, and Table 3.3 provides additional data on the trend in arms sales by seller nation. It shows that the world arms market has become steadily more oriented towards sales to the Near East and Southwest Asia. It also shows that Soviet and other Communist sales are steadily increasing relative to those of the West, and often in states which are serious potential threats to Saudi Arabia and the other Gulf oil exporters.

This conclusion is further reinforced by Table 3.4, which examines the trend in new arms agreements in the Middle East and the relative trends in total market share. The key shifts in Table 3.4 clearly reflect the growth of Communist arms transfers relative to those of the West. It is interesting to consider these data in light of the full range of information that the United States government has released on world arms sales:[1]

- The steady long-term growth of Middle Eastern arms imports has been part of a general global shift towards arms sales to Third World states. Sales to the Third World accounted for roughly 85% of all world arms imports in 1985, versus 78% in 1983 and 75% in 1973.
- Similar shifts are taking place in Third World defense expenditures. These have risen from 17% of the world total in 1973 to 21% in 1983 and 25% in 1985.
- Arms sales in North Africa, the Near East, and Southwest Asia in current dollars totaled nearly 42% of all world arms transfers, even though these nations accounted for only 8% of total world defense expenditures.
- Arms transfers to the Middle East from past sales almost totally dominated the arms transfers of developed states to developing countries. During 1980–1983, they represented 59% of U.S. transfers, 62% of major West European transfers, 40% of Soviet transfers, and 58% of other Communist transfers.
- Defense expenditures in North Africa, the Near East, and Southwest Asia in current dollars were only about 8% of the world total in 1983, the last year for which ACDA and the Central Intelligence Agency have hard data, but they were growing at a rate of over 11%. The defense expenditures of the OPEC states were growing even faster—at a rate of 13.3%.
- Arms sales to the Middle East during 1981–1983 increased at an annual rate of roughly 11.5%. This compares with a rate of about 1.1% for the developed world, 7.7% for the developing world, and 3% for NATO.
- This high rate of growth in arms sales to the Middle East compares with 2% for NATO, 2.1% for all OECD countries, 2.6% for the Warsaw Pact, 1.9% for all of North America, 2.5% for East Asia, 2.5% for all of Europe, 4.6% for South Asia, 5.0% for Latin America, and 8% for Africa.
- The expenditures of the OPEC states alone totaled 36% of all world arms imports. This compares with 11.4% of all world arms imports for NATO, 15.7% for all OECD countries, 6.1% for the Warsaw Pact, 1.9% for all of North America, 9.8% for East Asia,

Table 3.3: Middle Eastern Arms Imports by Importing Country: 1981–1985 (Current $ millions)

Importer		Seller Nation										
	Total	USSR	US	FR	U.K.	FRG	IT	CZ	PRC	R0	P0	Others
Southern Gulf												
Saudi Arabia	14,760	–	6,400	4,300	1,400	190	170	–	–	–	–	2,300
Kuwait	1005	90	230	360	20	210	80	–	–	–	–	15
Bahrain	115	–	20	10	5	60	10	–	–	–	–	10
Qatar	895	–	10	650	230	–	–	–	–	–	–	5
UAE	560	–	40	130	220	70	40	–	–	–	–	60
Oman	955	–	90	40	550	240	10	–	5	–	–	20
GCC	18,290	90	6,790	5,490	2,425	770	310	–	5	–	–	2,410
Sub-Total	48,650	7,860	6,790	11,790	2,695	1,470	800	210	3,680	–	645	13,900
Iran-Iraq												
Iran	6,435	370	–	–	100	–	–	30	575	–	20	5,340
Iraq	23,925	7,400	–	5,100	170	700	490	190	3,100	–	625	6,150
Sub-Total	30,360	7,770	–	6,300	270	700	490	210	3,675	–	645	11,490
Red Sea												
Sudan	560	–	140	30	10	120	–	–	80	–	60	170
Ethiopia	2,100	2,000	–	–	–	–	30	20	–	–	–	50
Somalia	365	–	70	10	5	–	140	–	30	–	0	110
YAR	1,675	850	90		–	10	–	–	–	–	150	575
PDRY	1,110	1,100	–	–	–	–	–	–	–	–	–	10
Sub-Total	5,810	3,950	300	40	15	130	170	20	110	–	210	905

Levant												
Israel	4,105	–	4,100	–	–	–	–	–	–	–	–	5
Syria	8,950	8,000	–	50	60	20	–	350	110	–	10	350
Jordan	3,805	525	850	1,100	1,200	–	–	–	10	–	–	120
Lebanon	630	–	450	140	–	–	10	–	–	–	–	30
Sub-Total	17,490	8,525	5,400	1,290	1,260	20	10	350	120	–	10	500
Other												
Turkey	2,120	–	1100	20	110	800	10	–	–	–	–	80
India	6,070	4,200	60	550	800	120	30	60	–	–	110	140
Pakistan	2,190	–	1300	340	20	40	30	20	350	–	–	90
Afghan.	1,590	1,500	–	–	–	–	–	70	–	–	–	20
Sub-Total	11,970	4,700	2,460	910	930	960	70	150	350	–	110	330

Source: Adapted from ACDA, *World Military Expenditures and Arms Transfers, 1989*, Washington, GPO, 1987, pp. 145–146.

Table 3.4: Trends in Western Market Share of Arms Sales to the Middle East—New Agreements (In Billions of Current U.S. Dollars)

Exporter	1977–1980			1981–1985		
	$Billions	% of All Sales in Middle East	% of Total World Wide Sales of Seller	$Billions	% of Total Sales in Middle East	% of Total World Wide Sales of Seller
Non-Communist						
U.S.	14.6	29	38	15.2	20	31
France	12.0	24	68	13.1	17	68
U.K.	4.7	9	52	4.5	6	51
FRG	1.5	3	22	1.6	2	21
Italy	3.1	6	51	1.6	2	33
Other NATO	0.6	1	14	–	–	–
Other Non-Communist	1.5	3	25	16.1	21	54
Communist						
USSR	9.2	18	20	18.4	24	33
Other Warsaw Pact	1.2	2	17	1.3	2	18
Sub-Total	10.4	21	19	19.7	26	31
PRC	1.4	3	63	4.3	6	78
Grand Total	49.8	100	34	76.1	100	40

Source: Adapted from Arms Control and Disarmament Agency, *World Military Expenditures and Arms Transfers, 1985*, Washington, GPO, 1986, pp. 42–47; and *World Military Expenditures and Arms Transfers, 1986*, Washington, GPO, 1987, pp. 145–146. Figures for 1981–1985 are not directly comparable since ACDA does not break out for other NATO, and other Warsaw Pact countries in the 1986 edition.

18.3% for all of Europe, 5% for South Asia, 7.6% for Latin America, and 15% for Africa.

- The rate of increase in arms imports in the Middle East was faster than in any other region except Latin America, which imports only about one-sixth as many arms. The other two high-growth regions were Africa (9%) and South Asia (9%).
- The total volume of arms imports by Middle Eastern and Gulf states rose from $29.0 billion in 1974–1978 to $65.4 billion in 1979–1983.

- Communist arms transfers to Middle Eastern and Gulf states rose from \$24.5 billion to \$44.4 billion. Soviet sales rose from \$7.5 to \$20.4 billion, and sales in other Communist states including the PRC and Vietnam rose from \$0.5 to \$4.3 billion.
- European arms sales to Middle Eastern and Gulf states rose from \$5.5 billion in 1974–1978 to over \$16 billion in 1980–1983. French arms sales led the increase, rising from \$1.8 billion to \$9.7 billion. German sales rose from \$1.0 to \$1.2 billion. Italian arms sales rose from \$0.6 to \$1.3 billion. Britain's sales rose from \$2.1 to \$5.1 billion.
- Growing competition from Asia, Latin America, and other European states increased the size of sales to Middle Eastern and Gulf states from \$1.9 to \$2.3 billion.
- More recent U.S. reporting includes South Asia and North African states with Middle Eastern and Gulf states. These reports show, however, that the Near East and Southeast Asia accounted for 75% of all new Third World arms agreements in 1982–1985, and 73% of all deliveries.
- The U.S. and Soviet Union now have a virtually equal share of sales to the Near East and Southeast Asia. The U.S. has 26% and the USSR has 25.7%. West Europe has 27.3%. The USSR leads, however, in actual arms deliveries, with 26.3% to 27.5% for the U.S. and 21.9% for West Europe.
- The U.K. has experienced the largest recent growth in market share, thanks to the Tornado sale. The British share rose from 1.6% in 1984 to 21.8% in 1985. The U.S. increased its sales to the Middle East, but lost market share. U.S. arms sales rose from \$28.5 billion in 1976–79 to \$30.5 billion in 1980–83. This was the worst performance of any major exporter except West Germany, which could not complete several key contracts because of political constraints on arms exports.
- Because of the high unit cost of Western arms, particularly those of the U.S., the Western share of actual weapons transfers is far lower than the share of arms sales measured in dollar terms. From 1978 to 1985, the USSR led in the delivery of tanks, self-propelled guns, artillery, APCs and AFVs, supersonic combat aircraft, subsonic combat aircraft, other aircraft, surface-to-air missiles, helicopters, major surface combatants, and submarines. West Europe has led in minor surface combatants and guided-missile patrol boats. The U.S. has not led in any category.

The latest data on total sales to the Middle East released by the U.S. Arms Control and Disarmament Agency also reflect the growth of Soviet influence. These data are shown in Table 3.4. The USSR increased its share of Middle Eastern arms sales from an average of 18% during 1977–1980 to 24% during 1981–1985. During these same periods, the U.S. share of the Middle East arms market dropped from 29% to 20%. If one analyzes the directly comparable four-year period of 1977–1980 with the four-year period of 1981–1984, Soviet arms sales to the Middle East rose from $9.2 billion to $15.9 billion, and from 18% to 23% of all sales. Total Communist arms sales to the Middle East rose from $10.4 billion to $19.6 billion, and from 21% to 28% of all new sales agreements.

The U.S., in contrast, suffered sharply from its inability to make new arms agreements in the face of pro-Israel lobbying groups and pressure on the Congress. New U.S. arms agreements dropped in value, even in current dollars, from $15.6 billion in 1973–1976 to $14.6 billion in 1977–1980, and to $11.9 billion in 1981–1984. In contrast, new Soviet arms agreements rose in value, in current dollars, from $9.2 billion in 1977–1980 to $15.9 billion in 1981–1984. The U.S. lost about 8–10% of its market share in the Middle Eastern arms sales—or nearly $10 billion dollars—during 1981–1984, and this drop preceded the virtual collapse of U.S. arms sales efforts that had begun in 1985. While West European sales helped offset the decline in U.S. influence, they did not rise enough to prevent a major growth in Soviet arms sales and military influence.

The U.S. share of the regional arms sales market has dropped for several reasons, but all of these reasons have had the net impact of increasing the relative influence of the Soviet Union throughout the region. The U.S. is not a major supplier to Iran or Iraq. Many Arab states are turning to other arms sellers because they feel the U.S. has failed to move Israel towards a peace settlement. The Southern Gulf countries and most Arab states no longer perceive the U.S. as a reliable arms supplier because of Congressional refusal to support the export of advanced arms. The Soviet Union and other Communist states have made arms sales a critical part of their hard currency exports and do not have to sell their arms at market prices.

Fortunately for the West, Western European and other non-Communist arms exporters have expanded their sales. They have done so for several reasons. One is price: The strong dollar has raised U.S. costs, and American arms prices have also risen more quickly in real terms than those of France, Italy, and most Asian, Latin American, and other suppliers. Further, nations with state-owned and state-supported arms industries often can undersell the U.S. because of subsidies and more favorable credit terms. Nevertheless, the main reason is political: No other non-Communist arms exporter now has significant military ties to Israel.

The relative position of specific Western states in selling military- and defense-related services and construction is more complex, but again Europe is increasing its importance relative to the U.S. The U.S. is steadily losing its share of military-related construction because many competing local and Asian companies can offer these services more cheaply. The U.S. share of the service and maintenance market has grown because of the added complexity of U.S. weapons systems, but European nations have positioned themselves to compete in much of this market. The expansion of British and French sales to both Saudi Arabia and the other Gulf countries will inevitably expand Europe's share of sales of military services, and Singapore, South Korea, and Taiwan are beginning to offer service and support for U.S. combat aircraft.

At the same time, the relative impact of Soviet arms sales is increasing and with it the vulnerability of Saudi Arabia, the GCC, and other moderate Arab states. This growth in Soviet military influence in the region is likely to continue through the next decade because of (a) U.S. inability to become a reliable source of arms and military assistance to any Arab state not formally at peace with Israel, (b) growing pressure on the USSR to sell arms as a main source of hard currency, and (c) advances in technology that will lead to major new arms exports, many of which the U.S. may be unable to provide for domestic political reasons.

THE IMPACT OF TECHNOLOGY TRANSFER AND THE PROBLEM OF CONVENTIONAL PROLIFERATION

The rate of technology transfer to the Near East and Southwest Asia has increased to the point where it is becoming a major problem. A decade ago, there was usually a 5–10 year lag between the initial deployment of major new weapons systems in U.S., NATO, Soviet, and Warsaw Pact forces, and any large-scale sale of such arms to the developing world. That lag is now being eliminated. Western Europe is selling aircraft, armor, and ships to developing nations at the same time it introduces such systems to its own armed forces. The USSR is not only selling its new MiG-29 fighters to India; it is selling coproduction rights. Soviet SS-21 missiles appeared in Syrian forces almost at the same time they became fully operational in Soviet forces in East Germany.

Both the West and the Southern Gulf states face a massive problem in terms of "conventional proliferation". For example, by the time that all of Saudi Arabia's new Tornados become fully operational, a wide range of new weapons technologies will be present in the forces of hostile nations. These technologies are summarized in Table 3.5.

Table 3.5: Key Near-Term Trends in the Technology of Arms Sales to the Middle East from 1988 to 1995

Weapon/Technology	Impact
Challenger, AMX-40, M-1, T-80	Advanced tanks with 3rd and 4th generation fire control systems, spaced and other advanced armor, and advanced 120 mm guns. Will be matched by advanced types of other armored fighting vehicles.
ITOW, HOT, AT-6, AT-7, and AT-8, Hellfire	Advanced anti-tank missiles with full automatic tracking or fire forget capability.
MRLS, BM-24, BM-25, ASTROS	Western and Soviet multiple rocket launchers capable of firing advanced submunitions and "smart" minelets at ranges beyond 30 km.
Night vision devices	Widespread use of night vision devices, "24 hour" infantry, helicopter, and armored combat.
Secure, switched, advanced communications	Conversion to advanced secure communications with automated tactical message traffic and battle management capabilities.
SA-10, Patriot, Improved Hawk, SA-12	Advanced surface-to-air missiles which cannot easily be suppressed with current weapons and electronic warfare means. Many will be netted with advanced sensor and battle management systems and linked to advanced short-range systems.
SHORADS: SA-14, Stinger-POST, etc.	Next generation short-range crew and man portable surface-to-air missiles and radar guided AA guns with far better tracking and kill capability and greater ranges. Many will be "netted" into an integrated battlefield and point defense system.
E-3A (Imp), E-2C (I), IL-76, SUAWACS	Airborne warning and control aircraft capable of managing large-scale air wars using radar and electronic support measures (ESM) equivalent to NATO-level capabilities.
F-15E, MiG-29, SU-27, Lavi, F-16C, F-20A, Mirage 2000, Tornado	Next generation air combat and attack fighters with far more accuracy and up to twice the range payload of existing fighters.
Aim 9L/M, Phoenix, Mica, AA-8, AA-X10, AA-X-P2, Super 530, Python III	Advanced short- and long-range multi-aspect air-to-air missiles which greatly improve the air-to-air combat capability of all modern fighters.

Table 3.5 (continued)

Durandal, Paveway ERAM, ACM, SUU-65 WASP, J-233	Advanced air-to-surface munitions including runway suppression, anti-armor, anti-hardpoint, anti-personnel, anti-radar, and other special mission point and area weapons with far more lethality than current systems. Many will use stand-off weapons like glide bombs or advanced dispensers for low-altitude single pass penetrations under radar.
RPVs, IMowhawk, MiG-25 (I)	Improved air borne sensor and reconnaissance platforms which can provide advanced targeting, intelligence, and battle management data.
PAH-2, AH-64, Mi-24	Next generation attack helicopters with much longer ranges, improved air-to-air missiles, 3rd or 4th generation launch and leave anti-tank guided missiles, and air defense countermeasures. Will be supported by steadily improved troop lift helicopters with improved protection and firepower.
Peace Shield, Project Lambda, Lion's Dawn and C^3I/BM Systems	Air sensor and battle management systems equivalent to NATO NADGE level systems for integrating fighter and SAM defenses. Many with advanced attack mission control capabilities.
Maritime Patrol Aircraft	More advanced versions of E-2C-type aircraft armed with ASW weapons and air-to-surface missiles.
FAC(M), Missile Saar 5, Lupo, F-2000, etc.	Next generation missile patrol boats and corvettes with frigates: Improved Harpoon and other moderate-range advanced ship-to-ship missiles.
Sea Skua, Harpoon II, Exocet II, Gabriel III/IV, AS-4, AS-6, AS-7	Advanced ship, shore, and air launched anti-ship missiles with advanced sensors and electronics, and far more lethal payloads. Can kill war ships and tankers far more effectively than today.
Coastal Submarines	Advanced diesel submarines with excellent silencing, moderate cruise ranges, and smart torpedoes.
SS-22/SS-23	Advanced surface-to-surface missiles with ranges up to 900 miles.
Nerve gas	Widespread stocking of single or binary nerve gas agents and limited CBW defense capabilities.

Western arms sales to the Southern Gulf states can have only a limited effect in helping friendly Gulf forces cope with these trends. They will almost certainly be unable to use technology to fully compensate for their limited ground and naval strength. While it may maintain some technical "edge" over the equipment in the forces of potential threats, this edge is unlikely to compensate for their superior mass. The only forces in which the Southern Gulf states can hope to use technology to compensate for their weakness in numbers are air and naval forces, and it is unlikely that they can achieve sufficient deterrent strength in these areas before the mid-1990s at the earliest.

THE IMPACT OF NEW SOVIET SYSTEMS

It is also important to note that advanced systems like the Tornado, F-15C/Ds, Mirage F-1s, and Mirage 2000s in the forces of the Southern Gulf states will not face a MiG-21, F-4, F-5, or MiG-23 threat in the late 1980s and 1990s. They will face a threat equipped with a wide range of new Soviet systems. Table 3.6 provides a brief technical summary of some of the new Soviet systems that will enter hostile air and air defense forces in the late 1980s and 1990s, and provides a clear indication of why Saudi Arabia is seeking to modernize its air forces.

The USSR is making much more rapid progress in fighter design than many experts previously predicted. The MiG-23 in service with Warsaw Pact forces marked the first major departure in twenty-five years from the relatively simple and mass-produced MiG-15, 19, 17, and 21 and the Sukhoi Su-7 through Su-22. Its Tumansky R-27 engine was the first Soviet afterburning turbofan and its Highlark doppler air intercept radar and laser range finder made it roughly comparable to the F-4J.

The Su-24, which appeared in 1974, was the first Soviet fighter optimized for the attack role, and seemed a close copy of many features on the F-111. Its new features included a swing wing design, a separate weapons control officer, and an all-weather and terrain-following capability, plus extensive internal fuel space and an external munitions carrying capability.

While the Fencer has not yet been deployed outside the Warsaw Pact, it seems likely that the Su-24C will appear in the forces of potential threats by the early 1990s. The Su-24C has a modern pulse doppler radar and advanced attack avionics, and a new inertial navigation system (INS) and electronic countermeasure (ECM) suite

Table 3.6: New Soviet Systems Affecting the Gulf and Red Sea Threat

Airborne Warning and Air Control Systems

- Il-76 Mainstay: An E-3A-like aircraft which entered developmental production in 1984, and which has a rotating radome saucer and refueling probe. Likely to enter Gulf and Red Sea threat forces in the early 1990s.

Fighters and Fighter Bombers

- MiG-23MF Export II: An improved version of the MiG-23 Flogger E with the full High Lark radar. This expands radar coverage of the export version with the Jay Bird radar from a search range of 18 miles to 53 miles and tracking range from 12 to 34 miles. Unlike the previous export MiG-23, this fighter has good look up capability and a limited look down capability. It has much more sophisticated avionics, built-in ECM, and possibly an optional IR pod. Unlike previous export versions, it can use advanced Soviet air-to-air missiles like the AA-7 (an IR/SARH missile with a 20 mile range) and AA-8 (a multi-aspect missile similar to the AIM-9J with a range of 3.5–4 miles). This aircraft may enter service in Syria and Iraq during 1985 or 1986.
- MiG-29 Fulcrum: First of a new generation of Soviet fighters which is roughly equivalent to the F-16. Entered service in 1984, and has a large pulse-doppler with moderate look down/shoot down radar with day/night and all-weather combat capability. Maximum speed is Mach 2.2, and combat radius is 500 miles. Primarily an air defense aircraft, but is equipped for dual-role attack mission. Carries six AA-10 missiles (similar to the AIM-9L and AIM-9M), and has wing and fuselage racks for bombs and rockets. To be coproduced in India. Likely to enter Iraqi, Syrian, and Ethiopian forces in the late 1980s.
- SU-27 Flanker: Similar to the MiG-29 in air defense and dual role capability, but somewhat slower and with much longer range. Maximum speed is Mach 2.35 and combat radius is 715 miles. Carries eight radar homing AA-8 missiles and up to 13,000 pounds of attack munitions. May enter Gulf threat forces in the late 1980s, but the early 1990s is more likely.
- MiG-31 Foxhound: The first Soviet long-range fighter with an advanced look down, shoot down capability. A two seat air defense oriented fighter with a superior pulse doppler radar and avionics display and the ability to carry eight air-to-air missiles. Maximum speed is Mach 2.4 and combat radius is 930 miles. Carries a new radar homing version of the AA-9. Likely to enter Gulf threat forces in the early 1990s.
- SU-22 Fitter: An improved version of the Fitter J attack aircraft. This aircraft has been supplied to Libya and may enter Gulf threat forces during the next three years. Performance data are uncertain.

(continued)

Table 3.6 (continued)

- SU-24 Fencer: An advanced long-range attack fighter which entered service in 1984. It has advanced attack avionics and a weapons officer who sits next to the pilot. It has an advanced variable wing system and roughly five times the range and payload of any previous Soviet attack fighter. Maximum speed is Mach 2.18. Combat radius in LO-LO-LO missions is over 200 miles. HI-LO-HI radius with 4,000 pounds of munitions is 1,100 miles. Unlikely to enter Gulf threat forces until the early 1990s. Maximum payload is 17,635 lbs of attack munitions.
- SU-25 Frogfoot: A new Soviet close support attack fighter which has demonstrated excellent performance in Afghanistan. Somewhat lighter and slower than the U.S. A-10, but more maneuverable. Carries a heavy caliber tank killing gun, rockets, and bombs. Maximum speed is 546 mph; combat radius is 345 miles.

Surface-to-Air Missile Systems

- SA-5 Gammon: Long-range high-altitude radar guided SAM defense system with speed above Mach 3.5, a slant range of 185 miles, and a ceiling of 95,000 feet. Already deployed in Syria.
- SA-8 Gecko: A mobile six launcher vehicle mounted SAM system with a range of 6–8 miles and a ceiling of 20,000 feet. Now being deployed in Jordan and Syria, and may be entering the Yemens and Ethiopia during the next year.
- SA-10: A single state advanced SAM system which became operational in the spring of 1984. The SA-10 can accelerate up to 100 "G" to a cruising speed of Mach 6, and has a range of 60 miles with an advanced radar and multiple target engagement capability. Will probable replace the SA-6 and SA-3 in the Near East and Gulf.
- SA-11: An advanced new short-range system now deployed alongside the SA-6 in Soviet forces with radar guidance and speeds in excess of Mach 3. Range is 18.5 miles. The four rail launcher is vehicle mounted and can be netted into an air defense system along with the SA-6. This system is likely to deploy to the Gulf area in the late 1980s.
- SA-12: A dual mode anti-aircraft and anti-cruise missile system just entering production. The dual launcher system has an estimate range of 60 miles. This advanced system is unlikely to deploy to the Gulf before the early 1990s.
- SA-13: A tracked vehicle mounted missile which replaces the SA-9, and is netted with the ZSU-23-4 tracked radar guided AA gun. It has a range of 5 miles and can hit targets at altitudes between 165 and 16,000 feet. It has low-altitude tracking capability far superior to previous light Soviet SAMs and very fast reload capability. Already deployed in Syria.

Sources : Various editions of *Aviation Week, Jane's, Air Force Magazine, Jane's Defense Weekly*.

roughly comparable to NATO avionics. It can deliver 2,000 kg of payload at a combat radius of up to 1,800 km, and has a combat radius of 322 km [(LO-LO-LO)-950 km (LO-LO-HI)] with 2,500 kg of payload. Its terrain-following capability means that only a fighter with advanced look-down/shoot-down capability, like the F-15C/D MSIP, can successfully defend against it.

The USSR's new first-line fighters continue this trend. The MiG-31 is an upgraded MiG-25 with an advanced "semi-look down" radar, an additional crew member, 6 advanced AA-9 Acrid air-to-air missiles with limited anti–cruise missile capability, a new engine with better subsonic performance, and considerable additional engine and air frame refinements over the MiG-25. It is specifically designed to work with the USSR's new Mainstay AWACS.

It is the Su-27 and MiG-29, however, that are most comparable to advanced Western fighters like the F-15, which are designed for high performance at subsonic speeds, and which Saudi Arabia must plan to face no later than the early 1990s. These two aircraft both have heads-up displays (HUDs), advanced infrared search/track sensors, full all-weather mission capability, digital data links, and true look-down/shoot-down capability. They have hard points and connections for sophisticated attack mission avionics, and can fire air-to-air missiles beyond visual range.

While the F-16, F-15, and Tornado are often treated as fighters which have no equal in the forces of potential threat nations, it is important to note that both the Su-27 and MiG-29 can outmaneuver the F-15C/D, F-16A/B, and Tornado F.3 in a number of ways in a substantial part of their flight envelope. Both Soviet fighters have a nominal maximum speed of Mach 2.3, and the MiG-29 seems to have a higher thrust-to-weight ratio than the F-16A/B while the Su-27 is both slightly larger than the F-15C/D and Tornado and has a Tumansky R-31 engine with a 13,600 kg thrust which is 25% higher than that of the F-15C/D.[2]

Equally impressive, the USSR is matching or surpassing the former U.S. advantage in fighter radar range and avionics sophistication. While the exact radar range of the Su-27 and MiG-29 is not yet public, various U.S. government sources have made it clear that the Foxfire radar on the MiG-25 can detect fighter sized targets at ranges of 60 nautical miles, which is beyond the detection range of Saudi Arabia's current APG-63 radar. They have also noted that the MiG-23's Highlark radar exceeds the detection range of the F-16A/B—which lacks the MiG-23's look down/shoot down capability. The long-range track-while-scan pulse Doppler radars of the MiG-29 and Su-27 match or outperform the radars on the F-15C/D and Tornado F.3.[3]

THE SOVIET MILITARY THREAT

The Soviet Union is also steadily improving the strength and readiness of the forces it can deploy in the Gulf area. These forces include the forces in the Soviet Southern Theater of Operations, or STVD, which is currently under the command of General of the Army Mikail Zaystsev. It includes a total force of some 32 combat divisions, 280,000 troops, 5,400 tanks, 22 fighter regiments, and over 1,000 tactical aircraft.[4]

This theater of operations was largely a military backwater before the Soviet invasion of Afghanistan. It has steadily increased in strength, readiness, and sustained logistic and support capability, however, and this build-up is regional rather than tied to the Soviet presence in Afghanistan. Since 1978, the number of divisions has increased from 25 to 32. Some 15–20% of the Soviet-based divisions, and all of the four divisions and independent combat units, in Afghanistan are fully combat ready. USCENTCOM estimates that all the divisions in the region could be built up to full combat strength in four to six weeks.

The forces in the STVD are being steadily expanded and modernized. In broad terms, the number of tanks, armored vehicles, and heavy artillery pieces in the region has doubled since 1978. The number of tanks has increased from 4,500 to 5,400, the number of armored personnel carriers has increased from 3,000 to 9,100, the number of artillery pieces has increased from 3,100 to 5,600, and the number of mortars has increased from 1,200 to 2,900.

During 1987, there was a slow but steady flow of new types of ground force equipment throughout the STVD. This included a continuing upgrading of combat battalions that used to be truck mounted with new BMP and BTR-70 armored vehicles. It included the latest long-range artillery pieces, such as the 152 mm field gun, with tactical nuclear capability. It also included SS-23 missiles, although these will be removed as part of the INF Treaty. The USSR has also greatly strengthened the forward air defense capability of its combat divisions and support forces. These are receiving a steady flow of SA-8 and SA-13 missiles.[5]

The build-up in air capability has increased the number of air regiments in the STVD from 18 to 22 since 1978. These regiments have also shifted to a largely offensive force. There were 9 air defense regiments in 1978, and 9 fighter-bomber regiments. There are now 4 air defense regiments, and 18 fighter bomber regiments. While the total number of fighter regiments has increased by 20%, the number of fighter-bomber regiments has doubled.[6] While no authoritative recent counts of aircraft numbers exist, earlier reports indicate a build-up in the 34th Tactical Aviation Army (TVA) from 400 active and 175 reserve aircraft in June 1979 to 900 active and 350 reserve aircraft in April 1985. A similar build-up is reported in the 6th TVA from 175 active and 75 reserve aircraft to 400 active and 100 reserve aircraft. These totals may have increased to some 1,195 combat aircraft by mid-1987, with 670 combat aircraft in the area headquartered in Tashkent and 525 in the Transcaucasus MD, headquartered in Tbilisi.[7]

While precise counts by type are not available, they now include the latest Soviet combat aircraft, including the Su-25 Frogfoot, the MiG-27 Flogger D/J, and the MiG-25 Foxbat D. They also include the long-range SU-24 Fencer fighter bomber. During 1987, the USSR introduced more of these all-weather aircraft, improving both ground attack and battlefield surveillance capabilities.

The USSR has also countered some of its problems in conducting mountain and desert warfare by steadily increasing the number of modern attack helicopters in the region, and steadily deploying MI-26 Halo helicopters. These are the world's largest transport helicopters, and can carry C-130 sized loads. Roughly 50% of the total Soviet inventory of MI-26s is now located in the STVD. The cut in air defense fighter numbers has also been more than offset by the introduction of SU-27 Flankers, the Soviet Union's most advanced air superiority interceptor.[8]

For all the problems involving morale and other factors, the Soviet experience in Afghanistan has also given the USSR invaluable practical combat experience in the region. The USSR has modified virtually all of its modern combat equipment in the region to reflect the lessons it is learning. It has developed almost completely new tactics for thrusting land forces through mountain terrain, or using helicopters and SPETZNAZ special forces to overcome defended barriers. It has refined its attack helicopter and SU-25 tactics, and adopted night ambush and a host of other innovative combat methods. For example, its SPETZNAZ units in Afghanistan increased from two battalions in 1981 to two brigades in 1987.[9]

A rough unclassified estimate of the total Soviet forces in the region is shown in Table 3.7. These figures reflect the Soviet forces in the Southern TVD, which includes the North Caucasus, Transcaucasus, and Turkestan military districts. They do not include the Soviet Union's massive Central Reserve in its Moscow, Ural, and Volga military districts, any naval assault forces that might be deployed from the Gulf, or the more than 116,000 men and massive new air facilities in Afghanistan.

While experts differ over the details of the Soviet force expansion in Afghanistan, most agree that there are now three motorized rifle divisions and one airborne division, and two independent motorized rifle

Table 3.7: The Soviet Land and Tactical Aviation Threat to Western Oil, Part One

Service	Forces	Personnel & Equipment
Land Forces		
	Total Combat Formations	
	Combat divisions	32
	Tank Divisions	1
	Motorized Rifle Divisions	27
	Airborne Divisions	2
	Other Units	2
	Total Troops	280,000
	Total Build-Up Capability	
	Fronts	2–3
	All Arms Armies	9
	Divisional Forces Not Including Afghanistan	
	1 Tank Division	11,000
	24 Motorized Rifle Divisions	165,000
	1 Airborne Division	8,000
	Forces in Afghanistan	
	Personnel	116,000
	2 Motorized Rifle Divisions	165,000
	1 Airborne Division	8,000
	1 Air Assault Brigade	
	Mi-24 Hind	140
	Mi-8 Hip	130
	M-6 Hook and Mi-2 Hoplite	
	Major Combat Equipment	
	Main Battle Tanks	5,400
	APCs/AFVs	9,100
	Artillery Weapons	5,600
	Mortars	2,900
	Combat Helicopters	400+
	FROG	115
	Scud	75
	Surface-to-Air Missiles	1,100

Table 3.7: The Soviet Land and Tactical Aviation Threat to Western Oil, Part Two

Service	Forces	Personnel & Equipment
Tactical Aviation	22 Fighter Regiments	18 Fighter-bomber 4 Air Defense 1,195 Combat Aircraft
	TVD Air Headquarters (HQ) Tashkent*	670 Combat Aircraft
	Trans-Caucasus MD Air Force (HQ Tbilisi)*	525 Combat Aircraft
	2 Tactical Air Armies*	3 Fighter Regiments 135 combat aircraft MiG-23 Flogger B/G MiG-29 Fulcrum
		8 Fighter-Bomber Regiments 360 Combat Aircraft Su-17, MiG-27 Flogger D/J, Su-24, Fencer, Su-25 Frogfoot
		1 Reconnaissance Regiment 30 Combat Aircraft Su-17 Fitter H MiG-25 Foxbat B/D
	Turkestan MD (HQ Tashkent) and 1 Air Army (HQ Kabul)*	3 Regiments 145 combat aircraft 90 Su-25 Frogfoot 45 MiG-23 Flogger B/G 10 Su-17 Fitter

*Data do not provide adequate detail of aircraft numbers and types to relate counts by aircraft type to totals by major command.

Source: Anthony H. Cordesman, *The Gulf and the Search for Strategic Stability*, Westview, Boulder, 1984, p. 818; IISS, *Military Balance, 1987–1988*, and Department of Defense, *Soviet Military Power, 1987;* General George B. Crist, *Status of the United States Central Command*, Defense Subcommittee of the House Appropriations Committee, February 22, 1988.

divisions and one air assault brigade in the country. There are 9,000–10,000 Frontal Aviation personnel in the country, and 145 combat aircraft, including 90 Su-25s, 45 MiG-23 Flogger B/Gs, and 10 Su-17 Fitters. There also are some 650 helicopters (including 2 attack helicopter regiments) with 140 Mi-24 attack helicopters, 130 Mi-8 Hips, and Mi-6 Hooks and Mi-2 Hoplites. The Soviets have built or improved major air facilities at Baghram, Kabul, Mazer-E-Sharif, and Jalalabad and have military air facilities at Herat, Shindand, Farah, Lashkar Gah, Serden Band, Askargh, and Qandahar.[10]

The USSR could only use a small portion of this pool of forces in most scenarios in the Gulf, unless it had months to deploy them or was invited into a friendly state. It lacks the airlift and sealift to rapidly deploy more than one or two divisions, and its only land routes into the Gulf pass through Turkey, Iran, and Afghanistan. Turkey can probably defend itself, and Iran is already at war with the Afghan people. As for Iran, the Khomeini regime occasionally flirts with the USSR, but also constantly describes it as the "lesser Satan".

A Soviet invasion of Iran would force troops to move through the Dareth Dagh mountains to the West and the Golul Dagh mountains to the East, or to try to move south from the Caspian Sea through the Elbruz Mountains. To reach the Gulf, Soviet forces would then have to drive through the Zagros Mountains. The Eastern-Central part of Iran consists of salt deserts with no major routes of communication.

In spite of many improvements in the road system over the years, there are still many choke points along all the routes through the mountains and in many parts of the "plains". There is only one rail route south, and both the railroad and the road system are dependent on passes and bridges. The USSR could certainly defeat Iran, but it would pay quite a cost to do so, and would then have to live with a hostile Iranian popular threat to its lines of communications and military presence.[11]

It seems unlikely, therefore, that the USSR will take the risk of directly invading any Gulf state, or of challenging the West in the Gulf area, in the near future. The Soviet Union seems likely to remain bogged down in Afghanistan for at least several more years, and seems unlikely to risk any adventures in Iran until friendly political elements gain far more strength. The fact remains, however, that it does have an immense military presence on the edge of the Gulf, and this gives it immense political and strategic leverage.

As will be described shortly, the USSR is also creating a growing presence in the Red Sea area. It has significant naval and air facilities in both Ethiopia and South Yemen. It already has immense strategic lift capability, and its advantage in proximity means that it can move forces and heavy equipment into the Gulf and Red Sea areas more quickly than the U.S. This Soviet strategic lift will expand in the early 1990s as the USSR deploys its new Condor military transport. The Condor will have about 125 metric tons lift capability compared to 92 metric tons for the U.S. C-5A, and will be the largest air transport in the world.[12]

Even in normal times, the USSR maintains a significant Indian Ocean and Red Sea fleet, and the Soviet Navy will be able to deploy a full carrier task force into the area by the early 1990s, with a new class of 65,000 ton carriers. These carriers will lack the catapults and advanced capabilities of Western carriers, and will rely on VSTOL aircraft with limited payloads, range, and sensor capabilities. They also will lack advanced air defense capabilities similar to those in U.S. battle groups. Nevertheless, they will be equipped with advanced surface-to-surface missiles and cruise missiles, and will be heavily supported by long-range land-based bombers and maritime intelligence and reconnaissance aircraft in Soviet Naval Aviation.

The Soviet fleet in the Indian Ocean area averaged 15-20 ships before the beginning of the crisis in the Gulf in 1987, and normally included nuclear submarines, destroyers, frigates, mine vessels, intelligence ships, support vessels, and transport ships.[13] During 1987, the crisis in the "tanker war" that led to a massive Western naval presence also led to an increase in Soviet forces. The number of Soviet naval combatants in the Indian Ocean squadron more than doubled.[14]

The Soviets began escort operations for three Kuwaiti chartered tankers in May, 1987. They escorted a total of 27 ships through the Gulf during the rest of the year. The Soviets also escorted 26 ships to Kuwait that were carrying arms for Iraq. They also began escorting general cargo ships, after the tanker *Marshall Chuykov* hit a mine, and cargo vessels were fired upon, in May, 1987. By the end of 1987, the Soviet forces in the Indian Ocean normally included two to three major combatants, such as destroyers and frigates, and two to five minor combatants, such as amphibious landing ships and minesweepers. The USSR established a mobile logistics base in the Gulf of Oman to provide command and control capability and support. It deployed several combat ships and auxiliaries in the Gulf.[15]

This naval presence is linked to the Soviet Union's military advisory and basing presence in the region. This presence is particularly strong around the Bab el Mandeb, at the southern entrance to the Red Sea. The USSR and Cuba how have 8,000 military advisors and combat personnel stationed in these two countries. The strategic importance of the Bab el Mandeb is exactly the same as the strategic importance of the Suez Canal at the other end of the Red Sea. Over 325 million tons of cargo, or roughly 10% of the world's commercial shipping, passes through the Bab el Mandeb each year. This is roughly 45–50 ships a day, and over 18,000 ships a year. It is also approximately one-third more cargo than passes through the Panama Canal.

The USSR has long had anchorages and access to facilities in the PDRY, and used them during part of its intervention during the Yemeni civil war in 1986. It deployed a Soviet Naval Aviation combat unit in Aden. IL-38 May anti-submarine warfare and surveillance aircraft from this unit flew an average of one reconnaissance mission a week against U.S. naval forces in the North Arabian Sea during 1987.

The USSR has provided Ethiopia with more than $4 billion in military hardware and 1,700 Soviet military advisors. It has established a small naval support base at Dahlak, off the coast of Ethiopia. This base is equipped with a dry dock, MCMV, naval infantry detachment, and staging facilities for IL-38 maritime reconnaissance aircraft. The USSR made steady use of these facilities throughout 1987.

The USSR is the Seychelles' largest arms supplier and has delivered some $18 million in arms. It keeps up a steady pattern of ship visits, regularly transits military transports through the main airport of Mahe, and has sent naval vessels to help prop up the regime.[16]

These developments mean the USSR will steadily expand its strategic leverage in the Gulf and Red Sea areas. While USCENTCOM and the U.S. Navy will remain the primary deterrent to any Soviet attempt to use such forces for overt aggression, the Gulf states must still develop their own forces to provide some level of deterrence and self-defense capability, and to avoid a situation where they are so weak that the Soviet Union, or some Soviet-backed state, could intervene before USCENTCOM or other Western forces could arrive.

There are five additional variables which could radically increase the Soviet threat to Saudi Arabia and the Gulf during the next decade:

- *Oil:* It is likely that the USSR will become radically more dependent on oil imports during the next decade. The USSR presents extraordinary problems in estimating proven reserves in terms of economically producible oil, and some experts now feel its undiscovered oil reserves may ultimately prove as large as its proven reserves. Nevertheless, Soviet production fell from 616 million metric tons in 1983 to 612 million metric tons in 1984, and 595 million metric tons in 1985.[17] This was the first drop in production since World War II, and it occurred in spite of massive infusions of capital and the priority given a resource that accounts for two-thirds of Soviet foreign exchange earnings. The USSR now maintains an extremely high ratio of production to proven reserve.

(The ratio is 10–14:1 versus over 100:1 for Saudi Arabia and Iraq and 250:1 for Kuwait.) Many of the USSR's reserves are also in the Volga-Urals and Tyumen oil fields—both of which have high lifting costs and low recovery from conventional extraction methods. There are growing reports of overproduction of Soviet wells, and some estimates indicate that as many as 20% of the wells in the Tyumen field (which produces half of all Soviet output) are now shut down due to lack of spares. The USSR has experienced growing problems in meeting its oil production quotas, and is virtually certain to fall short again in 1986. Soviet oil production is now at about 11.9 MMBD, whereas Soviet plans call for production of well over 12.5 MMBD. It would be extremely dangerous to assume that the USSR will be able to maintain its past levels of production and exports—much less increase production—and its development of alternative energy resources now lags far behind Soviet goals. These trends could greatly increase the pressure on the USSR to find some means of dominating a Gulf oil nation.[18]

- *Afghanistan:* A Soviet victory in Afghanistan may still be possible if the USSR is ruthless enough to continue its tactic of striking at the Afghan population as a substitute for its inability to locate and suppress the Afghan freedom fighters. Such a victory would not only give the USSR vastly increased military credibility and strategic leverage in the Gulf area; it would also give it greatly enhanced capability to threaten Iran and Pakistan. Even if the USSR withdraws from Afghanistan, it will have a far more experienced force in fighting in the region than the West.
- *Iran:* There does not seem to be any immediate prospect that Khomeini's death, or the other currents of Iranian politics, will bring a pro-Soviet or Marxist movement to power. The USSR, however, made significant improvements in its relations with Iran during 1987, and the Iranian revolution is building up a deadly legacy of alienation in terms of economic mismanagement, the murder and suppression of minorities, and losses from the Iran-Iraq War. One cannot to rule out the possibility that some Marxist element will come to power during the next ten years or that some faction or strong element in Iran will seek Soviet aid and intervention to give it internal power. Such Soviet intervention could radically change the entire defense structure of the Gulf almost over night.
- *The Yemens:* As will be discussed later, there is at least a possibility that the tensions in North Yemen (YAR) and South Yemen (PDRY) could explode into another civil war and result in a

united and anti-Western Yemen. Such unity could be achieved only with Soviet arms and backing, and could transform the Soviet advisory and basing presence in the Yemens into a major military capability. The USSR has already demonstrated that it can conduct massive military airlifts to the PDRY in a matter of days, and is helping both the PDRY and YAR expand their current air and naval bases and forces.

- *The Soviet Threat to Africa:* The USSR has steadily expanded its efforts to win power and influence in the Horn of Africa, and this region seems likely to be convulsed by economic and political crises throughout the next decade. There is a possibility that the USSR could expand its presence in Ethiopia to include Soviet-backed regimes and basing facilities in the Sudan and Somalia.

SOVIET ARMS SALES AND ADVISORY EFFORTS

Even if none of these forces lead the USSR to directly intervene in the region or support some anti-Western state or movement—and the cumulative near-term probability of some such Soviet intervention is significant—the USSR is likely to try to expand its arms sales and military advisory efforts in every nation that poses a current or potential threat to the Gulf and Saudi Arabia. The threat posed by such arms sales is much more serious than that implied by the dollar data which are issued by CIA and ACDA, and which are used in most press and academic reporting.

The dollar trends shown in Tables 3.2 and 3.3 do not portray the volume of arms involved. They disguise the fact that the USSR provides more weapons per dollar than the U.S., and that states like Saudi Arabia must make a massive investment in munitions and military support equipment that generally is not included in the cost estimates made for Soviet client states.

Unfortunately, no unclassified data are available on the specific dollar cost per weapon delivered for particular nations in the region. These ratios are reflected, however, in the broad regional patterns shown in Table 3.8.

The data in Table 3.8 show that the USSR alone has exported as many major land combat systems to the Middle East since 1981 as the U.S. and Western Europe combined. Similarly, the USSR has exported 625 supersonic jet combat aircraft to the Middle East versus 373 for the U.S. and Western Europe. Although the U.S. and Europe have sold about twice as many arms to the Middle East as the USSR over the last

Table 3.8: Major Arms Sales to the Middle East by Major Supplier: 1981–1985

Weapons Category	Total	USSR	Other Warsaw Pact	U.S.	France	U.K.	Other NATO	PRC	Other Developed	Other Developing
Tanks	8,763	1,790	860	1,228	45	300	30	2,000	2,510	–
Other Armored Vehicles	10,730	3,210	510	3,270	585	75	515	995	1,570	–
Field Artillery	10,082	825	460	1,317	165	70	4,590	645	1,810	200
Supersonic Combat Aircraft	1,288	625	25	208	155	10	–	95	170	–
Subsonic Combat Aircraft	140	85	–	–	15	25	–	–	15	–
Other Aircraft	193	35	80	8	10	10	20	–	20	10
Helicopters	496	340	10	16	30	10	80	–	–	10
Surface–to–Air Missiles	14,122	5,385	5,000	1,437	220	420	55	30	1,430	145
Anti–Air Artillery	3,494	340	405	1,317	165	70	4,590	645	1,810	200

(continued)

(Table 3.8, continued)

Weapons Category	Total	USSR	Other Warsaw Pact	U.S.	France	U.K.	Other NATO	PRC	Other Developed	Other Developing
Major Surface Combatants	20	4	–	–	5	3	4	2	2	–
Minor Surface Combatants	148	17	–	17	3	22	18	12	27	32
Submarines	10	2	–	–	–	–	–	4	4	–
Missile Attack Boats	40	9	–	–	3	6	10	6	6	–

Adapted from ACDA, *World Military Expenditures and Arms Transfers, 1986*, Washington, GPO, 1987, pp. 151–153; and Richard F. Grimmett, *Trends in Conventional Arms Transfers to the Third World by Major Supplier, 1979–1986*, Congressional Research Service 87-418F, May 15, 1987.

Table 3.9: Soviet Bloc Military Advisors and "Technicians" in the Middle East, the Gulf, and Africa

	DoD Estimates as of 1985–87			CIA 1987 Estimates		
	Soviet Advisors	East European Advisors	Cuban Advisors	Soviet Bloc Advisors	Military Trained in Soviet Bloc 1986	Military Trained in Soviet Bloc 1955–86
Algeria	850	250	15–170	1,065	110	3,780
Iran	–	–	–	100	–	875
Iraq	1,000	–	–	1,100	300	5,810
Kuwait	5	–	–	–	–	–
Libya	1,820–2,300	?	(3000?)	2,260	1,800	10,130
Morocco	–	–	–	–	–	145
Syria	2,480–4,000	210	–	3,000	–	9,615
North Yemen	500	–	–	350	200	4,510
South Yemen	1,000	?	1,200	1,100	100	1,785
Ethiopia	1,700	700	13,000	1,700	680	680
Other	–	–	–	50	–	–
Afghanistan	4,000	–	–	3,025	500	12,425
India	150	–	–	600	250	4,870
Angola	1,000–1,400	500	36,000	1,200	1,500	3,260
Benin	80	–	50	–	–	120
Burundi	35	–	–	–	–	–
Cape Verde	25	–	20	–	–	–
Congo	300	15	3,000	80	405	1,665
Equatorial Guinea	40–50	200	25	–	–	200
Ghana	–	–	–	–	75	395
Guinea	85–150	–	50	70	55	1,495
Guinea Bissau	200	–	40–50	70	–	–
Madagascar	–	–	–	55	–	–
Mali	180	–	–	60	65	1,100
Malagasy Republic	400	–	55	–	–	–
Mozambique	850	100	100–215	950	1,000	580
Nigeria	50	–	–	50	240	1,255
Rwanda	15	–	–	–	–	–
Sao Tome	–	–	100	150	–	–
Sierra Leone	–	–	90	–	–	–
Somalia	–	–	–	–	–	2,635
Sudan	–	–	–	20	–	610
Tanzania	60	15	100	90	–	3,230
Zambia	150	–	–	315	–	685
Other		–	–	1,115	400	2,780

Sources differ significantly. The first three columns are taken from *Soviet Military Power* and other Department of Defense sources. The other columns are updates of estimates in from CIA, *Handbook of Economic Statistics, 1987*, September, 1987, CPAS 87-10001, pp. 120–128.

decade, *such transfers—when measured in dollar terms*—reflect the sale of substantially fewer major weapons.

A detailed analysis of recent trends indicates that the number of Soviet bloc weapons transfers to the Middle East has increased sharply relative to those of the West since 1982, and that European weapons transfers are increasing faster than those of the U.S. This reflects the impact of massive arms purchases by Iraq, Syria, Algeria, Libya, and North Yemen; increasing European efforts to capture the Western share of the arms markets; and rising U.S. unit costs coupled with political problems in selling to the Arab world.[19]

Equally significant, the USSR has a massive military advisory presence in the region. This presence is shown in Table 3.9, and it is important to note that the USSR backs its direct military advisory effort to the nations which surround Saudi Arabia with efforts to infiltrate host-nation communications and educational, police, intelligence, and internal security forces.

It is true that the Soviet threat is only part of the problem that Saudi Arabia and the other Southern Gulf states face. It is also true that the Soviet threat is more likely to be indirect, and work through the support of radical states in the region, than to take the form of a direct invasion. Nevertheless, the USSR is clearly committed to a massive effort to improve its military role and influence in the region. There are also an ample number of radical states for the USSR to work with, and several present a serious risk to the West and Southern Gulf states whether or not they have Soviet support.

NOTES

1. All of the following arms transfer data are based on the data base developed by the CIA and reported by ACDA. The primary written sources are ACDA, *World Military Expenditures and Arms Transfers, 1986*, Washington, GPO, 1985, and Richard F. Grimmett, *Trends in Conventional Arms Transfers to the Third World by Major Supplier, 1978–1985*, Washington, CRS Report 87-99F, May 9, 1987. While the individual data from SIPRI are sometimes useful, the data base is so uncertain and variable in quality as to be useless for reporting on regional trends. The reader should be aware that the ACDA data were used in preference to the more up-to-date data available from Richard F. Gimmett because the Grimmett data include India and Pakistan in the same totals as for the Middle East.

2. See Jim Bussert, "Can the USSR Build and Support High Technology Fighters?" *Defense Electronics*, April, 1985, pp. 121–130; Bill Sweetman, "New Soviet Combat Aircraft", International Defense Review, 1/1984, pp. 35–38; Bill

Gunston, *Modern Soviet Air Force*, ARCO, New York, 1982, pp. 84–88 and 112–116; Martin Streetly, "Su-24 Fencer C; Major Equipment Change", *Jane's Defense Weekly*, June 22, 1985, pp. 1226–1227; James B. Schultz, "New Strategies and Soviet Threats Spark EW Responses", *Defense Electronics*, February, 1985, pp. 17–21; Daniel P. Schrage, "Air Warfare: Helicopters and the Battlefield, *Journal of Defense and Diplomacy*, Vol. 3, No. 5, pp. 17–20; "Helicopter Special", *Defense Update*, No. 60, March, 1985.

3. Soviet MiG-29s have already been delivered to India, Iraq, and Syria.

4. Unless otherwise specified, estimates are taken from the statement by General George B. Crist, *Status of the United States Central Command*, Defense Subcommittee of the House Appropriations Committee, February 22, 1988, and from various annual editions of IISS (London), *Military Balance*.

5. Taken from unclassified briefing materials provided by USCENTCOM, and from General George B. Crist, *Status of the United States Central Command*, Defense Subcommittee of the House Appropriations Committee, February 22, 1988.

6. Taken from unclassified briefing materials provided by USCENTCOM, and from General George B. Crist, *Status of the United States Central Command*, Defense Subcommittee of the House Appropriations Committee, February 22, 1988.

7. Estimate based on figures in various editions of IISS, *Military Balance*. The IISS land and air data are serious flawed because of a failure to properly define the readiness and size of forces in given command areas, and because of errors in the description of the organization, major combat unit numbers and types, and aircraft numbers and types in the region.

8. Taken from unclassified briefing materials provided by USCENTCOM, and from General George B. Crist, *Status of the United States Central Command*, Defense Subcommittee of the House Appropriations Committee, February 22, 1988.

9. Taken from unclassified briefing materials provided by USCENTCOM, and from General George B. Crist, *Status of the United States Central Command*, Defense Subcommittee of the House Appropriations Committee, February 22, 1988.

10. "Soviet Air Force in Afghanistan", *Jane's Defense Weekly*, July 7, 1984, pp. 1104–1105, and G. Jacobs. "Afghanistan Forces: How Many Soviets Are There?" *Jane's Defense Weekly*, June 22, 1985, pp. 1228–1233; *Department of Defense, Soviet Military Power*, 1986, pp. 136–138.

11. Typical invasion scenarios are described in some depth in the author's *The Gulf and the Search for Strategic Stability*, Westview, Boulder, 1984. Also see Joshua M. Epstein, *Strategy and Force Planning*, Brookings, Washington, 1987, pp. 49–69. The Epstein book uses some severely dated material to describe some of the terrain and LOC problems, but is generally an excellent analysis.

12. DoD, *Soviet Military Power, 1985*, pp. 83–84, and *Soviet Military Power, 1987*, pp. 98–99.

13. Ibid, pp. 126–125; IISS, *Military Balance, 1987–1988*, p. 30.

14. Taken from unclassified briefing materials provided by USCENTCOM, and from General George B. Crist, *Status of the United States Central*

Command, Defense Subcommittee of the House Appropriations Committee, February 22, 1988.

15. Taken from unclassified briefing materials provided by USCENTCOM, and from General George B. Crist, *Status of the United States Central Command*, Defense Subcommittee of the House Appropriations Committee, February 22, 1988.

16. DoD, *Soviet Military Power, 1986*, pp. 132 and 135–136, and *Soviet Military Power, 1987*, pp. 134–136.

17. *Chicago Tribune*, July 3, 1986, pp. 3–9, and July 20, 1986, pp. 7–10; *Economist*, July 5, 1986, pp. 62–63; William L. Randol and Ellen Macready, *Petroleum Monitor*, First Boston Corporation, October, 1987, p. 14.

18. For a good summary of recent trends see Gary Lee, "Soviet Oil Output Shows a Decline", *Washington Post*, April 3, 1985. The broad trends in Soviet oil production are well summarized in William L. Randol's "Petroleum Monitor", First Boston Corporation, Vol. 4, No. 4, April, 1985. Also see Mikhail S. Bernstam, Soviet Oil Woes", *Wall Street Journal*, January 10, 1986; Ernest Conine, "Soviets Sit on Oil's Power Keg", *Los Angeles Times*, February 17, 1986; and "Mother Russia's $9 Billion Headache," *Economist*, July 5, 1986, pp. 62–63.

19. Richard F. Grimmett, *Trends in Conventional Arms Transfers to the Third World by Major Supplier, 1978–1985*, U.S. Congressional Research Service, Report 86-99F, May 9, 1986, pp. CRS 26-27, and 44.

4
The Regional Threats to the Southern Gulf States

The major near-term threats to the Gulf and Western oil supplies come from within the region. The USSR may exploit them, or act as a catalyst in triggering them, but it is the radical anti-Western states in the region which are the primary threat. The events of the last few years have shown that these threats are far from theoretical. The Iran-Iraq War and the Iranian revolution present a constant risk to Iraq, the Southern Gulf states, and the West.

The radical Red Sea states also present a continuing threat to Saudi Arabia and Oman. After several years of relative quiet, the PDRY is again internally unstable, and could resume its expansionist threat to Oman and the YAR at any time. The PDRY, Ethiopia, and possibly the Sudan can all be future threats to the Gulf states and Western interests.

THE THREAT FROM THE NORTHERN GULF: IRAN AND IRAQ

Iran and Iraq have the wealth to build up massive air and ground forces, are politically unstable, have a history of aggressive ambitions in the region, and now have military forces with nearly a decade of intensive experience in combat. As a later chapter discusses in more detail, the Iran-Iraq War is unquestionably the most immediate military threat to both the West and the Southern Gulf states.

Iran has become steadily more aggressive in challenging the Southern Gulf states. As a later chapter will discuss, Iran has stepped up its terrorist attacks on Gulf states, attempted to launch a naval raid on Saudi Arabia, started a bloody riot during the pilgrimage in Mecca, and has actively begun to attack Gulf shipping in international waters.

While Table 4.1 shows that Iran's lack of a major source of military equipment since the Shah's fall has given Iraq a major edge in terms of equipment and technology it can import, Iran can still launch major spoiler attacks at Saudi Arabia and the other Southern Gulf countries,

Table 4.1: The Trends in Iranian and Iraqi Military Forces, 1980–1988

Force Category	1980/81		1987/88	
	Iran	Iraq	Iran	Iraq
TOTAL ACTIVE MILITARY MANPOWER SUITABLE FOR COMBAT	240,000	242,250	654,000–1,000,000	750,000–1,035,000
LAND FORCES				
Regular Army Manpower				
Active	150,000	200,000	305,000	955,000
Reserve	400,000+	256,000	NA	(480,000)
Revolutionary Guards	–	–	350,000	–
Basij/People's Army	–	–	130,000	650,000
Hezbollah (Home Guard)	–	–	2,500,000	–
Arab Volunteers	–	–	–	6,000?
Division Equivalents				
Armored (Divisions/Brigades)	6+4	12+3	4?	5
Mechanized	3	4	3–4? (a)	3
Infantry and Mountain	0	4	7–11 (a)	10+9 (b)
Special Forces/Airborne	–	–	1/1 (a)	11
Pasdaran/People's Militia	–	–	9–20	–/15
Major Combat Equipment				
Main Battle Tanks	1,740	2,750	900–1,150	4,500–6,150
Other Armored Fighting Vehicles	1,075	2,500	1,190–2,000	3,550–5,000
Major Artillery	1,000+	1,040	750–1,000	3,000–3,500
AIR FORCES				
Air Force Manpower	70,000	38,000	35,000	40,000
Combat Aircraft	445	332	60–118 (c)	500–592 (d)
Combat Helicopters	500	41	45	150–170
Total Helicopters	750	260	120–370	360–433
Surface-to-Air Missile Batteries (e)	–	–	12	70
NAVY				
Navy Manpower	26,000	4,250	14,500	5,000
Destroyers	3 (f)	0	3(f)	0
Frigates	4 (g)	1 (h)	4(g)	5 (h)
Corvettes	4	0	2	6 (i)
Missile Patrol Craft	9 (j)	12(k)	8–11(j)	8(k)
Major Other Patrol Craft	–	–	4–13	7–12
Mine Warfare Vessels	–	5	1	8
Hovercraft	14	0	6	0
Landing Craft and Ships	–	17	8	7
Maritime Patrol Aircraft	6 P–3F	0	1–5 P–3F	0

Table 4.1 (continued)

(a) Estimates differ sharply. One detailed estimate of the regular army shows 7 mechanized divisions with 3 brigades each and a total of 9 armored and 18 mechanized battalions as well as 2 special forces divisions, 1 airborne brigade, plus 8 Revolutionary Guard divisions and large numbers of other brigades and battalions. A recent Israel estimate says there are about 10 regular divisions and 20 Pasdaran divisions. The latest JCSS estimate shows 4 corps with 4 armored and 29 infantry divisions, plus 3 independent special forces brigades, and 2 airborne divisions. This is equivalent to 13 regular army and 20 Pasdaran divisions.
(b) Includes 5 infantry divisions and 4 mountain divisions. There are 2 independent special forces divisions, 9 reserve brigades, and 15 People's Volunteer Infantry Brigades.
(c) Includes 20–50 F-4D/E, 17–50 F-5E/F, 10–14 F-14A, and 3 RF-4E. Large numbers of additional combat aircraft are in storage due to lack of parts. Some Argentine A-4s and PRC or North Korean F-6s and F-7s may be in delivery. The number of attack helicopters still operational is unknown.
(d) Includes up to 7–12 Tu-22 and 8–10 Tu-16, 4 FGA squadrons with 20 Mirage F-1EQ5 (with Exocet), 23 Mirage F-1EQ200, 4 FGA squadrons with 40–60 MiG-23BM/MiG-27, 3 with 75–95 Su-7 and Su-17/20, and 1 training unit with 12–15 Hunter FB-59/FR-10. There is 1 recce squadron with 5 MiG-25 and 5 interceptor squadrons with 25 MiG-25, 40 MiG-19, 150–200 MiG-21, and 30 Mirage F-1EQ. Figures for Mirage strength vary sharply according to assumptions about delivery rates and combat attrition. Typical estimates of combat helicopters are 40–50 Mi-24, 50–70 SA-342 Gazelle (some with HOT), 30 SA-316B with AS-12, and 44 MBB BO-105 with SS-11.
(e) The number of operational SAM units on each side is unknown. Many of Iran's 12 Hawk batteries are not operational. Iran also has extensive holds of SA-7s and some RBS-70s. Iraq has shown very limited ability to use its Soviet-made SAMs and some sites do not seem to be fully operational. Counts of Iraq's missile strength are controversial but Iraq seems to have roughly 20 SA-2 (120 launchers), 25 SA-3 (150 launchers), and 25 SA-6 batteries. It also has SA-7 and SA-9 units and some 60 Roland fire units.
(f) Three are equipped with Standard Arm SSMs. One Battle-class and two Sumner-class are in reserve.
(g) Equipped with Sea Killer SSM.
(h) 5 Lupo class frigates with 8 Otomat-2 missiles and 1x8 Albatros/Aspide missiles, plus 1 helicopter. There is also 1 Yugoslav training frigate.
(i) 6 Wadi-class Italian-made 650 ton corvettes. Each has 1X4 Albatros/Aspide. Two have 2 Otomat-2 and 1 helicopter each; 4 have 6 Otomat 2 SSMs.
(j) Equipped with Harpoon surface-to-surface missiles. No missiles are currently available.
(k) Equipped with Styx missiles.

Source: Adapted from various editions of IISS,*The Military Balance*; JCSS, *The Middle East Military Balance*; and work by Drew Middleton for the *New York Times*.

as well as encourage Shi'ite separatism and religious tension. There is also some risk that Iran can use its air or naval power to threaten Gulf shipping, or achieve a coup d'etat in Bahrain. Further, any Iranian victory over Iraq would not only make Iran the Gulf's dominant military power; it would pose a direct threat to Kuwait which no combination of the current or projected forces of the Southern Gulf states can deal with.

There is no way to predict the outcome of the fighting or how long Iran's clerical regime can rule, with or without victory and with or without Khomeini. While Iraq has won the latest rounds of fighting, as it did during most of 1984, it also made serious mistakes in responding to Iran's February, 1985, and February, 1986, offensives and suffered a serious defeat when Iran successfully invaded the Iraqi city of Faw, just opposite to Kuwait. Iran achieved significant initial successes in both sets of offensives, and Iraq suffered high casualties. Iraq experienced similar problems during Iran's attacks on Basra in early 1987. Iraq does retain superior equipment strength and technology, and the advantage of excellent defensive positions, but any major tactical mistakes could still cost it the war.

Iran, in turn, continues to experience internal political problems, and the war does seem to be producing a rising social and political backlash. Iran has also run down much of its military inventory to the point where it will be virtually forced to rearm once the war is over. The PRC cannot provide Iran with competitive arms and technology. While Western Europe may be willing to sell Iran the arms it will need, Iran might also turn to the USSR. The ambitions of its new class of military leaders remain obscure (as is the case with Iraq's new class of generals). While a clerical regime seems most likely over the next five years, it is impossible to rule out some form of "man on horseback" once Khomeini is gone, and it is impossible to predict his ambitions and politics.

Recent Trends in Iranian Forces

Iran has suffered badly since the fall of the Shah from a near cutoff in Western supplies of major military equipment. At the same time, the war and revolution have led to a near halt in many forms of advanced military training, and Iran's inability to find reliable and consistent sources of ammunition and spare parts has often created near chaos in the Iranian logistic and support system. At the same time, Iran now has hundreds of thousands of experienced troops, and is steadily converting to a revolutionary force equipped with weapons and supplies obtained from China, North Korea, and the Third World.

Estimates of Iranian manpower are now very uncertain. In the beginning of 1988, Iran seemed to have about 650,000–700,000 full-time active personnel in its Revolutionary Guard Corps, and regular army, air force, and navy forces, and anywhere from 150,000 to 250,000 more men which it mobilized for short periods for key offensives. Its regular army had around 350,000 active men, with 250,000 conscripts. While it is unclear that the regular army still has an active reserve system, it does have a pool of up to 350,000 reserves organized into Qod (ex-service battalions).

The regular army had at least three major army headquarters along the front, and probably four. It had three to four mechanized divisions (each with three brigades of nine armored and eighteen mechanized battalions), seven infantry divisions, one to two airborne brigades, and one Special Forces division with three to four brigades.

Since the revolution, Iran has also created 300,000 to 350,000 man full-time Revolutionary Guard Corps (Pasdaran Inqilab) with eleven regional commands loosely organized into battalions of no fixed size. These forces had at least eight divisions and many independent regiments and brigades of very different composition. Some estimates combined these units into a total strength of twenty Pasdaran divisions. The units were specialized into border, infantry, armored, special forces, and paratroop units, and now have their own engineering, artillery, and air defense support units.

These Pasdaran forces are supported by numerous Popular Mobilization Army (Basij) volunteer units, many of which are organized into roughly 500 lightly equipped 300–350 man battalions. These battalions have different strengths, but many have a nominal strength of three companies, each with four platoons and support forces. They have a total organized strength of around 130,000. There is also a 45,000–70,000 man Gendarmerie and border guard—equipped with patrol boats, Cessna 185/310s, AB-205s, and AB-206s, 96 coastal craft, and 40 harbor craft. Various elements of the Sevama (secret police) are scattered throughout both the regular armed forces and the Guards to ensure suitable orthodoxy and loyalty. Finally, the Kurdish Democratic Party (Persh Merga) is reported to provide up to 12,000 men for Iranian paramilitary training and equipment as anti-Iraqi guerrillas.

Iran's ground troops have a very mixed selection of equipment, reflecting Iran's isolation from the West and its shift to the PRC, North Korea, and the Third World for many of its arms. The ground forces now have around 1,000 Soviet-made T-54/55s, 260 Chinese-made T-59s, some Soviet-made T-62s and T-72s, some British-made Chieftain Mark 3/5s, and some U.S.-made M-47s, M-48s, and M-60A1s. Iran has more operational PRC- and North Korean-made T-54, T-55, and T-59 tanks than Western ones.

Iran also now has roughly 500 operational BTR-50, BTR-60, and Urutu armored personnel carriers (APCs), and 180 BMP-1 and 130 EE-9 Cascaval armored infantry fighting vehicles (AIFVs) versus around 250 M-113s which remain from the days of the Shah; it is also heavily dependent on Soviet bloc, PRC, and North Korean artillery. It has more than 750 major artillery weapons, including M-46 130 mm guns, 30 self-propelled M-107 175 mm guns, M-101 105 mm howitzers, 36 Oto Melara 105 mm howitzers, M-109A1 155 mm howitzers, M-110 203 mm howitzers, Chinese-made Type 63 12 X 107 mm multiple rocket launchers, and 65 Soviet-made BM-21 40 X 122 mm multiple rocket launchers. It has Scud surface-to-surface missiles, and claims to be manufacturing a 40-kilometer range-free rocket called Oghab.

The air defense weapons strength in Iran's ground forces is unclear. There are around 1,500 ZU-23 23 mm towed, ZSU-23-4 23 mm self-propelled, 35 mm towed, 37 mm towed, and ZSU-57-2 57 mm self-propelled anti-aircraft guns. Iran's ground forces also retain some Improved Hawk surface-to-air missiles, and have SA-7, RBS-70, and some captured Stinger light manportable surface-to-air missiles. While Iran has been able to buy and manufacture enough parts to keep some of its helicopters flying, most of its once massive force of over 400 helicopters—including AH-1 Cobra attack, CH-47C heavy transport, and Bell 214A, AB-205A, and AB-206 light helicopters—are either non-operational or only partly operational.

It is important to note that while some friction exists between the largely apolitical army and the Revolutionary Guards, it has been nine years since the fall of the Shah, and most of Iran's current manpower has no Western training or history of loyalty to the former monarchy. The high command has been chosen on the basis of loyalty and proven success both in suppressing Khomeini's domestic rivals and in the Iran-Iraq War. The Pasdaran units are, however, getting large shares of new equipment from the PRC, North Korea, and the Third World. While this is partly explainable on the basis that such equipment is simpler to operate and easier to maintain, and that it makes sense to have the regular army rely more on trying to maintain the Western equipment acquired under the Shah, the regular army units tend to be starved for supplies unless they are actually committed to an offensive. The regular ground units are being treated as a dwindling asset which must conserve irreplaceable skills and key equipment items, while the Pasdaran are being treated as the forces of the future.

The Iranian Air Force still had about 35,000 men in early 1988, although it is becoming difficult to distinguish the air force from air elements of the Pasdaran. The Pasdaran now man many of the air force's ground-based air defense units, and are being trained to use Chinese-made J-6 fighters, and to use light aircraft in suicide missions. There are reports that the air units of the Pasdaran have up to 22 J-6 fighters, but these are unconfirmed.

Iran still has extensive stocks of U.S.-made combat aircraft, and excellent sheltered air bases at Bandar Abbas, Bushehr, Ghaleh-Marghi, Isfahan, Kharg Island, Khatami, Mehrabad, Shiraz, Tabriz, and Tehran. It now, however, has only about 60 to 90 operational fighters. These are organized into four fighter ground attack squadrons with some 20–35 F-4Es, four fighter ground attack squadrons with some 20–45 F-5Es, one interceptor squadron with 10–12 F-14s and possibly 10 Chinese-made J-6s, and one reconnaissance squadron with five F-5s and three RF-4Es.

Many of these aircraft can fly only limited numbers of sorties, and have at least partially nonfunctional avionics. Most of Iran's more sophisticated AWG-9 and APQ-120 fire control radars are not operational. While the F-4E forces has recently improved in readiness because of deliveries of spare parts from various unidentified sources, none of the F-14As seem to be able to use their Phoenix missiles. Iran's only fighter-carried missile is the AGM-65 Maverick. This has a range of 12.95 to 25.91 nautical miles, depending on launch altitude, but has a very small warhead. Operational stocks of the missile are believed to be low, and most F-4E aircraft no longer can successfully fire the missile. Iran may have up to 50 PRC and North Korean copies of the MiG-19 and MiG-21 in delivery, but none have yet been seen in combat.

The land-based air defense holdings of the Iranian Air Force are equally uncertain. The U.S.-supplied radars and Improved Hawk missiles were in the ground forces and many no longer function. The same is true of Iran's five British-made Rapier squadrons and 25 Tigercat launch units. This is making Iranian air units increasingly dependent on new deliveries of Chinese-made CSA-1s, which are variants of the obsolete Soviet-made SA-2, on various versions of the SA-7 man portable short-range missile, and on "curtain fire" by large numbers of AA guns and automatic weapons.

Iran retains a fairly strong airlift capability. It has two tanker-transport squadrons with 10 B-707s and seven B-747s. It also has five transport squadrons with up to 26 C-130E/Hs, 9 F-27s, two Aero Commander 690s, and four Mystere-Falcon 20s. Its operational helicopter strength is unknown, but it has 10 HH-34Fs (S55), 10 AB-206As, 5 AB-212s, 39 Bell 214Cs, 10 CH-47s, and two S-61As.

The Iranian Navy had about 14,500 men in early 1988. Like the army and air force, it also was being supplemented by a naval branch of the Revolutionary Guard. The navy has its main naval headquarters at Bandar Abbas. It had bases at Khorramshar, Bushehr, Bandar Abbas,

Bandar Khomeini, and Bandar Lengh. It based its destroyers and frigates at Bandar Abbas and its patrol boats at Bushehr.

The navy's operational strength is increasingly uncertain. At the start of the war, Iran had two U.S. Sumner-class destroyer and 1 British Battle-class destroyer. The Sumner-class ships were armed with four two-missile Standard surface-to-surface missile launchers. The Battle-class ship was armed with the Standard ship-to-ship missile on a four-missile launcher and the Sea Cat ship-to-air missile.

Iran also had four Vosper Mark 5 Saam-class frigates armed with one five-missile Sea Killer surface-to-surface missile launcher, two U.S. PF-103 corvettes, and eight Kaman-class (Combattante II) fast attack boats armed with Harpoon missiles. Iran also had 3 MSC 292/268 and two Cape-class minesweepers, 14 BH-7 and SRN Hovercraft, 4 modern Larak-class amphibious support ships, and about 50 small patrol craft.

Iran seems to have suffered serious damage to at least two of its destroyers and possibly three. Three of its SAAM-class frigates and one of its US PF-103 corvettes seem to be nonoperational. It has lost two Kaman class patrol boats and two more have been seriously damaged—two were evidently hit during Karballa 5. None of the Kaman-class ships seem to have operational Harpoon missiles. It has lost two of its minesweepers, and the only operational mine ship is in the Caspian. Its amphibious strength consists of three marine battalions, four Hengam LSTs, four ex-Dutch LSMs, and six BH-7 Mark 4 Hovercraft. It has one ocean-going replenishment ship, one repair ship, and two fleet supply oilers.

Iran seems to have severe shortages in both anti-ship and anti-air missiles and may have converted some of its Standards to an air-to-surface role. Most of the larger Iranian warships have serious problems in maintaining their combat readiness and electronics. This includes most of their missile systems such as the RIM-66 Standard (anti-aircraft), Sea Cat (anti-aircraft), and RGM-84 Harpoon (anti-ship). All of the Western missiles in the navy's inventory have long exceeded their maximum reliable storage life, and Iran's stocks of Harpoons may be limited to seven missiles for its Kaman-class fast attack craft.

The Iranian Navy has lost significant amounts of its trained maintenance personnel in various purges and upheavals since the revolution, and many of its key radar and electronic systems are no longer operational. This included the Contraves Sea Hunter, SPG-34, and Mark 37, 51, and 61 fire control systems; the WM-28 tactical and fire control radars; the Plessey AWS 1 and SPS6 search radars; and the SPS-37 air surveillance radars.[1]

Iran, however, is still able to operate many of its British-made

Sa'am-class fast-attack craft. According to some reports, it can operate up to 2 frigates, 1 destroyer, 1 corvette, 9 fast-attack craft (FAC), 7 large patrol boats, 40 coastal patrol boats, a maximum of 14 Hovercraft, and 57 amphibious assault ships, logistic ships, and small patrol boats. This still gives it a total force of more than 80 vessels. It also has significant stocks of U.S. Mk. 65 and Soviet AMD 500, AMAG-1, and KRAB anti-ship mines, and may have bought both PRC-made versions of the Soviet mines. It also claims to make its own non-magnetic acoustic free-floating and remote-controlled mines.

The navy's air capability consists of one to two operational P-3F Orion maritime patrol aircraft out of a force of five. None of the P-3Fs have operational radars and their crews use binoculars. There are up to 12 Sikorsky SH-3D ASW helicopters, and 2 RH-53D minelaying helicopters. It also has had up to 7 Augusta-Bell AB-212 helicopters equipped with Italian-made Sea Killer missiles. It has used Air Force AH-1J attack helicopters, equipped with French AS-12 missiles, in naval missions, and has adapted Hercules C-130 and Fokker Friendship aircraft for minelaying and patrol missions. At least some of these aircraft have been lost, but the numbers are uncertain.

Iran has also created a new branch of its Naval Guards force, with a strength of around 20,000 men. These include frogman, minelaying, and suicide attack units, and are equipped with at least 29 to 50 Swedish-built fast interceptor craft, Marin small launches equipped with anti-tank guided missiles, and at least 30 Zodiak rubber dinghies to carry out rocket, small arms, and recoilless rifle attacks. The Swedish fast interceptor craft are particularly important. They are built by Boghammer Marine, and can reach speeds up to 69 knots and have a range of up to 926 kilometers with a 1,000 pound equipment load. The Boghammers are equipped with heavy machine guns, grenade launchers, and 106 mm recoilless rifles. The Swedish boats and Zodiacs are extremely difficult to detect by radar in anything but the calmest sea state, and are based at a number of offshore islands and oil platforms with key concentrations at Al Farisiyah, Halu Island (an oil platform), Sirri, Abu Musa, and Larak.

The Naval Guards units are believed to have some "suicide boats" designed to ram with high explosives and other boats it could fill with fast-drying concrete to block key ports or shipping channels. The Guards operate Iran's PRC-supplied Silkworm surface-to-ship missiles, and have three missile units with three to six missile launchers each. They have extensive stocks of Scuba equipment, and an underwater combat center at Bandar Abbas. They also have a facility at Nowshahr Naval Academy on the Caspian Sea, where some Guards are reported to have had training in suicide attacks. Other branches of the Guards operate

Iran's new Silkworm anti-ship missiles, and possibly PRC-made F-6 and F-7 fighters and small private aircraft. These units trained in suicide missions against ships.[2]

Recent Trends in Iraqi Forces

In contrast to Iran, Iraq has steadily strengthened its relations with the West since the late 1970s, while preserving its ties to the USSR. As a result, it has been able to buy virtually any major weapons it has sought since the start of the Iran-Iraq War, and has obtained a constant flow of replacements, critical spares, and munitions to support its logistic pipeline.

Iraq's superior access to arms has allowed Iraq to gradually dominate the skies, conduct its oil war against Kharg and Iranian tanker traffic with comparative impunity, and launch air and missile raids against Iranian cities and rear areas with only limited fear of retaliation. It also has tended to make the land war one in which the Iranians can constantly score minor gains they cannot fully exploit because of their lack of armor. Iran has lost much of its ability to fight in the open or to maneuver outside the areas it can hold with masses of well-dug-in infantry, and has limited ability to rapidly extend and support offensives much beyond the initial area of attack.

Iraq had about 1 million men and women under arms in early 1988. Its army had about 955,000 men, including some 480,000 active reserves. These were organized into 7 to 8 corps with 5 to 6 armored divisions, and 3 motorized/mechanized divisions.[3] It had up to 31 infantry divisions, including People's Army and reserve brigades. Smaller formations consisted at least of 6 special forces brigades and a Presidential Guard force with 3 armored, 1 infantry, and 1 commando brigade. Some estimates have gone as high as 14 special forces brigades, including all Presidential Guard units.

The Iraqi Army was very well equipped. It had some 4,500 main battle tanks. Up to 2,900 of these were Soviet supplied T-54, T-55, and T-72; up to 1,500 were PRC and North Korean supplied T-59s, and about 150 were captured Iranian Chieftain, M-47, and M-60 tanks. The army also had about 100 PT-76 amphibious tanks, some 1,000 Soviet-made BMP AIFVs, some French CT-TH with HOT ATGMs, and 2,800 to 4,000 BTR-50, BTR-60, BTR-152, OT-62, OT-64, M-113A1, EE-11 Urutu, and Panhard M-3 APCs.

The army had roughly 3,000 to 5,700 artillery pieces—depending on what calibers were counted and whether the total included weapons in reserve or storage. These included a wide mix of Soviet bloc towed weapons and multiple rocket launchers, and Soviet and French self-propelled artillery weapons. There were 80 GCT 155 mm howitzers on order. Unlike Iran, Iraq has been able to buy modern anti-tank weapons relatively freely. It now has large stocks of HOT, AS-11, AS-12, AT-2, and AT-3 missiles.

The army's land-based air defense weapons strength consisted of some 4,000 anti-aircraft guns, including ZSU-23-4 23 mm self-propelled guns, M-1939 37 mm guns, ZSU-57-2 self-propelled 57 mm guns, and 85 mm, 100 mm, and 130 mm cannons. Its surface-to-air missile strength included 120 SA-2 launchers, 150 SA-3 launchers, 60 Roland fire units, and SA-6, SA-7, and SA-9 launchers.

The Iraqi Air Force had a strength of roughly 40,000 in early 1988, including some 10,000 air defense personnel. It had excellent air bases at Basra, H-3, Habbaniyah, Kirkuk, Mosul, Rashid, Shaiba, and 13 additional military air strips.

Estimates differ, but the Iraqi Air Force seems to have about 500 to 590 combat aircraft, including some initial deliveries of MiG-29 fighters, and at least 20 French Mirage 1EQ5 fighters with Exocet. It has 2 bomber squadrons with 7-12 Tu-22 Blinders and 8–10 Tu-16 Badger. It has 11 fighter-ground attack squadrons: four with 40–80 MiG-23BMs, 4 with 23 Mirage F-1EQ-200s with Exocet and 30 Mirage F-1EQ5s, and 3 with 60 Su-7s and Su-20s and up to 10 Su-25s.[4] It has 24 FROG-7 and 12 Scud-B surface-to-surface missile launchers. It has 5 interceptor squadrons with 25–30 MiG-25s, 40 MiG-19s, 30 Mirage F-1EQs, 10 MiG-29s, and up to 200 MiG-21s.[5] It has 1 reconnaissance squadron with 5 MiG-25s.

Iraq has been able to obtain a wide range of modern air ordnance from the West. Its air-to-air missile inventory includes Soviet-made AA-2s, AA-6s, AA-7s, and AA-8s. Its French-made inventory includes R-530s and R-550 Magics. Iraqi inventories of air-to-surface missiles include French-made AS-30 Lasers, Armatts, and AM-39 Exocets. It also includes AS-4 Kitchens and AS-5 Kelts.

The Iraqi Army Air Corps adds at least 150 combat helicopters to this air strength, including 40 Mi-24 Hind with AT-2 Swatter, 50 SA-342 Gazelle with HOT, 30 SA-316B Alouette III with AS-12, 56 MBB BO-15 with SS-11, and 10 SA-321 Super Frelons. Some of the Super Frelons are equipped with AM-38 Exocet and some with AS-12 missiles. The air corps also has 26 Hughes 530F, 30 Hughes 500D, 30 Hughes 300C, 10 Mi-6, 100 Mi-8, 20 Mi-4, and 10 SA-330 helicopters.

Iraq's land-based air defenses are more uncertain. It has made major improvements since the Israeli attack on its Osirak reactor, but the overall readiness and proficiency of both its major surface-to-air missile units and sensor/battle management system is poor. The army forces described earlier are organized into air defense units which are often associated with the air force, and which have 20 operational SA-2 batteries, 25 SA-3 batteries, and 25 SA-6 batteries.

Iraq's transport aircraft include two squadrons with 10 AN-12s, 6 AN-24s, 2 AN-26s, 19 IL-76s, 19 IL-14s, and 1 DH Heron. The Iraqi air force has large reserves of training aircraft, including MiG-15s, MiG-21s, MiG-23Us, 16 Mirage F-1BQs, 50 PC-7s, and 21 EMB-312s.

Iraq's major weakness is its 5,000 man navy. Ironically, Iraq had allowed its navy to run down before the war because it was awaiting the arrival of frigates and corvettes which were under construction in Italy. At the start of the war, it had only 1 training frigate, 8 fast-attack craft (FAC) armed with Styx SSM, 4 FAC armed with torpedos, 3 large and 8 coastal patrol boats, 2 Polnochy class-landing craft, and some inshore patrol vessels. Iraq seems to have lost 3 missile FAC, 2 torpedo FAC, its 2 Polnochy landing craft, 5 patrol boats, and many small craft in the fighting.

Iraq now has 4 Lupo-class frigates held in Italy. These ships have 8 Otomat 2 ship-to-ship launchers, 1 Albatross/Aspide surface-to-air missile launcher with 8 missiles, and 1 helicopter. Iraq's only other frigate, an obsolescent Yug training ship, is laid up in port. Iraq also has 6 Assad frigates, each with 1 Albatross/Aspide surface-to-air missile launcher with 4 missiles. Two have 2 Otomat 2 launchers with 2 missiles each and a helicopter, and 4 more are held in Italy. These have 4 Otomat 2 launchers with 6 missiles each.

Iraq's smaller warships consist of 6 Osa II and 2 Osa I guided-missile fast-attack craft with 4 Styx missiles. Iraq also has 4 old P-6 fast-attack craft which do not seem to be operable. It has 3 large SO-1 patrol craft, 5 Zhuk coastal patrol craft, 2 Soviet T-43 and 3 Yevgenya-class oceangoing minesweepers, 3 Polnocny-class LSMs, and 3 LSTs. It has 1 modern support ship held in Italy, 2 old Poluchat-class torpedo support ships, 1 tanker, and 1 small support ship.

Most of Iraq's warships are trapped in port by Iran's mining and blocking of the Shatt al-Arab. Iraq's only major naval bases are at Basra (which is closed) and Um Qasr (which only has a small channel). Iran has captured its small naval facility at Faw. While the Iraqi Navy does operate in the Northern Gulf, and has attacked Iran's Khor Musa convoys and done some mining, it has only token capability. There also is no Iraqi port where it could base its 4 new Lupo class frigates and 6 Assad-class corvettes, although most are completed and several are completing trials.

The Southern Gulf states not only face these uncertainties regarding an Iranian victory and Iran's future politics; they must also cope with important uncertainties regarding Iraq. They are all too conscious that

Iraq is becoming a massive military power, and that any peace or cease-fire between Iran and Iraq will leave many critical problems unresolved regarding both Iranian and Iraqi attitudes and actions:

- Iran faces at least a decade of political turmoil and further religious conflict. There is no way to predict how, and how fast, Iran will rearm. It is important to note that Iran successfully operated about 450 modern combat aircraft at the time of the Shah's fall and retains the basing and infrastructure to support a first-line fighter force roughly twice the size that Saudi Arabia can hope to field.
- Iranian acquisition of several hundred modern Soviet fighters, or Mirage F-1/Mirage 2000 equivalents, would allow it to challenge the Southern Gulf states even if they were equipped with all the aircraft it is now requesting, particularly because of their lack of air and sea power in the Eastern or lower Gulf.
- Iran has also operated a force of about 1,700 main battle tanks, or four times the present strength of Saudi forces. Its current army manning is close to 20 times that of Saudi Arabia, and its navy represents a continuing threat. The Iranian Navy has 3 modern missile-equipped destroyers, 4 modern missile-equipped frigates, and large numbers of Hovercraft and landing ships.[6]
- Iraq's trend towards political maturity, and friendly relations with the Southern Gulf states, began in the mid-1970s, before the Iran-Iraq War. This trend has been reinforced by the painful lessons of that conflict, but there is no guarantee that the present Ba'ath regime will remain in power. There is a good chance that Iraq's considerable military forces could come under hostile or radical control during the late 1980s to mid-1990s.
- Iraq already has about 1.2 million men under arms, and some 3,700 tanks, 532 combat aircraft, and 409 helicopters. It has 60 Mirage F-1s and 165 MiG-23, MiG-25, and SU-20 fighters. It will replace at least 200 of its current fighters with advanced types such as the Mirage 2000 and MiG-29 by the early 1990s. This will give it not only massive land superiority over Kuwait and Saudi Arabia, but also air superiority—unless Kuwait and Saudi Arabia can obtain the modern fighters and air defense systems they are now requesting. Iraq also has ordered a significant number of new naval vessels, including four Italian missile equipped *Lupo* frigates and six 650-ton corvettes.[7]
- Iraq and Iran are likely to emerge into a world "oil glut" with a massive need for revenue and a substantial surplus oil export capacity—both can probably export at around 4 MMBD within a

year to two years after the war versus current exports of under 2 MMBD. This raises the specter of major pressure on the Southern Gulf states to cut their production. While "oil wars" are not a direct military threat, they certainly are viewed as one of the most serious strategic threats the Kingdom faces.

- Iran must eventually find some massive new source of modern arms, and this may mean it will eventually turn to the USSR for arms which are competitive in technology and performance with those Iraq is obtaining from Western Europe.
- Iraq already has a major capability to strike at tankers and targets in the Gulf. By the late 1980s, it will have modern strike fighters equivalent to those in the French and Soviet air forces with a range of up to 800 miles and much heavier and more lethal air-to-ship and air-to-surface missiles. Iraq also has vast supplies of modern land force armor, advanced munitions, and C^3I equipment on order. Like Iran, it is certain to steadily expand its land and air strike capability against the Southern Gulf states throughout the next decade.[8]

No currently foreseeable outcome of the Iran-Iraq War can free the Southern Gulf states or the West from the prospect of a continuing threat from at least one of the Northern Gulf states. The strength of Iranian and Iraqi forces, in terms of the key measures of force strength and force quality, reflect military realities that no Western strategist can ignore.

The Southern Gulf states also have only one conceivable advantage over Iran and Iraq that allows them to create a significant deterrent: This advantage is the ability to use their oil wealth to buy superior technology and to take advantage of the geography of the Gulf to use air power and advanced technology to compensate for their weakness in land forces and air strength. The comparative figures on manpower, tank numbers, and aircraft numbers discussed in the previous chapters reveal inherent weaknesses which grow out of the size of GCC forces, and which the Southern Gulf states simply cannot overcome.

THE RED SEA THREAT

While the Gulf is of critical importance in terms of oil supplies, and the Red Sea can be bypassed by sailing around the Cape, the Red Sea is also of great strategic importance. Over 325 million tons of cargo, or roughly 10% of the world's commercial shipping, passes through the Suez Canal and Bab el Mandeb each year. This is roughly 45-50 ships a day, and over 18,000 ships a year. It is also approximately one-third more cargo than passes through the Panama Canal.

At least two Gulf states—Saudi Arabia and Oman—face serious potential threats from the PDRY. Oman has chosen to deal with this threat by concentrating its best ground troops in bases along its southern border and steadily building up its ability to operate in the desert in its southwest. These forces are led by cadres of British SAS advisors. Saudi Arabia also deployed a significant portion of its forces to deal with the threat from the PDRY, and to secure its border with the YAR. Saudi Arabia is also far too large to use most of its bases and forces to support or cross-reinforce those on other fronts. Saudi air forces cannot operate against the Red Sea from bases in Hafr al-Batin, Dhahran, or Riyadh. Forces at Tabuk cannot support forces at Sharurah or Khamis Mushayt.

Saudi Arabia must disperse its limited ground forces to forward bases throughout the country, and this has left it with an exceptionally poor force density on any given front. Its growing naval forces are limited in readiness and capability and will remain so for the next decade, while Northern Gulf and Red Sea naval threats can be expected to grow steadily throughout this period. This, in turn, has made air power Saudi Arabia's only means of compensating for the weakness and dispersal of its land and naval forces.

Saudi Arabia can use air power in such a role, however, only if (a) its limited first-line fighter strength has the range and refueling capability to mass quickly, (b) its air units can maintain a decisive technical and performance edge over threat forces, (c) it can provide sufficient air defense capability to provide air cover for Saudi ground forces, naval forces, and key targets, (d) it can provide sufficient dual capability in the attack mission to offset its limited ground strength and give it time to reinforce its army units, and (e) its air units are cumulatively strong enough to provide at least limited coverage of the Northern Gulf or Red Sea front while facing an active threat on the other front.

The PDRY

The virtual economic collapse of the PDRY, accompanied by a continuing internal struggle for power following Ismail's replacement by Ali Nasr, allowed Kuwait, Saudi Arabia, and the UAE to "buy off" the threat from South Yemen in the early 1980s. The PDRY exchanged ambassadors with Saudi Arabia and the U.K. in 1983, agreed to accredit a nonresident ambassador from Oman, condemned the mining of the Red Sea in 1984, and played a relatively moderate role in trying to heal the splits in the PLO.

It is far from clear, however, how long the PDRY will stay bought. Saudi and other Gulf aid has been limited since 1982, with Abu Dhabi providing the bulk of the external aid coming from the Southern Gulf states (about $30 million annually). Coupled with a disastrous set of floods in the spring of 1982, this meant that the PDRY lacked the external funding (70%) it needed for its 1981–1984 five year plan.

The PDRY's more recent development efforts have fared even worse. Before the recent civil war, debt service rose from $3.4 million annually in 1980 to $252 million in 1986. South Yemen now owes well over $1 billion, nearly half of which is owed to the USSR and Soviet bloc states. The PDRY has also seen its annual balance of payments deficit grow from $39 million in 1978, to $368 million in 1984. Part of this was formerly offset in part by worker remittances, but these have dropped significantly since 1984.[9]

Equally important, the long-standing rivalry between President Ali Nasr Muhammad and his more radical deputy, Brigadier General Ali Antar, led to a twelve-day civil war that began on January 13, 1986, and which resulted in a official estimate of 4,300 dead and some 35,000-60,000 refugees fleeing the country.[10] While the civil war caused the death of Ali Antar and Ismail, the more radical and pro-Soviet faction of the ruling National Front (NLF) won the struggle, and the civil war ended in forcing Ali Nasr Muhammed and most of the PDRY's "moderate" radicals to leave the country.[11]

The coup attempt against Ali Nasr had some of the elements used in Soviet-backed coups in Afghanistan, although the Soviets do not seem to have been involved in plotting the coup or even to have been aware it was going to be triggered. Ali Antar and Ismail had spent some time preparing for a rebellion before the civil war started. Many of the forces loyal to Ali Nasr were dispersed away from the capital because false intelligence was provided that Israel would carry out a retaliatory strike against Palestinian bases in the PDRY. Nevertheless, Ali Nasr's forces won the first phase of the fighting because the pro-Antar regular army was poorly positioned, and the navy, much of the air force, and the local militia remained loyal.

At some point, however, the USSR seems to have decided it could not afford to let Ali Nasr win. The Soviets rushed in a radio transmitter for the pro-Antar forces, which it set up at Lahej. The Soviets supplied arms to the rebels during its evacuation of Soviet dependents from Aden. The USSR also helped organize a rump session of the Central Committee of the NLF, and ousted Ali Nasr although only 15 of the 77 members were present.

The USSR then used its advisory role with the military to persuade key units to back a new radical government, headed by Prime Minister

Haider Al Attas, who was then in Moscow. Soviet pilots flying from Socotra and Ar-Rayan in the eastern PDRY intervened on the side of the forces that had supported Ali Antar. A battalion of Cuban troops was airlifted to the PDRY. This Cuban unit spearheaded the military drive from Abyan Province that led to the expulsion of Ali Nasr's force into the YAR.[12]

The new ruler of the PDRY, President Haydar Abu Bakr al Attas, is a pro-Moscow hardliner—as are his other colleagues and rivals. They have so far had to take a relatively moderate line because of their weakness, and the civil war had led to some tension between the PDRY and Ethiopia, but this moderation seems unlikely to continue once the new regime consolidates its strength. While many of the factions driven into North Yemen still remain in that country, the PDRY may still attempt to force unification with the YAR. The new leadership of the PDRY conducted a showpiece trial of some 108 former officials in the Ali Nasr government during 1987, and sentenced Ali Nasr and 34 other officials to death in December, 1987, in spite of appeals not to do so by the YAR. Although only 64 of the prisoners under trial were still in the PDRY and present in the courtroom, 16 were sentenced to death. These included Ahmed Hussein Mussa, the former head of the air force, and Mubarak Salem, the former head of Ali Nasr's security forces.[13]

The PDRY began to take a progressively harder line towards Saudi Arabia and Oman beginning in late 1987. It resumed border clashes with Oman in October, 1987, and the planned exploitation of oil deposits along the PDRY/YAR border makes new clashes with the YAR almost certain.[14]

The PDRY did lose nearly 140 tanks, other army equipment, many of its ships, and some of its aircraft during the civil war. These were rapidly replaced by the USSR, however, and South Yemen remains a fairly large military power by Arabian and Red Sea standards. The PDRY's active military manpower has expanded from 20,000 in 1979 to over 27,000 men, and its annual defense budgets have risen from $115 million to around $200 million. Its arms imports dropped from a peak of $900 million in 1981, to $330 million in 1983, and to only $90 million in 1985, but have risen sharply in the last year now that a more pro-Soviet regime has taken power. Its total arms imports during 1981-1985 totalled $1.11 billion, all but $10 million of which came from the USSR.[15]

Many of the officers and technicians in the PDRY's military forces were loyal to Ali Nasr and have been purged. Its desertion rate is high, and its overall military proficiency is low—especially in operating aircraft, armor, ships, and military electronics. Nevertheless, the PDRY still has sufficient officer and technical cadres, and sufficient ground forces, to pose a threat to Oman, Saudi Arabia, or North Yemen.

Table 4.2: The Military Build–Up in the PDRY, 1978–1987

	1978	1988[16]
LAND FORCES		
Armored Brigade	–	1
Mechanized Brigades	–	1–3
Infantry Brigades (Regiments)	10	8–10
Independent Tank Battalions	8	8
Mixed Field Artillery Brigade	–	1
Medium Artillery Brigade	–	1–2
Various Artillery Battalions	7	10
Luna and Scud SSM Brigades	–	2 each
Tanks	355–365	470–500
Other Armored Vehicles	240	310–400
Artillery	176	340–350
SSM Launchers	–	18
NAVY		
Total Combat Ships	20	20–23
BM–21 (Osa II) Missile Boats	2	6–8
AIR FORCE		
Combat Aircraft	167–177	62–220
Advanced Combat Aircraft	0	27–113
Transport Aircraft	10–11	130
Helicopters	26–33	45
AIR DEFENSE		
SA–2/SA–3 Launchers	–	48–54
SA–7 Launchers	36–40	200+
Chelika	–	6
Warning Radar Sites	18	20

Source: Adapted from various editions of IISS, *Military Balance*, and JCSS, *Middle East Military Balance*.

It has also made some recent improvements in its training efforts, although there are no signs that the PDRY is improving its training and organization at the same rate as the Saudi, YAR, and Omani forces.

The PDRY's build-up in combat units and major military equipment is summarized in Table 4.2. It had about 27,500 men under arms at the beginning of 1988, plus about 15,000 Cuban-trained People's Militia, and a growing 30,000-man Public Security Force.[17] The PDRY army had over 700 tanks before the civil war began, and now has at least 470 to 550. It still has 12–24 FROG-7 and 6 SCUD-B missile launchers, roughly 400 to 450 APCs and armored cars, and some 220–350 artillery pieces of varying types. The army is receiving an increasing number of modern BMP-1 fighting vehicles and AT-3 Sagger anti-tank guided missiles.

The PDRY air force has 62-110 operational combat aircraft, depending on the source. These include 20 MiG-23/27s, 35 MiG-21s, and 25 Su-20/22s.[18] Its naval forces include one Soviet corvette, six OSA-II missile-equipped patrol boats with SS-N-2B surface-to-surface missiles, two P-6 large patrol craft, two Zhuk patrol craft, one Ropucha-class LST, three Polnocny-class LSMs, and five T-4 LCMs.

This force strength compares with a total of 63,500 regular military for Saudi Arabia, 550 medium tanks, and 226 combat aircraft. Even a relatively small state like the PDRY has much larger heavy armored forces than Saudi Arabia, and the PDRY can concentrate all of its strength on a single front.

The PDRY will probably replace most of its MiG-21s and SU-20/22s, much of its armor, and most of its ships by the early 1990s. The recent history of Soviet force modernization and support indicates that the PDRY may get first-line Soviet export fighters as replacements.[19] While the PDRY's military build-up has recently been delayed by the bloody power struggle between the tribal elements of its Marxist government, it is likely to build up to a total strength of about 700–800 tanks and 120–140 combat aircraft by the mid-1990s. The PDRY also has a significant Soviet military presence which seems to have been reinforced since the civil war in January, 1986.

The PDRY now has a main air base at Khormkar (Aden), which it shares with the USSR. It also has bases at Al Dali, Mukayris, Lawdar, Bayshan, al Qisab, Ansab, Ataq, Mukalla, Lawdar, Zamakah, and In Fadhl. The navy has bases at Aden, Mukulla, Socotra, and Perim and may be building a joint base with the USSR in the Bay of Turbah. The national dockyards at Aden have a 4,500 ton floating dock and 1,500 ton slipway, and perform maintenance on Soviet military vessels.

The Soviet Military Presence in the PDRY

The 1,000 man Soviet military advisory effort in the PDRY is supported by a substantial number of combat troops and security advisors in the police, internal security, and intelligence branches. While such estimates are controversial, some British and Egyptian experts estimate that there are now as many as 40,000 Soviet combat troops, specialists, and technicians assigned to South Yemen.[20] There are other training missions from Cuba, East Germany, and North Korea. Limited foreign training of officers and NCOs takes place in the USSR and East Germany.[21]

The Soviet naval squadron in the Indian Ocean and Red Sea area normally totals 15–20 units, including surface ships and cruise missile and attack submarines, and often uses Socotra and Aden as a port or anchorage. Recent anchorages have included one submarine, one to two principal combat ships, one to two minor surface combatants, one amphibious ship, and at least six to eight support ships.[22]

The Soviet Union bases a Soviet Naval Aviation combat unit with IL-38 May aircraft in Aden. Since January, 1986, it has installed a fleet command headquarters in the port. It has naval facilities, radars, and SIGINT units on the island of Perim, between the Indian Ocean and the Red Sea, and a similar base on Dahlak Khebir further north in the Red Sea.

Reports that Socotra has Soviet submarine pens are incorrect, but it does have Soviet operated or commanded radars, air defenses, and a 3 kilometer (1.8 mile) long runway and two secondary runways. There is a small naval base at Socotra, with surface-to-air and anti-aircraft guns, and there may be a Scud or surface-to-ship missile unit. The Soviets maintain up to four AEW aircraft in the PDRY.[23]

Recent Soviet Pacific Fleet developments indicate that true Soviet carrier task forces may begin to visit the Indian Ocean by the late 1980s, and that the PDRY is the most logical center for such operations. The USSR maintains extensive intelligence and communications facilities in the PDRY, and Soviet IL-38 May maritime patrol aircraft fly regularly from Al-Anad airfield. The USSR has delivered more than $3.5 billion worth of military equipment to the PDRY since 1968.[24]

The YAR

The situation in the YAR is considerably more favorable to the West and the Southern Gulf states, but it still presents some problems. The YAR generally supports moderate Arab positions, and has provided some 3,000 troops to Iraq—the only nation in the Gulf–Red Sea area to provide such assistance.[25] Its foreign policy is moderate and it does not support any radical or military movements that threaten the moderate Gulf states. The YAR is not dependent on either the East or West. It obtains military and economic aid from Saudi Arabia, the U.S., and USSR, although the USSR provides nearly 70% of all of the YAR's military equipment.

The YAR has made limited discoveries of oil, and is gradually beginning to export oil from the Azal field. Estimates differ over the size of YAR oil reserves and production capability, however, and the YAR is still heavily dependent upon Saudi Arabia for both aid and special treatment in the hiring of Yemeni workers and the handling of worker remittances and foreign trade flows.[26] This special treatment is important because the YAR's income from its foreign workers is one of its largest sources of income. Saudi influence in North Yemen has slowly diminished during the last five years, however, and may continue to do so. Saudi Arabia has slowly lost much of its influence over the YAR's northern tribes, just as their conservative leaders have lost influence over an increasingly expatriate work force.

The YAR and Saudi Arabia are also divided by several boundary disputes, and the search for oil in the region led to armed clashes between YAR and Saudi forces in 1983, 1984, 1986, and 1987—largely as the result of disputed claims over the Marib-Jawl area in the northeastern YAR. These disputes are typical of the minor conflicts in the region, and do not pose a serious risk of war at the present time. They could, however, lead to more serious problems if the YAR came under a more radical regime.[27]

While President Ali Abdullah Saleh has been in power for more than nine years, and seems to be as secure as any YAR head of state can be, there were reports of a coup attempt in the YAR in late September, 1984, just before President Saleh left to sign a 20 year treaty of friendship with the USSR. These reports surfaced again since a series of Cabinet reshuffles began in December, 1984, and after the civil war in the PDRY in January, 1986. These details of reports are somewhat dubious, but they have occurred with sufficient frequency to act as valid indicators of the continuing tension in the country.

There also is a long heritage of unification talks between the YAR and the PDRY, during which the PDRY has generally tried to undermine the government of the YAR and seize power.[28] Ali Nasr Muhammed of the PDRY visited North Yemen for relatively cordial talks before his overthrow, and a Supreme Yemen Council was set up in 1983 that created a Joint Ministerial Committee (JMC) chaired alternatively by both presidents. This council met sporadically in alternative capitals—most recently in Sanaa in December, 1985.[29]

In spite of the recent civil war in the PDRY, there is still the possibility that North and South Yemen may eventually be combined under a single radical regime. Some experts believe that the existence of the JMC—coupled with agreements on reduced trade barriers, joint development projects, and the joint manning of border posts—may continue to meet the PDRY's demand for "unity" without leading to the kind of political union that could weaken the YAR. Others see the Council as creating an ideological push towards unification which could lead to the merger of the two countries through some future coup.

As for internal stability, North Yemen's economic development has never fully recovered from the effects of a December, 1982, earthquake. Even though it introduced major austerity measures in 1983, its economy has weakened to the point where there are few near-term prospects of development other than the possibility of oil. In 1987, it had an annual deficit of some $900 million, and per capita income was probably below $440.

Recent oil discoveries in the YAR have, however, been promising. The YAR opened its first major export terminal in late 1987. It has constructed a 263 mile pipeline from from the Alif and Azal oil fields in its eastern desert over the mountains to Hodeidah on the Red Sea. As a result, the YAR may earn up to $600 million in oil revenues in 1988.[30]

Recent estimates have also raised its proven reserves from 200 million barrels to 500 million, and this could raise the YAR's output to 200,000 barrels a day by 1988. This level of oil production would earn up to $6,000 million a year, but it would not solve the YAR's economic problems. This figure is close to the recent drops in annual remittance payments from Yemeni labor working in foreign countries.[31]

In the interim, the YAR's massive trade deficit continues to grow. Its external debt has risen at a rate of over 30% annually, and remittances from workers in other nations have steadily diminished since 1980. While Saudi Arabia has deliberately subsidized labor from the YAR to help maintain North Yemen's stability, and remittances approached $1.3 billion in 1985, they have dropped significantly since 1986.

Industrial and mercantile development remains low. The government's efforts to cut the growing dominance of Qat over the YAR's agriculture since 1972 have fallen far short of the government's goals. Economic growth has averaged about half of the 7% called for in the YAR's 1982-1986 development plan, and the YAR has not approached its investment goal of $6.4 billion.

While the central government has steadily improved its control over the country, it still does not fully control the countryside outside the major cities and key lines of communication. Military and security units control the roads, and there are extensive checkpoints and garrisons, but the area is often under tribal control. The YAR is also unable to properly pay its forces, and desertion rates in the Saudi armed forces remain high.[32]

The YAR could, however, be a major military threat if it came under radical control. Like the PDRY, the YAR has managed to build a major military force by local standards. It is able to draw on the manpower output of a nation of 6 to 8 million people, which is roughly equivalent to the native population of Saudi Arabia.[33] The YAR's annual defense budgets have risen from $150 million in 1978, to around $570–$670 million annually in 1987 and 1988. Its arms imports have fluctuated sharply from year to year, ranging from $100 to $460 million annually. Its total arms imports during 1981–1985 totalled $1.59 billion, of which $850 million came from the USSR, $150 million came from Poland, $90 million came from the U.S., and $575 million came from a variety of other countries.[34]

The YAR had 36,800 men in its armed forces in early 1988. It had five to six armored brigades, one to three mechanized brigades, six to nine infantry brigades, one commando and one special forces brigade, and one central guards unit. It is important to note that these formations were very irregular in organization, and were little more than battalions by Western standards.

The YAR army had 775–800 main battle tanks in early 1988. This is roughly 150% of the heavy armor strength of Saudi Arabia, although only 514–558 of the YAR's tanks were T-55s, T-62s, and M-60s. The YAR had roughly 295–320 artillery weapons, 65 BM-21 122 mm multiple rocket launchers, some 440–500 APCs, and armored cars. The overall readiness and maintenance of most YAR army equipment was poor.

The YAR had an air force with an active strength of 1,000–1,500 men and 73–99 combat aircraft. These combat aircraft included 30 MiG-21s, 15–20 SU-20/22, and 11–15 F-5E, although 34–49 of its combat aircraft were old and poorly maintained. The YAR plans to expand its force to 130 aircraft by 1990. At least 42 of its existing fighters will have to be replaced with new aircraft by the early 1990s, and the YAR may receive up to 70–100 modern Soviet fighter types over the next decade. The military air bases at Hodeida and San'a are acceptable, but not of high quality. A secondary base seems to have been set up on Kamran Island.

The small 500–800 man navy had only 2 real combat ships: 2 obsolescent Osa II guided-missile patrol boats. It also had 2 Soviet supplied Yevgenya-class mine warfare ships, 6 patrol boats (2 Zhuk-class, 1 Poluchat-class, 3 ex-U.S. Broadsword-class), and 4 landing craft (2 T-4 and 2 Ondatra). Its only naval base was at Hodeida, although it had an anchorage at Kamran Island.[35]

While the YAR can scarcely be called pro-Soviet, the USSR has played a steadily increasing role as an advisor and supplier of its military forces. The YAR signed a 20 year friendship treaty with the USSR on October 9, 1984, and the USSR has provided the YAR with roughly one-third of the foreign loans it has needed to survive in recent years. The USSR has a 500 man advisory group in the YAR, which is far larger than the U.S.-Saudi groups, and Soviet technicians now maintain most of North Yemen's complex military equipment.

The initial Soviet-YAR arms agreement, which was signed in 1979, has expanded in value from an initial level of $1 billion to over $2.5 billion—with $2 billion worth of actual deliveries since 1962, and more than $1.1 billion since 1981. These deliveries include new major deliveries of armor, including T-62 tanks. The USSR signed a new friendship treaty with North Yemen in 1984, and while President Saleh continues to support economic alignment with the GCC states and the West, U.S. military aid plays a token role at best.[36]

This means that Saudi Arabia and Oman must continue to plan their military and political strategy to cope with a threat from both Yemens. Oman can afford to concentrate on defending its southern and southwestern borders, but Saudi Arabia will have to divide its forces between coverage of the Gulf, Red Sea, and the southeast. In fact, the need to divide Saudi forces is already reflected in the fact that Saudi forces at Sharurah and Khamis Mushayt are being given a considerably higher readiness and force improvement priority than those at Tabuk, and that the military cities (army, air force, and air defense) at Khamis Mushayt have been steadily expanded.

Ethiopia

The Ethiopian regime is hostile to both the West and the Southern Gulf states, and harbors deep-seated resentment of Saudi Arabia because of Saudi financing of the Eritrean rebels and Saudi support of hostile regimes in the Sudan and Somalia. As the tables in Chapter 3 have shown, Ethiopia has become a major military power in the Red Sea area. It has received more than $4 billion in Soviet arms since 1975, and has signed agreements for $1 billion more.

The extent of Soviet involvement in Ethiopia is indicated by the fact the USSR made new major arms deliveries to the port of Aseb to help Ethiopia in its 1985 campaign against the Eritrean rebels, including T-55 tanks, APCs and AFVs, and more MiG-23 fighter bombers. Direct Soviet logistic and advisory support was the only reason that Ethiopia was able to deploy some 50,000 troops in less than three weeks in August,

1985, and could capture the key town of Barentu in spite of the fact it was the rainy season.[37]

Ethiopia's annual defense budgets have risen from $131 million in 1978, to around $400–$430 million annually in 1987 and 1988. Its arms imports have recently fluctuated sharply from year to year, ranging from $300 to $570 million annually. Its total arms imports during 1981–1985 totalled $2.1 billion, of which $2 billion came from the USSR, $30 million came from Italy, $20 million came from Czechoslovakia, and $50 million came from a variety of other countries.[38]

Ethiopia had up to 320,000 men in uniform, and another 169,000 men in paramilitary units, in early 1988. The Ethiopian Army had about 313,000 men including 150,000 People's Militia. It had some 22 infantry divisions (including 3 motorized, 4 mountain, and 3 light), 8 paracommando brigades, 37 artillery battalions, 6 air defense gun battalions, 3 SA-2 battalions, and 3 SA-3 battalions. Although Ethiopia's terrain is often unsuited to armored warfare, it has some 750–1,010 main battle tanks. The army also has some 835 other armored fighting vehicles, and 700 artillery pieces.

The 4,000 man Ethiopian Air Force has only about 138–160 combat aircraft. The USSR is conducting extensive MiG-21 and MiG-23 training, however, and Ethiopia is likely to build up to 200–220 combat aircraft by the early 1990s. Most of these fighters will be MiG-23 or follow-on generation fighters. Ethiopia had 24-30 armed helicopters in early 1988, and is steadily increasing its overall helicopter strength. These helicopters do not pose a direct threat to the Gulf states when they operate out of Ethiopia. The Mi-24 has a range of 99 miles, and the new Mi-28 has a range of 150 miles. This is insufficient range to operate against Saudi Arabia from across the Red Sea. Such helicopters could, however, rapidly deploy to, and operate out of, the Yemens. The new Soviet Mi-26 heavy transport aircraft can carry up to 100 troops, or even armored vehicles, for ranges of 497 miles and is well equipped for night operations.

Ethiopia has a 3,000 man Soviet-supplied navy of two Petya II–class frigates, eight missile-equipped Osa-II patrol boats, four Mol-class fast-attack craft, five large and seven coastal patrol craft, and eight landing craft. It has one support ship and one training ship.[39]

Ethiopia also provides the only true Soviet naval base in the Red Sea, on the island of Dahlak. The USSR has an 8,500-ton floating dry dock, floating piers, helipads, fuel and water storage, a submarine tender, and other repair ships. Soviet guided-missile cruisers and nuclear submarines routinely call at Dahlak for service, and Soviet IL-38 May aircraft operated from Dahlak until they were destroyed by Eritrean rebels in May, 1984.[40]

Ethiopia's forces are still "pinned down" by its civil war. In spite of President Mengitsu's efforts to declare Ethiopia a Democratic Republic in September, 1987, and to offer limited autonomy to various rebel groups, the Eritrean Liberation Front (ELF) and Tigray People's Liberation Front (TPLF) opened new attacks in October. Nevertheless, the rebels in the Northern Provinces have been hit hard by the current famine and massive new Soviet support, and Ethiopia may well emerge as a major radical threat in the 1990s.

Ethiopia also has political and strategic links to the PDRY and Syria, and may have sold its stocks of U.S. arms to Iran. Ethiopia is a member of the Tripartite Alliance with the PDRY and Libya, which was signed in 1981. This alliance has not been particularly active in recent years. Libya withdrew its People's Bureau (embassy) from the PDRY on September 25, 1984, and Ethiopia backed Ali Nasr in the PDRY's civil war in January, 1986. A number of Ali Nasr's supporters remain in Ethiopia, and relations between it and the new regime in the PDRY remain poor. Nevertheless, this kind of coalition may be a more serious threat in the future.[41]

The Sudan

The Sudan also may become a radical and hostile regime during the next decade, although it is far too soon to predict this as a trend. Nimeiri's fall in an April, 1985, coup led to the election of Sadiq al-Mahdi in April, 1986. Unfortunately, Al-Mahdi has proved to be a notably weak and incompetent leader in virtually every respect. He had done nothing to reduce the tensions between the Arab north and Christian and Animist south, or to end the abuse of Islamic law that Nimeri instituted as part of his last bid to retain power. The civil war in the Sudan continues to grow. The Sudanese People's Liberation Army (SPLA) scored major new gains during the 1987 rainy season (May–October). It expanded its area northward, and seized the Sudanese Army outpost at Jaku, along the Sudanese-Ethiopian border. It is periodically able to seize other border outputs at Kumuk and Gizen, and regularly controls some 75 isolated outposts and villages in the Southern Sudan. The SPLA is able to obtain steadily better access to arms, and it is unclear whether the current government can consolidate power in the north, much less over the entire country.

The Sudan is also covertly supporting the Libyan military effort in Chad, in return for military support. Libyan forces and Chadian dissident forces use Darfur province in the Sudan as a safe haven and line of supply and communications. This is done with the tacit permission of the al-Mahdi government.

The nation's economic development has virtually halted, it faces a crushing debt burden of over $12 billion, and it suffers from a continuing drought and famine in much of the country. There are no real signs of reform, the agricultural sector has degenerated into near chaos, and virtually every aspect of the nation's economic development activities has halted.

Libya has actively courted the new military government since Nimeiri's fall, and supported it in several battles against the rebels in the south. The USSR has also increasingly courted the Sudan. The U.S., Egypt, and Saudi Arabia have all experienced growing problems in dealing with the Sudan's new officials. If the Sudan does become hostile, the combination of the threats in the Red Sea and in related areas in Africa could become far more serious.

THE MILITARY TRENDS IN THE RED SEA AREA

The military efforts of the states in the Red Sea area are summarized in Table 4.3, and any major new radical combination of these states could greatly increase the threat to oil shipments from the Gulf, the movement of pilgrims and cargos to Saudi Arabia, and Saudi Arabia's ability to ship oil and gas directly from its Red Sea ports.

While most of the radical threats in the region are now of peripheral importance, the fact remains that Saudi Arabia and the other Southern Gulf states may face not only a more active and hostile Ethiopia and PDRY in the future, but also stronger linkages between all the radical states in the region. Given Ethiopia's large military forces, and the potential ability of the PDRY and Ethiopia to control the lower Red Sea, the Red Sea area may present steadily greater threats to both the movement of Gulf oil and all shipments moving through the Suez Canal.

INSTABILITY IN THE WESTERN BORDER AREA

Saudi Arabia and Kuwait must also take account of possible threats to their western borders. While direct military threats are now unlikely, Syria has long put political pressure on Saudi Arabia and Kuwait. No one familiar with the Near East is likely to place much faith in "intentions" as a test of force requirements. The Southern Gulf states are forced to take account of "capability" as well as "probability".

Table 4.3: Comparative Military Efforts of Red Sea and Key African States Affecting Red Sea Security

	Defense Expenditure in 1986 ($Millions)	Arms Imports '81–'85 ($Millions)	Military Manpower	Battle Tanks	Combat Aircraft
Saudi Arabia	16,200	14,760	73,500	470–500	62–220
Oman	1,510	955	21,500	39	53
North Yemen	414	1,675	36,800	775–800	73–99
South Yemen	230	1,110	27,500	550	120
Sudan	440	560	58,500	175	43
Ethiopia	450	2,100	320,000	750–1,020	138–160
Somalia	146	365	65,000	293	71
Egypt	4,950	7,120	460,000	1,750	427
Libya	5,100	10,455	76,500	2,280	544
Israel	5,110	4,105	141,000	3,900	676
Chad	49	65	17,000	(65)	2
CAR	17	10	7,000	4	2
Zaire	45	150	51,000	50	20
Uganda	11	135	20,000	13	(6)
Kenya	258	300	13,350	76	26

Source: Adapted from IISS, *Military Balance*, 1987/1988; ACDA computer data base for *World Arms Transfers and Military Expenditures, 1986*. Military data differ from text which is adapted to include information from other sources. Figures in parentheses indicate country has similar equipment in form of lighter AFVs or armed training aircraft.

Egypt and Jordan

Neither Egypt nor Jordan seem likely to become a significant threat to the Southern Gulf states during the next decade. If anything, they are likely to contribute to their defense. Jordan provides a number of officers, NCOs, and technicians for Gulf armies, and may provide reinforcements in an emergency.

Egypt is an important source of equipment and military aid. It reached a military accord with Kuwait in December, 1987, to "reinforce with limits the conditions of military cooperation between the two countries," and is discussing taking the place of the Pakistani advisors and combat manpower in Saudi Arabia.[42]

Egypt already contributes to the defense of Iraq. Thousands of Egyptians have been recruited into the Iraqi army out of the one to two million working in Iraq. Egypt is also exporting to Iraq at a rate of nearly $1 billion a year, and is selling tanks, armored vehicles, munitions, artillery, and spare parts to the Iraqi forces. It has assembled and exported about 80 Brazilian Tucano light strike aircraft. Egypt is also actively seeking to revive the Arab Organization for Industrialization and to sell arms to the Southern Gulf states.

Nevertheless, the internal stability of Egypt and Jordan is threatened by the tensions growing out of the Arab-Israeli conflict, serious internal economic problems, and religious extremism. Egypt remains the largest Arab state, and controls the strategic lifeline through the Suez Canal. Although its military forces have declined sharply in total effectiveness since Egypt's break with the USSR, it remains a far more powerful military power than Saudi Arabia, and a radical or hostile Egypt could be a major threat.

Similarly, Jordan has long stabilized Saudi Arabia's western flank, and limited the political and military leverage Syria could exert on the Kingdom. Jordan's forces have not modernized at the same rate as those of other Arab states, but they remain the most professional of all Arab armies and any coup in Jordan could change the Kingdom's security position overnight.

Syria

Syria has not been a recent threat to Saudi Arabia and the other GCC states, and some Saudi leaders, such as Crown Prince Abdullah, have close ties to Syria. Nevertheless, Saudi Arabia and Kuwait have paid dearly in economic and military aid to ensure that Syria's ruthless political opportunism has never been directed against them, and the future attitudes of a post-Assad Syria are unclear.

While it seems likely that any successor regime will be more concerned with Israel and internal conflicts than with any political ambitions in the Gulf, the fact remains that Syria's pressure on Iraq has gravely increased the risks created by the Iran-Iraq War, and that the Kingdom cannot rely on Syrian stability or friendship. Once President Assad is gone, and his hold on the country seems to be weakening, Syria may well come under more radical rule or be destabilized by the tensions between its Sunni majority and its ruling Alawite minority.

Further, the growth of Syrian military forces has given it ample capability to threaten Saudi Arabia as well as Jordan. Syria has ordered over $19 billion worth of arms from the USSR, and there are nearly 4,000 Soviet advisors in country—more than in any other Third World nation. There are an additional 1,000 Soviet economic technicians, and Syria is now heavily dependent on Soviet economic credits as well as Arab aid.

Syria has more than 300,000 active men in its army, plus some 270,000 reserves. It has more than 4,000 main battle tanks, and roughly 2,000 of these are high-quality newer types. It has some 490 combat aircraft, and about 300 of these aircraft are medium- to high-performance types. It also has some 255 helicopters, 110 of which are attack helicopters.

As Table 4.4 shows, Syria's forces match or exceed those of Israel, approach those of Iraq, and vastly exceed the forces of any combination of the GCC states. Syria seems to have been the first Arab country to get the advanced version of the MiG-23 discussed earlier, and is now taking delivery on the MiG-29. It has recently received SA-13 surface-to-air missiles and was the first country outside the Warsaw Pact to get the SA-5 and SS-21 missile systems. It will be able to pose a major threat to Saudi Arabia and Kuwait for the foreseeable future.[43]

Libya and Radical Threats

The threat posed by Libya, and the various radical movements it finances, is more complex. Libya does not have significant power projection capability. It does, however, have a large oil income and it has continuously used it to try to enforce its radical political views on other Arab states and to support anti-Western and extremist movements. It also has imported vast amounts of arms. These imports totalled $10.5 billion during 1981–1985 alone, and $5.6 billion of these imports came from the Soviet bloc.[44]

There is little question that Libya's rhetoric about Israel disguises the fact that it is far more of a threat to moderate Arab regimes than it will ever be to Israel. Libya has sponsored virtually every radical opposition movement in the Gulf area; it has attempted to subvert the Saudi armed forces and has sponsored assassination attempts; it has supported Iran, the PDRY, and Ethiopia in hardline or radical actions and in their military build-up. It is supporting extremist left-wing Islamic elements in the Sudan, Lebanon, and possibly in Syria.

Libya has sponsored subversive movements in Saudi Arabia and the other Gulf states for years, and has sponsored several assassination attempts against members of the royal family and efforts to get members of the armed forces or internal security forces to back a coup. Libya probably supported or carried out the mining of the Red Sea in 1984. Libya also supplied approximately 72 Scud B missiles to Iran for use against Iraq.

Over half of Libya's massive inventories of armor and combat aircraft are now in storage and could be used to rapidly supply a hostile Gulf or Red Sea state. This now includes nearly half of the $10 billion worth of arms that Libya has received from the USSR. While Libya is not a Soviet satellite, it does have about 2,000 Soviet military advisors, and 1,200 East European advisors and technicians. Libya can, therefore, be a powerful catalytic threat to Saudi Arabia and every other moderate Arab state.[45] It lacks serious military capabilities to intervene in the Gulf, but it uses its oil income and weapons stockpiles to increase the risk of political instability and the emergence of hostile radical regimes.

THREATS AND THE PROCESS OF DETERRENCE

Like the threat from the USSR, the importance of the regional threats to Saudi Arabia and the other members of the GCC must be kept in careful perspective. With the exception of Iran, no regional state poses an immediate threat of invasion to any of the Southern Gulf states. Most of the regional hostilities and tensions are far more likely to take the form of border wars, subversion, and the backing of coup d'etats, than the form of large-scale war. Further, many of the more radical and hostile states surrounding the Gulf are as likely to attack each other, or degenerate into civil war, as pose a threat to Saudi Arabia, its neighbors, or the West.

The practical problem, however, is that the region is filled with a history of unpredictable wars which have been fought to intensities far beyond the level justified by their strategic purposes. The Iran-Iraq War is only the latest of these conflicts. The Yemens have been a constant source of civil and external conflict, and tensions caused by the Arab-Israeli conflicts may yet lead Syria to turn to the south or the east in order to satisfy its ambitions.

The sad fact remains that the ability to use force is still the only way to achieve security in the region, and that the GCC states will be secure only to the extent that they and friendly states have sufficient force to halt any challenge. The fact that border wars are normally low-level wars can also be extremely misleading. Even high-technology forces normally have to deploy 10 to 20 men per guerrilla to secure a border area. Securing desert territory and urban areas can be equally difficult, and Iran has shown that far more sophisticated and higher thresholds of air and naval forces may be used. The threats in the region may have many political and military limitations, but they are all too real.

Table 4.4: The Military Build-Up to The West of the Southern Gulf States

Category	Israel		Syria		Jordan		Egypt	
	1982	*1988*	*1982*	*1988*	*1982*	*1988*	*1982*	*1988*
Total								
Defense Spending ($billions)	6.1	5.1	2.4	4.0	0.4	0.8	2.1	4.6
Manpower (1,000s)								
Active	174.0	141.0	222.5	407.5	72.8	80.3	452.0	445.0
Conscript	120.3	110.0	120.0	180.0	-	-	255.0	250.0
Mobilizable	500.0	504.0	345.0	272.5	107.8	35.0	787.0	825.0
Army								
Manpower (1,000s)								
Active	135.0	104.0	170.0	300.0	65.0	70.0	320.0	320.0
Conscript	110.0	88.0	120.0	180.0	-	-	180.0	180.0
Moblizable	450.0	406.0	270.0	272.5	-	-	620.0	500.0
Tanks	3,600	3,900	3,990	4,200	569	986	2,100	2,250
APCs	8,000	10,300	1,600	3,900	1,022	1,370	3,030	3,250
Artillery/MRLs	960	1,000	2,100	2,800	274	247	2,000	1,800
Air Force and Air Defense Forces								
Manpower (1,000s)								
Active	30.0	28.0	50.0	105.0	7.5	10.0	113.0	105.0
Conscript	7.0	19.0	-	-	-	-	60.0	60.0
Mobilizable	35.0	37.0	-	-	-	-	133.0	90.0

Total Combat								
Aircraft	634	676	450	492	94	109	429	441
Bombers	0	0	0	0	0	0	14	9
Attack/Int.	432	391	0	0	0	0	0	102
Attack	174	130	205	217	29	66	218	83
Interceptor	28	0	244	275	45	32	152	161
Recce/EW	28	29	0	10	0	0	45	24
OCU	0	0	0	0	20	11	0	0
Armed Helicopter	42	76	16	110	0	24	24	53
Major SAM Bns/								
Bty/Sites	15	15	75	95	5	23	151	125
<u>Navy</u>								
Active Manpower	9.0	9.0	2.5	2.5	0.3	0.3	20.0	20.0
Conscripts	3.3	3.3	-	-	-	-	15.0	10.0
Mobilizable	10.0	10.0	5.0	2.5	-	-	35.0	34.0
Submarines	3	3	0	3	0	0	12	12
Guided Missile	27	22	18	24	0	0	19	30
Destroyer/Escort/								
Frigate/Corvette	2	6	2	2	0	0	8	9
Small Combat	44	38	12	24	9	6	56	88
Amphibious	7	12	?	3	0	0	20	17

Source: Adapted from various editions of IISS, *Military Balance*, and JCCS, *Middle East Military Balance*. Note that these figures do not show equipment in storage, use estimates which often do not reflect actual readiness of manpower and equipment, and use dated material. For example, the IISS data for Israel reflect virtually no updating between 1982 and 1986, and grossly exaggerate the size of Egypt's operational forces and equipment holdings in both 1982 and 1985. In fact, virtually no Soviet-supplied equipment in Egyptian forces is combat operational and the true strength of Egyptian forces is less than one-third the totals shown. These figures are deliberately presented as a contrast to the trend curves shown earlier which reflect more substantial adjustments by the author.

NOTES

1. Larry Dickerson, "Iranian Power Projection in the Persian Gulf", *World Weapons Review*, August 12, 1987, p. 7.

2. These assessments are based on various editions of IISS, *Military Balance*, the *Jaffe Center Middle East Military Balance*, and *Jane's Defense Weekly*, July 11, 1987, p. 15.

3. The armored or mechanized divisions all differ in structure. A nominal unit has one armored and one infantry brigade.

4. Some estimates of the Mirage strength go as high as 70 aircraft. The same estimates show 60 Su-20s in the FGA units, and 40 F-6s and 40 Su-7s in other combat units.

5. They may have some F-6s.

6. See IISS, *Military Balance, 1980–1981* and *1984–1985*, for details.

7. See Aharon Levran and Zeev Eytan, *The Middle East Military Balance, 1986*, Jaffe Center for Strategic Studies, Tel Aviv University, 1985, and IISS, *Military Balance, 1987–1988*, for details on Iraq's current force strength.

8. For an analysis of the threat to Gulf tankers, see Dr. Raphael Danziger, "The Persian Gulf Tanker War", *Proceedings of the Naval Institute*, May, 1985, pp. 160–176, and the author's article in the February, 1988, edition of *Seapower* magazine.

9. Economist Intelligence Unit, *EIU Regional Review: The Middle East and North Africa, 1985*, Economist Publications, New York, 1985, pp. 267–271; ibid., 1986 edition, pp. 277–284.

10. *Washington Post*, December 14, 1987, p. A-27. For a good summary of the history of the political rivalries in the region see J.E. Petersen, "Legitimacy and Political Change in Yemen and Oman," *Orbis*, Winter, 1984, pp. 971–998.

11. The PDRY reported after the fighting that 4,230 had been killed and that it had done $115 million worth of damage. Other sources reported nearly 10,000 dead. Ali Nasr first fled to his native Abyan Province and then surfaced in a guest house in Syria in late 1986. He has since joined his supporters in the YAR. Up to 20,000 troops and other supporters fled to the YAR and Ethiopia. Much of the navy fled to Ethiopia and only part of it returned when the new government promised to protect the rank and privileges of its members. The government tried 141 of Ali Nasr's supporters for treason, plus 94 in absentia.

12. Department of Defense, *Soviet Military Power, 1987* pp. 140–141; *Washington Times*, January 15, 1986, p. 1; *Intelligence Digest*, January 14, 1987, p. 2; and Mohammed Abdullah Kerim, "Soviets and South Yemen", *Jane's Defense Weekly*, February 15, 1986.

13. *Washington Post*, December 14, 1987, p. A-27, and January 1, 1988, p. A-24; *Washington Times*, December 14, 1987, p. A-2.

14. Ibid.; and Department of Defense, *Soviet Military Power, 1987*; pp. 140–141.

15. Arms Control and Disarmament Agency, *World Military Expenditures and Arms Transfers, 1986*, pp. 98, 140, 146.

16. Initial estimates of the cost of the civil war ran as high as 145 tanks, 80% of naval equipment, and 90% of air equipment. Data on resupply is uncertain, but

the USSR seems to have provided substantial deliveries. Some equipment is in storage.

17. The ranges shown cover the high-low range between the estimates of the JCSS and IISS.

18. These force strength estimates, and those that follow throughout the rest of this chapter, are taken from Aharon Lehvran and Zeev Eytan, *The Middle East Military Balance,* Jaffe Center for Strategic Studies, Tel Aviv University, 1986, and IISS, *Military Balance, 1987–1988.*

19. Egyptian press sources have claimed that British intelligence sources said there were as many as 10,000 Soviet advisors and 44,000 troops stationed in the PDRY. These reports seem to be incorrect. See *Jane's Defense Weekly,* May 23, 1987, p. 996.

20. *Insight,* December 7, 1987, pp. 46–47; and *Washington Times,* November 30, 1987, p. A-10.

21. IISS, *Military Balance, 1987–1988,* p. 45.

22. *Insight,* December 7, 1987, pp. 46–47; and *Washington Times,* November 30, 1987, p. A-10.

23. Department of Defense, *Soviet Military Power, 1985,* GPO, Washington, 1985, pp. 123 and 128; *Soviet Military Power, 1986,* GPO, Washington, 1986, pp. 135–136; and *Soviet Military Power, 1987,* pp. 140–141

24. See Note 16.

25. *Washington Post,* December 11, 1987, p. A-24.

26. Worker remittances are sometimes reported as $1 billion annually, but peaked at $440 million annually in 1982, and have dropped to less than $160 million since 1984. The Alif oil field in the YAR seems to have up to 400 million barrels in reserve and a maximum production potential of around 250,000 BPD.

27. Department of Defense, *Current News, Supplemental,* April 16, 1987, p. W; *Washington Post,* April 12, 1986, p. A-18; *Wall Street Journal,* April 15, 1987, p. 34; *Financial Times,* November 26, 1987; Fred Halliday, "North Yemen Today", MERIP Reports, February, 1985, pp. 3–9.

28. There is also a history of military disunity. The two Yemens have fought three times in the last fifteen years.

29. Economist Intelligence Unit, *EIU Regional Review: The Middle East and North Africa, 1986,* Economist Publications, New York, 1986, p. 268; *Economist,* December 26, 1987, p. 50.

30. *Wall Street Journal,* December 11, 1987, p. A-24, and December 15, 1987, p. 28; *Washington Post,* January 2, 1988, p. A-24; *New York Times,* January 5, 1988, p. D-4.

31. Ibid.; and *Economist,* December 26, 1987, p. 50; Economist Intelligence Unit, *EIU Regional Review: The Middle East and North Africa, 1986,* Economist Publications, New York, 1986, pp. 267–275; and *Washington Post,* December 11, 1987, p. A-24, January 2, 1988, p. A-24.

32. EIU, *Regional Review: The Middle East and North Africa, 1985,* pp. 257–265

33. Estimates of the YAR's total population differ sharply. The country is virtually entirely Muslim, but is approximately 50% Sunni, 48% Zaydi Shi'ite, and 2% Isma'ili Shi'ite. The latter two sects have not sided with Khomeini, however,

and the YAR's key ethnic divisions are tribal, not religious.

34. Arms Control and Disarmament Agency, *World Military Expenditures and Arms Transfers, 1986*, pp. 98, 140, 146.

35. Eytan, *Middle East Military Balance, 1985*, pp. 387–392.

36. Department of Defense, *Soviet Military Power, 1985*, pp. 123–129; *Soviet Military Power, 1986*, GPO, Washington, 1986, p. 136; and *Soviet Military Power, 1987*, pp. 140–141.

37. *Soviet Military Power, 1986*, GPO, Washington, 1986, p. 132; and *Soviet Military Power, 1987*, pp. 134–135.

38. Arms Control and Disarmament Agency, *World Military Expenditures and Arms Transfers, 1986*, pp. 98, 140, 146.

39. IISS, *Military Balance, 1987–1988; Air Force Magazine*, March, 1985, p. 108.

40. Department of Defense, *Soviet Military Power, 1985*, pp. 123–129; *Soviet Military Power, 1986*, GPO, Washington, 1986; *Soviet Military Power, 1987*, pp. 134–135.

41. For good discussions of Soviet and Ethiopian strategy in the Horn see Lt. Colonel David R. Mets, "The Dilemmas of the Horn", *Proceedings of the Naval Institute*, April, 1985, pp. 49–57; and Samuel Makinda, "Shifting Alliances in the Horn of Africa", *Survival*, January/February, 1985, pp. 11–19.

42. *New York Times*, December 24, 1987, p. A-8.

43. Syria signed a preliminary July, 1986, arms agreement that would provide such advanced technologies as the MiG-29, SA-8, SA-11, SA-13, SA-14, and SS-23. *Washington Times*, July 21, 1986, p. 3A; and *Jane's Defense Weekly*, July 26, 1986, p. 92.

44. Arms Control and Disarmament Agency, *World Military Expenditures and Arms Transfers, 1986*, p. 143.

45. Department of Defense, *Soviet Military Power, 1986*, GPO, Washington, 1986, p. 134; and *Soviet Military Power, 1987*, pp. 140–141.

5
The Southern Gulf States and Western Power Projection Capabilities

It is not easy for the West to deal with these threats or to secure its access to Gulf oil. The West's ability to play a military role outside Europe, the Mediterranean, and the Atlantic has changed radically since the early 1960s. The steady elimination of the last vestiges of colonialism has led most Western states to create forces with sharply limited range and support capabilities.

Only four Western countries are still capable of playing any significant power projection role in the Gulf: Turkey, France, Britain, and the United States. All of these states face major limitations on the power they can project. Turkey can project power only near its borders, and France and Britain no longer have major power projection forces. The U.S. must project its power nearly halfway around the world and faces major limitations in terms of strategic lift and forward basing facilities.

TURKEY

Turkey has common borders with Iran, Iraq, Syria, and the southwestern USSR. Up to half of its armed forces could play an out-of-area role in defending eastern and southern Turkey. Turkey has 45,000 men in eastern Turkey, and its Third Army has more than six bases around Erzerum, with six more strung out along the 160 mile highway towards Agri. It is conducting active operations against the Kurds in the area around Hakkair and Chirze.

Turkey sent some 30,000 men to reinforce units on the Syrian and Iraqi borders in the spring of 1985, and it has steadily built up the capability of the Third Army since.[1] Turkey is expanding its strategic road net near both the Syrian and Iraqi borders, and is steadily increasing its ties to the Northern Gulf states. It has developed steadily stronger trade relations with the Middle East, and with Iraq and Iran in particular. It deployed forces equivalent to two divisions in Iraq in 1983, with Iraq's consent, in a military operation against the Kurds.

While Turkey would normally have to deploy most of its forces to meet the threat of Soviet or Greek action, it could use at least 6 mechanized and/or armored divisions in operations near its borders with Iraq and Iran or Syria (up to 500 tanks, 800 AFV/APCs, and 400 artillery weapons). It also has 1 commando and 1 parachute brigade it could commit, and could deploy at least 1 mobile Gendarmerie Brigade with V-150 and UR-416 APCs suitable for light paramilitary action. The Turkish Air Force could deploy up to 5 fighter squadrons, and 1 reconnaissance squadron, and 2–4 Rapier/Redeye units (24 fire units), and is improving its C^3I net, military facilities, and air bases in Eastern Turkey.[2]

It is important to note, however, that Turkey is severely short of mobile combat and service support assets. It has a very poor communications net, lacks the logistics and infrastructure for extended power projection much beyond its border areas, and badly needs modern anti-tank weapons, portable artillery, and other combat gear for mountain and urban warfare. Turkey would experience severe operational problems the moment it shifted from border or territorial defense to operations beyond its border. It has no real power projection capabilities and would require full-scale external support for any operation outside its border areas.[3]

Turkey has also carefully avoided taking any political or military action, thus implying that its forces could play any regional role other than self-defense. It is still deeply committed in Cyprus, and faces a massive Soviet and Bulgarian threat on its northern borders. It would probably avoid taking any contingency action that did not directly threaten its territory, and would almost certainly avoid playing any role in an Arab-Israeli contingency even if it involved a substantial Soviet presence in Syria. Turkey's willingness to intervene in the event of the collapse of Iraq or a Soviet move into Iran is unclear. It has publicly declared that its facilities in eastern Turkey are not U.S. or NATO contingency facilities for action in the Gulf, although such declarations are deliberately ambiguous.

FRANCE

Britain and France are the only Western European nations that still deploy substantial out-of-area forces in the Gulf, Red Sea, and Indian Ocean. It is important to note, however, that these British and French forces in the region are normally relatively small, and are limited to roles in very low-level conflicts. Although both nations joined the U.S. in intervening in the Gulf in 1987, they do not have the strength or sustaining power to engage in armored, air, or naval combat in the Gulf, except against low-level threats like the Dhofar rebel units in the

PDRY or as part of a regional or U.S.-led force. They have limited surface-to-air (SAM), electronic warfare (EW), and reconnaissance aircraft capabilities, but lack advanced sensor and intelligence systems other than maritime patrol aircraft, photo reconnaissance, and land-/air-/ship-borne electronic support measures (ESSM).

French capabilities and trends are summarized in Table 5.1. France's main contingency forces in the region normally consist of its forces in Djibouti, its Indian Ocean Command (ALINDIEN), and its South Indian Ocean Command. France could also deploy substantial land and air forces, including elements of the Forces d'Action Rapide (FAR), and up to a two-carrier task force to the Gulf in an emergency. France has some particularly well trained elite units for such operations, including the Foreign Legion.

The FAR, which was organized in its current form in June, 1983, has a nominal strength of 47,000 men, the majority of whom are professionals. Its main combat units include the 6th Light Armored Division (Nimes), the 4th Air Mobile Division (Nancy), the 9th Marine Infantry Division (Nantes), the 11th Paratroop Division (Toulouse), and the 27th Alpine Division (Grenoble).

Two of these divisions have a new structure. The 6th Light Armored Division is a 7,400 man unit with 2 infantry regiments with armored vehicles, and 2 armored regiments with 36 AMX-10 reconnaissance vehicles and Milan anti-tank guided missiles. It has a command and support regiment, a 155 mm artillery regiment with 24 TR-155 mm towed guns, and an engineer regiment. Its total equipment includes 340 VAB armored personnel carriers, 24 VAB/Mephistos, 72 AMX-10RCs, 48 Milans, and 48 Mistral anti-aircraft missiles. The 4th Air Mobile Division has 250 helicopters, of which half are combat helicopters. It is organized into four combat helicopter regiments, two infantry regiments with anti-tank guided missiles, two engineering companies, and a command and support regiment.[4]

The FAR is not, however, dedicated solely or even principally to out-of-area missions. It has both a role in European combat and a potential role in out-of-area contingencies. Further, some elements are clearly organized for combat in the center region. The 4th Air Mobile Division, for example, is organized to halt Soviet armored penetrations at ranges up to 250 kilometers from its main position. The 27th Alpine Division is also best suited to a particular defense mission in Europe. Only the 6th Light Armored Division is specially configured for overseas deployment. The other two main out-of-area units would be the 9th Marine Infantry and 11th Paratroop Divisions. The 9th Marine Infantry Division has four infantry, one armored cavalry, one combat engineer, and one headquarters and support regiment. It is being equipped with

Table 5.1: French Contingency Capabilities in the Middle East, 1987–1988, Part One

Forces Normally Deployed in the Area:

Mediterranean Fleet:
2 SSNs
9 Submarines
2 carriers
14 Destroyers/frigates/escorts
5 mine countermeasure ships
5 amphibious ships

Peacekeeping:
Lebanon (Unifil): 1,380: 1 infantry and 1 logistic battalion
Sinai MFO: 40: 2 Twin Otter and 1 C-160 aircraft

Djibouti:
Total manpower: 3,800 men
Permanent and Prepositioned Forces:
- 10th BCS (Command and Services Battalion)
- 5th RIAOM (Overseas Regiment)
- 13th DBLE (Demi-Brigade of the Foreign Legion)
  —ALAT (Army Light Aviation Unit) with 5 attack and 5 medium transport helicopters.
  —CDMB (Engineering Company)
- 1 motorized company
- Army equipment includes AMX 13 light tanks, AMX with SS-11, AFVs, 105 mm battery, 1 AA artillery battery
- One Mirage IIIC squadron with 10 fighters, and air elements with 1 C-160 transport, and two Allouette helicopters.
- Naval elements with 1 Atlantic MPA.

Rotated Units (every four months): 1 motorized company

ALINDIEN: Indian Ocean Inter-Service Overseas Command
Total Manpower: 1,400
5 frigates, 3 minor combatants, 1 amphibious, and 1 support ship
South Indian Ocean Joint Service Command: La Reunion and Mayotte:
Total manpower: 2,200 men
Permanent and Propositioned Forces
- 53rd BCS
- 2nd RPIMA (Marine Parachute Regiment)

Rotated Units: 1 parachute company

Table 5.1: French Contingency Capabilities in the Middle East, 1987–1988, Part Two

Probable Maximum Out-of-Area Contingency Forces in France

Rapid Action Force
- 1 Parachute Division (13,500)
- 1 Air Portable Marine Division (8,500)
- 1 Light Armored Division (7,400)
- 1 Air Mobile Division (5,100)
- 1 Signals Regiment
- Foreign Legion Force (8,500): 1 armored, 1 parachute, 4 infantry, and 2 engineer regiments
- Independent Army Elements 1 Support Brigade and 2 Mixed Regiments

Naval Forces
- 2 Carriers with 20 Super Etendard, 4 Etendard IV or 7 F-8E
- 5 ASW and 2 AA Destroyers with Exocet, Crotale, and Malafon
- 8 Frigates with Exocet
- 2 Amphibious Assault Ships with 9 LCM or 2 LST
- 2–5 Diesel or Nuclear Submarines
- 11 Support Ships
- 5 Maritime Patrol Aircraft

Air Forces
- 3 Mirage and 3 Jaguar Squadrons
- 1 Mirage III Reconnaissance Squadron
- 6 Crotale Batteries (8 fire/4 radar units)
- 1 DC-8 and 2 C-160 transport squadrons
- 60–80 armed Alouette II, Puma, and Gazelle army helicopters, some with HOT ATGM.
- 2–5 transport Helicopter squadrons

Source: Adapted from IISS, *Military Balance, 1987–1988*; and John Chipman, *French Military Policy and African Security*, Adelphi Papers No. 201, IISS, London, 1981, p. 20.

the Panhard Sagaie armored car with a 90 mm gun to give it air portable anti-armor capability. The 11th Paratroop Division has six parachute regiments, one parachute combat engineer, and one HQ and support regiment.[5]

In practice, most FAR units must be kept in Europe unless some other nation provides strategic lift, support, and sustaining capabilities. This is particularly true of any combat involving armored forces, or substantial enemy air power, since most FAR units have only limited air defense capability and are equipped to fight only against heavy armor in a defensive role. It seems unlikely that France could deploy more than one light division to any area in the Middle East except Djibouti without several months of build-up, unless the U.S. provided most of the support and sustaining capabilities—both of which are areas where USCENTCOM already faces serious shortages.[6] In practice, France would probably be limited to a maximum of about three squadrons for any short-term deployment, although it could certainly build up much more substantial deployment and support capabilities over a period of three to six months. France would also need U.S. strategic lift to rapidly move air basing and air defense assets. [7]

France can deploy considerable air and naval power as well, but would again face the problem of strategic lift, support, and sustaining capabilities. France has not been able to buy the C-141 Starlifters it once sought from the U.S., cannot afford the C-5A, and has had to rely on limited numbers of DC-8F jet freighters, C-160 Transall twin turboprops, and old piston engine Noratlases. It is badly short of tankers for refueling, and although it plans to buy more strategic transports in the 1990s, it is now limited to a maximum of one regiment worth of long-range air lift. Its tank landing ships are obsolete and most have been withdrawn from service, and funds are not available for the three 10,000 ton amphibious combat ships that France has sought since the early 1980s.[8]

The French Navy has redeployed most of its major surface forces to the Mediterranean since 1976, including its two carriers, the *Foch* and *Clemenceau,* and many of its best ASW and air defense cruisers. The French carriers have a 22,000 ton standard displacement and 32,780 ton full load. Each normally carries 16 Super Etendards, 3 Etendard IVPs, 10 F-8E Crusaders, 7 Alize, and 2 Alouette helicopters. Unlike Britain and Italy, France retains true attack carriers, but French carriers scarcely compare with the full fleet carriers of the U.S., which can displace up to 85,000 tons.[9]

As the French deployments in the Gulf in 1987 demonstrated, France can conduct naval operations the size of a carrier task force. French carriers and ships do, however, lack the air defenses and anti-missile capabilities for operations against threats with modern missile-equipped strike aircraft unless they can receive support from ground-based fighters. France also would find it very difficult to sustain air operations from its carriers for any length of time because of its carriers, small complement of aircraft and supply, and lack of air defense aircraft. Further, it lacks the replenishment capability to operate more than one carrier group in the Indian Ocean.

France spent some 2.5 billion French francs on out-of-area operations in 1985, of which 555 million francs were spent overseas. This included costs for the French forces that were in Lebanon, the protection of the embassy in Beirut, the French presence in Chad, French operations in New Caledonia, and the creation of a new base there. France played a critical role in ending Libya's military adventures in Chad in 1986 and 1987. French naval forces have played an important role in the West's recent naval intervention in the Gulf.

France has also shown on many occasions that it can project small forces effectively—although it has generally needed U.S. assistance in strategic lift for significant long-range troop movements.[10] For example, the U.S. provided a C-5A for the recent operations in Chad because French C-160 Transalls could not carry large weapons systems such as the Hawk missiles France used for air base protection.[11]

France also showed it could quietly provide significant internal security assistance to the Southern Gulf states when it had to suppress radicals that seized the Grand Mosque in Mecca in 1982. France has, however, faced increasing political problems in sustaining out-of-area operations in cases like Chad. It is increasingly doubtful that the French public would support any intervention that produced significant casualties or required a sustained military presence in the face of hostile forces.

Any significant out-of-area deployments mean cutting France's capability to support the defense of the Atlantic and Mediterranean, although SACEUR and SACLANT have already declared that NATO has major deficiencies in ship strength to meet critical defense missions. The U.S. would have to provide substantial airlift and resupply for combat operations, and would have to provide intelligence support and air cover in the face of a Soviet or well-equipped Third World air threat with long-range air or ship-to-ship missiles.[12]

France is also experiencing steadily greater budget problems in out-of-area operations. While France now seems likely to remain in the Indian Ocean and Djibouti through 1995, it may well have to phase out one of its three carriers. Since the Super Etendards, Etendards, and F-8Es on French carriers, and much of the C^3I and air/missile defense gear on French ships, are already obsolete or obsolescent, French carrier task force contingency capability is likely to drop to a one-carrier "demonstrative deployment" capability sometime between 1990 and 1995.[13]

France should, however, still be able to fund the deployment of more advanced maritime patrol aircraft, modern fighter-reconnaissance aircraft, and better ESSM systems. It plans to upgrade the surface-to-air missile defenses on many of its ships, including better close-in protection. It is seeking to buy long-range airlift, and will deploy improved SHORADs, although it has no heavier land-based SAMs that it could deploy to the Gulf. French carriers are nuclear armed, and both French Mirage 2000s and Mirage IVs could provide theater nuclear support.

UNITED KINGDOM

British military intervention capability in the Gulf is more difficult to estimate. Britain can still deploy a large portion of its forces for the Atlantic and the Mediterranean to the Gulf. Although it officially no longer deploys military forces "east of Suez", it still provides a major advisory presence to many of its former Trucial states, including Kuwait. It also provides contract naval and air officers to Oman, as well as small SAS units. Britain also has increased its naval deployments in the Mediterranean, and has played an active role in patrolling the Gulf and Straits of Hormuz since the beginning of the Iran-Iraq War.

Like the U.S., Britain has had forces in the Gulf since the start of the Iran-Iraq War. The Royal Navy joined the U.S. Navy in the Gulf area in a demonstrative exercise in the Indian Ocean in October, 1980, which was clearly designed to demonstrate Western determination to keep the Gulf open in spite of the war.

Britain maintained its presence at a demonstrative level until early 1984, when Britain agreed to join the U.S. in convoying ships through the Straits of Hormuz after renewed threats by Iran. Britain moved the frigate *Brazen* and destroyer *Glamorgan* to the Straits. By July, 1984, Britain had deployed the *Glasgow*, a Type 42 destroyer, and the *Charybdis*, a Leander-class frigate, plus the Royal Fleet Auxiliary ship *Appleleaf* in demonstrative sailings through the Gulf of Oman, the Straits of Hormuz, and into the Gulf. The U.K. also sent the Air Defense Troop of the 3rd Commando Brigade of the Royal Marines to the Gulf to provide air defense for British ships with shoulder-launched Shorts Javelin surface-to-air missiles. Britain has kept a naval task force in the Gulf ever since, and it too played a major role in the Western intervention in the Gulf in 1987 and 1988.[14]

British Conservative politicians have often suggested that Britain should strengthen its out-of-area role, and Britain has regularly deployed ships in cooperation with the U.S. in the Gulf during the Iran-Iraq War. Britain also demonstrated excellent out-of-area contingency capabilities during the Falklands conflict. In practice, however, Britain has never obtained the defense budgets necessary to maintain or expand its out-of-area contingency forces. Both its presence in the Gulf and its total power projection capabilities have declined steadily since 1968, and are likely to continue to decline further.[15]

An estimate of British contingency capabilities is provided in Table 5.2. Both Britain's 5th Airborne Brigade (2nd and 3rd Battalions) and the Royal Marines 3rd Commando Brigade (40, 42, and 45 battalions) are now specifically earmarked for out-of-area operations, and the Parachute regiment and a Gurkha battalion have out-of-area experience. The 5th Brigade has been specially equipped for such missions, and has been converted from an infantry to an airborne unit during the last two years. It has a special leading battalion group with artillery and combat engineer support designed for very rapid long-distance deployment. Nevertheless, all such British forces have been affected by Britain's defense budget problems and are short of the air defense, armor, artillery, and lift required for operations against armored or mechanized Middle Eastern forces.[16]

Britain now has very limited carrier power, and would find it difficult to project more than one light VSTOL carrier to the Gulf. Britain would be dependent on small numbers of comparatively short-legged Harrier and Sea Harrier aircraft. The latest British carrier, the *Ark Royal,* was commissioned in 1985. It is only 680 feet long, however, and displaces only 20,000 tons. This makes it far closer to a U.S. LPH-class assault ship than a U.S. fleet carrier, which is more than 1,000 feet long and displaces over 65,000 tons. The *Ark Royal* also falls far short of a French carrier in terms of on-board aircraft. The deck has a ski jump for VSTOL aircraft and it cannot operate with modern fixed-wing naval fighters. The *Ark Royal* normally carries only fourteen aircraft: five Sea Harrier fighters, three airborne early warning Sea King Helicopters, and six ASW Sea King Helicopters.

Even the coming refit of the *Invincible* will mean that Britain's largest "carrier" can deploy a total of only eight Harrier VSTOL fighters and twelve Sea King AEW and ASW helicopters. The *Invincible* is being expanded to allow the storage of missiles for more extended operations and is being given a new 12 degree ski jump to extend Harrier range, a modern Type 996 three-dimensional radar, and three Goalkeeper 30 mm guns for terminal missile defense. Nevertheless, a British carrier task force would require U.S. air/missile defense coverage against a Soviet or sophisticated Third World threat, although the Sea Harrier now has improved air defense avionics and BVR air-to-air combat capability.

The future of British amphibious capability remains uncertain. The *Atlantic Causeway,* the sister ship to the *Atlantic Conveyor* which was sunk in the Falklands, was converted to an amphibious helicopter ship capable of carrying several hundred Royal Marines and heavy lift helicopters. It now, however, is laid up along with the *Hermes,* Britain's only dedicated commando carrier. This leaves Britain with a

Table 5.2: British Contingency Capabilities in the Middle East, 1987–1988, Part One

Forces Normally Deployed in the Area

INDIAN OCEAN SQUADRON
2 destroyers/frigates
1 support ship

OMAN
SAS detachment
RAF contract pilots
RN seconded and contract officers

DIEGO GARCIA
1 Naval detachment
1 Marine detachment

CYPRUS
UNFICYP: 750 men
Army: (3,250)
- 1 Infantry battalion less 2 companies
- 1 Infantry battalion plus 2 companies
- 2 Armored reconnaissance squadrons
- 2 Engineer and 1 logistic support squadrons
- 1 Helicopter flight

RAF: (1,347)
- 1 Helicopter squadron

Table 5.2: British Contingency Capabilities in the Middle East, 1987–1988, Part Two

Probable Maximum Out-of-Area Contingency Forces in the U.K.

LAND FORCES
2 Armored reconnaissance regiments
10 Infantry battalions
2 Paratroop battalions
1 SAS regiment
- Scimitar, Ferret and Fox AFVs
- FV-432, Saracen, MCV-80, Spartan APCs

2–4 artillery regiments with 105 mm guns on AFVs or towed

155 mm howitzers.
1 SAM regiment with Rapier
2 SAM batteries with Blowpipe and Rapier
30–40 Helicopters: Gazelle AH-1 and Lynx AH-1, some with TOW

NAVAL FORCES
1 Commando brigade, 2 SBS assault squadrons with Blowpipe,
105 mm guns, Milan ATGMs.
2 Carriers with 10–15 Sea Harrier and Harriers, 9 Sea King
5 Destroyers with Sea Slug, Sea Cat, and Sea Dart SAMs
8–10 frigates with Exocet, Sea Wolf, Sea Cat
4–6 Nuclear and diesel submarines
2 Assault ships with 4 LCM and 4 LCVP, and Seacat SAMs.
4 Landing ships
15 Tankers, 6 store and 1 helicopter support ships.
5 Nimrod maritime patrol aircraft; AEW-2 Sea Kings
14 Commando and 20 ASW Sea King, 20 Lynx and Wasp helicopters

AIR FORCES
2–5 Tornado, Jaguar, and Buccaneer squadrons
1–2 Tornado and Jaguar reconnaissance squadrons
1 Nimrod ECM aircraft
1 VC-10 and 2–4 C-130 transport squadrons
20–30 Wessex, Chinook, and Puma helicopters

Source: IISS, *Military Balance*, 1985–1986, pp. 41–43; and Peter Foot, *Beyond The North Atlantic: The European Contribution*, ASIDES, No. 21, Spring 1982, p. 28.

brigade-sized amphibious lift capability which is dependent on two amphibious assault ships, five Sir Bedivere class logistic land ships (LSLs), two chartered merchant ships performing the same role, a series of flat-bottomed landing craft, self-propelled pontoons, and Wessex and Sea King helicopters. These would be supported by requisitioned Ships Taken Up From Trade (STUFT) which include RO-RO ships, troop-carrying ferries and liners, tankers, and break bulk and cargo ships.[17]

Britain has begun to replace its two aging amphibious assault ships (also known as landing platform docks, or LPDs), the *Fearless* and *Intrepid*, but the *Fearless* and *Intrepid* were designed in the late 1950s, were laid down in 1962, and were commissioned in 1965 and 1967. They are now twenty years old, and if they are not replaced, they would need a massive service life extension program (SLEP) to fully rebuild and modernize both vessels. Such amphibious ships are critical to British out of area operations. They are the only ships with the command and control, helicopter landing, and amphibious craft loading capability to support the rapid assault operations essential to successful amphibious actions.

Britain also lacks modern amphibious landing ships (LSLs). This can create serious problems. LSLs are critical to any rapid landing of armor and heavy equipment across beaches. Britain experienced major difficulties in trying to land in bad weather in the Falklands and could not conduct amphibious landings off the Danish coast in a 1984 Bold Gannet exercise because of bad weather. The existing LSLs need major modernization and added air defense and missile capability.[18]

Britain is not organized or equipped to conduct major land operations east of Suez, and lacks the ability to rapidly deploy more than light armor in Gulf contingencies. This situation is unlikely to change. Britain would experience significant problems in conducting even brigade-sized operations in the Gulf against an opponent with extensive heavy armor and air power. British out-of-area forces are relatively lightly equipped and do not have main battle tanks or long-range surface-to-air missile defenses.

Britain would take at least two to three weeks to assemble the sea lift and forces to deploy a mechanized brigade, and could use strategic airlift for such an operation only if it were made available from the U.S. Even high-speed movement of a reinforced battalion of British mechanized forces would now require supporting U.S. airlift or sea lift. British land forces will, however, steadily improve their AFVs, holdings of Milan and Improved TOW ATGMs, and modern SHORADs like the Improved Blowpipe and Improved Rapier.

UNITED STATES OF AMERICA

Given this background, it is clear that the United States is the only Western power that can intervene in the Gulf in a moderate- to high-level conflict. The U.S., however, will be critically limited by the availability of bases and facilities in the area. In fact, the size of the forces the U.S. can commit to the region are determined more by the limitations on its strategic air and sea lift, forward basing and support facilities, and the risks inherent in redeploying U.S. forces from other regions, than by the total combat forces available from the U.S. order of battle.

The U.S. has steadily improved the forces it can commit to the Middle East which are tailored to regional needs and contingencies since the fall of the Shah of Iran, and has stepped up these efforts since it began its convoy activity in the Gulf in early 1987. Nevertheless, the U.S. must still rely heavily on prepositioned equipment and stocks in the Gulf to minimize the strain on its strategic lift, and on access to friendly bases so that it can allocate its lift resources to moving U.S. combat forces, rather than creating bases and support facilities.

As is described later in this book, the U.S. built up massive naval forces in the Gulf during 1987. These forces, however, are exceptional. Under normal peacetime conditions, the U.S. can draw on two groups of forces to intervene in the Gulf: the forces already in the Mediterranean area and those in the U.S.

The U.S. forces already in the Mediterranean normally include 470 army personnel in Greece, 3,950 in Italy, and 1,250 in Turkey. These are largely support personnel, but many could support out-of-area operations. The U.S. Air Force has 5,300 men in Spain, and a tactical wing with three squadrons of 72 F-16A/B fighters. One tactical fighter wing with F-4E fighters is deployed in the U.S. on an "on call" rotational basis. The air force has 5,800 men and two air base groups plus one Ground Launched Cruise Missile (GLCM) unit in Italy; 2,700 men and two air base groups in Greece; and 3,800 men and two air base groups in Turkey.

The U.S. Sixth (Mediterranean) Fleet has a nominal strength of 27,000 men. It typically has two nuclear attack submarines (SSNs), two full fleet carriers, twelve major surface combatants, eleven support ships, one Amphibious Ready Group (3–5 ships and a battalion-sized landing team or Marine Amphibious Unit), and three stores ships with prepositioned combat equipment. It has major base facilities at Rota, Spain (3,600), and at Gaeta, Naples, Sigonella, and La Maddalena in Italy (5,250). The U.S. Marine forces afloat normally total 1,900 men or one Marine Amphibious Unit (MAU). An MAU has a reinforced infantry battalion group, including tank and artillery elements, a composite air group with AV-8B fighters and helicopters, and an additional logistics unit.[19]

The key aspect of U.S. amphibious capability, however, is the world-wide pool of amphibious forces, which are not just an assembly of ships but which include fully ready and functional carrier task groups. Even one U.S. carrier provides more air power than the total air strength of the far smaller carriers that could be deployed by other NATO countries. The size of a typical air wing in a carrier task group is shown in Table 5.3. Not only is one U.S. carrier task group considerably stronger than the total naval forces of any other NATO nation; it is also capable of self defense, major air operations, and limited forced entry through amphibious or helicopter assault.

Table 5.3: Strength of Typical U.S. Navy and Marine Air Wings

Aircraft Type	Function	Squadrons	Aircraft
A. Carrier Air Wing			
F-4, F-14 (TARPS)	Fighter (Reconnaissance)	2	20–24
A-7, F/A-18	Light Attack	2	20–24
A-6	Medium Attack	1	10–20
KA-6D	Tanker	1	0–4
S-3A	ASW (Fixed Wing)	1	10
SH-3	ASW (Rotary Wing)	1	10
EA-6B	Electronic Warfare	1	4–5
E-2C	Airborne Early Warning	1	4–5
Totals		9	86
B. Marine Corps Air Wing			
F-4, F-18	Fighter	4	48
A-4	Light Attack	1	9
AV-8B	Light Attack	2	40
A-6	Medium Attack	1–2	20
KC-130	Tanker/Transport	1	12
EA-6B	Electronic Warfare	1	8
RF-4 or F/A-18	Reconnaissance	1	8
OV-10	Observation	1	12
AH-1	Attack Helicopters	1	24
CH-53	Transport/Utility Helicopters	6–7	48
UH-1	Helicopters	–	24
Totals		27–30	313

Source: DoD, *Annual Report, FY1988*, p. 188.

The other Western forces available for a Gulf contingency do not meet this test. No other NATO nation has a single carrier task group capable of self-defense against a modern air force, and no allied carrier group is normally equipped to support land or air operations against an enemy equipped with modern combat aircraft and tanks.

The second part of U.S. regional contingency capabilities is easier to define: It includes the forces allocated to USCENTCOM. These forces are shown in Table 5.4, and once again it is important to contrast them with the other forces NATO can deploy. Britain, France, and Turkey can play an important role in Indian Ocean–Red Sea–Gulf scenarios, but all their forces combined lack the muscle to sustain even moderate-level military operations.

The USCENTCOM forces have massive naval and air power by regional standards, although they lack anything approaching Soviet land force strength and would have serious problems in defending against Iraq or Iran. These forces are also far better tailored to their mission than they were in the late 1970s, and are now specially trained and equipped to operate in the region—something that no longer is true of most British air and land forces, and many French power projection units. The scale of the training involved is illustrated by the annual Bright Star exercise in 1987, which involved 12,700 men ashore and 15,000 afloat, 4 sea lift ships, and 205 air lift missions. The U.S. also conducts major special operations, field training, mobility, and command post exercises.

This does not mean that USCENTCOM would find it easy to operate in the Gulf region, or that it is without flaws. Further, the details of the shortfalls in U.S. capabilities, which are infinitely greater in terms of support, readiness, supply, and C^3I than those of British and French forces, illustrate two main issues: Combat forces are not the key problem at this kind of power projection distance, and the U.S. would have to sacrifice support of its own forces to support any European forces. In short, European reinforcements would limit maximum U.S. build-up and combat capabilities, in addition to adding major interoperability problems.

The U.S. Army combat units for USCENTCOM are combat ready, but more than 50% of the combat support and service support units are reserve units. This includes 49% of combat support units and 59% of service support units. Although these forces were rapidly improved in 1987, because of the crisis in the Gulf, the full-time active combat support units in USCENTCOM have 83% unit readiness and the reserve units have 62% readiness.

The full-time active service support units in USCENTCOM have 78% unit readiness and the reserve units have 57% readiness. There are serious shortfalls in specialized skilled manpower, in total manning, and in senior NCOs. USCENTCOM support units all have equipment problems. The full-time active combat support units in USCENTCOM have 89% of their equipment on hand and the reserve units have 83%. The full-time active service support units in USCENTCOM have 93% of their equipment on hand, and the reserve units have 80%. While the navy, marine corps, and air force support units are at high readiness, these shortfalls mean that U.S. Army units would take far longer to bring to full readiness, and to sustain in combat, than U.S. contingency plans call for.[20]

Table 5.4: USCENTCOM Forces in FY1989

Force and Element	Manpower
U.S. Central Command Headquarters	1,100
U.S. Army–Central Forces Command	131,000
Headquarters, U.S. Army Central Command (Third U.S. Army)	
XVIII Airborne Corps Headquarters	
82nd Airborne Division	
101st Airborne Division (Air Assault)	
24th Infantry Division (Mechanized)	
6th Armored Cavalry Brigade (Air Combat)	
lst Corps Support Command	
U.S. Navy Forces–Central Command	123,000
Headquarters, U.S. Naval Forces Central Command	
3 Aircraft Carrier Battle Groups[a]	
1 Surface Action Group	
3 Amphibious Groups	
5 Maritime Patrol Squadrons	
U.S. Middle East Task Force (Bahrain)	
U.S. Marine Corps Forces	70,000
1 Marine Expeditionary Force (MEF), including	
1 Marine Division	
1 Marine Aircraft Wing[b]	
1 Force Service Action Group	
1 Marine Expeditionary Brigade (MEB), including	(16,000)[c]
1 Marine Regiment (reinforced)	
1 Marine Air Group (composite)	
1 Brigade Service Support Group	
U.S. Air Force, Central Command Air Forces (9th Air Force)	33,000
7 Tactical Fighter Wings[d]	
3 1/3 Tactical Fighter Wings (available as attrition fillers)	
2 Strategic Bomber Squadrons[e]	
1 Airborne Warning and Control Wing	
1 Tactical Reconnaissance Group	
1 Electronic Combat Group	
1 Special Operations Wing	
Unconventional and Special Operations Forces	3,500
TOTAL	291,600

The U.S. Air Force combat units for USCENTCOM are all active combat units. The USAF Reserve, however, provides two-thirds of the C-130 squadrons and more than 50% of the in-theater communications aircraft. There are serious shortfalls in some aspects of the USAF units committed to USCENTCOM. Shortfalls in funding for spare parts and war readiness spares will leave these units 19–24% short of their planned sortie support.

Further, there are serious shortfalls in the ability of assigned aircraft to fly all-weather and night missions, and these will not be corrected until the mid-1990s, when the F-111D is upgraded with PAVETACK and the F-15E and F-16 with LANTIRN are deployed. The tactical air control equipment used to coordinate between land and air units is 25 years old, bulky, and limited in information-processing capability. While munitions are being upgraded with improved BLU 109/8 2,000 pound bombs, HARM anti-radiation missiles, and IR Mavericks, recent budget cuts mean a serious delay in such modernization.

U.S. Navy and Marine Corps forces are at a high state of readiness because of the U.S. intervention in the Gulf in 1987. Marine forces are 100% combat ready, and a fully combat ready 16,500 man Marine

Table 5.4 (continued)

Source: Data furnished by USCENTCOM; see also the Department of Defense, *Annual Report, FY1986*, p. 212, and *FY1988*, pp. 223–233.

[a]A typical active Navy carrier wing consists of nine squadrons (approximately 86 aircraft): two fighter squadrons, two light attack squadrons, one medium attack squadron, plus supporting elements for airborne warning, antisubmarine and electronic warfare, reconnaissance, and aerial refueling operations.
[b]An active Marine Corps air wing typically consists of 23–25 squadrons (338–370 aircraft) with four fighter attack squadrons, two or three light attack squadrons, one or two medium attack squadrons, plus supporting elements for electronic warfare, reconnaissance, aerial refueling, transport, airborne assault, observation, and tactical air control.
[c]The MEB is currently the only element of the MEF which has prepositioned equipment. The full MEF has prepositioned equipment on ships although the other two sets will be located for missions in NATO and Asia.
[d]Each Air Force Wing typically contains three squadrons of 24 aircraft each. (Combat support units, such as those composed of EF-111 electronic warfare aircraft, are generally organized into squadrons of 18 to 24 aircraft.) By the end of FY1989, the U.S. will have the equivalent of 40 tactical fighter wings—27 active and 13 Air National Guard and Reserve.
[e]There are a total of 7 B-52G squadrons assigned to general purpose as well as nuclear missions. These have a strategic reconnaissance and anti-shipping mission as well as a conventional land bombing role.

Expeditionary Brigade can now rapidly deploy using prepositioned assets. The Landing Craft Air Cushioned (LCAC) is being deployed, and the MV-22 Osprey tilt-rotor aircraft is reaching final development and will replace the Corp's aging CH-46 helicopters. This will soon give the Marines high-speed over-the-horizon amphibious capability and enhanced lift.

On a broader level, however, USCENTCOM is short of the assets it needs to off-load ships and carry them across beaches. Unless it has ready access to a port in the rear combat zone, it will have only 44% of the U.S. Army logistics over the short (LOTS) assets it needs. The army needs air cushion vehicles, lighters, better landing craft, and ferries. These will be available by FY1996, but USCENTCOM can now move only 9,300 short tons a day versus a goal of 21,000.

The special forces elements of all services are generally ready, but much of their combat equipment is aging and the air force units are only at 75% of readiness because of a lack of spare parts for the HH-53 helicopter and AC-135 gunship. They lack long-range air lift and vertical lift capability.

The U.S. now has only 57% of the strategic air lift it needs to deploy and support USCENTCOM. Current plans to improve U.S. strategic airlift will provide only about 50 million ton miles per day worth of capacity in FY1990, versus the 66 million ton miles per day that is USCENTCOM's requirement, and it is unlikely USCENTCOM will get the airlift it needs before the year 2000—if ever. The U.S. is even more short of intra-theater lift. It has only 55% of its goal of 13,500 tons per day, and there are serious limits on cargo size and speed of handling. This will begin to change in the early 1990s, if the C-17 is deployed, but the situation will not be corrected until after the year 2000. Efforts to upgrade the C-5 and reach high levels of operability have not been fully successful. This not only leaves the U.S. dependent on the C-17 to improve its strategic lift, but the C-17 is planned to provide 60% of intra-theater lift by the year 2000.

Over 95% of USCENTCOM tonnage moves by sea lift. Fortunately, the navy is closer to meeting its goal of 1 million short tons worth of capacity for USCENTCOM, and will be at about 89% of its goal during FY1986–FY1992. It has reached its goal for fast sea lift ships and maritime prepositioning ships.[21]

Nevertheless, USCENTCOM is far short of its total goals for shore-based prepositioning. It now has a goal of 300,000 short tons worth of prepositioned supplies and equipment, and only 20% of this is available afloat and another 13% ashore. It has only 77% of its goal of 150,000 tons worth of prepositioned ammunition, all of it afloat. It has 41% of its goal of 12 million barrels worth of prepositioned petroleum, oil, and lubricants (POL), with 5% prepositioned on shore in the area, 8% afloat, and 28% stored outside the USCENTCOM area.[22]

USCENTCOM would also face major sustainability problems. If one considers its total theater requirements for prepositioned sustainability, it has only 60% of its requirement of supplies and equipment; only 50% of its requirement for petroleum, oil, and lubricants (POL); and only 36% of its requirement for munitions. This situation will improve if equipment continues to move into Oman, if Bahrain and Saudi Arabia quietly allow limited prepositioning as agreed on September 26, 1987, and if new facilities are provided in Somalia. USCENTCOM will then have 81% of its requirement of supplies and equipment, 73% of its requirement for POL, and 39% of its requirement for munitions. Most of this will be located on land, although 20% of its requirement of supplies and equipment, 25% of its requirement for POL, and 31% of its requirement for munitions will be afloat. The POL situation will also be much better than it seems because the U.S. Navy has an additional new 1.6 million barrel storage facility in Bahrain.

USCENTCOM faces even more serious sustainability problems if one considers its total theater requirements for both prepositioned and rear area sustainability. It has only 46% of its 1 million short ton requirement of supplies and equipment; 37% of its 16 million barrel requirement of petroleum, oil, and lubricants (POL); 43% of its 495,000 tons worth of munitions; and 59% of its threat-oriented munitions. It will have only 76% of the items it needs to meet its goal for F-15 sorties and 81% for F-16 sorties. It will only have 48% of its radar-guided air-to-air munitions. Roughly 19% of its requirement of supplies and equipment, 18% of its requirement for POL, 25% of its requirement for level-of-effort munitions, and 8% of its threat-oriented munitions will be afloat.

Communications are a particularly serious problem. The Defense Communication System (DCS) effectively stops at Turkey in the West and the Philippines in the East. There is only limited commercial capability, and it would be hopelessly inadequate in combat. The DCS is slowly being extended into the area, and an initial operational capability was achieved in June, 1987. It will be some years, however, before an adequate system is deployed. Fortunately, tactical satellites are now available to cover the 7,000 miles between the theater and the U.S., and automated secure communications are steadily improving. Efforts are also being made to implement a new theater-wide intelligence architecture, and deployable intelligence data handling systems will be deployed in FY1989. There have been intelligence satellite imagery processing facilities in the theater since 1986.[23]

Medical readiness will also be a problem. USCENTCOM had only 60% of the hospital beds it needed in FY1988, although it will have

95% in FY1991. The readiness of medical personnel is a problem: Readiness is 85% for active personnel and 78% for reserve personnel. Only about 36% of its goal for prepositioning key medical items has been met, and recent budget cuts mean this situation will improve slowly, if at all. These problems could be particularly serious in the event of chemical warfare—something that seems all too possible in the Gulf. USCENTCOM has 79% of the needed individual protection gear, 50% of the collective protection gear, 83% of the decontamination capability, and 90% of the detection systems.

Even when all currently programmed steps are completed in the early 1990s, it will take several weeks for the U.S. to deploy the equivalent of a two-division force, and a month to six weeks to deploy three full divisions.[24] These forces will still be light on armor compared to the major powers in the Northern Gulf.[25] The U.S. also faces the practical reality that carriers are vulnerable in Gulf waters, or in any other place where they cannot operate a long-range air and missile defense screen, and enemy operations can strike from nearby air bases or the terrain masking provided by land.

Further, the U.S. Navy will remain badly short of mine-clearing capability well into the 1990s. While the U.S. has mobilized and deployed mine forces to the Gulf as a result of its convoy effort, it normally has only six active ocean-going mine-clearing vessels in the regular U.S. Navy and fifteen in the reserve fleet. Its only other mine warfare ships are seven minesweeping craft for ports and harbors.[26]

None of of these mine warfare resources are normally deployed in Southwest Asia, and the U.S. experience in the Gulf in 1987 showed that such capabilities normally take so long to mobilize and deploy that it is unlikely they would arrive in a contingency where time was critical. The U.S. Navy also has three mine-clearing helicopter squadrons, but they too take several weeks to deploy and would compete with other U.S. forces for available strategic airlift.[27]

The U.S. Air Force will be almost totally dependent on access to friendly air bases. Its effectiveness will also be heavily dependent on having sheltered, defended, well stocked, and interoperable facilities with suitable C^3I/BM capabilities. Such bases now exist only in Spain, Morocco, Italy, Greece, Israel, Egypt, Turkey, Oman, and Saudi Arabia, and contingency access is uncertain and scenario dependent.

Any U.S. effort to conduct major land force operations will be heavily dependent on friendly locals and good staging facilities. Strategic sea and air lift will be critical. This means free access to critical NATO and Middle Eastern staging facilities such as those in the Azores, Morocco, Egypt, and Oman. The struggle to conduct successful operations half a world away from the U.S. will be difficult at best.

ACCESS TO BASES IN THE SOUTHERN GULF

These limitations on U.S. capabilities help explain why Western access to bases and facilities in the Southern Gulf could be so critical in a conflict, and why the quality of U.S. military relations with Saudi Arabia is so important to protection of the Gulf's oil supplies.

Table 5.5 shows the full range of U.S. contingency bases in the region. The U.S. spent some $1.1 billion just on military construction in the USCENTCOM area—and en route support facilities in Morocco, Lajes, and Diego Garcia—between FY1980 and FY1988. These facilities have also largely been completed, and spending has been drastically reduced since FY1985. It has dropped from a peak of around $250 million annually in FY1980–82, to less than $50 million in FY1985–88. The only major projects now under way are an Intermediate Staging Facility (ISF) in the Horn of Africa, and a theater war reserve facility (TRF). The U.S. Army portion of this program will be completed in FY1989, and the USAF portion in FY1990 and FY1991.

Yet, for all the bases listed in Table 5.5, the U.S. has only four sets of bases or basing facilities it can use to defend the Gulf and must rely on bases in Oman and Saudi Arabia. The only other Western bases in the entire region are the French facilities in Djibouti in the Red Sea, and the British advisory presence in Oman.

The only permanent fully active Western base in the Indian Ocean and Gulf area is on the British island of Diego Garcia in the Southern Indian Ocean—which the U.S. now leases from Britain and where the U.S. now prepositions much of USCENTCOM's equipment. This base is so far to the South, however, that it is nearly as far away from the key strategic areas in the Upper Gulf as is Dublin, Ireland. The U.S. is also helping Turkey strengthen its bases in Eastern Turkey, but Turkey has firmly stated that it will not provide contingency bases for USCENTCOM, and that it must make defense of its territory against the USSR its primary concern. The bases in Turkey are also useful primarily for contingencies involving a Soviet invasion of northern Iran.

The U.S. has contingency bases in Oman—the one Gulf state whose internal and external politics allow it to grant such facilities with minimal risk to its security. Oman provides important staging facilities on the island of Masirah, allows the U.S. to fly maritime patrol aircraft from its soil, and has supported contingency arrangements to allow U.S. tankers to stage out of Omani air fields and refuel U.S. carrier aircraft flying from the Indian Ocean. Oman, however, is too far

Table 5.5: U.S. Military Contingency Facilities in the Near East

Base	Status
NORTH AFRICA AND STAGING POINTS	
Morocco: Total Cost of Military Construction in FY1980–88 was $58.6 Million	
Slimane	Agreement signed in May, 1983. A former B-47 base closed in 1963 is now being modernized to support C-141 and C-5 operations.
Navasseur	This base or Rabat may be given similar modernization later.
Liberia	
Monrovia	Agreement signed in February, 1983, to allow U.S. to make contingency use of international airport for stage air operations. U.S. will fund expansion of airport to allow use of C-5s, C-17s, and C-141s.
Portugal	
Lajes	Negotiations were completed in 1983–84 to keep Lajes as a major air staging point for U.S. air movements. The fuel, runway, and other facilities at this base in the Azores are being upgraded. The total cost of military construction in FY1980–88 was $66.6 million.
EASTERN MEDITERRANEAN AND RED SEA AREA	
Egypt	
Suez Canal	U.S. has been granted tacit permission to move warships through the Canal.
Cairo West	The U.S. shares an unnamed air base with Egypt, and normally deploys about 100 men on the base. It has been used for joint F-15 and E-3A AWACS operations.
Ras Banas	Still under negotiation. Ras Banas would provide basing capabilities for C-5 aircraft, and for unloading and transit of SL-7 and other fast sea lift ships.
Djibouti	Access agreement and arrangements with French allow port calls and access to maritime patrol aircraft.

Table 5.5 (continued)

Turkey	
Mus Batman Erzurum	The U.S. has informal arrangements to use three Turkish air bases near the Soviet border, Iran, and Iraq. These bases are NATO bases and are being funded to allow the deployment of U.S. heavy lift aircraft and fighters.
GULF AND RED SEA	
Diego Garcia	Used through a long-term 50-year lease from the U.K. signed in 1965. The base provides 12,000 foot runways and facilities suitable for B-52 and heavy air lift facilities, and is where seven U.S. prepositioning ships in the Gulf are now deployed. The total cost of military construction in FY1980–88 was $542.9 million.
Seychelles	Satellite tracking and communications base with NASA and air force personnel.
Kenya	
Mombassa Nanyuki Airport Kenya Naval base	Provides a potential staging point, maintenance facilities, and port call. Access agreement signed in mid-1970s. Facility expansion program was completed in 1983. The total cost of military construction in FY1980–88 was $66.6 million. U.S. spent $30 million dredging harbor to allow it to be used for carrier port calls.
Somalia	
Mogadishu Airport Berbera	Staging facilities for U.S. air and sea movements. Limited repair capability. Expansion was completed in 1983. The total cost of military construction in FY1980–88 was $24.4 million. Somalia is 1,400 miles from the Gulf and facilities would be used for sea control and intermediate staging.
Oman	Total cost of military construction in FY1980–88 was $270.3 million.
Al Khasab	Small air base in the Musandem Peninsula near Goat Island and Straits of Hormuz. Limited contingency capability. Largely suited for small maritime patrol aircraft.

(continued)

Table 5.5 (continued)

Masira	Island being expanded to a major $170 air and naval staging point, with limited deployment of prepositioning ships. Some $121 million worth of equipment is prepositioned, including food, trucks, air traffic electronics, artillery shells, and air-to-air missiles.
Thumrait & Seeb Air bases	Contingency air base facilities. Now used by U.S. maritime patrol aircraft.
Saudi Arabia	No formal basing agreements, but the U.S. has deployed F-15s, KC-10 and KC-135 tankers, and E-3As to Saudi air bases in emergencies, and operates E-3As from Riyadh. All Saudi major air bases have the sheltering and facilities to accept extensive U.S. air reinforcements and/or support U.S. deployment of heavy lift aircraft.
Bahrain	U.S. Middle East Force deploys in Bahrain, although formal agreement has lapsed. A 65 man U.S. support unit is present. The U.S. has spent $2.6 million on military construction.

Sources: Adapted from material provided by USCENTCOM, and from Barry M. Blechman and Edward N. Luttwak, *International Security Yearbook, 1983/84*, St. Martin's Press, New York, 1984, pp. 154–159.

east to allow U.S. forces to efficiently defend most of the Gulf oil fields and key oil nations like Kuwait.

The U.S. also has de facto access to Bahrain for a number of contingency purposes. The U.S. Middle East Force is home ported in Bahrain, and there is a small USCENTCOM headquarters on a ship in this force. This was of considerable importance during the U.S. intervention in the Gulf in 1987, and Bahrain provided important support and repair capabilities for the U.S. Navy during its operations in the Gulf.

Bahrain has agreed to provide large fuel-storage facilities, and some prepositioning facilities. It allows U.S. personnel and military cargos to transit into the Gulf via its international airport, and allows relatively free use of its ports and anchorages by U.S. naval forces. It also has allowed the U.S. to deploy two large barges just outside its territorial waters to act as bases for special forces, attack helicopters, radars, air defenses, and intelligence sensors. Nevertheless, Bahrain is a small and divided state. It cannot base or support large air or land forces and it is vulnerable in both political and military terms.[28]

Kuwait is larger than Bahrain, and could base a substantial number of U.S. troops or aircraft, but it has no strategic depth and its air, naval, and cooperable support facilities are acutely limited if the U.S. should have to deploy for a low-level war involving Iran or Iraq. It would take several weeks to build up enough U.S. forces in Kuwait to sustain large numbers of combat sorties or deploy land troops to help Kuwait defend its territory.

Kuwait did agree to allow the U.S. to deploy a charter barge as an offshore base in Kuwaiti territorial waters in late 1987. This set an important precedent, and indicates that Kuwait might accept larger U.S. forces in the face of an immediate threat to Kuwait. Neverthless, Kuwait is in a highly vulnerable position in terms of any threat from the Northern Gulf states. Any Western base or deployment, except in the face of a direct and publically apparent threat to Kuwait's security, would make it highly vulnerable to political attacks from radical Islamic states and Arab movements. Kuwait might well accept deployment of U.S. forces in an absolute emergency, but only for its national defense. Even if it did, however, it might delay until U.S. reinforcements would be difficult to deploy in time, and the U.S. would need other bases in the region to secure its access to the Gulf in virtually any contingency—including an attempt to defend Kuwait.

All of these factors make Western access to Saudi bases like the ones at Dhahran and Hafr al-Batin critical in any major defense of the Gulf against a threat from Iran, Iraq, or the USSR. In fact, USCENTCOM probably cannot function in its most critical contingency roles without Saudi cooperation and wartime access to Saudi bases and facilities. While Diego Garcia, Djibouti, Turkey, and Egypt provide useful contingency facilities on the periphery of the Gulf and lower Red Sea, they cannot make up for the range and reinforcement problems the West would face in defending its critical oil facilities in the upper and central Gulf.

Saudi Arabia has never formally agreed to provide the U.S. with contingency bases. The politics of the Gulf preclude Saudi Arabia from overtly granting base facilities without a clear and immediate threat. It would be accused of neocolonialism, of supporting an ally of Israel, and of having brought superpower confrontation into the region. Saudi Arabia has, however, quietly consulted U.S. defense planners and senior USCENTCOM officers regarding U.S. use of Saudi facilities in an emergency, and it has made USCENTCOM "over the horizon" reinforcement capabilities one of the mainstays of its defense planning.

Saudi Arabia has, however, sought U.S. deployments in past contingencies, and these have recently included detachments of U.S. minesweeping forces, USAF F-15s, and USAF E-3A AWACS aircraft. Saudi

Arabia now has U.S. E-3A AWACS aircraft deployed on its soil (these were first requested early in the Iran-Iraq War) and a small USAF detachment which operates in a joint headquarters at Dhahran, and it shares data on the Iran-Iraq War and other developments in the Gulf.

The Saudi Air Force's past reliance on U.S. equipment has also given it the capability to conduct joint operations with U.S. forces, to support USAF reinforcements, and to provide C3I and support facilities. The Saudi Air Force now operates 57 F-15C/Ds, and nearly 114 F-5E/Fs and RF-5Es. It uses U.S. training and maintenance standards, and many Saudi squadrons now approach USAF proficiency and qualification levels. Saudi Arabia also has the largest and most modern air bases in the Middle East. These air bases have extensive shelter facilities, and those in the Gulf and lower Red Sea areas are equipped to support U.S.-made F-15 and F-5 aircraft. They are also defended with Hawk missiles, and will soon have the new Peace Shield command, control, communications, and intelligence (C3I), and air control and warning (AC&W) system using U.S. E-3A AWACS aircraft and advanced ground radars and electronics.

Such Saudi air facilities could base up to two wings of USAF fighters, and give them full munitions and service support. The U.S. has large numbers of contract personnel servicing Saudi equipment in the air force, army, national guard, and navy, and large numbers of Saudi military and civilian personnel have had U.S. training and can operate with, or support and service, U.S. military equipment.

If Saudi Arabia had been successful in obtaining the additional 40 improved F-15C/Ds and modern U.S. attack munitions to supplement its air-to-air weapons that it requested in 1985, it would have further developed and equipped a network of bases in the Gulf (Riyadh, Dhahran, and dispersal facilities at Hafr al-Batin), and in the Red Sea area (Taif, Khamis Mushayt, Sharurah, Jiddah, and Tabuk), that could have allowed large amounts of U.S. air power to deploy in 48 to 72 hours to the most threatened areas in the Gulf.[29]

Further, the U.S. experience in deploying naval forces to the Gulf in 1987 has shown that Saudi bases are located in areas where U.S. carriers cannot operate their aircraft effectively without moving into the Gulf. The Gulf is a highly vulnerable area for such operations. Hostile states like Iran have anti-ship Harpoon missiles and can launch suicide air attacks with only limited warning, and Soviet attack fighters and bombers can launch air-to-ship missiles after taking advantage of the terrain masking provided by the mountains in Iran and with far less chance of detection than in the open sea.

Saudi Arabia has equally modern naval facilities and ground bases. These bases have extensive stocks of parts and munitions, and service and support equipment, which can be used by USCENTCOM forces. The Saudi base at Hafr al-Batin (which is located in the critical border area near Kuwait and Iraq) will also have two full brigades equipped with U.S. armor.[30] Saudi army and naval bases have some of the most sophisticated infrastructure and service facilities in the world, and can both speed the deployment of U.S. forces and make them more effective once they arrive.

WESTERN AND U.S. SECURITY ASSISTANCE PROGRAMS TO THE GULF STATES

Power projection capabilities are closely linked to a wide range of facilities in the region, and our access to these facilities is closely linked to military assistance and advisory efforts. By and large, however, most Western nations no longer provide military assistance to the region. The exceptions are French assistance to Djibouti and British assistance to Oman, most of which is paid for by the governments involved.

The U.S. was a major exception during the 1970s and early 1980s, in large part because it was seeking to create a broad range of regional basing infrastructure and service facilities. Since 1985, however, the U.S. Security Assistance effort in the region has slowly collapsed. The total U.S. foreign military sales credit and military assistance program has shrunk from $1,735 million in 1985 to $1,703 million in 1986, $1,677 million in 1987, and $1,599 million in 1988. The amount of this aid going to countries in the Gulf region, however, has shrunk by 83.4%. It dropped from $235 million in 1985, to $148 million in 1986, $65 million in 1987, and $39 million in 1988.[31]

This drop has had nothing to do with U.S. military needs and strategic commitments. It has been forced upon U.S. military planners by a mixture of cuts in the total federal and defense budgets and by U.S. domestic politics, which protect Egypt and Israel—the two largest consumers of U.S. grant aid—from any share of the reductions in U.S. security assistance.

These cuts are shown in Table 5.6. They are particularly serious in the case of Jordan, which had a $1.9 billion aid package withdrawn from the Congress in 1986, and where cuts in 1987 and 1988 funding have left funds only for minimal sustainment of older U.S. systems. The virtual end of aid to Kenya may deprive the U.S. of key staging facilities. The loss of aid has left no funds for modernization, and even basic maintenance activity has had to be cut back. The cut in aid to Oman, coupled with the fall in its oil revenues, has left it with very limited

Table 5.6: Trends in U.S. Security Assistance Impacting on the Gulf Region: FY1985 to FY1988

Grant Aid, FMS Credits, FMS Sales, MAP Aid, FY85/FY88 ($ in Thousands)

Country	FY75–86**	FY85	FY86	FY87	FY88	Change
Bahrain**	339,165	–	–	382,718	?	NA
Djibouti	8,347	2,500	1,914	1,200	1,000	–60%
Jordan	1,197,028**	90,000	81,345	39,941	26,500	–71%
Kenya	255,967**	20,000	19,140	10,500	5,000	–75%
Kuwait**	1,475,252	–	–	57,060	?	NA
Oman	231,813**	40,000	9,140	0	0	100%
Qatar**	2,497	–	–	635	?	NA
Saudi Arabia	88,962,000**	–	–	1,100,000**	?	NA
Somalia	180,807**	33,000	19,140	7,500	5,500	–83%
Sudan	404,646**	45,000	16,140	5,000	0	–100%
UAE**	750,503	–	–	90,659	?	NA
YAR	359,072**	5,000	1,914	1,000	1,000	–80%

*Number of Military Personnel Trained by the U.S., FY85/FY88****

Country	82	83	84	85	86	87	Change in $
Bahrain		4	1	6	25	42	NA
Djibouti	4	7	6	7	8	11	+7%
Jordan	–	381	370	405	381	491	–7%
Kenya	73	90	126	138	169	80	–29%
Kuwait	263	108	227	297	298	266	NA
Oman	–	–	15	24	13	29	–3%
Qatar	–	13	2	4	6	22	NA
Saudi Arabia	563	866	447	928	688	717	NA
Somalia	26	32	46	72	97	35	–12%
Sudan	229	112	149	150	55	57	–37%
UAE	–	93	44	30	127	48	NA
YAR	–	97	87	91	63	69	27%

Source: Adapted from briefing materials provided by USCENTCOM.

* Includes MAP and FMS credits. Bahrain, Kuwait, Qatar, Saudi Arabia, and the UAE are self-financed.

**Includes a small amount of International Military Education and Training (IMET) aid.

*** Change in IMET aid only.

funds for modernization, and may ultimately produce the kind of backlash that could deprive the U.S. of its most important staging site and prepositioning point in the region—particularly since Mauritius and India are stepping up their campaign to persuade Britain to push the U.S. out of Diego Garcia.[32]

U.S. aid to Somalia has been cut to the level where the Somali armed forces have been virtually deprived of their ability to modernize, or even maintain existing equipment. The cuts in U.S. aid to the Sudan have been so serious that they are forcing the U.S. to reconsider the effect of leaving the country open to Libyan and Ethiopian pressure, although nothing has yet been done to change the situation.

The cuts in U.S. aid to the YAR have left a program so small that the U.S. aid program between 1974 and 1988 has totalled only 2 percent of the aid the Soviet Union has provided the YAR since 1980. The MAP program was cut another 49% in FY1987, and was left at that level in FY1988. These cuts have virtually abandoned the military aid effort to the USSR.

The political and military effect of these cuts in U.S. aid has been deferred by (1) the fact that many of the nations involved have nowhere else to turn to in the West or the Gulf, (2) the fact that Bahrain, Kuwait, Qatar, Saudi Arabia, and the UAE finance all U.S. arms sales and military assistance, (3) the impact of cuts in U.S. training has been minimized by selective use of aid or direct funding, and (4) their reluctance to react at a time when the Iran-Iraq War is so important.

Nevertheless, these cuts may ultimately have a critical impact on U.S. contingency capabilities. They are building up a degree of resentment that could be expressed in virtually any future crisis in the region, and they are leaving a power vacuum for the USSR to exploit. For example, the USSR has provided over ten times as much military assistance to Ethiopia over the past ten years as the U.S. has to Kenya, the Sudan, Djibouti, and Somalia, the four surrounding states.

In sum, the unique character of American power projection capabilities is a critical factor in any discussion of Western strategic relations with the Gulf. These capabilities may only be important *in extremis*, but the events of 1987 and 1988 have shown just how real and immediate the West's need for such capabilities can be. Europe can help the Southern Gulf states modernize their forces, but it cannot help them or the U.S. compensate for the failure to standardize the key aircraft and munitions used by USCENTCOM. Further, the fact that Britain and France help maintain close military relations between the Southern Gulf states and the West is not an effective substitute for the kind of U.S. relations with Bahrain, Kuwait, Oman, and Saudi Arabia that would increase their willingness to provide the U.S. with forward basing facilities or develop informal links between USCENTCOM and Saudi and other GCC military forces.

NOTES

1. "Guarding Turkey's Eastern Flank", *The Middle East*, April, 1986, pp. 9–10.

2. This estimate is based on working data from Defense Marketing Services, various editions of *Jane's Defense Weekly*, and IISS, *Military Balance, 1987–1988*.

3. Andrew Borowiec, "Turks Seek Aid To Upgrade Army", *Washington Times*, May 16, 1986, p. 7.

4. Jonathan Marcus and Bruce George, "French Rapid Deployment Force", *Jane's Defense Weekly*, April 28, 1984, pp. 649–650; Giovanni de Briganti, "Forces d'Action Rapide", *Armed Forces Journal*, October, 1984, pp. 46–47. Personal note by General Fricaud-Chagnaud, "La Force d'Action Rapide", July 2, 1986.

5. Ibid.

6. Ibid.

7. It should be noted that any discussion of French and British deployment capabilities is necessarily time sensitive. Both nations could build up a major out-of-area presence over a period of months using sea lift and by drawing down on support equipment and assets normally reserved for European defense. Provided that sufficient domestic political support could be sustained, both nations could also reconfigure existing units and forces or simply assemble new units out of the force elements of other major combat units. The tacit assumption made here is that most out-of-area action will require very rapid reaction, involve serious political and financial constraints, and occur in a climate where the defense of Europe must still be given clear priority.

8. Giovanni de Briganti, "Forces d'Action Rapide", *Armed Forces Journal*, October, 1984, pp. 46–47; and "France's Special Operations Forces", *Defense and Foreign Affairs*, June, 1985, pp. 32–33.

9. John Jordan, *Modern Naval Aviation and Aircraft Carriers*, Arco, New York, 1983, pp. 22–25; and *Defense News*, June 26, 1986, p. 7.

10. When France conducted Operation Mantua in Chad in 1984, it had to lease 24 Boeing 747s for 22 days. It chartered C-5As for the operations in Chad in 1986. *Jane's Defense Weekly*, March 15, 1986, p. 454.

11. The Crotale is effective only to about 12,000 feet versus up to 52,000 feet for Hawk. As a result, the SAMs transportable on French lift aircraft cannot reach medium and heavy bombers. Although France deployed only 900 men, it also had to move its 24 armored vehicles by land from the Cameroons since it lacked air lift for such vehicles. *Jane's Defense Weekly*, March 15, 1986, p. 454.

12. For a good recent description of the goals of the French navy, see Pierre Lachamade, "The French Navy in the Year 2000", *Jane's Naval Review*, London,

1985; Jane's, pp. 79–90. This description precludes recent budget cuts but still indicates the probable future strengths and weaknesses of French naval out-of-area capabilities.

13. Jean de Galard, "French Overseas Action: Supplementary Budget", *Jane's Defense Weekly*, December 14, 1985, p. 1281.

14. Ibid.; and *Jane's Defense Weekly*, July 28, 1984, p. 93.

15. For illustrative British views see Lt. General Sir Geoffrey Howlett, "NATO European Interests Worldwide—Britain's Military Contribution, *RUSI Journal*, Vol. 130, No. 3, September, 1985, pp. 3–10; Simon O'Dwyer-Russel, "Beyond the Falklands—The Role of Britain's Out of Area Joint Forces", *Jane's Defense Weekly*, January 11, 1986, pp. 26–27; "Battle Continues to Preserve British Amphibious Warfare Capability", *Jane's Defense Weekly*, February 8, 1986, p. 185; "UK's Amphibious Dilemma", *Jane's Defense Weekly*, April 12, 1986, pp. 661–662; Professor Neville Brown, "An Out of Area Strategy?" *Navy International*, October, 1982, pp. 1371–1373; and Keith Hartley, "Can Britain Afford a Rapid Deployment Force?" *RUSI Journal*, Vol. 127, No. 1, March, 1982, pp. 18–22.

16. See Simon O'Dwyer-Russel, "Marines Fear Loss of Capability", *Defense Attache*, No. 3, 1984, pp. 60–68; and "Beyond the Falklands—The Role of Britain's Out of Area Joint Forces", *Jane's Defense Weekly*, January 11, 1986, pp. 26–27.

17. *RUSI News Brief*, May, 1986, pp. 5–6.

18. It should be noted that the Dutch Royal Marines no longer have amphibious lift of their own and use British amphibious ships as part of the UKNL Amphibious Force.

19. Estimate based on IISS, *Military Balance, 1987–1988;* data furnished by DoD Public Affairs, January, 1986, Department of Defense, *Annual Report, FY1988*, pp. 172–176; Office of the Joint Chiefs of Staff, *Military Posture, FY1988*, pp. 61–66.

20. All readiness and strength data on USCENTCOM in the text of this section are taken from USCENTCOM briefing material dated February, 1988, and pertain to strength and readiness at the end of calendar 1987.

21. The seventeen prepositioning ships now deployed carry some 165,000 short tons of ammunition and supplies. To put this into perspective, this is equivalent to more than 6,100 C-141 sorties.

22. These and the following shortfall statistics are taken from the FY1987 USCENTCOM command briefing as provided by USCENTCOM.

23. General George B. Crist, *Status of the United States Central Command*, February 22, 1988, pp. 83–84.

24. The U.S. Navy construction program calls for 14 Avenger-class ocean-going mine sweepers, 17 coastal minesweepers, 32 RH-53E helicopters, and 15 RD-53D helicopters. These program are experiencing serious cost and delivery problems and will not be complete before 1997. Many of the current forces are over 30 years old.

25. For an excellent recent treatment of U.S. build-up capabilities in the region, see Thomas McNaugher, *Arms and Oil*, Washington, Brookings, 1985, pp. 53–89.

26. *Defense News*, September 28, 1987, p. 35.

27. David M. Ransom, Lt. Colonel Lawrence J. MacDonald, and W. Nathaniel Howell, "Atlantic Cooperation for Persian Gulf Security", *Essays on Strategy*, Washington, D.C., National Defense University, 1986, p. 102.

28. USCENTCOM briefing; and *Christian Science Monitor*, Janurary 19, 1988, p. 1.

29. Saudi Arabia originally planned a base at Al Kharj, near Riyadh, for its E-3As. This base expansion has been cancelled for funding reasons, but may be reinstated because of the problem of securing Riyadh airport against terrorism.

30. This base was originally designed for three brigades, and has considerably better equipment storage and support facilities than its normal deployment strength of two brigades indicates. Saudi Arabia has also discussed the possibility of converting one brigade to a "Gulf brigade" which would include forces from other GCC countries.

31. All data on U.S. aid are supplied by USCENTCOM. These data differ in time period and/or definition from some of the similar data issued by DSAA and the State Department.

32. *New York Times*, January 4, 1988, p. 10.

6
The Military Capabilities and Vulnerabilities of the Southern Gulf States

It should be clear from the previous analysis of potential threats and Western capabilities that the West cannot secure its strategic interests in the Gulf without the support of friendly Gulf states. Similarly, the West cannot assume all the security burdens in the region. The Southern Gulf states must provide for their own internal security, create sufficient military forces to deter regional threats, and provide the facilities and support that are necessary for Western reinforcement.

The key statistical details of the military forces of the Southern Gulf states are shown in Table 6.1, along with those of the key "threat" states in the Northern Gulf and Red Sea areas.[1] There are several aspects of these statistics which are critical to an understanding of what the Southern Gulf states can and cannot do. One is the relative importance of Saudi Arabia. Saudi Arabia so dominates the overall capabilities of the Southern Gulf that close strategic relations with the Kingdom are vital to any partnership between the West and the Southern Gulf states.

The interaction between these statistics and geography is another critical factor. The geography of the Gulf spreads the smaller Southern Gulf states along the entire coast of the southern Gulf. Kuwait, Bahrain, Qatar, and the UAE all lack strategic depth and are vulnerable to Iranian attacks. Each of the smaller Gulf nations must use all of its air and naval forces to defend its own airspace and waters in any confrontation with Iran or Iraq. Kuwait, Bahrain, and the UAE face particularly serious problems because of the small size of their military forces relative to the threats they face and the size of the border area and territory they must defend.

The smaller conservative Gulf states may spend a great deal on defense, but each faces special problems in building up an adequate deterrent or defense capability. Bahrain is small, relatively poor, and ethnically divided. Kuwait is highly vulnerable to both Iranian and

Table 6.1: The Size and Military Capabilities of the Southern Gulf States in 1987–1988

Country	Size (Sq. Km.)	Population (1,000s)	GDP ($B)	Defense ($B)	Active Military Manpower	Tanks	Combat Aircraft	Combat Ships
Bahrain	676	440	4.5	.14	2,800	60	12	6
Kuwait	17,818	1,800	21.5	1.4	15,000	260	80	8
Oman	212,380	1,310	9.8	1.5	21,500	39	53	8
Qatar	11,000	309	3.1	.25	7,000	24	23	9
Saudi Arabia	2,149,690	9,600	93.7	16.2	73,500	550	226	25
UAE	83,600	1,350	25.7	1.6	43,000	136	65	12
Iran	1,648,000	49,800	163.5	6.1	980,000	1,050	60-85	24
Iraq	434,924	16,900	22.5	11.6	845,000	4,500	500+	28
YAR	194,250	6,500	3.4	.41	36,800	683	73	8
PDRY	322,968	2,300	1.1	.24	27,500	470	62	10

Sources: Adapted from IISS, *Military Balance, 1987–1988*; JCSS, *Middle East Military Balance, 1986*; John Aronson and Inge Lockwood, "Background Materials on the Persian Gulf Region", Washington, Middle East Institute, Sultan Qaboos Center, December, 1987; CIA, *The World Factbook, 1986*, 041-015-00163-9, Washington, GPO, 1986.

Iraqi attacks, and is an extraordinary prize since its small territory and population make its military vulnerable although it has massive oil and gas resources. Oman is acutely limited in the amount of modern heavy weaponry it can buy and operate effectively. Qatar is small, and has too small a native population. The defense effort of the UAE is so divided because of tensions among the individual sheikdoms that it is making little progress in coalescing into an effective force. Further, the individual sheikdoms in the UAE have taken very different stands about whether to organize to defend against Iran or appease it.

At the same time, Saudi Arabia's combination of force levels and geography tells a somewhat different story. Table 6.1 shows that Saudi Arabia is the only Southern Gulf state with sufficient military forces to cross-reinforce the other Gulf states. Its geography also makes it the only state with the lines of communication and strategic depth both to make such reinforcement possible and to deploy at least some of its forces where they are safe from attack.

Even Saudi Arabia, however, has a low ratio of forces to the space which is covered by its critical defensive areas. Further, the problems inherent in insufficiency of force in the Southern Gulf states are compounded by the fact that the forces of the individual Southern Gulf states have little standardization and poor interoperability. While they are gradually improving in individual military capability, many are still "showpiece" forces which cannot operate effectively except in carefully planned exercises, have few native combat troops, and have wholly foreign manned combat units with little loyalty to the nation or regime. In many cases, they have bought weapons for their prestige, rather than for their deterrent or combat capabilities.

Most of the military forces in the smaller GCC states have inadequate warning sensors, and weak command and control systems. Most armies lack modern communications, battle management, and target acquisition systems. There is little heliborne or amphibious capability to rapidly move troops. There are few airborne early warning (AEW) and no air control and warning assets. Most ships have inadequate air and no anti-missile defense. The smaller GCC navies have no mine warfare capabilities, and poor ability to conduct combined operations. There are few modern reconnaissance and intelligence assets. The various states and military services differ sharply in sheltering and passive defense capabilities, and only Oman has pipelines that allow it to avoid dependence on ship movement through the Gulf.

The military expansion and modernization of the Southern Gulf states have also been sharply affected by the recent changes in oil prices. The total oil revenues of the GCC states shrank from about $150–

163 billion in 1981, to $45–55 billion in 1985. According to some estimates they were around $40–43 billion in 1986, and although they were probably above $50 billion in 1987, the rapid drop in the value of the dollar in the fall of 1987 brought their purchasing power back down to something approaching the 1986 level. These drops in income are still cushioned by nearly $160–200 billion in investment abroad, but virtually all of these reserves are held by Saudi Arabia, Kuwait, and Abu Dhabi. Since oil accounts for over 90% of the foreign exchange earnings of the GCC states, Bahrain and Oman are dependent on Saudi Arabia and Kuwait to finance their military modernization.[2]

These individual weaknesses could be of critical importance to both Western strategic interests since a hostile regime or takeover of any one of the Southern Gulf states might cripple both regional defense efforts and Western ability to deploy reinforcing units.

Fortunately, most of the Southern Gulf states seem to be relatively secure against immediate internal threats to their political security of the kind that could overthrow their present regimes, or turn them into hostile radical states. All, however, are vulnerable to outside pressure and threats unless they can count on strong outside assistance. These vulnerabilities become clearer when the evolution of the Gulf Cooperation Council (GCC) and the current security situation in each of the Southern GCC states are examined individually.[3]

THE IMPACT OF THE GULF COOPERATION COUNCIL

The creation of the Gulf Cooperation Council in February, 1981, is gradually leading the Southern Gulf states towards a more effective form of military cooperation. The GCC was created in large part as a reaction to the dangers posed by the Iran-Iraq War, and all six Gulf states—Bahrain, Kuwait, Oman, Qatar, Saudi Arabia, and the UAE—are members. While it has scarcely catalyzed an effective collective security effort, it has led to a long series of meetings and, more substantively, to improved cooperation in defense planning and procurement.

A now regular series of meetings of the chiefs of staff of the Southern Gulf states began on September 20, 1986. This meeting of the chiefs of staff involved the first GCC-wide discussions of each member country's defense plans, and there have been continuing low-level discussions at the expert level. Meetings of the GCC defense ministers followed on January 25, 1982, and October 10, 1982. They led to reports that the GCC countries would try to coordinate a total of $30 billion in defense expenditures, move towards the creation of a joint command, and discuss the creation of a joint defense network.[4]

The resulting follow-up efforts, however, did not lead to serious feasibility studies until June, 1985. The basic problem has been the reluctance of the smaller Gulf states either to become netted into a system using Saudi Arabia's E-3A AWACS, or to cooperate in a system Iran might feel was provocative. The UAE has been particularly reluctant to integrate its Lamda C^3I/battle management system into the rest of the GCC systems, although this reluctance also reflects serious system development problems and rivalries among the UAE's individual sheikdoms. Similarly, Kuwait refused to integrate its Thomson radar array and I-Hawk missile system until the events of 1987 made it clear that Kuwait faced an Iranian threat it simply could not face alone.

At the third meeting of the GCC Supreme Council in Bahrain in November, 1982, the GCC did begin to move towards collective defense, and agreed to support a $1 billion aid program for Bahrain, which was to pay for a modern air base and fighter aircraft. In July, 1983, the Supreme Council agreed to a $1.8 billion twelve-year aid program for Oman, to improve its ability to defend the Straits of Hormuz. Neither of these programs was fully funded because of the ensuing drop in oil prices and revenues, and the problems raised by the Iraq-Iraq War. Nevertheless, they did lead to some aid transfers and established an important principle.

The meeting of the defense ministers in Doha in March, 1983, led to agreement in principle to Saudi Defense Minister Prince Sultan's call for a GCC defense industry to reduce the Gulf's dependence on arms imports, with a projected initial funding of $1.4 billion. While little initial progress took place, Saudi Arabia initiated covert discussions with Egypt regarding the revival of the Arab Military Industrial Organization, and set up a General Institute of Military Industries in late 1985. It immediately initiated open discussions with Egypt about joint efforts at military industrialization following the Arab League Summit meeting in November, 1987.[5]

The GCC Supreme Council meeting in November, 1982, also led to the decision to hold the Southern Gulf's first joint military exercises, Peninsular Shield I, during October 7–16, 1983. This was successful enough to make the exercise a regular military activity, and the Peninsular Shield II exercises were held near Hafr al-Batin during October 10–23, 1984. These exercises brought fairly large forces together for the first time (10,000 men, 3,200 armored and other vehicles, mobile artillery and tanks, and air units using F-5 and F-15 fighters and C-130 transports).

Other efforts in the summer and winter of 1984 led to a series of bilateral and multilateral exercises to strengthen air defense and maritime surveillance cooperation. Saudi Arabia and Oman began a limited number of joint naval patrols, and even Bahrain and Qatar participated in joint exercises in spite of a boundary dispute. The GCC carried out further multilateral exercises in 1985, 1986, and 1987, although the expansion of the Iraq-Iraq War into the Gulf in 1987 shifted the focus of the key GCC states from exercises to actual operations.

The GCC ministers of state and defense tentatively agreed at their meeting in Abha on September 18, 1984, to create a 10,000 man joint strike or "intervention" force that would be based at the Saudi facility at Hafr al-Batin. This agreement was ratified at the Supreme Council meeting in Kuwait on October 18, 1984. Less than 1,000 men were actually deployed in this "Peninsular Shield Force" by the beginning of 1986, however, and virtually all came from Kuwait and Saudi Arabia—the two states closest to the base and the ones that most needed such a force as a hedge against Iranian action. While the Supreme Council of the GCC again approved the need for such a force at its meeting in Abu Dhabi in November, 1986, serious differences existed within the GCC over how any such force should be created. Both Oman and the UAE took the position that it would interfere with their internal military modernization, and quietly made it clear that they did not want their forces to become involved in a conflict with those of Iran or Iraq.[6]

The increased threat to the Southern Gulf states from Iran's "tanker war" during 1986 led to expanded discussions about creating a maritime reconnaissance and strike force in 1986. When the foreign and defense ministers of the GCC met at Abha on August 26, 1986, they discussed three ideas: using Saudi Arabia's E-3As to create a GCC-wide air control and warning system, integrating naval and ground radar systems, and creating a regional capability to convoy ships through the Gulf.7

The first phase of these discussions centered on the future role of the five E-3As and eight KE-3A tankers that Saudi Arabia had bought from the U.S. Studies by the staff of the GCC and its member countries concluded that the GCC needed an integrated airborne warning and control system, but as Kuwait's Chief of Staff Major General Abdullah Farraj Ghanim pointed out, as "only one country needs to buy the aircraft, the GCC would cooperate and share the cost."[8]

By the time the Supreme Council of the GCC met in Abu Dhabi on November 2, 1986, the defense ministers and military staffs had agreed on plans that would have used the Saudi aircraft to provide coverage of the entire coastal area, with downlinks to each of the GCC countries. The plan, however, ran into several problems. The first was the fact that the intensity of the Iran-Iraq War increased rapidly, and the

Saudi air force was still having problems in getting sufficiently trained crews to operate the aircraft. Although two E-3As were delivered in 1986, and the rest were due to be delivered in 1987, it became clear that USAF personnel would have to jointly crew the aircraft for several more years. Further, it became clear that it would take at least two years to convert the communications and IFF systems of the other Gulf states, particularly those of the UAE, to be compatible with those of Saudi Arabia.

Somewhat similar issues arise regarding the maritime patrol mission. The E-3As had advanced maritime reconnaissance systems, but were not suitable platforms to carry anti-ship weapons or for visual inspection of ships, and were too expensive to commit to many day-to-day maritime missions. This led GCC officials to discuss the purchase of 10 advanced maritime patrol aircraft. The GCC defense ministers approved this idea at their meeting in Muscat in October, 1986, and it was endorsed by the Supreme Council in November.[9]

As a result, the GCC informally approached Washington regarding the purchase of ten P-3Cs, in what were reported to be a $1 billion deal.[10] This initiative, however, ran into funding problems and faced the difficulty that the sale would have to have been made to Saudi Arabia, and would have presented the almost certain prospect of another U.S. domestic political debate over arms sales to the Arab world. Moverover, the aircraft were not readily available from inventory. This may lead the GCC states to buy the Atlantique from France or the Fokker F-27 MPA variant from the Netherlands.

The GCC also showed during 1987 and early 1988 that it could remain unified in the face of growing pressure from Iran, although this unity was sometimes tenuous and came at the cost of substituting declarations for action. It was only at the GCC foreign ministers' meeting in Riyadh in late December, 1987, that the GCC states could agree to take a hard line of supporting an arms embargo of Iran. Even then, the agreement to support an embargo was coupled with a new peace initiative towards Iran, and the original wording condemning Iran for its recent actions was softened to remove part of the attack on Iran because of the fears of Oman and the UAE. It was also far from clear what the ministers meant at their December meeting, when they agreed to a joint security plan for increased cooperation in internal security and defense.[11]

This progress in military cooperation is real, but it must be kept in careful perspective. All of the GCC's collective defense efforts are still more political gestures than military realities, and the GCC often substitutes rhetoric about military cooperation for serious planning. Discussions of creating a common rapid deployment force, military standardization, common support facilities, and common military production facilities have led to words and studies, rather than to actions. They also have led to a situation where the lower and upper Gulf states find it difficult to cooperate because of their different political and military objectives, and the smaller states tend to use the GCC to ask Saudi Arabia and Kuwait for money. This has been true of Bahrain and Oman since the start of the GCC, and Saudi Arabia has evidently recently agreed to fund the improvement of Omani naval facilities and the new UAE naval base at Taweelah as "GCC facilities".[12]

The creation of the GCC has also slowly led to improved cooperation in intelligence and internal security. Many of these efforts were triggered by the uprising at the Grand Mosque in Saudi Arabia in 1979 and the coup attempt in Bahrain in 1981. The need for such cooperation has since been reinforced by the long series of pro-Iranian bombings and assassination attempts in Kuwait and Saudi Arabia since 1985.

The GCC ministers of interior discussed ways to improve their cooperation in internal security at three meetings during 1981 to late 1983. They produced a draft agreement at the first meeting and seem to have agreed on a common text at the third. The agreement covered a wide range of areas including the movement of arms, terrorism, illegal immigration, smuggling, crime, standard identity codes, documentation rights, and rights of hot pursuit and extradition.

Kuwait initially refused to join the other states in implementing this agreement because of its fear that it would lead to internal tensions with its Palestinian workers and friction over the extradition of political figures who were allowed to speak out in Kuwait, but not in some other Gulf states. An Iranian-assisted assassination attempt on the Emir of Kuwait in May, 1985, helped lead Kuwait to reconsider its position, however, and further terrorist incidents in 1985 and 1986 led it to sign the GCC multilateral security agreement in February, 1987. Although progress has been much slower in sharing military intelligence data, considerable progress has now been made in sharing data on subversive and terrorist groups, and the situation is now far better than in 1980.[13]

THE MILITARY CAPABILITIES OF THE SOUTHERN GULF STATES

Impressive as some of the progress within the GCC has been, it also reinforces the point that the Southern Gulf states are not a collective entity either in military terms or in terms of internal security. The West must shape its strategic relations with the region in ways which allow it to deal separately with each Southern Gulf state, and which take account of their long-standing rivalries as well as recent moves towards cooperation.

The West must also recognize the fact that Bahrain, Kuwait, Oman, Qatar, and the UAE are fundamentally different from Saudi Arabia. This is not simply a matter of force numbers and geography. As the following country-by-country analysis shows, it is a matter of deep-seated differences in the ability of Saudi Arabia and the smaller Gulf states to act in region-wide and strategic terms, as distinguished from having to concentrate purely on national survival.

Bahrain

Bahrain is the smallest, and perhaps the most vulnerable, of all the Southern Gulf states. It consists of one main island, which is some 596 square kilometers in area, and 32 smaller islands. As Table 6.2 illustrates, Bahrain is deeply divided in ethnic and religious terms and is the weakest GCC state in terms of social coherence. Its royal family has done an uncertain job of sharing the national budget, of demonstrating its concern for the poor, and of showing its respect for Islamic custom. Bahrain is divided on religious grounds. The ruling Khalifa family is Sunni, although Bahrain has a 60% Shi'ite majority, and 37% of its population consists of expatriates from other countries, of whom at least 8% are still Iranian.[14]

Unfortunately, the Khalifa family has also reacted to a 1981 Iranian-sponsored coup attempt by increasing its centralization of power, and it does not rule or administrate as well as it should.[15] The regime has made efforts to improve social services, but it still tends to be discriminatory in dealing with the nation's Shi'ite majority and has not extended equal privileges to its large "foreign" population—a problem which is of considerable importance in a nation where less than 12% of the population can trace their roots back to families who were living in Bahrain in 1921.[16]

The current ruler, Sheik Isa ibn Salman al Khalifa, does hold frequent meetings of his Majlis, but he is not widely popular, and the government's favoritism and corruption have produced some internal hostility. The royal family is able to move freely with only minimal security measures, but two new arms caches were discovered in 1984, and as many as 30 people may have been arrested.

Bahrain now has only token oil resources, and has experienced a sharp contraction of its offshore banking operations as a result of the recent "oil glut". This has created an unusual amount of unemployment

Table 6.2: Population, Ethnic, and Religious Differences Within the Southern Gulf States

Country	Population			Labor Force			
	Military Age Males (1,000s)	Annual Growth (%)	Foreign Population (1,000s)	Domestic (%)	Foreign (%)	Ethnic Composition	Religious Structure
Bahrain	73	3.5	130	42	58	Bahraini (63%) Asian (13%) For. Arab (10%) Iranian (85) Other (6%)	Sunni (30%) Shi'ite (65%)*
Kuwait	266	3.5	850	30	70	Kuwaiti (39%) For. Arab (39%) So. Asian (9%) Iranian (4%) Other (9%)	85% Muslim 15% Other**
Oman	162	3.4	?	50	50	Arab (90%) Baluchi Zanzibari Indian	Ibahdi (75%) Sunni Shi'ite Hindu
Qatar (95%)***	70	4.2	210	15	85	Arab (40%) Pakistani (18%) Indian (18%) Iranian (10%)	Muslim
Saudi Arabia	1,760 (106 annually)	3.2	3,000	50	50	Arab (90%) Afro-Asian (105)	Sunni (90-93%) Shi'ite (7-10%) ****
UAE	381	3.1	700	20	80	Emirate (19%) For. Arab (23%) So. Asian (50%) Other (8%)	Muslim (96%) Other (4%) *****

Iran	6,629 (462 annually)	3.1	100	-	-	Persian (63%) Turkic (18%) Kurd (3%) Arab & Jew (3%) Other Iranian (3%)	Shi'ite (93%) Sunni (5%) Other (2%)
Iraq	2,105 (177 annually)	3.2	1,230	69	21	Arab (75%) Kurd (15-20%) Turkic, Assyrian, Other (5-7%) Other (5-10%)	Shi'ite (50%) Sunni (43%)
YAR	664 (69 annually)	2.9	*** ***	-	-	Arab (90%) Afro-Arab (10%)	Sunni (48%) Zaidi Shi'ite (52%)
PDRY	276	2.9	-	-	-	Arab (95%) Indian, Somali, European (5%)	Sunni (97%) Christian & Hindu (3%)

* Bahraini citizens are 100% Muslim and 65% Shi'ite. Foreign workers are 96% Muslim and 20% Shi'ite. Some estimates show Bahrain's total population as 70% Shi'ite and 30% Sunni.
** Kuwaiti citizens are 100% Muslim and about 30% Shi'ite. Foreign workers are 90% Muslim and about 11% are Shi'ite.
*** Qatari citizens are 100% Muslim and 15% are Shi'ite. The foreign population is 95% Muslim and 35% is Shi'ite.
**** Saudi citizens are 100% Muslim and 7% Shi'ite. The foreign population is about 98% Muslim and 10% "twelver" Shi'ite, although Ismaili and Yemeni Shi'ites make up a significant additional percentage.
***** UAE citizens are 100% Muslim and 10% are Shi'ite. Foreign residents are 70% Muslim and 17% are Shi'ite.
****** Roughly one-third of the YAR labor force works abroad. The remainder is virtually agricultural .

Sources: Adapted from: John Aronson and Inge Lockwood, "Background Materials on the Persian Gulf Region", Washington, Middle East Institute, Sultan Qaboos Center, December, 1987; and the CIA, *The World Factbook, 1986*, 041-015-00163-9, Washington, GPO, 1986.

among the native labor forces, particularly among the younger Shi'ites. Saudi Arabia does, however, seem likely to continue to subsidize Bahrain's 250,000 BPD BAPCO refinery by providing 70% of its oil. Bahrain is developing its own "tourist" industry and shipyards, and diversifying into industries using its still extensive stocks of gas feedstock.

Bahrain increased its annual military expenditures from around $40 million in 1979, to $150–$280 million in 1982 through 1985. Its recent arms imports have ranged from $10 million to $30 million annually, and its military manpower slowly increased from 2,000 in the early 1980s to 3,000 in 1985. Bahraini defense expenditure was officially set at 50.7 million dinars ($135 million) in 1986, and 51.5 billion dinars ($137.5 million) in 1987, but seems to have risen to an average of around $200 million annually, including Saudi aid.[17]

Bahrain spends about 4 to 8% of its GDP on defense, and 10 to 20% of its central government expenditures. Its total arms imports during 1981–1985 totalled $115 million, with $20 million coming from the U.S., $10 million coming from France, $5 million from the U.K., $60 million from the FRG, $10 million from Italy, and $10 million from other states.[18]

Bahrain's defense minister, the ruling Sheik's son Crown Prince Hamad bin Isa al Khalifa, is competent and was educated at Sandhurst and Leavenworth. Bahrain avoided much of the waste that has taken place in other Gulf states, but Bahrain has only token military forces and little prospect of building up a real military capability at any time in the future. Bahrain had only 2,800 men in its armed forces in early 1988, and only 2,300 men in its army, many of whom are expatriates. Military service is not a popular career, particularly among its ruling elite. Many of its officers, NCOs, and technicians are foreign, and it is heavily dependent on foreign contract personnel. While some of these are from Jordan, Oman, and North Yemen, they seem to offer only a marginally more loyal and stable cadre than other expatriates.

Bahrain's order of battle includes one small mechanized brigade which is little more than an under-strength regiment in terms of effective fighting power. Sources disagree as to whether Bahrain also is forming an independent mechanized battalion and a mechanized company.[19] It has just deployed 54 M-60A3 tanks into its force structure.[20] Its other mechanized weapons are 10 AT-105 Saxons, 90 M-3 Panhard armored cars, and 45 obsolete armored cars (8 Saladin, 8 Ferret, and 20 AML-90). Its artillery strength consists of 8 towed 105 mm howitzers, and 7 M-198 155 mm howitzers.

The army's only other major weapons include 60 BGM-71A TOW ATGM launchers, and 40–50 light RBS-70 short-range surface-to-air missile launchers. Bahrain has repeatedly attempted to order Stingers from the U.S. since the early 1980s, but was rebuffed until late 1987 because of opposition from the U.S. Congress triggered by pro-Israeli lobbying groups. The Reagan Administration was finally able to get Congressional consent because of the situation in the Gulf, but only 70 missiles and 14 launchers were approved. These weapons will have to be returned to the U.S. either 18 months after they are delivered, or once a substitute U.S. air defense system like the Improved Hawk is approved.[21]

Training and proficiency levels are inadequate if the army has to go into combat against a foreign force. The army is also poorly equipped for point defense of its own forces or Bahraini territory. The main mission of Bahrain's army is, however, internal security and the army has been organized and trained to suppress any uprising or coup by the nation's Shi'ite majority. It has a British-organized intelligence and internal security elements, few Shi'ites, and virtually no Iranian expatriates. All of the services are under the direction of the royal family, and all have good pay and privileges. The army is likely to side with the royal family, and with any Saudi forces that come to the aid of the royal family, in any coup attempt or civil crisis.

Bahrain is building up its naval forces. Its navy is part of the Ministry of Interior. It had 300 men in early 1988, and was equipped with two Lurssen 62-001 63 meter missile corvettes with two MM-40 Exocet or Harpoon missile launchers, dual-purpose OTO Melara 76 mm guns, and two triple 234 mm torpedo tubes. Bahrain had two Lurssen TNC(FPB)-45 missile fast patrol boats with four single-cell MM-40 Exocets. Bahrain also had two Lurssen FPB-38 gun boats with two 40 mm twin Bofors, and twenty-two other small patrol boats ranging from 11 to 15.3 three meters in length. Its smaller vessels included one Hovercraft and two utility landing craft. Bahrain has fourteen British landing craft on order, and two more broadbeam FPB-45s on order.[22]

While Bahrain has some good native naval personnel, it will take Bahrain some time before it can effectively man its existing ships and those it has on order. Bahrain's navy also suffers from abuses like the order of a 100 foot "warship" equipped with 20 mm anti-aircraft guns and air defense missiles, but which is also equipped with an oval bed, a 22-karat gold-plated sunken bathtub in a blue marble bathroom, television and stereo, and a carpeted bridge lined with grey suede.[23]

The Bahraini navy will remain dependent on expatriate support to maintain and operate most of its more advanced sensors, weapons systems, and communications gear. Bahrain is, however, getting considerable support from the U.S. Middle East Force, which is no longer officially headquartered in Bahrain, but which uses the island's harbor and wharfage facilities. Bahrain has benefited significantly from the

U.S. build-up in the Gulf in 1987. Its naval base at Jufair is also well equipped, and it has excellent commercial shipbuilding and repair capabilities.

Bahrain has a separate 180 man coast guard. This force is also under the Ministry of Interior. It is equipped with twenty-three coastal patrol craft, two landing craft, and one Hovercraft.

Bahrain acquired its first modern combat aircraft—four F-5Es and two F-5Fs—in 1986. It now has eight F-5Es and four F-5Fs.[24] The operation of these aircraft is dependent on foreign technical support, but Bahrain is training some good native pilots. Bahrain also has one Gulfstream transport, two Hughes 500D helicopters, and 10–14 Augusta-Bell AB-212 and BO 105/117 helicopters. Three of the latter helicopters are equipped with Bendix 1400 long-range maritime radars.

Saudi Arabia has helped finance the construction of a military air base in Bahrain, and Bahrain plans the acquisition of Improved Hawk MIM-23B missiles, Cossor SSRs, and Plessey Watchman air traffic control radars. It has ordered a $400 million arms package from the U.S. that includes 12 F-16C/D fighters, Sidewinder air-to-air missiles, AGM-65 Maverick air-to-surface missiles, AN/A LE-40 chaff dispensers, and spares, support, and training. It also has bought AN/ALQ-131 electronic countermeasure pods.[25]

If all of these plans go forward, Bahrain will have acquired a total of 12 F-5s, 12 additional fighters, and Improved Hawk defenses by the early 1990s. It will, however, then be far more dependent on expatriate military personnel and technicians if it is to operate such systems. Bahrain's present command and control system is also very poor. It relies largely on voice links and Teletype, and Bahrain generally does a mediocre job in using these links both internally and in cooperating on air and maritime traffic data with Saudi Arabia and the other GCC states. Bahrain plans to improve this system, but this again will require foreign personnel.

Given the recent drop in Gulf oil income, and in Bahrain's income from its various service industries, it will have to depend on Saudi Arabia to help finance its armed forces, and on the West and Saudi Arabia to reinforce its air, naval, and land defenses for the foreseeable future.[26] The completion of the 25 kilometer causeway between Jasrah in Bahrain and Al-Aziziyah in Saudi Arabia in December, 1985, has, however, made it far easier for Saudi Arabia to reinforce Bahrain in an emergency.[27]

Bahrain's internal security forces include a 2,000 man police force, which is part of the Ministry of Interior, and which has two Bell 412 and two Westland Mark-1 helicopters. It has significantly improved its internal security efforts since a Shi'ite-Iranian coup attempt in

December, 1981. It has quietly consulted with both Britain and the United States regarding both internal security assistance and defense against Iran, and has discussed contingency plans for Saudi military assistance. It has also improved its treatment of its Shi'ite majority, its controls over foreign labor, and its surveillance of the relatively limited PFLOAG elements in the country.

None of the current internal or external pressures on Bahrain seem immediately threatening, but it is uncertain how long it can go on without an improvement in the leadership from the Khalifa family and some broadening of power. Saudi Arabia is now providing Bahrain with the economic support it needs, but it cannot provide it with political cohesion or leadership.

The fact that Bahrain and Qatar have recently revived their absurd rivalry over the Hawar Islands and the coral reef of Fasht-e-Dibal, which divide the two countries, is also a grim warning regarding Bahrain's ability to develop effective collective security arrangements with the other GCC states. This rivalry owes far more to a heritage of feuding between two royal families than to any serious issue over oil wealth, and the issue was scarcely of serious importance at a time when GCC solidarity was of particular importance because of the Iranian victory at Faw.

Nevertheless, a clash took place between Bahrain and Qatar on April 26, 1986, in which Qatari helicopters fired on construction crews working on Fasht e-Dibal, and Qatari forces landed and seized 30 Bahraini workers. Bahrain claimed these work crews were building a GCC facility to monitor tanker traffic and Qatar claimed it was a Bahraini coast guard base.[28] Both nations then called military alerts and deployed troops. Bahrain reinforced Hawar and Qatar reinforced Fasht e-Dibal. The countercharges included claims that Qatar might seek Iranian aid. While Saudi and GCC attempts at mediation finally succeeded, and a GCC observation team was sent to end the disagreement, the resulting message was scarcely one of Gulf unity and strength.[29]

Kuwait

Kuwait mixes geographic vulnerability with an awkward combination of the deep ethnic divisions shown in Table 6.2 and the immense oil wealth shown in Table 6.3. It has steadily experienced more serious security problems as the Iranian seizure of Faw and expansion of the tanker war have increased Iran's ability to use force to try to make Kuwait cease supporting Iraq. It also has had sharply growing internal security problems.

Table 6.3: The Relative Oil Resources of the Southern Gulf States[30]

Total Production Capacity and Recent Production Levels

Country	Oil Reserves (Billions/Bbl)	Maximum Sustainable Production (MMBD)	Average Oil Production (MMBD) 1986	1987	Gas Reserves (Trillion TCF)
Bahrain	0.1	.44	.04	–	7.0
Kuwait	94.5	2.3	1.25	–	41.0
Oman	4.0	.62	.55	–	8.0
Qatar	3.2	.6	.31	–	152.0
Saudi Arabia	169.2	10.3	4.72	–	130.0
UAE	33.1	2.4	1.4	–	105.0
Iran	48.8	3.7	2.36	–	28.0
Iraq	47.1	3.5	1.69	–	41.0
YAR	0.5	0.025	0.01	0.02	?
PDRY	?	–	–	–	?

Dependence on Oil Exports Through the Persian Gulf (Ave. MMBD in 1987)

Country	Production	Domestic Consumption	Available for Export: Through Straits of Hormuz	Via Pipeline
Bahrain	*	*	*	*
Kuwait	1.3	0.2	1.1	-
Neutral Zone	0.3	-	0.3	-
Qatar	0.3	*	0.3	-
Saudi Arabia	4.9	0.9	3.5	0.5
UAE	1.3	0.1	1.2	-
Iran	1.9	0.7	1.2	-
Iraq	1.7	0.3	-	1.4

* Less than 100,000 BPD.

Kuwait has stronger military forces than Bahrain, but its geographic position is even more vulnerable. It is experiencing serious problems in military modernization, and it has no prospects of creating an independent deterrent or defense capability. Kuwait increased its annual military expenditures from around $770 million in 1979, to $1,100–$1,400 million in 1982 through 1985. It spent $1.39 billion on defense in 1986/87 and $1.42 billion in 1987/88. Its recent arms imports have ranged from from $110 million to $390 million annually, and its military manpower slowly increased from 12,000 in the early 1980s to 15,000 in 1985.

Kuwait spends about 4 to 8% of its GDP and 10 to 20% of its central government expenditures of defense. Its total arms imports during 1981–1985 totalled $1,005 million, with $90 million coming from the USSR, $230 million coming from the U.S., $360 million coming from France, $20 million from the U.K., $210 million from the FRG, $80 million from Italy, and $15 million from other states.[31]

Kuwait's 15,000 man military forces have a well-trained officer corps, many members of which have attended Sandhurst. Officers and some NCOs and technicians are also trained in the U.S., Pakistan, and Jordan. The officer corps is recruited from the ruling family and loyal tribes, but recruitment and promotion are dominated by favoritism rather than performance.

Kuwait has tried to recruit its citizens for its armed forces, combining excellent pay and privileges with the threat of a draft. This draft now involves two years of service, except for university students—who have to serve only one year. There are many exemptions, however, and Kuwait is forced to use a large number of troops who are not really citizens. It has not done a good job of personnel management, or of using its trained personnel. There are, however, U.S., British, and French military and contractor support missions for virtually all of Kuwait's more advanced and Western-supplied military equipment.

Kuwait's total army manpower was only about 13,000 at the start of 1988, and while Kuwait's order of battle has two armored brigades and two mechanized brigades, and one surface-to-surface missile battalion, its total army manpower is equivalent to only one Western brigade slice.

Kuwait has limited ability to effectively employ its strength of 260 main battle tanks (of which 160 are first-line Chieftains), in anything other than a set piece defense. Its Chieftain tanks are also under-powered and experience continuing overheating and maintenance problems.[32] It has more capability to use its 200 M-113 APCs, 100 AT-105 Saxons and Saladins, 100 Saracens, and 60 Ferret armored cars. It has Scorpion armored fighting vehicles on order.

The Kuwait Army has 18 M-109A2 and 40 AMX Mark F-3 155 mm

howitzers, but it has no combat training in the use of such artillery beyond set piece and firing range exercises. Kuwait's surface-to-surface missile battalion has 4 FROG-7 launchers, but these have little more than symbolic importance. Kuwait has bought a wide range of anti-tank weapons, including the AT-4, BGM-71A Improved TOW, HOT, M-47 Dragon, and Vigilant, and it has 56 M-901 ITV armored TOW carriers. It has 4,000 Improved TOW missiles on order. This is a good mix of anti-tank weapons, but there is an uncertain training and support effort.

Kuwait is gradually developing improved army air defenses, although it has too many diverse types and poor training in operating them. It has U.S.-supplied Improved Hawk surface-to-air missiles, Soviet-supplied SA-7s, SA-8s, and ZSU 23-4s, and may have SA-6s. The U.S. has refused to sell it the Stinger, but it has more SA-7s and SA-8s on order, and may have ordered some Crotale or Sea Wolf light surface-to-air missile systems. Britain, however, has been reluctant to sell Kuwait a key system in service in the British navy because of its fear of loss of the details of the technology to the USSR.[33] Kuwait does have reasonably good weapons numbers, but there are too many types and it is uncertain how quickly its forces will be able to absorb such systems into its force structure.

Kuwait has a massive $100 million military complex about twenty miles from Kuwait City, and a new $29 million naval base. These facilities, however, owe more to political convenience, and an effort to maintain high living standards, than to military effectiveness. They are vulnerable to air attack and over-centralize both the deployment of Kuwait's forces and their support functions in fixed locations. Further, most of Kuwait's army support personnel are foreign civilians, and Kuwait lacks the ability to sustain its forces in the field without such civilian support. Kuwait has, however, recently concluded an agreement with Turkey to provide for more advanced training, and this may improve some aspects of its military proficiency.[34]

Kuwait is just beginning to create a real navy. It has created an 1,100 man naval force to replace its coast guard. This force, however, remains part of the Ministry of Interior.[35] The core of the Kuwait navy consists of eight Lurssen guided-missile patrol boats. Two of these boats are FPB-57s, and six are TNC-45s—which have 76 mm OTO Melara guns, twin 40 mm guns, and four Exocet MM-40 missile launchers each. It should be noted that these patrol boats have some important limitations common to virtually all GCC naval vessels. They lack serious air defense capability, and while their voice communications are good, they cannot be integrated into a data link exchange network. They also require nearly 60% of Kuwait's native naval manning.[36]

Kuwait received four 55-meter South Korean missile patrol boats in

August, 1987, and these are based in Kuwait's offshore islands. They have anti-ship missiles, helicopter pads, and a Hovercraft docking facility. The ships are not fully combat ready, but they increase Kuwait's shallow water defense capability.

Kuwait also has 47 11–24 meter patrol craft, 4 modern British Cheverton LCTs, 3 LCUs, 3 LSUs, 4 tugs, 6 launches, and some light coastal vessels and support craft. The air force provides support in the form of Super Puma helicopters equipped with Exocet.

Finally, Kuwait has 6 SRN-6 Hovercraft, some Exocet-capable SA 365N Dauphin II helicopters, 20 Magnum Sedan patrol boats, 2 Italian 18.4 meter patrol boats, 2 20-meter Italian patrol boats, and more South Korean patrol boats on order. It also is negotiating with the Netherlands to buy two Alkmaar-class minehunters. The Dutch parliament has approved the loan of two such vessels until new production is available.

These orders will inevitably increase the Kuwaiti Navy's manpower problems and dependence on foreign technicians. Kuwait has modern naval bases at Kuwait City and Ras al-Qulayah, and major civil ship-repair facilities at Kuwait City's Shuwaikh harbor, including a 190 meter floating dock with a 35,000 DWT repair capability.

Kuwait's 2,000 man air force is slowly improving in effectiveness, and it now has 80 combat aircraft and 23 helicopters. Kuwait's combat aircraft include 9 BAC-167 Strikemasters and 12 Hawk in the COIN, training, and liaison roles. It has 34 A-4KU/TA-4KU attack fighters, which are in full combat-ready service.

The air force is still experiencing problems, however, in absorbing 34 Mirage F-1BK/CK fighters. The Mirage F-1 aircraft have proved hard to maintain, Kuwait has lost several of the aircraft to accidents, and the 55-kilometer air intercept range of their radars has proved too short to meet Kuwait's operational needs. It is forced to use its A-4 attack aircraft in the combat air patrol role when it needs to create an air defense screen. Kuwait has, however, contracted for Pakistani service and support crews.[37]

The air force has 9 transport aircraft, including 1 B-707-200, 6 C-130-30s, and 2 DC-9s.[38] The air force also operates 46 helicopters. These include 23–30 SA-342K Gazelle attack helicopters, and 23 of these are equipped with TOW. They also include 5–6 AS-332 Super Pumas equipped with Exocet, and 10–12 SA-330 Pumas. It has 6 AS-332F Super Pumas on order.

Kuwait's air weapons inventory includes AIM-9 Sidewinders, Matra Super R-530s, and R-550 Magique air-to-air missiles. It has AS-11 and AS-12 air-to-surface missiles, and 12 AM-39 air-to-ship missiles on order. It has also ordered the French SA-365N maritime attack system.

Kuwait does benefit from data exchanges with the E-3As flying in Saudi Arabia, but the quality of the data links between them and Kuwait's French-designed air control and warning system is uncertain. It is far from clear that it can react quickly and effectively enough to deal with Iranian or Iraqi intruders into Kuwait's air space.

The Kuwaiti Air Force also has four batteries of Improved Hawk surface-to-air missiles with twin launchers, and SA-8 surface-to-air missiles. It may have two Shahine batteries on order. Kuwait has had serious difficulties in absorbing its more sophisticated surface-to-air missiles, however, and this presented serious problems in 1987, when efforts were made to re-site the missiles to defend against attacks by Iran's Silkworm missiles. The U.S. refusal to sell Kuwait Stinger missiles in June, 1984, however, led Kuwait to delay the purchase of Hawk systems, and Kuwait responded by buying some $327 million worth of Soviet arms—none of which can be netted into an effective air defense system. While Kuwait is still oriented towards buying arms from the West, it may not end up buying a ground-based air defense system which can be made fully compatible with that of Saudi Arabia and the other GCC states.

Kuwait has separate National Guards, Palace Guards, and Border Guards, which are equipped with a total of 20 V-150 and 62 V-300 Commando armored personnel carriers. It has had to steadily improve its internal security because of the Iran-Iraq War. Kuwait has been a strong supporter of Iraq, has provided Iraq with massive economic aid, and is a major transshipment point for Iraq's civil and military imports. This has caused serious problems with some of Kuwait's native Shi'ites and with expatriate Iranians and Lebanese Shi'ites working in Kuwait.

Roughly 20–35% of Kuwait's population is Shi'ite, if expatriate Iranians and Lebanese are included (4–6% of total). Many of these Shi'ites have family ties to Iran, and they face discrimination. Shi'ite mosques get little or no funding, while Sunni mosques get 100% government suupport. There are Shi'ite quotas at Kuwait University, and Shi'ites are rarely promoted to senior government positions or management positions in Kuwaiti firms. There are no Shi'ite newspapers or community centers.[39]

This has already led to serious internal security problems. Suicide bombers attacked the U.S. and French embassies in 1983. Pro-Iranian Shi'ites made an attempt on the Emir's life on May 26, 1985, and a series of sea front bombings on July 11, which killed 12 and injured 89, dramatized just how serious Kuwait's problems really are. So did the discovery of other bombs targeted against key oil facilities, members of the ruling elite, and the U.S. Embassy.[40] As is discussed later in this book, Iran kept up its pressure on Kuwait in 1986 and 1987, and made a long series of efforts to force Kuwait to halt the execution of pro-Iranian terrorists and to free others.

Kuwait has attempted to deal with these problems with its Shi'ites through a series of mass deportations—which totalled some 400 immediately after the bombings and involved 26,898 people by November 1, 1986.[41] It also has systematically purged its military and security forces of Shi'ite personnel, organized much stronger anti-terrorism forces, and procured a great deal of advanced security equipment. Nevertheless, it has become clear that a substantial number of the terrorists are Kuwaiti citizens, and Kuwait's problems are far from over.

Kuwait also has a problem with infiltration from the outside. After the Iranian seizure of Faw in early 1986, Kuwait began to actively procure sensors and weapons to guard against naval infiltration. It banned fishing in the Gulf war zone to improve its control of local naval traffic, and built a series of four man-made platforms or floating islands to protect the country against seaborne infiltrators. These islands were completed in July, 1986, and were equipped with surveillance radars, communications equipment, missiles, artillery, and helipads; they were also part of a surveillance barrier supported by Kuwait's patrol boats.[42]

It is unclear, however, how well this mix of internal security measures and perimeter defenses will really work. Kuwait has long been caught between Iraq and Iran. Syria has backed anti-Iraqi groups operating in Kuwait, probably including those that have been responsible for much of the recent terrorism. Iraq has put pressure on Kuwait for war aid tantamount to blackmail, and has evidently resorted to threats against members of the royal family. Iran's actions have varied from direct and critical attacks on Kuwait's support of Iraq to attempts to calm the situation, but Iran remains an obvious threat.

Kuwait also faces problems in defending Bubiyan and Waribah—its two large main islands in the salt marshes north of Kuwait. Iraq made new attempts to obtain or lease the islands in 1984, and even sent troops across the border. This led to a sudden visit to Baghdad by Kuwait's Prime Minister Saad Sabah on November 10–13, 1984. This, in turn, may have led to a substantial aid payment by Kuwait to Iraq, which then contented itself with creating a Hovercraft base across the river from Waribah.[43]

Kuwait faced a different threat when Iran seized Iraq's Faw Peninsula in early 1986. Kuwait deployed part of its 35th Armored Brigade on the island and expanded its military facilities by filling in more of the marsh. This activity continued through November, 1987, when Iran launched a series of Silkworm strikes on Kuwait City that led the government to take the risk of the Iranian invasion of Bubiyan far more seriously.[44]

As for the other aspects of internal stability, Kuwait also faces growing uncertainties. Until recently, it did a good job of reacting to the shifting political currents and demands of its native Kuwaiti population, and a good job of controlling foreign labor. It maintained a relatively free press and a lively public assembly until mid-1986, when the ruling Emir, Sheik Jaber al-Ahmed Al Sabah, dissolved the 50 member National Assembly and clamped down on the press.

The Emir took this action because of the election of a more radical parliament in October, 1985. This rapidly led to repeated speeches and articles attacking Iran and Syria, attacks on government ministers for being insufficiently Islamic, and criticism of the royal family for failing to do enough to preserve Kuwait's security and for suppressing its involvement in Kuwait's long-standing stock market scandal. What remains to be seen is whether Kuwait will be more stable without these relief valves.[45]

Kuwait, however, does a good job of distributing its oil wealth to Kuwaiti nationals, and the Al-Sabah regime's conservation policies, conspicuous "non-alignment" between East and West, use of technocrats, and investment in a Fund for the Future have all helped to stabilize the political situation.

The only other long-term internal threat to Kuwaiti security is likely to be the fact that 70% of Kuwait's labor force and 61% of the population is still foreign. Kuwait also has a high birth rate (6.2%), and its young population is becoming politicized more rapidly than the Al-Sabah family can coopt it or liberalize the government. The Al-Sabah family lost popularity since the Suq al-Manakh collapse in 1982, although it has recovered some of this popularity since that time—in part because of its firmness in dealing with terrorism and Iranian threats. It is coming under increasing attack both for failing to listen to the nation's younger technocrats and for being insufficiently Islamic. It did not prove as able to influence the outcome of the 1985 National Assembly elections as in the past, and it has since experienced growing problems with Sunni fundamentalists and native Shi'ites.

Qatar

Qatar is another small nation which is primarily dependent on Saudi Arabia to provide for its external defense. As Table 6.3 shows, it is a relatively small oil power, although it has very large reserves of natural gas. It is its geography—which divides the Gulf below

Bahrain—that gives it its primary strategic importance. As Table 6.2 shows, Qatar is 15% Shi'ite and is extraordinarily dependent on foreign workers. It is is relatively homogeneous in terms of its domestic population, but it is dependent on a diverse labor force. This labor force includes enough Asians that most of the workers can be counted on to want to return to their homelands, but it also includes a significant number of Iranians and illegal workers.

Qatar has attempted to create modern military forces. Qatar's military manpower slowly increased from 5,000 in the early 1980s to 6,000 in 1985. It spends about 10% of its GDP on defense, and 20% of its central government expenditures. Its total arms imports during 1981–1985 totalled $895 million, with $10 million coming from the U.S., $650 million coming from France, $230 million from the U.K., and $5 million from other states.

Qatar increased its annual military expenditures from around $475 million in 1979, to $604–$780 million annually in 1982 through 1985. Its recent arms imports have ranged from $200 million to $270 million annually, although massive financial mismanagement during 1985, forced it to temporarily reduce arms imports to a minimum and sharply reduce its defense expenditures.[46]

Qatar now has about 7,000 men in its armed forces. The armed forces are commanded by members of the royal family, and Major General Shaykh Hamad ibn Khalifa al-Thani combines the jobs of the heir apparent, Minister of Defense, and Commander in Chief. Two other members of the al-Thani family serve as Commander of the Air Force and Chief of the Royal Family.

Officers and enlisted men are recruited from members of the royal family, and from the leading desert tribes. Qatar has drawn heavily in the past on nomadic tribes that have crossed the Qatari-Saudi border, but has been forced to increase its intake of urbanized Arabs. Pay and privileges are good, and there have been no visible signs of disaffection within the military.

Qatar's main military problem is total manpower. It lacks the native manpower to field significant military forces, and is dependent on foreign Arab and Pakistani recruits to fill out its combat units, on some British officers, and on British, Egyptian, French, Jordanian, and Pakistani advisors. Qatar does, however, have a growing pool of native personnel, who train in Britain, France, Jordan, Pakistan, and Saudi Arabia. An increasing number of young Qataris have joined the armed forces in recent years, and a steadily rising number of competent young native officers are being trained.

The Qatari army has a nominal strength of 6,000 men, and its order of battle includes an armored battalion, a guard/infantry regiment, five infantry battalions, one artillery battery, and one Rapier surface-to-air missile battery. These are extremely small combat units by Western standards; the total manpower in the Qatari army is too small to fill out even one Western regimental formation plus support.

The army is still lightly equipped. It has 24 AMX-30 tanks, and a diverse mix of other armored fighting vehicles, including 30 AMX-10Ps, 136 VABs, 8 V-150 Mark 3 commando armored combat vehicles, and 10 Ferret and 25 Saracen armored cars. Its artillery is very limited. It has 6 modern French AMX Mark F-3 155 mm self-propelled howitzers and 8 obsolete 25 pound (87 mm) howitzers. It has 18 Rapier and Blowpipe short-range air defense systems. It also has well-equipped modern barracks and casernes.

Qatar's air force has only a little over 300 men, and is based at the military airfield at Doha. Many of its pilots and officers are from Qatar, but it is heavily dependent on French and other foreign support. It has 17 combat aircraft (3 Hunter FGA-78/T-79s, 1 T-79, 6 Alphajets, and 14 Mirage F-1C/Bs). Qatar has AM-39 air-to ship missiles for its Mirages.

Only its Mirages are really combat capable, and it has had serious maintenance and advanced training problems. Qatar also has a fairly large helicopter force, including 2 SA-342 Gazelles, 6 AS-332 Super Pumas, 3 Westland Commando Mark 2/3s, 3 Westland Lynxes, 2 Whirlwinds, and 4 Westland Sea Kings. It has 29 SA-330 Pumas on order.

Qatar has five Tigercat surface-to-air missiles. It is considering an order of Hawk MIM-23B surface-to-air missiles, and may have ordered two Shahine batteries. Qatar is also negotiating to buy a squadron of modern fighters like the F-16 or Mirage 2000, but funding is now uncertain. Qatar will have to rely on Saudi Arabia for most of its air and naval defense through the year 2000.

Qatar's navy has about 700 men, many of whom are expatriates. Its main combat ships consist of three Combattante III missile fast patrol boats, equipped with Exocet missiles. It also has six Vosper Thorneycroft 33.5 meter patrol boats. These ships are operated with a reasonable degree of professionalism by largely expatriate crews. They are adequate for local missions, but lack effective air defense and sensors for surveillance and target acquisition.

Its other ships consist of two 22.5 meter patrol boats, two 13.5 meter patrol boats, 25 Spear-class patrol boats, two Fairey Marine Interceptor-class rescue and assault boats, and nine P-1200 patrol boats. It also has two tugs. Qatar has a small coastal defense unit with three MM-40 Exocet launchers with four missiles each. Its main base is at Doha, but it is building a naval base at Halul Island.

Qatar has a large police force with some paramilitary elements. They are equipped with three Lynx and two Gazelle helicopters. It has not experienced major internal security problems of late. The royal family has improved the sharing of its oil wealth in recent years, and its society is gradually becoming more modernized without becoming radicalized. The Al-Thani family has done a relatively good job of maintaining living standards and private sector opportunities in spite of declining oil revenues, and its recent cuts in government budgets and development activity have so far been healthy.

While there were rumors of a Libyan- and Iranian Shi'ite–backed coup attempt in September, 1983, these reports have never been confirmed. What seems to be more serious are constant rumors of feuding within the ruling Al-Thani family. There is a long-standing dispute within the royal family because the current ruler, Sheik Ahmad bin Ali al-Thani, named his son his heir in 1977, rather than his younger brother Sheikh Suhaim. This produced sporadic tension until Suhaim's death from a heart attack in 1985. Suhaim provided arms to some of the more distant members of the royal family, and sought Saudi support for his claims with at least limited success. This has led to some tension between Qatar and Saudi Arabia.

These rivalries surfaced again during Qatar's absurd 1986 border confrontation with Bahrain. There were rumors that Sheik Nasr bin Hamed, another younger brother of the ruler, had been shot in a family quarrel. The information minister, Isa Ghanim al Kawari, issued a public denial of reports that he himself had been shot. It is obvious that some tensions still remain within the royal family over the control of key ministries, if not the succession.

Qatar's internal security problems with Shi'ite minorities and foreign workers seem limited. Although the nation is about 15% Shi'ite, there have been few signs of support for Khomeini. The fact that 90% of Qatar's labor force is expatriate, and over 60% of its total population is expatriate, has not presented serious problems except for workers who stay or come without work permits. There are few reports of troubles between Qatar and its comparatively large population of expatriate Iranians. Qatar has, however, greatly strengthened its security controls—particularly of Iranians and Shi'ites.

The United Arab Emirates

The United Arab Emirates (UAE) now presents more problems in terms of creating an effective GCC-wide collective or cooperative defense effort than any other member state. It is still an awkward amalgam of seven small Sheikhdoms—Abu Dhabi, Dubai, Sharjah, Fujairah, Umm Al-Quwain, Ajman, and Ras al-Khaimah—that were forced together into the UAE during 1968–1972, after Britain decided to leave the Gulf.

The UAE is a federation and under its charter, each Emirate is supposed to contribute 50% of its oil-related income to help finance the national budget. Nevertheless, it is still divided by rivalries among its senior Sheikhs, particularly those of Abu Dhabi, Dubai, and Ras al-Khaimah. Ras al-Khaimah has also pursued its own low-level border dispute with Oman against the opposition of the Sheikhs in the UAE, and has exhibited some separatist ambitions towards creating a Qasimi state that would include Sharjah. The UAE's individual Sheikhdoms have also pursued different policies towards Iraq and Iran, with Abu Dhabi backing Iraq and Dubai, and Sharjah tilting towards Iran. The UAE has found it difficult to cooperate collectively with any of the GCC efforts to integrate Southern Gulf defenses, although Abu Dhabi has generally supported such efforts.

The UAE is relatively wealthy, although the recent cuts in oil prices have seriously affected its budget. It has not been able to issue a budget on time for the last six years, and its 1987 budget contained a deficit of 3.35 billion dirhams, or $914.8 million. This was much higher than the 1.19 billion dirham deficit in 1986. The 1987 budget was, however, an increase over the 1986 budget. It was 14.42 billion dirhams, a rise of 2.8%. This occurred although 90% of its budget revenue comes from oil and natural gas, and its annual revenue was down from 12.83 billion dirhams in 1986 to 11.06 billion.[47]

The UAE has collectively spent a great deal on defense. It increased its annual military expenditures from around $1,197 million in 1979, to $1,930–$2,100 million during 1982 through 1985. Its arms imports ranged from $60 million to $190 million annually, and its military manpower slowly increased from 25,000 in the early 1980s to 44,000 in 1985.

The UAE now spends about 6.7 to 7.4% of its GDP on defense, and 38 to 41% of its central government expenditures. It spent $1.88 billion on defense in 1986 and $1.58 billion on defense in 1987. Its total arms imports during 1981–1985 totalled $560 million, with $40 million coming from the U.S., $130 million coming from France, $220 million from the U.K., $70 million from the FRG, $40 million from Italy, and $60 million from other states.[48]

Unfortunately, much of the UAE's defense effort has been wasted, or spent on internal rivalries. The individual Sheikhdoms that make up the UAE remain deeply divided, and most of its 43,000 man military forces are expatriate and under the de facto command of individual Sheikhs. According to some estimates, nearly 90% of its forces are Omani. The command structure puts the President of the UAE, Sheikh

Zaid bin Sultan al-Nhayyan, in control of the Union Security High Command. Rashid bin Said al-Maktum, the ruler of Dubai and Vice President of the UAE, acts as commander in his absence.

The Chief of Staff of the Union Defense Forces is now Major General Muhammed Saeed Al Badi, a UAE citizen. He was appointed about seven years ago to replace a Jordanian, Lt. General Awad Khalidi. He is also an advisor to the President. This appointment came, however, only after Sheikh Zaid of Abu Dhabi tried and failed to install his 18 year old son, Colonel Sultan bin Zaid, as Commander in Chief in February, 1978. Sheikh Zaid did not consult with either Sheikh Rashid, or his son the Defense Minister, before making the appointment. Dubai responded by putting its forces on emergency alert, and forced Zaid to withdraw the appointment.

Arms purchases have supposedly been centralized since 1976, but this simply is not the case. Dubai, for example, bought Italian-made tanks to have a supply of arms that Abu Dhabi could not influence. As a result of a minor power struggle between Abu Dhabi and Dubai, the Minister of Defense is Sheikh Muhammed bin Rashid, the son of the ruler of Dubai. The Minister of Defense is supposed to be responsible for personnel, logistic, and support matters. He is not formally in the chain of command, and it is not clear what role he has in procurement.

The UAE is still heavily dependent on foreign mercenaries at every level from mid-level officer down. It recruits from Jordan and Oman, and Pakistan. This has had the advantage that many troops were trained before joining the armed forces, but it has raised obvious issues regarding loyalty. The UAE suddenly raised military pay by 50% after the ruler of Ras al-Kaimah attempted an abortive military confrontation with Oman in late 1977. There are still some British and many Pakistani contract pilots, and contract British officers. The UAE has talked about conscription, but has not implemented it.

The UAE Army has a nominal force strength of 40,000 men. Its order of battle includes one armored brigade, one mechanized brigade, and three infantry brigades—including a royal guard brigade. In practice, these units are organized into formations controlled by Abu Dhabi, Dubai, and Sharjah—often without any standardization of equipment, training, or personnel career structures. Abu Dhabi has the Western Command with 22,000 men. Dubai has the Central Military Region with 5,000 men, and Ras al-Kaimah has the Northern Military District with 1,700 men. Each military district is under the command of a son of the ruling Sheikh of the individual Sheikhdom involved, and the smaller Sheikhdoms have their independent commands.

As in Qatar and Kuwait, the UAE's army cannot really operate its main battle tanks effectively, although its ratio of tanks and other major weapons to troops makes more sense than that of Kuwait. The UAE now has 100 AMX-30 and 36 OF-40 Lions, plus 80 Scorpion armored fighting vehicles.[49]

The AMX-30s are in two battalions in Abu Dhabi, and sources disagree as to whether there are 64 AMX-30s plus 4 recovery vehicles, or all 100 AMX-30s plus 6 recovery vehicles, in active service. The AMX-30s are worn and obsolete, and Abu Dhabi is examining possible replacements in cooperation with Saudi Arabia. Candidates include the AMX-40, EE-T1 with either a 105 mm or 120 mm gun, the Challenger, the M-1, and possibly the T-72.

Abu Dhabi is also examining the option of joining Saudi Arabia in accepting a West German proposal to up-engine the AMX-30, install a new fire control, and make other improvements. Dubai took delivery on the first 18 OTO Melara OF-40s in 1981, and then on 18 more plus 3 armored recovery vehicles. It has converted all of its OF-40s to the improved Mark 2 version.[50]

The UAE's other armor includes about 430 armored fighting vehicles. These include 90 AML-90s and VBC-40s in the armored reconnaissance role, 30 AMX-10P MICVs, and a wide range of APCs, including 30 AMX-VCIs, 66 Engesa EE-11 Urutus (some with TOW), 300 M-3 Panhards, and VAB/VBCs. It has too many types of these vehicles for cost-effective training and support. In addition, it has 235 armored cars, including AML-60s, AML-90s, Ferrets, Saladins, Saracens, and Shoreland Mark 2s. Once again, it has too many types of armored cars for efficient support and training, although 70 Saladins, 60 Ferrets, and 12 Saracens are in storage.

To add to this confusion, various sources report that the UAE has 100 M-113A2s, more EE-11s, and M-998 Hummer light reconnaissance vehicles on order. There are also rumors of a massive deal between the UAE and Embraer, Avibas, and Engesa to provide large numbers of EE-11 Urutus, EE-9 Cascavals, EE-T-1 Osorio tanks, EMB-127 Tucano aircraft, and EDT-FILA air defense systems. It seems likely that there are no M-113A2S, and that the only Brazilian armored vehicles are the five sent to the UAE for evaluation purposes.[51]

The UAE is relatively light in artillery strength. It has 20 155 mm AMX Mark F-3 self-propelled howitzers, up to 50 ROF and M-102 105 mm towed howitzers, 18 M-56 105 mm pack howitzers, and 12 120 mm mortars. It lacks more artillery sensors and fire control aids. The UAE is effective by Gulf state standards in set piece firing exercises, but has problems in target acquisition and maneuver.

The UAE has 24 BGM-71A Improved TOW launchers, aging Vigilant ATGMs, BAT 120 mm recoilless rifles, and 84 mm Carl Gustav M-2 light recoilless rifles. Its short-range air defenses include Crotales, Rapiers,

RBS-70s, and Tigercats. There are also 60 M-3 VDA self-propelled twin 30 mm guns, and 30 GCF-BM2 self-propelled twin 20 mm guns These are all adequate weapons systems, and the UAE has done relatively well in operating the Crotale. Nevertheless, the UAE has too many types of anti-tank and light air defense weapons, too few of any given system, and poor overall training.

The UAE now has 1,500 men in its air force, including the police air wing. This is a reasonable overall manning level for an air force with 65 combat aircraft, 32 transport aircraft, and 59 helicopters, but overall training levels are low. The UAE is heavily dependent on foreign pilots and technical support for all operations.

Its fixed-wing combat aircraft include 36 Mirage 2000 in the air defense role, which are in the process of delivery. It has 27 Mirage V AD/RAD/DADs with attack and multi-role missions, 3 Alphajet attack aircraft, and 10 aging Hunter FGA-76/T-77s in the light attack role. There seem to be 8 MB-326K and 2 MB-339A light attack aircraft operating in the COIN role.

The UAE Air Force is having trouble absorbing its major equipment deliveries, especially its new Mirages. Abu Dhabi refused to take delivery on the first 18 Mirage 2000s in early 1987 because they were not equipped to a special standard as specified. This dispute had begun in March, 1986. In spite of an agreement by Dassault-Breuget to modify the aircraft, Dassault proposed to deliver but could not fire the same U.S. ordnance as other Gulf aircraft, and did not have compatible communications, IFF, and data links. Dassault sought to deliver first, and then to modify.

The UAE finally agreed to accept the aircraft in November, 1987, but only if Dassault later completed the necessary changes and paid penalties. The French were forced to compromise because they were competing with McDonnell Douglas F/A-18s for a $500 million follow-on order for 25 more aircraft, after the F-16 and Tornado were rejected as candidates. The UAE Foreign Minister visited Paris and threatened the French Prime Minister and Minister of Foreign Affairs with excluding Dassault.[52]

Four CASA C-212s are employed in an electronic warfare role, and the UAE has 10-12 SA-342K Gazelle attack helicopters and 7 SA-316 Alouette IIIs with AS-11. Combat-capable training and liaison aircraft include 21 Hawks, and 20 Pilatus PC-7s. There also are 6 SF-260TPs and two MB-339As. The UAE is now considering purchase of either 25 more Mirage 2000s or F/A-18s. Three more Alphajets, 24 Hawks, 1 G-222, and 12 Aermacchi MB-339 may also be on order.

The UAE has actively investigated ordering C-130s equipped for electronic warfare, E-2C Hawkeyes for the AWACs and maritime surveillance role, and 2 BN-Defender AEW aircraft. Its advanced air ordnance includes R-550 Magique air-to-air missiles, and AS-11, AS-12, and AM-39 Exocet air-to-ground missiles. It has some Beech MQM-107A RPVs, which it uses as target drones.

The UAE's transport force includes 3 B-707s, 1 B-737, 1 VC-10, 1 L-100-20, 5 C-130Hs, 1 Gulfstream II, 1 Falcon 20, 5 Islanders, 1 G-222, and 8 Caribou and Buffalo. Its medium transport helicopters included 8 AB-205/Bell 205s, 3 AB-212s, 3 AB-214s, 10 AS-332F Super Pumas (4 in the naval attack role), and 90 SA-330 Pumas. Its light transport helicopters include 6 AB-206s, 3 Alouette IIIs, and 3 BO-105s. It has 30 A-129 Mangustas and Lynx helicopters on order.

The air force's main air bases at Abu Dhabi and Jebel Ali (Dubai) are sheltered and have light anti-aircraft defenses. It also has military fields at Batin in Abu Dhabi, Dubai, Fujairah, Ras al-Khaimah, and Sharjah. Abu Dhabi is creating a major modern air base at Suwaihan.

The UAE Air Force merged with the air defense force in January, 1988. This merger occurred because of growing coordination problems between the fighter force and land-based air defenses.[53] The UAE has ordered 7 MIM-23B Improved Hawk units with 42 launchers (342 missiles), and Skyguard twin 35 mm anti-aircraft guns.[54] It will take at least three years, however, to finish deploying these ground-based air defense units as fully effective forces.

The overall command and control, and air control and warning, capabilities of the UAE Air Force are barely adequate. It lacks effective battle management and warning capability, effective secure voice and data links to the Saudi Air Force, and a compatible IFF system. Its proficiency in manning and operating its Hawk missiles is also poor.

The UAE's small 1,500 man navy is also divided by Sheikhdom. Its only major combat ships are 6 Lurssen TNC-45 guided-missile patrol boats with two twin MM-40 Exocet launchers, and 6 Vosper Thorneycroft 33.5 meter patrol boats. Abu Dhabi, however, seems to have ordered Lurssen 62 meter patrol boats with Exocet, 76 mm guns, and Goalkeeper close-in defense systems.

The UAE also has three 17 meter patrol boats, two 15.3 meter patrol boats, six Keith Nelson or Dhafeer-class 12.3 meter patrol boats, and 6 Fairey Marine Spears. It has four 20 foot police boats and 19 P-1200s. A small coast guard, which is part of the Ministry of Interior, operates many of these patrol boats. The UAE navy's only support ships are two Cheverton tenders, although it has two Crestitalia 30-meter diver support vessels on order.

Four of the air force's Super Pumas are equipped with AM-39 Exocets and possibly with depth charges. Abu Dhabi is considering buying a 60 meter missile boat from Lurssen and the Goalkeeper close-in weapons system.

The navy is highly dependent on foreign personnel and advisors. There are small naval bases at Ajman, Mina Zayd and Dalma (Abu Dhabi); Mina Rashid and Mina Jebel Ali (Dubai); Mina Sakr (Ras al-Kaimah); Fujairah; and Mina Khalid and Mina Mina Khor Fakkan (Sharjah). There are plans, however, to create a $1 billion naval base at Taweela if funds become available. A naval facility is also under construction at al-Qaffay Island.

In regard to the internal stability of the UAE, the individual Sheikhs who lead the UAE continue to struggle for power. Only Abu Dhabi's forces regularly participate in GCC exercises, although Dubai has sometimes contributed. Dubai has resisted Abu Dhabi's support of Iraq, and pushed for relations with the USSR.

The individual Sheikhdoms in the UAE also have considerable internal instability. The crown princes in several Emirates are either ill or are considered to have uncertain capability to rule. The long tradition of internal coups within these individual Sheikhdoms has scarcely ended—as became all too clear during June, 1987.

Sharjah was the more prominent emirate in the Gulf until Abu Dhabi and Dubai proved to have much larger oil wealth. The al-Qasimi family—which also rules Ras al-Khaimah—has had a long tradition of internal rivalry, as well as one of rivalry with Abu Dhabi and Dubai which dates back to a time when the Qasimi dominated the area. The Qasimi family in Sharjah is divided into two main branches, and these have long been rivals. The British overthrew one branch in favor of the other in 1971, and in 1972, the ruling Sheikh Khalid bin Muhammed was killed in a coup attempt by the other branch of the family.

The June 17, 1987, coup attempt occurred, however, within one branch of the royal family. Sheikh Abd al-Aziz bin Muhammed al-Qasimi attempted to overthrow his younger brother, Sheikh Dr. Sultan bin Muhammed, on the grounds that his younger brother had misused the nation's oil income and had hurt Sharjah's economy by failing to cut the budget in response to the fall in oil revenues.

Abd al-Aziz had been commander of the royal guard since 1972, as well as of the local defense command since 1972. When his brother was in London, closed the airport, seized control of the radio, TV, and newspapers, and announced his brother's "abdication." He also threatened to have his troops fire on anyone who challenged his rule.

Abd al-Aziz's coup triggered considerable debate within the UAE, particularly because a number of the other rulers felt that Sultan bin-Muhammed had created millions of dollars of unnecessary debt. Abu Dhabi, in fact, had previously refused to provide further support to

Sharjah in repaying its loans and showed every sign of recognizing Abd al-Aziz. Dubai, however, forced the issue to the UAE's Supreme Council. It did so partly because of a friendship between Dubai's ruler and Sultan bin-Muhammed, partly out of rivalry with Abu Dhabi, and partly because of fears that the precedent might present problems for Dubai's crown prince, who was not felt to be equal to his brothers.

After three days, the Council decided that the other Sheikhdoms could not accept the precedent of a military overthrow of a ruling Sheikh. It avoided any fighting, however, through a compromise whereby Sultan bin-Muhammed stayed as ruler, but Abd al-Aziz became Crown Prince and joined Sultan bin Muhammed on the UAE's Supreme Council.[55]

This kind of internal and external instability is not reassuring. The UAE's one strong leader—Sheikh Zayid bin Sultan al Nuhayyan of Abu Dhabi—has no strong successor, although Sheikh Khalifa, the Crown Prince and Deputy Supreme Commander of the Armed Forces, may be emerging as a possibility. The sons of his principal rival, Sheik Rashid of Dubai, are divided about the merits of increasing federalism within the UAE. This rivalry between Abu Dhabi and Dubai reached the point in 1985 where Dubai created its own air line as a rival to Gulf air, and tensions with Sharjah and the smaller Emirates are growing as the decline in oil income forces more budget and project cutbacks. As is discussed in later chapters, the Sheikhs have also split over the UAE's attitudes towards the Iran-Iraq War, with Abu Dhabi favoring Iraq and Dubai and Sharjah favoring their traditional trading partner, Iran.

There is no inherent reason that any of these political and economic problems should destabilize a nation with such wealth and strong strategic incentives to remain unified. Nevertheless, the overall level of economic management in the UAE is poor, and only Abu Dhabi and Dubai have done a reasonable job of sharing the nation's oil wealth.

The visible divisions among the Sheikhs have also weakened the UAE's efforts at economic development. This has led to a growing disenchantment among its technocrats, and to the politicization of the UAE's youth. The resulting resentments and tension now lack a clear political focus, but could become much more serious if the Sheikhs should quarrel to the point where they return to the paralytic divisions of the 1970s, or fail to deal with the cut in oil revenues by spreading more widely their oil and gas income.

Finally, Table 6.2 has shown that the UAE has a problem with foreign labor and ethnic divisions. By some estimates, only 19% of the UAE's population is now Emirian, and only 42% is Arab. Over 80% of the labor force is expatriate, and there is a significant Shi'ite minority. Abu

Dhabi, Dubai, Sharjah, and the other states still do not coordinate adequately on internal security, much less on military affairs.

Oman

Oman is one of the few Gulf states whose forces have had any experience with modern war. Unlike the other British Trucial states, its forces had to be organized to fight in order to deal with the Green Mountain and Dhofar rebellions, and its army developed a cadre of high-quality infantry. Oman has experienced growing problems as it has had to convert from an infantry force to more modern military services, but it has still had the best managed military modernization effort in the Southern Gulf.

Like the other Gulf states, Oman has suffered from the drop in oil prices, although this has had limited impact on development and local living standards. The 1988 budget is projected to be 1.60 billion riyals ($4.13 billion), just slightly less than the 1987 budget of $1.61 billion riyals. Oil revenues are projected to be $1.07 billion rials. This leaves a deficit of $252 billion rials ($650 million), when other sources of revenue are taken into account. Oman estimates it will borrow $150 million in 1988 to help bridge the gap.[56]

Oman increased its annual military expenditures from around $780 million in 1979, to $1,513–$2,111 million in 1982 through 1985. It spent $1.56 billion on defense in 1986 and $1.51 billion on defense in 1986. It spends about 23 to 28% of its GDP on defense, and 47 to 49% of its central government expenditures. Its military manpower slowly increased from 15,000 in the early 1980s, to 25,000 in 1985. It has since reduced its manpower to remove some low-grade tribal forces.

Oman's recent arms imports have ranged from $100 million to $350 million annually. Its total arms imports during 1981–1985 totalled $995 million, with $90 million coming from the U.S., $40 million coming from France, $550 million from the U.K., $240 million from the FRG, $5 million from the PRC, $10 million from Italy, and $25 million from other states.[57]

Its main military strengths are a good cadre of Omani personnel and excellent British advisors. It appointed an Omani officer to the post of Commander of the Army for the first time in 1984.[58] It is spending almost $2 billion annually (or 40% of its total budget) on defense, and the other GCC states pledged to provide $1.8 billion in aid over 12 years in September, 1983.[59] Nevertheless, Oman will be hard-pressed to create even a minimal deterrent to the threat from Iran by the early 1990s and could have problems if it was attacked by South Yemen.

Oman has benefited significantly from U.S. and Saudi aid. It is rapidly converting its officer corps to native personnel, but still has nearly 500 British officers and NCOs seconded to the Omani armed forces. It also has a limited number of personnel from Jordan and Egypt, and large numbers of Pakistani Baluchis. Oman has established both specialized secondary schools to train its military intake and a central training center near Muscat. It also trains officer and technical personnel in Britain, the FRG, France, Jordan, and Saudi Arabia. Unlike other members of the GCC, it conducts exercises with outside powers, including Britain, Egypt, France, and the U.S.

While Oman did establish relations with the USSR in late September, 1985, Oman provides the U.S. with critical contingency and prepositioning facilities. These include airfields and prepositioned equipment at Masirah Island, and airfield facilities at the international airport at Seeb, at Hasb, and at Thamrait in Western Oman, plus additional storage and naval facilities at Masirah and Ghanam.[60] The U.S. Army Corps of Engineers has upgraded the old 2,000 foot direct runway at Khasab with a 6,500 foot surface air base. British forces make frequent use of Omani facilities, and use the Omani base at Goat Island in the Straits of Hormuz and a new intelligence post at Qabal in the Musandam Peninsula for a variety of reconnaissance functions.

The base at Goat Island and other facilities in the Musandam Peninsula are particularly important because they guard the Straits of Hormuz and are only 26 miles from Iran. More than 50 large ships, 60% of them tankers, passed through these waters every day during 1986 and 1987. The Musandam Peninsula is a small enclave with a population of about 12,000, and is separated from the rest of Oman by a 40 mile strip of the UAE. Oman has been assisted in developing this region since 1976 by a U.S. firm called Tetra Tech International. It has spent nearly $5,000 per person in recent years to develop the region.[61]

Oman has conducted some impressive public military exercises since 1985. Its March, 1985, exercise was called Codename Thunder and involved roughly 10,000 men. It was the largest and most effective exercise that any GCC state has conducted, and Oman has regularly played an important role in GCC exercises ever since. Oman also exercises regularly with British forces. Refueled British RAF Tornados have flown nonstop to air bases in Oman, and Oman held a joint exercise with Britain in December, 1986, called "Swiftsword" that involved over 5,000 British servicemen. This exercise included a 400 man landing by British marines, air drops, and air reinforcements. Some 5,000 Omani troops acted the role of defenders, while another 2,000 played the role of an enemy that was clearly Iran.[62]

At the end of 1987, Oman had about 21,500 regulars and 17,500 highly trained regulars in its army. Omani soldiers and officers are respected throughout the Gulf and often form an important portion of the total military manpower of other GCC states, especially those of the UAE. Omani Army forces, however, are not equipped with large numbers of modern weapons and are not highly sophisticated.

The Omani army consists largely of light mechanized infantry organized into a royal guard brigade, two partly motorized infantry brigades, an armored battalion, an armored reconnaissance battalion, and one airborne/special forces battalion.[63]

The army is now equipped with about 39 main battle tanks (33 Chieftains and 6 M-60A1s), with 20 more Chieftains on order. It has 30 Scorpions and 6 VBC-90 light tanks/armored reconnaissance vehicles. It has only a very small pool of other armored vehicles. These include 20 V-150 Commandos, 6 VAB/VCIs (2 VCACs with Milan, 2 VDs with 20 mm AA guns, and 2 APCs), and 15 AT-105 Saxons. Partly because of cost and partly because of its own rough terrain, Oman is the least mechanized of all the GCC armies.

The Omani army has 12 FH-70 155 mm guns, 12 M-109A2 self-propelled howitzers, 12 Type 59/M-1946 130 mm guns, 36–39 M-102/ROF 105 mm howitzers, 18 obsolete 25 pound howitzers, 12 120 mm mortars, and 12 107 mm mortars. It has Palmaria 155 mm self-propelled howitzers on order. It also has 60 mm, L-16 81 mm, 107 mm, and 12 M-30 4.2-inch mortars. Its artillery lacks mobility and sophisticated target acquisition and fire management systems, but is generally better organized and more effective than that of the other small GCC states.

The army is equipped with 10 BGM-71 Improved TOWs, and Milan anti-tank guided missile launchers. It has Blowpipe surface-to-air missiles, and 12 Bofors L/60, 16 X 23 mm, and 4 ZU-23-2 23 mm anti-aircraft guns. It is not well equipped with either anti-tank or light air defense weapons, but is acquiring modern digital land lines similar to the British Army Fastnet communications system.[64]

The Omani Air Force has 3,000 men, and is gradually beginning to build up a cadre of native pilots. It also has a technical school and is trying to improve the technical base of its air force manpower. Its combat and technical training is still limited, however, and although the Omani air force is very small for a nation the size of Oman, it is still heavily dependent on foreign support and technicians.

Oman currently has 53 combat aircraft, 37 transport aircraft, and 39 helicopters. The combat aircraft include 23 Jaguar S(0) Mark 1, Mark 2, and T-2s. These are good attack aircraft, but have only limited visual air-to-air combat capability using guns and AIM-9-P4 infra-red missiles.[65] The other combat aircraft include 16 Hawker Hunter FGA-6/FR-10/T-67 light attack trainers. There are 13 BAC-167 Strikemasters in the COIN/training role, and AS-202 Bravos and PC-6 Turbo-Porters that could be armed.

Oman's air ordnance is relatively unsophisticated and includes R-550 Magique and AIM-9P air-to-air missiles, and BL-755 cluster bombs. Some 300 more AIM-9Ps are on order.

Oman has sought to buy up to two squadrons of F-16 or Tornado fighters, and is expanding Thamrait air base into a fully modern facility to base them. It has bought an improved air defense system, and has received U.S. aid in mechanizing its army.[66]

So far, however, it has lacked the funds for such major procurement efforts in spite of the fact Britain has offered preferential terms for a Tornado sale. It has ordered only eight Tornados as part of a $340 million arms package, and this order is now on an indefinite hold.[67] The Tornado may well prove to be too sophisticated and expensive for the Omani air force in its current stage of development, and Oman will be slow to absorb any advanced fighter aircraft, even though it has obtained British and Saudi help.

Oman does not have any major surface-to-air missile systems in service or on order. The air force does have two squadrons, 28 Rapier surface-to-air missile launchers, and some Tigercats in service. It has ordered 28 radars for its Rapiers, most of which have now been delivered.

Oman has three transport squadrons. Its transport aircraft include 3 BAC-111s, 3 C-130H Hercules, 1 Mystere-Falcon 20, 7 Britten-Norman BN-2 Defender/Islanders, and 15 Short Skyvan 3M STOL aircraft—which Oman has found to have great value in mountain and desert operations. One more C-130H and 2 DHC-5D Buffalos are on order. The royal flight has 1 Gulfstream, 1 DC-8, 1 VC-10, and 2 AS-202 Bravos.

The helicopter force includes 3 AB-206 Jet Rangers, 20 AB-205s, 4 AB-212B/Bell 212s, 11 AB-214,s 2 AS-332 Super Pumas, and 4 SA-330 Super Pumas. Oman has 6 Bell 214STs and possibly some Westland Sea King ASW/SAR helicopters on order.

Oman shelters its main air bases at Masirah and Thamrait, which have been greatly improved with U.S. aid. It has additional military airfields and strips at Khasab on the Musandam Peninsula, colocated at the modern international airport at Seeb near Muscat, Nizwa, and Salalah. It also has Marconi Martello three-dimensional radars, and has ordered two S-713 three-dimensional radars. This will give it the nucleus of a ground-based air control and warning system.

Oman's navy now confronts a more direct challenge than its other services. It must defend a 2,900 kilometer coastline, including the main shipping routes through the Straits of Hormuz.

Oman now has 2,000 men in its navy and a long seafaring tradition.

Its navy personnel lack technical expertise, however, and are still dependent on British support. Oman is steadily converting to all-Omani naval forces, but the recent growth of the navy's size and technical sophistication has meant that this process must take longer than Oman originally planned.

Oman has 5 major combat ships, 4 gunboats, 17 patrol boats, and 6 landing craft. It has a ship maintenance and repair facility at Muscat, and naval bases at Jazirat Ghanam, Mina Raysut near Salalah, Khasam, and Muscat. A new base is under construction at al-Masnaa al-Wudam Alwa.

Oman's principal ships consist of three 400-ton Province-class fast-attack boats, armed with 76 mm OTO Melara L/62 guns, twin 40 mm Breda Compact mountings, and MM-40 Exocet missiles. Two of the ships have 2 X 4 Exocet missiles each, and one has 2 X 3. They were handed over to Oman in 1982–1984. The radar and fire control system includes a Racal-Decca TM1226C surveillance radar and a Phillips 307 director. One more of the larger Province-class ships has been on order since 1986.

Oman also has four Brooke Marine (Al Wafji) 37.5 meter fast patrol boats, at least two of which are now armed with Exocet. None of these ships, however, has adequate air defense or sophisticated modern sensor packages.

The Omani Navy has six 23–27 meter patrol boats, and eleven 8–13 meter patrol boats. It is one of the few GCC navies equipped for amphibious operations and has one 2,000 ton Landingship Logistic (LSL) and one 2,200 LSL in service. These ships have good command and control facilities, a helipad, and the ability to carry two landing craft each with up to 188 men and eight tanks. They are well suited for missions like seizing an oil platform or small island.

Oman also has two 75–85 DWT landing craft utilities (LCU), and three 30–45 DWT LCUs. Oman's auxiliary craft include the royal yacht, a coastal freighter, and a survey craft. An 8,000 ton auxiliary with amphibious warfare capability and a helipad is on order from Bremer-Vulcan.

The main limitations to the Omani Navy are its lack of capital ships that could directly challenge Iranian ships, a lack of mine warfare capability, and a lack of maritime surveillance capability, although it has several transports it uses for visual airborne surveillance. Like ASW, however, these are missions which the Omani Navy can probably safely leave to the U.S. and U.K. until it is ready for more sophisticated and more expensive missions.

Oman provides the U.S. with extensive contingency facilities and has a strong contingent of British advisors, officers, NCOs, and technicians that help it as well as the West. Both British and U.S. ships and maritime patrol aircraft routinely support Oman in patrolling the approaches to the Straits of Hormuz, and the U.S. can rapidly supplement Omani forces with over-the-horizon reinforcements. The modernization of Omani forces should allow them to deal with low-level contingencies and create some deterrent capability against medium-level conflicts even without U.S. reinforcements.

There are also 4,500–5,000 men in the tribal forces, most of very low capability, and 7,000 men in the police, coast guard, and border forces. The latter units operate some light aircraft, helicopters, and patrol boats. The police coast guard has 15 AT-105 armored personnel carriers, 11 coastal and 3 inshore patrol craft, 13 support craft, and 28 speed boats. The air wing of the police has 1 Learjet, 2 Do-228-100s, 2 Merlin IVAs, 2 DHC-5 Buffalos, 5 AB-205s, and 3 AB-206s. There is a small 85 man security force on the Musandam Peninsula called the Shikuk Tribal Militia.

In spite of its recent past, Oman no longer faces major internal security problems. It is one of the best managed states in the Near East and Southwest Asia. While its population is becoming increasingly politicized and less tolerant of Britain's role in Oman's government, the Sultan has coopted a large number of former rebels, modernized his government, and increased the rate of Omanization. Oman's second five year plan has been relatively successful, and the country has succeeded in increasing oil sales in spite of the decline in the world market. This has helped to offset the serious decline in its oil revenues, which have provided nearly 90% of government revenues.

Oman has recently produced around 600,000 barrels a day, of which it has consumed about 50,000. It cut back production to 550,000 BPD in late 1986, however, as part of the OPEC quota effort to raise oil prices. This left 500,000 BPD for export.[68]

While Oman will experience economic problems until the oil market recovers, it seems capable of sustaining its most critical development efforts without any political backlash. Oman's proven oil reserves expanded from 2.5 billion barrels in 1983 to 4 billion barrels in 1985. This is still small by Gulf standards, although its agriculture, light industry, and infrastructure are capable of sustaining slow and steady growth in spite of sharply diminished funds for its new third five year plan. About 25% of Oman's population is expatriate, mostly Pakistani and Indian, but it is not politically active. The succession issue, however, does present a constant and growing immediate risk, although Sultan Qaboos is young enough so that this is now largely a risk of assassination.

As for foreign relations, Oman has increasingly tried to defuse some of the political costs of its ties to the U.K. and U.S. It established diplomatic relations with the USSR in 1986, and has made a point of trying to maintain good low-level relations with Iraq without causing a confrontation with Iran. It has been involved in minor border squabbles with Sharjah, but its only serious problem remains the PDRY. In spite of the establishment of diplomatic relations with South Yemen, ambassadors were never exchanged because the PDRY refused to settle several border disputes, and the new regime in Aden may present Oman with serious problems.

THE SOUTHERN GULF AND THE 360° THREAT

The GCC states are not particularly unstable by Third World standards, but there are serious internal pressures within them. They all face a long and uncertain period of transition from family rule to some form of a more modern state with broader public participation. They all must deal with serious problems in terms of religious tensions, problems in modernization, and problems in dealing with a young and politically awakening population.

No one can predict which of these pressures within the GCC states will become most dangerous to Saudi Arabia and Western interests in the region. The problem that the West faces in dealing with these states is not the vulnerability of any one country, but rather that there is at least a moderate *cumulative* probability that one of the more vulnerable conservative Gulf states will become radicalized over the next decade.

If such an event takes place, the West and/or the other GCC states will virtually be forced to act—just as they would in the event of an Iranian or any other attack on such states. The risks inherent in the radicalization or defeat of one of the smaller GCC states is simply too great. If one should fall, the geography of the Gulf would ensure that Saudi Arabia and the remaining conservative GCC states would face a radically different land, air, and naval threat.

Any shift in Kuwait's position would remove a critical buffer between Saudi Arabia and Iraq, and possibly Iran. A radical Bahrain could place hostile aircraft within less than five minutes' flying time of key oil, power, and water facilities in the other Southern Gulf states. The same would be true of Qatar. If any element of the UAE became hostile, the air force and navy of the other Southern Gulf states would experience major problems in trying to operate in the Eastern Gulf, and any combination of a radicalized Southern Gulf state and a radical Iran would seriously undermine the GCC's political and military security structure. A hostile Oman would threaten all naval and shipping traffic through the Gulf, make a British military role in the Gulf far more difficult, and force a massive restructuring of USCENTCOM in order to be able to deploy without the stocks and bases in that country.

There is no easy way to summarize the interface between the military development of the Southern Gulf states, and the potential threats to the West's strategic interests in the Gulf. The problem is never one of a single threat from a single direction, or of a threat that can easily be defined in military and political terms. While the Northern Gulf states and the USSR obviously have the most forces, other states border directly on Oman and Saudi Arabia and can pose an equally great, if not greater, political threat.

The key point is that both the West and the Southern Gulf states face a "360° Threat". Military forces cannot be shaped quickly. New combat units take a decade to mature, and new major military bases take up to five years to build. Governments and political attitudes can change overnight. Many of the surrounding governments are already hostile, and most of the remainder are at least partially unstable. The steady flood of arms into surrounding states is expanding the potential military threat on every border.

With luck, diplomacy and politics can contain this situation without a conflict of the kind that has already taken place between Iran and several of the GCC states. The chances of such luck continuing for the next decade, if Saudi Arabia does not build up its own forces and does not have convincing strategic ties to the West, seem slim indeed. Luck is also scarcely a wise basis for securing Western access to half of the world's proven oil reserves.

NOTES

1. It is important that the reader understand that there is no consistency in the statistical data provided on the Middle East. The author has used a wide range of sources throughout this book, and has often had to make his own estimates. The data on the GCC countries are particularly uncertain, and the author has often had to change sources to get consistent or comparable data on a given point. This leads to the use of contradictory data for the same measurement, often because of differences in definition or time of estimate, but sometimes simply because accurate data are not available. The reader should be aware that such statistical information is better than no information, but must be regarded as approximate and should be checked with at least three to four different sources before being used for specialized analytic purposes.

2. These estimates are made by the author, and are based on working data

from Wharton, the EIU, and the U.S. Department of Energy. Such estimates differ sharply according to source.

3. The military data used in this section are based on IISS, *Military Balance, 1987-1988;* Eytan, *The Middle East Military Balance,* Tel Aviv, JCSS, 1986, and the DMS data base on foreign military markets, Middle East, and Africa. The naval data also draw on James Bruce and Paul Beaver, "Latest Arab Force Levels Operating in the Gulf", *Jane's Defense Weekly,* December 12, 1987, pp. 1360–1363; and Michael Vlahos, "Middle Eastern, North African, and South Asian Navies", *Proceedings,* March, 1987, pp. 52–64. The data on ethnic and religious divisions are taken from the CIA, *World Factbook, 1986;* and *Middle East Review, 1985* and *1986,* World of Information, Saffron Walden, England, 1985. For additional background see the author's *The Gulf and the Search for Strategic Stability,* and Thomas L. McNaugher's "Arms and Allies on the Arabian Peninsula", *Orbis,* Vol. 28, No. 3, Fall, 1984, pp. 486–526.

4. For a full discussion of the GCC's defense role, and an excellent description of the overall impact and history of the GCC, see Erik Peterson, *The Emergence of the Gulf Cooperation Council.* The author reviewed Mr. Peterson's work while it was still in draft.

5. Saudi Arabia opened new arms production facilities at Kharg in late 1986 that now make about 60% of its small arms and their ammunition. *Jane's Defense Weekly,* November 8, 1986, p. 1088.

6. *Jane's Defense Weekly,* November 8, 1986, p. 1085.

7. *Jane's Defense Weekly,* September 6, 1986, p. 434.

8. *Jane's Defense Weekly,* November 1, 1986, p. 995.

9. *Los Angeles Times,* October 6, 1987, p. I-15; *Washington Times,* October 6, 1987, p. 6A; *Jane's Defense Weekly,* August 9, 1986, p. 198.

10. *Washington Post,* November 8, 1986, p. A-36; *Middle East Economic Digest,* October 25, 1986, p. 2, and November 13, 1986, p. 26.

11. *Washington Times,* December 29, 1987, p. A-10; *New York Times,* December 30, 1987, p. A-3; *Washington Post,* December 30, 1987, p. A-18.

12. *Middle East Economic Digest,* October 25, 1986, p. 2.

13. Economist Intelligence Unit, *EIU Regional Review: The Middle East and North Africa, 1986,* Economist Publications, New York, 1986, p. 16.

14. All demographic statistics used for Bahrain and the rest of the GCC states are based on CIA estimates in the relevant country sections of *The World Factbook, 1986,* Washington, GPO, 1985.

15. The coup involved arms shipped from Iran to Bahrain, the use of the Iranian embassy to support the coup attempt, and the training of Bahraini citizens in Iran. While some expatriates were arrested, the core force in the coup was some 73 Bahraini youths, all Shi'ite.

16. This definition of residency was the criterion for being able to vote for Bahrain's parliament before it was dissolved in 1976. Only 3% of the male population qualified.

17. Estimate based on Bahraini data and reporting in *Jane's Defense Weekly,* March 15, 1986, p. 452.

18. Arms Control and Disarmament Agency, *World Military Expenditures, 1986,* Washington, GPO, 1987.

19. The JCSS shows the independent formations; the IISS does not. The IISS seems out of date since it shows that the brigade has one infantry battalion, one armored car squadron, and one artillery and two mortar batteries. This does not reflect Bahrain's tank or AFV holdings.

20. These tanks were deployed at a cost of $90 million. *Jane's Defense Weekly*, November 30, 1985.

21. *Washington Post*, December 2, 1987, p. A-25; and December 18, 1987, p. A-25; *Defense Electronics*, February, 1988, p. 11.

22. *Jane's Defense Weekly*, February 13, 1988, p. 247.

23. *New York Times*, September 11, 1987, p. A-3.

24. These aircraft cost $114 million, and are part of a package including 60 AIM-9-P3 missiles. DMS data base.

25. *Jane's Defense Weekly*, February 7, 1987; *New York Times*, January 28, 1987, p. 2; *Washington Post*, January 21, 1987, p. A-15; *Defense News*, February 2, 1987, p. 15..

26. Saudi Arabia promised up to $700 million in military aid during GCC planning meetings in early 1984. It is unclear how much of this aid will be available in the near term, given the growing cash squeeze in Saudi Arabia.

27. The causeway consists of 11 km of embankment, 12.5 km of bridging in five separate units, and 1.5 km over the small island of Umm Nassan. The Saudi government paid for the causeway at a cost of $1.2 billion. It has presented serious social problems for Bahrain and Saudi Arabia because of the fact Bahrain permits the sale of alcohol and Saudi Arabia does not, and because of the different tariff and price structures in the two countries. *Washington Times*, August 19, 1986, p. D-3; *Economist*, January 4, 1986. p. 34; *Los Angeles Times*, April 28, 1986, p. IV-1.

28. *Defense and Foreign Affairs Weekly*, May 26–June 1, 1986, p. 4.

29. *Jane's Defense Weekly*, June 14, 1986, p. 1087; and *Defense and Foreign Affairs Weekly*, May 26, 1986, p. 4.

30. Kuwait is considering up-engining the tanks with new British or German engines. *Jane's Defense Weekly*, February 28, 1987, p. 323.

31. Arms Control and Disarmament Agency, *World Military Expenditures, 1986*, Washington, GPO, 1987.

32. *Jane's Defense Weekly*, January 30, 1987, p. 151.

33. *Jane's Defense Weekly*, February 28, 1987, p. 314, and March 7, 1987, p. 359.

34. I am indebted to Lt. Commander Jerry Ferguson, one of my students at Georgetown University, for much of the research, and many of the insights, on Gulf naval and air forces presented in this chapter.

35. The 76 mm and 40 mm guns can provide some air defense, but with little lethality. The TNC-45s have very complicated electronics, virtually all of which are maintained by foreign technicians. The voice network system used by the TNC-45 is so slow that it is virtually hopeless for air defense operations and generally creates confusion and increases delay and vulnerability if any attempt is made to use it.

36. The A-4s lack an air intercept radar, and can only engage in visual combat

using guns or Sidewinder missiles. There are 12 Lightnings and 9 Hunters in storage.

37. Sources differ. The data from JCSS are shown. The IISS says 2 DC-9s, 4 L-100-30s.

38. *Christian Science Monitor*, December 7, 1987, p. 9.

39. *Los Angeles Times*, August 12, 1985; *Washington Times*, August 19, 1985.

40. Figures stated by Brigadier Mohammed al-Qabandi, Assistant Under Secretary for Police Affairs at the Kuwait Ministry of Interior.

41. *Washington Times*, August 20, 1986, p. 6A; *Jane's Defense Weekly*, June 21, 1986, p. 1141, November 6, 1986, p. 1085; *Reuters*, October 17, 1985.

42. *Washington Post*, December 19, 1987, p. A-27.

43. *Jane's Defense Weekly*, August 30, 1986, p. 347

44. *Wall Street Journal*, July 7, 1986, p. 16. The 1982 collapse of Kuwait's highly speculative "camel market" left six of Kuwait's banks with $4.1 to $4.4 billion in bad loans.

45. These figures draw largely on DOE sources to ensure comparability between reserve and production data and are somewhat different from the commercial estimates used in Chapter 2.

46. Arms Control and Disarmament Agency, *World Military Expenditures, 1986*, Washington, GPO, 1987.

47. *Wall Street Journal*, December 22, 1987, p. 16.

48. Arms Control and Disarmament Agency, *World Military Expenditures, 1986*, Washington, GPO, 1987.

49. The Italian-made OF-40s are the only ones operational in the world in this configuration.

50. France demonstrated the AMX-40 in trials in both Qatar and the UAE. *Jane's Defense Weekly*, June 6, 1987, p. 1092.

51. These reports are dubious. Brazilian firms often leak false orders while trying to win new business. The EDT-FILA is a three-radar system with a scanning area of 2 25 kilometers.

52. *Defense News*, January 5, 1987, p. 19; *Jane's Defense Weekly*, October 17, 1987, December 5, 1987, p. 1302.

53. *Jane's Defense Weekly*, February 20, 1988, p. 301.

54. The U.S. temporarily delayed the sale of a $170 million upgrade package for the Hawks in June, 1987, because of the coup attempt in Sharjah.

55. *Defense and Foreign Affairs*, June 29–July 5, 1987, p. 7; *Economist*, June 27, 1987, p. 41;

56. *Wall Street Journal*, January 5, 1988, p. 21.

57. Arms Control and Disarmament Agency, *World Military Expenditures, 1986*, Washington, GPO, 1987.

58. The new army commander, Major General Naseeb Bin Haman Bin Sultan Ruwaihi, is almost certainly qualified for the post. It is important to note, however, that he was appointed at the end of 1984 when General Sir Timothy Creasy was replaced as Chief of Defense Staff by Lt. General John Watts. There are rumors this replacement occurred partly because of Creasy's insistence on an exemplary jail sentence for Robin Walsh, a British MOD official accused of misappropriating $8,700 in MOD funds. Walsh died in an Omani jail in

October, 1984. This was followed by broader accusations that up to $74 million annually was being wrongly appropriated by the Ministry of Defense, and that both British and Omani officials knew of the problem. Economist Intelligence Unit *The Middle East and North Africa, 1986*, Economist Publications, New York, 1986, pp. 186–187.*EIU Regional Review:*

59. Oman has refused to support the creation of a Gulf Defense Force at al-Batin. This is partly because it is under Saudi command, and partly because the force is primarily oriented towards the defense of Kuwait, and Kuwait has pressed for freer movement of Gulf labor than Oman prefers and refuses to sign intelligence agreements that would provide more data on the movement of politically sensitive individuals. This reflects a high level of vestigial rivalries and issues between the GCC countries. For example, Abu Dhabi and Dubai refused to fully cooperate during the initial Peninsular Shield exercises.

60. The U.S. has spent over $300 million onm upgrading these facilities. *Washington Post*, July 19, 1985, p. A-29.

61. There are two major radars at Goat Island. It is garrisoned by 250 Omani soldiers and marines and 10 Britons. *New York Times*, December 22, 1986, p. A18; *Defense News*, December 1, 1986, p. 6; *Washington Post*, March 24, 1986, p. A-13; *Christian Science Monitor*, October 30, 1979; *Time*, December 2, 1985, p. 58.

62. Richard Green, editor, *Middle East Review, 1986*, London, Middle East Review Company, 1986, pp. 168–167; *New York Times*, December 22, 1986, p. A18; *Defense News*, December 1, 1986, p. 6; *Jane's Defense Weekly*, November 22, 1986, and December 6, 1986; *Chicago Tribune*, November 18, 1985, p. I-12.

63. The two brigades are not standardized and are mission-tailored commands. They contain a total of one armored regiment with two tank squadrons and one self-propelled artillery battery, three artillery regiments (two light and one medium), a light anti-aircraft battery, a reconnaissance regiment with two armored car squadrons, eight infantry regiments (battalions), one signals regiment, one field engineer regiment, and one paratroop regiment.

64. The army acquired digital switching for some 6,000 telephone lines as part of a two million pound contract with Plessey. *Jane's Defense Weekly*, November 1, 1986, p. 998.

65. Oman bought some 300 AIM-9P4s in late October, 1985. *Baltimore Sun*, October 11, 1985, p. 9A.

66. *Washington Times*, May 9, 1985, p. 7.

67. Economist Intelligence Unit, *EIU Regional Review: The Middle East and North Africa, 1986*, Economist Publications, New York, 1986, p. 193.

68. *Washington Post*, September 1, 1986, p. 33.

7
The Development of Saudi Forces

The preceding analysis of threats to the region, of Western power projection capabilities, and of the capabilities of the smaller Gulf states has shown that the West has no ideal options in the Gulf—although it is equally true that no Gulf nation has any ideal options in the West. The practical problem the West faces is to build the best possible strategic relations with the Southern Gulf states in spite of these difficulties, and to do so in a form that offers both the West and its regional allies maximum security at an acceptable cost. Seen from this perspective, the military capabilities of Saudi Arabia present a unique opportunity to the West that merits analysis in considerably more depth than has been presented in discussing the other Gulf states.

This opportunity must be kept in perspective. Saudi Arabia is not strong enough to serve as a proxy for Western military forces or as a "pillar" of Western security. It has many of the vulnerabilities of its smaller neighbors, and it can achieve security only through a combination of cooperative defense efforts with its neighbors and the West. At the same time, Saudi Arabia has the wealth and the population to act as the core of the GCC's efforts to build regional security. Further, it is large and strong enough so that Western military forces can remain over the horizon in many contingencies, and limited amounts of Western reinforcement should be adequate in most contingencies.

Further, Saudi Arabia's recent military development has reflected its leadership's understanding of two key facts: the scale and complexity of regional threats, and the nation's dependence on Western military equipment, technology, and military advice. These facts help explain the central thrusts behind Saudi military modernization.

SAUDI DEFENSE EXPENDITURES

Saudi Arabia's greatest strength is oil wealth; its greatest weakness is manpower. While there are many varying estimates of Saudi Arabia's population, all agree that it has a severely limited pool of military aged and technically trained manpower. The most recent CIA estimates indicate that Saudi Arabia's total native and foreign manpower pool is about 3,079,000 males, of whom 1,760,000 are fit for military service. A maximum of 106,000 males reach military service each year. As much as half of its manpower fit for military service consists of foreign workers. Further, in spite of massive recent investments in education, literacy is only a little over 50%.[1]

This ensures that Saudi Arabia cannot build up large ground forces to match those of the Northern Gulf states. It is forced to rely on technology as a substitute for manpower, and on Western support for its combat forces while it creates its own technicians and support manpower. This reliance on technology, however, forced Saudi Arabia to make extremely high levels of defense expenditure during most of the 1970s and early 1980s in a "brute force" effort to create the capability to support high technology forces. This has been a driving factor behind Saudi requests for the transfer of advanced weapons and technology.

Few developing countries have spent as much on defense as Saudi Arabia, but few countries have had to begin with so limited a military and technical infrastructure, and faced so many challenges. Saudi Arabia's defense expenditures have had to respond to a massive regional arms build-up, and to rapidly expanding threats on all its borders. Saudi Arabia has had to buy advanced technology, and to pay the extraordinarily high cost of creating high technology forces without a base of trained manpower and modern military infrastructure.

Since the late 1960s, this effort has been driven by the impact of Britain's withdrawal from the Gulf, the massive arms race and the war between Iran and Iraq, and the arms race and fighting among the Red Sea states. Saudi Arabia had to begin this effort with only token military forces, and has had to conduct a "brute force" effort to use money to make up for constraints in manpower, facilities, and time.

This "brute force" effort has concentrated on support and infrastructure, rather than on combat forces. It has been shaped and driven by various U.S. plans and studies over a period of nearly two decades, and it has been relatively successful. Saudi Arabia has slowly built up division- and brigade-sized military cities to guard its borders. It has built up some of the most modern air bases in the world, created several new military ports, and established extensive training, logistics, and headquarters facilities. There is no easy way to cost this, but the bulk of the Saudi modernization effort has been planned and supervised by the U.S. Army Corps of Engineers, and much of it is reflected in the U.S. FMS budget. Tables 7.1 and 7.2 provide a picture of the overall trends in foreign military sales to Saudi Arabia, and how they relate to Saudi defense spending.

Table 7.1: Saudi Military Imports and Spending (Millions of Constant 1983 U.S. Dollars)

Year	Total Defense Spending	Defense Spending as a Percent of GNP	Defense Spending as a Percent of Central Govt. Expend.	Defense Effort Spending Per Capita	Defense Effort Mil.[a] Per 1,000	Total Arms Imports	Arms as a Percent of Total Imports
1974	4,956	7.1	10.9	701	10.6	434	5.9
1975	11,040	7.3	17.4	1,516	10.3	1,021	7.2
1976	15,020	7.7	19.1	1,943	9.7	1,696	7.5
1977	14,300	8.3	15.3	1,727	9.1	1,853	6.4
1978	14,950	8.7	15.9	1,711	8.6	1,584	4.9
1979	17,810	9.1	18.1	1,962	8.7	2,178	6.0
1980	19,760	9.4	14.4	2,098	8.4	3,247	8.2
1981	22,450	9.8	12.9	2,301	8.1	3,243	7.6
1982	25,110	10.1	15.7	2,486	7.9	3,600	9.2
1983	27,020	10.4	23.4	2,587	7.7	2,513	7.7
1984	21,480	10.8	21.3	1,990	8.8	2,339	8.1

[a]Military personnel

Source: Arms Control and Disarmament Agency, *World Military Expenditures and Arms Transfers, 1986*, Washington, GPO, 1987, pp. 79, 121, and 134.

Table 7.2: Recent Sources of Saudi Arms Imports, 1976–1985 ($ Current Millions)

Source	1976–1980 Total Value ($ Millions)	1976–1980 Percent of Total	1981–1985 Total Value ($ Millions)	1981–1985 Percent of Total
Major Western Countries				
France	700	15	4,300	29
FRG	350	7	190	1
Italy	150	3	170	1
U.K.	975	21	1,400	9
U.S.[a]	2,000	43	6,400	43
Major Communist				
Czechoslovakia	0	0	0	0
Poland	0	0	0	0
PRC	0	0	0	0
Romania	0	0	0	0
USSR	0	0	0	0
All Other	520	11	2,300	16
Total	4,700	100	14,760	100

[a]Includes some U.S. Corps of Engineers construction equipment

Source: Arms Control and Disarmament Agency, *World Military Expenditures and Arms Transfers, 1971–1980*, and *1986*, Washington, GPO, 1980 and 1986.

The trends in Tables 7.1 and 7.2 are complex, but a close examination reveals several consistent patterns. The first is that Saudi Arabia's massive defense expenditures have normally been a relatively small proportion of its oil wealth. Large as they have been, they have generally represented less effort in terms of percentage of GNP and total national manpower than those of its northern Gulf neighbors and the vast majority of Third World states. Second, Saudi Arabia has bought remarkably few major arms in comparison with its total defense expenditures. It has spent its money on construction, operations and maintenance, and personnel.

Saudi Arabia has been the leading cash customer for U.S. military sales and support services in both the Gulf and the world. It has received only $38.5 million dollars in U.S. grant aid since the Korean War, and all grant aid terminated in 1971. Similarly, it has received only $250 million in FMS credits. None were concessional, and all were terminated in 1971. Between FY1974 and FY1986, however, it spent $50.2 billion on FMS cash sales, and it spent $1.1 billion on such sales in FY1987.

This point becomes clearer in Table 7.3, which compares the patterns in U.S. FMS sales to Saudi Arabia from 1950 to 1985. It shows that the U.S. had sold Saudi Arabia nearly $20 billion worth of construction agreements, and delivered about $13 billion worth. This compares with about $28 billion in other FMS agreements, of which about $15 billion were delivered.

Table 7.3 also shows just how important the U.S.-Saudi connection has been to the Kingdom's military development. U.S. sales of military construction services have approached the level of all other military sales, and they exceeded all Saudi arms imports well into the late 1970s. The construction totals in Table 7.3 also show that the West has done more than provide Saudi Arabia with arms and military advice: U.S. modernization plans and the construction activities of the U.S. Army Corps of Engineers have dominated development of Saudi military bases, facilities, and lines of communication.

Even today, Saudi Arabia is the largest customer for U.S. FMS aid. It is receiving deliveries of around $1 billion annually. It ordered $712 million worth of FMS aid in FY1986, and $637 million worth in FY1987. The next largest customer was the PRC, which ordered $552 million in FY1987.[2] The largest NATO customer was the FRG, which ordered $511 million. This illustrates Saudi Arabia's continued dependence on U.S. military advice and arms sales even though much of its infrastructure is complete and in spite of the political resistance of the U.S. Congress.

Table 7.3 also understates the role the U.S. has played in shaping Saudi Arabia's military modernization since the mid-1970s. Many major

Table 7.3: Saudi Arabia's Main Source of Western Military Support—U.S. Military Assistance and Foreign Military Sales ($ Current Millions by U.S. Fiscal Year) [a]

Year	Foreign Military Sales Agreements	Foreign Military Sales Deliveries	Military Construction Agreements	Military Construction Deliveries	Commercial Exports
1950–1975	3,683.1	705.7	6,901.6	663.8	68.3
1976	1,826.6	461.4	5,451.5	465.5	92.7
1977	1,122.8	1,066.1	589.9	483.9	44.1
1978	1,312.2	1,129.2	647.0	1,193.3	166.3
1979	5,449.6	940.5	1,021.0	1,193.3	166.3
1980	2,954.2	1,124.2	1,590.8	1,457.9	29.0
1981	1,042.6	1,435.8	877.4	1,491.6	71.5
1982	5,347.3	2,072.8	1,888.2	1,775.6	50.0
1983	891.4	3,860.9	716.5	2,153.2	251.8
1984	2,840.3	1,950.0	263.0	1,470.8	359.5
1985	2,791.4	1,904.2	743.0	901.0	186.5
1950–1985	29,261.4	16,650.8	20,698.8	13,556.8	1,364.1

[a]Unlike most states receiving U.S. foreign military sales, Saudi Arabia has paid for virtually all of its purchases from the U.S. It has never received MAP merger funds, and has not received any Military Assistance Program funds since 1976. Its total MAP funds from 1950 to 1975 totalled only $239 million. Saudi Arabia also received $1.8 million in excess defense articles during 1950–1975, and $12.5 million in International Military Education and Training (IMET) funds during 1950–1975. It received $234 million in financing before 1976, all of which it repaid.

Source: Department of Defense, *Foreign Military Sales, Foreign Military Construction Sales, and Military Assistance Facts*, Washington, D.C., DoD, September 30, 1985.

Saudi military projects which have not been funded through FMS have been the result of U.S. plans or studies, and have been designed and/or supervised by the U.S. Army Corps of Engineers. The U.S. has been the principal architect behind most of Saudi Arabia's transition from a low-level infantry force to a relatively modern military force which has gradually built up one of the most sophisticated basing and military infrastructure systems in the developing world.

This "brute force" period in Saudi military development is also reaching its end. While nearly $10 billion worth of U.S. military construction services are still in the pipeline, many have already been fully paid for and many are associated with the completion of the high technology aspects of Saudi Arabia's force structure, such as its air base sheltering program. Its key military "cities", ports, air bases, training and support facilities, and headquarters are nearing completion.

The completion of Saudi Arabia's basic military infrastructure is easing both the burden of Saudi defense expenditures and the Saudi transition to dependence on British, French, and other non-U.S. military equipment. Saudi defense expenditures have had to drop sharply since the early 1980s because of the major decline in its oil revenues. Nevertheless, the completion of Saudi Arabia's military facilities should allow it to proceed with the key elements of its military modernization in spite of this decline.

As has been discussed in Chapter 2, Saudi Arabia has seen its recent oil revenues drop to about one-fifth of their 1981 level of around $100 billion. Its financial reserves have dropped to less than half their 1981 level of $190 billion. The Saudi riyal has been devalued by 12% over the last three years, from 3.75 to the dollar to 3.35.

Saudi Arabia has had to cut its original FY1986, FY1987, and FY1988 budget estimates, and accept annual deficits ranging from $10 billion to $20 billion.[3] Its FY1988 budget is projected to be 141.2 billion riyals ($37.7 billion), down some 17% from the 1987 level of 170 billion riyals. Oil revenues were unofficially projected at 65.2 billion riyals. The 1988 deficit was projected to be 35.9 billion riyals ($9.57 billion) versus deficits of about 50 billion riyals in the previous four years. Saudi Arabia sought to reduce the deficit through utility surcharges, 12–20% import duties, hospital charges, airline fare increases, and foreign borrowing in the form of some $8 billion in bonds, and by imposing taxes on foreigners up to 30%.[4]

While precise estimates are impossible, and Saudi Arabia often adjusts its flow of defense expenditures without reporting them in its budget documents, the FY1985 defense budget seems to have been about $21–23 billion, and FY1985 was a year in which a combination of aid to Iraq and Syria may have driven foreign aid expenditures up to over $5 billion.[5]

The FY 1986 Saudi defense budget was planned to be $17.7 billion (64.6 billion riyals), or 32% of the total budget. The oil revenue deficit led to minor cuts and spending of about 64.09 billion riyals ($17.3 billion). The FY1987 defense budget was about 60.7 billion riyals, or $15.78 billion. The FY1988 defense budget, which includes the National Guard and the Interior Department and its police forces, is planned to be about 50.8 billion riyals or $13.21 billion. This is a cut of 9.9 billion riyals, or $2.57 billion, from FY1988.[6]

At the same time, it is clear that the level of Saudi defense expenditures in FY1986 to FY1988 stills represent a high level of defense spending for a nation with Saudi Arabia's military forces, and it seems unlikely that Saudi defense spending will ever drop much below $13 billion unless there is a massive further cut in world oil prices. It also seems likely that Saudi Arabia will continue high levels of indirect defense spending in the form of foreign aid. The mix of threats it faces seem likely to force it to continue its high level of defense effort regardless of the present decline in its oil revenues.

Even if Saudi defense budgets should drop below $13 billion, the Kingdom probably can sustain reduced spending levels for several years without harming its basic defense program. It has largely completed a $30 billion dollar investment in military facilities and infrastructure, and retains a relatively modern major weapons mix. Expensive as first-line major weapons and combat equipment now are, Saudi Arabia can buy its essential needs with a budget far lower than its peak past budgets ($22 billion in 1983/84, and $22.7 billion in 1984/85).

The Kingdom can draw on its extensive capital holdings or easily manipulate its delivery schedule on major weapons and projects so that the key cost component comes after 1988, when most energy experts predict oil prices will begin at least a limited recovery. Saudi Arabia also has slowed down many of its direct payments for defense purchases, and is shifting to barter deals and sales which require a high level of offsets.

SAUDI MILITARY MANPOWER

The key military problem Saudi Arabia faces is manpower. While Saudi Arabia often exaggerates its population and military manpower for political purposes, it seems fairly clear that Saudi Arabia now has a total native population of only 7–9 million and only about 84,000 full-time uniformed actives in its armed forces—10,000 to 15,000 of whom are in its paramilitary Royal Guards and National Guard. It has roughly 73,500 men in its regular armed forces.

This manpower situation compares with an Iranian population of roughly 50 million and peak military manning levels of over 1 million men, with an Iraqi population of roughly 16 million and nearly 1 million men under arms, with a Syrian population of about 11 million and 400,000 men under arms, and with a total population in the Yemens of about 9–11 million and 63,000 under arms.[7]

Saudi Arabia's regular military forces now comprise about 63,500 men. By Western standards, it would take about 75,000–100,000 men to adequately man the Kingdom's current force structure. Even a full-scale draft would probably fail to give the manpower to meet its limited force expansion plans. Saudi Arabia solves this problem by:

- Depending heavily on foreign support and technicians (now over 10,000 personnel);
- Using small elements of Pakistani and possibly Turkish forces in key speciality and technical areas—such as combat engineers—to "fill in" the gaps in Saudi land forces, and some 10,000 Pakistani forces to fill out one brigade (the 12th Armored Brigade) at Tabuk;[8]
- Using French and British internal security experts;
- Selective undermanning while it builds its training and manpower base;
- Concentrating on building a fully effective air force as a first-line deterrent and defense; and
- Reluing, *de facto*, on over-the-horizon reinforcement by the U.S., France, Pakistan, or some other power to deal with high-level or enduring conflicts.

These are all intelligent methods of reducing the manpower problem, but they still leave many gaps and weaknesses in Saudi forces. The limitations in Saudi military manpower are also forced on Saudi planners by Saudi demographics, by civil competition for skilled manpower—which still makes it extremely hard to retain army personnel in spite of the contraction of the Saudi economy—and by the need to maintain a 10,000 man National Guard for internal political and security reasons.[9]

None of these manpower constraints will change significantly during the next decade, and Saudi Arabia can only hope to reach and maintain a technical edge over regional threats by concentrating on the modernization and "Saudization" of its combat arms while continuing to rely on foreign support. The Kingdom must allocate virtually all of its increasing output of skilled military manpower to operational forces and command roles, and it cannot hope to replace Western technical support.

There is no way that the Kingdom can hope to compete with most of its larger neighbors in sheer manpower or land forces. While it has talked about conscription for more than a decade, this can be only a token solution, and any full-scale program would have unacceptable political and economic costs. Although Saudi Arabia's total population is uncertain, it is reasonably clear that the Kingdom now has a maximum of about 1.8 million males eligible for military service. Only 60,000–90,000 new males fit for military service reach draft age every year, and it is unlikely that this number will climb above 200,000 before the year 2000.[10]

The demand for these males in the civil economy will continue to exceed supply almost regardless of the shift in oil revenues, and most will strongly prefer civil careers. Coupled with the fact that it generally takes at least 24 months to turn a conscript into an effective soldier in most developing states, this scarcely implies a political and economic climate where the Kingdom can adopt more than a limited conscription plan. While Saudi Arabia has indicated on several occasions in recent years that it may introduce a draft during the next year, such a draft would also be politically unpopular and extremely difficult to enforce.[11] If conscription is adopted, it is more likely to be a selective effort that will provide a limited increase in active manning levels than a broad popular draft that results in a major increase in Saudi total manning.

The National Guard adds to Saudi Arabia's manpower problems in filling out its regular forces, although it also aids internal stability. The Kingdom's internal politics indicate that Crown Prince Abdullah will continue to concentrate on seeking to convert the portion of the Guard that is actually fit for military service (about 8,000–10,000 men) into a fully mechanized force rather than allow its manpower to be absorbed into the regular military.

This modernization of the Guard is producing a much more effective internal security force, but it also increases the strain on Saudi manpower rather than reducing it. It also is likely to raise the level of competition among the various services, particularly if the Ministry of Interior continues to build up its own internal security and to counter terrorist forces. These Ministry of Interior forces now total about 8,500 men.

THE IMPACT OF FOREIGN MANPOWER

These internal manpower problems are a key reason why Saudi Arabia must depend on substantial numbers of U.S., Western European, and other advisors and technicians well into the late 1990s. About 1 million foreign workers now live in Saudi Arabia, some 35,000–40,000 of whom are Americans.

The Saudi military is equally dependent on foreigners. It must continue to concentrate most of its own military personnel in combat arms and combat support, and must rely heavily on outside help even to create a purely defensive set of military capabilities.

As a result, Saudi Arabia will continue to need something approaching its current 2,000–3,000 European technicians and advisors, 5,000 U.S. technicians and experts, and 500 U.S. military advisors well into the 1990s.[12] The U.S. portion of this manpower is only part of the foreign advisory effort, and Saudi Arabia depends heavily on Britain for air training and support, and France for naval and army training and support. Nevertheless, the breakdown of U.S. military and contract personnel supporting FMS purchases in Saudi Arabia shown in Table 7.4 provides a good illustration of the wide mix of companies involved and of the kind of role Western states are likely to play in the future.

It is important to note that the bulk of the U.S. personnel working on defense-related contracts in Saudi Arabia are not working on weapons-related projects but on construction and are normally listed as civil personnel. Further, Table 7.4 shows that a small—but growing—number of U.S. personnel are working for Saudi or Arab firms.

Nevertheless, Saudi Arabia must find U.S. or European technicians in order to ensure that it can get the maintenance and upgrade support it needs for its new weapons systems, as well as suitable military advice and training. The new E-3A Peace Sentinel and Peace Shield programs will raise the total significantly. So will the nearly 2,000 British technicians and advisors needed to help Saudi Arabia convert to and operate its Tornados, Hawks, and Pilatus PC-9s, although Britain already carries out much of the basic training for the Saudi Air Force and helps to support Saudi Arabia's remaining Lightning fighters.

This interdependence between Saudi Arabia and the West is deliberately given a low political profile. Like the role of U.S. plans and the U.S. Army Corps of Engineers in shaping Saudi military modernization, it is best left as a quiet reality, rather than given a high profile in a way that leads hostile local states, radical political movements, and the USSR to accuse Saudi Arabia of giving the U.S. "bases", or of betraying the Arab cause by close ties to Israel's principal ally. A shift to more reliance on Britain may also help defuse this issue, although it is too early to tell how various hostile Arab states and Arab radical groups will react.

It is also too early to tell just how much the Saudi shift to new British aircraft will affect Saudi relations with the U.S. The tensions arising out of the Iran-Iraq War have led to increasing visits of USCENTCOM personnel to Saudi Arabia, to the quiet upgrading of the personnel at the U.S. mission's headquarters in Dhahran, and to an increase in the number of U.S. personnel supporting the joint Saudi-U.S. section at Riyadh.

The Saudi Air Force now is heavily dependent on USAF advisory teams at the Saudi Air Force headquarters, at Riyadh Air Base, at King Abd al-Aziz Air Force Base at Dhahran (F-15 and F-5 support), at the King Faisal Air Academy, and at each Saudi air base.

Senior Saudi Air Force officers have been pleased with the quality of the USAF "Elf-1" unit that is now operating four to six E-3As, KC-10 and KC-135 tanker aircraft, and TPS-43 gap-filler radars in Saudi Arabia while the Saudi Air Force trains and prepares to operate its own E-3As. They also are pleased with the new "Peace Shield" C^3I/BM system they are buying from the U.S.[13] The 500-plus man Elf-1 unit is based at Dhahran Air Force Base, and has already provided Saudi Arabia with key training in operating an air defense system and using the E-3A AWACS. In addition, the Saudi Air Force has made increasing use of special U.S. task teams assigned for specific missions on a temporary duty basis. These U.S. advisory efforts have been expanded significantly since the U.S. intervention in the Gulf in 1987, but the details are classified.

Further, the Saudi Air Force has generally been pleased with its civilian contractor teams from Northrop, Boeing, and McDonnell Douglas. These are generally supported by contractor teams from each of the major U.S. aircraft manufacturers, and Northrop and McDonnell Douglas personnel have played a key role in training and supporting Saudi personnel as part of the Peace Hawk (F-5) and Peace Sun (F-15) programs, although Saudi Arabia is increasingly supporting its F-5s on its own and already maintains its own C-130s. The Kingdom has also made use of independent U.S. defense consulting firms like Braddock, Dunn, and McDonald (BDM). Saudi Arabia is now adding a major new British advisory presence to its existing 2,300 man British Aerospace team. British firms provide a wide range of training services in addition to supporting its remaining Lightnings. It also is adapting British training, support services, C^3I, and combat techniques to blend them with those it has learned from the U.S.

The Saudi Air Force's dependence on Western aid offers broad advantages. It underpins virtually every aspect of Saudi Arabian modernization in a way most developing states lack. Its predominantly civilian character gives it a lower political profile than a military effort, and still allows Saudi Arabia to concentrate on the "Saudization" of its combat crews, maintenance crews, and intermediate service crews. While this does mean that the Saudi Air Force depends on Britain and the U.S. for service and resupply in any extended training or combat use of its forces, it allows Saudi Arabia to eliminate dependence on U.S. personnel in combat.

The training and support effort the Saudi Army is receiving from the West draws less praise, both from the Saudis and from former U.S. and French personnel. There seems to be a general consensus that the Saudi Army is modernizing more slowly than the air force, and has weaker teams and poorer contractor support. The French and U.S. advisory effort supporting Saudi Arabia's ground-based air defenses has also been less effective than the support for the Saudi Air Force.

The quality of contractor advice seems to have been low to mediocre,

Table 7.4: U.S. Personnel Supporting FMS Purchases in Saudi Arabia

U.S. Civilian Contract Personnel

Function	Personnel	Function	Personnel
AEA Electromechanical Systems	4	Pacific Architects and Engineers	41
A.R. Ramlah Corp	2	Prefab Eld Co. Ltd.	4
Boeing	16	Rabya	1
Demauro Shuwayer Co. Ltd.	6	Raytheon	283
Ericsson	1	Ret-Ser Engineering Agency	1
FMC	7	Samwhan Corp	4
General Technical Services, Inc.	1	Saudi Computer Service	5
HCA	312	Saudi Medical Service	37
Al Henaki	2	SBT/Erectors	1
Al-Husseini ADA	1	SHIN WHA	1
Honeywell	6	SIBIC BASIL	661
Hughes Aircraft	4	SIYANCO	283
Hyundai Construction	17	Transworld Services	20
ICCI/AI Muraibid Est	1	Vinnel Corporation	404
Johnson Brothers International	2	Wallace International Ltd.	3
Landarun and Brown	14	Westinghouse	2
Litton	162	Weimer and Trachte	1
Lockheed	709	BDM	26
McDonnell Douglas	599	Saudi Arabian International Stores	1
Almusaadiah-Pepper	2	SOGREB/BELL/HAZAR	3
Northrop	617	Computer Science Corporation	45
Obeid and Al Mulla	10	BASI	2
OPS	23	Day and Zimmerman	9

U.S. Government Civilian and Military Personnel

Function	Personnel
U.S. Military Training Mission	187
APO	13
MAC	8
ACC	36
Commissary	4
Medical Clinic	4
C-130 Crews	19
Technical Advisory Field Team (TAFT)	103
MTT	16
OPM/SANG	89
AFDC-LSG	109
Corps of Engineers	519
TDY (largely ELF-1 AWACS)	530

Source: Based on the "Response to Section 36(A)(7) of the Arms Control Act on U.S. Government and Civilian Personnel resident in Saudi Arabia on assignments relating to FMS as of May 4, 1985", and data furnished by the Department of Defense.

and the French and U.S. Army advisory efforts often seem to have accepted problems rather than tried to solve them. The French advisory effort at Tabuk has, however, served as a tacit deterrent to Israeli attacks, although the French presence at Tabuk now consists of only 20–30 men, and the bulk of the French army effort in Saudi Arabia has shifted to Taif and Khamis Mushayt.

The improved C^3I system for Saudi Arabia's Hawk and fixed SHORAD defenses—which is part of the comprehensive U.S. Peace Shield C^3I/Battle Management system—may improve Saudi ground-based air defenses. Unfortunately, any overall integration effort will be complicated by the fact that Saudi Arabia's $4.5 billion deal with France to buy an extensive net of Shahine short-range air defense missiles owes as much to politics—and to the need to give France guaranteed arms purchases that would ease the burden of refinancing Iraq's military debt—as to well-planned requirements.

The land-based air defenses operated by the Saudi Air Defense Corps are unlikely to approach the air force in providing the readiness and combat capability that Saudi Arabia needs. The Air Defense Corps has also experienced major problems in managing its procurements and maintenance contracts. This is a key factor explaining Saudi interest in the Stinger man portable surface-to-air missile—which does not need sophisticated C^3I or battle management.

The U.S. Navy has advisors headquartered at Dhahran, and has training teams at the Saudi Navy bases at Jiddah and Jubail and at the main training base at Dammam. The U.S. has improved the quality of the personnel assigned to the Saudi Naval Expansion Program, but serious problems in the past quality of the U.S. advisory team helped lead Saudi Arabia to give France the bulk of the advisory role.

The problems in the U.S. Navy advisory effort stem from several causes. The U.S. initially failed to select high-quality personnel, but Saudi Arabia also advanced over-ambitious expansion plans. Further, the U.S. Navy did not operate or procure the kind of heavily armed and high-technology small vessel Saudi Arabia sought. France has been in a position to provide such vessels, but has had major problems in providing training, maintenance, and follow-on technical support. The Saudis have expanded their use of U.S. naval advisors since early 1987, however, and are considering seeking additional aid in naval training and modernization from the U.K.

The U.S. Army Corps of Engineers generally draws praise from the Saudis, and has helped to build up the relationship between U.S. and Saudi forces. It has also generally been successful in creating effective military bases and facilities, although often at an extremely high cost.

Unfortunately, neither the Corps nor any Saudi has had the authority to tightly manage construction efforts, and the Corps' role in the Kingdom is gradually coming to an end. The Saudi government has also resisted the Corps' efforts to standardize design and construction patterns—which might otherwise have significantly reduced the cost of major Saudi efforts.

The U.S. has tightened up the quality of the regular uniformed U.S. personnel assigned to advising the National Guard, and has improved the analytic and contracting support provided to the mission in both Saudi Arabia and the United States. Nevertheless, the National Guard remains a critical problem. The Guard's modernization continues to lag and much of the input of its foreign advisers is ignored. These problems have grown to the point where it has been necessary for the U.S. to impose strict security limits on any discussion of these problems by current or former advisors.

Prince Abdullah has made it clear that the official U.S. military role in advising the Guard must be kept more limited than that for the regular services. There are only about 50 military and 100 civilian advisors to the Guard, and the 404 men in the Vinnell Corporation have a more independent role than that of other contract advisors. This limits the role the U.S. government can play in ensuring that the Guard gets the advice and support it needs. Prince Abdullah has also indicated his intention to seek non-U.S. advisors and training personnel, although it is unclear how far this effort will progress.

The Guard also has an acute shortage of educated and technically trained personnel, and this has hampered the U.S. advisory effort. It also seems to have delayed the Guard's effort to learn how to maintain its Cadillac Gage armored vehicles, and Saudi Arabia has evidently been slower in converting to self-maintenance of this equipment than any of the other 30-odd developing nations that have bought it.

The Kingdom's future need for the deployment of foreign military combat units is more uncertain. Saudi Arabia does seem likely to remain dependent on the U.S. Air Force Elf-1 team to operate E-3As and part of the new air defense system it is buying from the U.S. through the early 1990s. Saudi Arabia also takes advantage of warning and intelligence data from U.S. vessels in the Gulf and has held several low-level exercises with U.S. naval and air forces.

What is less clear is Saudi dependence on foreign security and land force units. Rumors that Saudi Arabia has obtained the support of Thai and South Korean security forces cannot be confirmed. There is, however, considerable evidence that Saudi Arabia has sought contingency support from France and Pakistan, and that some French and Pakistani security forces are in the country.

There are also reports that the French reinforcement role may be more significant than providing internal security advisors. Some sources have hinted that the equivalent of a French mechanized brigade set of equipment is prepositioned at Sharurah in special storage, and there are contingency plans for French forces to stage out of Djibouti or fly directly to Khamis Mushayt to reinforce Saudi Arabia in an emergency. These reports are very doubtful, but the French navy seems to have conducted low-level exercises with the Saudis, and could play a key role in areas like minesweeping—where neither the Saudi Navy nor the U.S. Navy has adequate capability.

Western officials have indicated in the past that a Pakistani training "brigade" plus some other engineer elements close to the North Yemeni border, were deployed at the Saudi army base at Tabuk. These forces built up the army's strength in the early 1980s, and filled out most of the 12th Saudi Armored Brigade. There were around 11,000–15,000 Pakistani troops and advisors in Saudi Arabia at the end of 1987. They brought Pakistan some $300 million annually, and were under Saudi command, although a Pakistani Major General head the Military Liaison Office in Riyadh.

Growing friction between the Saudis and Pakistanis began in 1986, however, and Saudi Arabia did not renew its contracts with Pakistan when it became clear that Pakistan would not remove the Shi'ite elements from these troops or allow them to be used in situations where they might have to fight Iranian forces. These elements made up roughly 10% of the Pakistani manpower in the country.

As a result, the 10,000 Pakistanis in the armored brigade left the country in late 1987 and early 1988.[14] Up to 3,000 more men, recruited individually or seconded to the air force, are reported to be staying. According to some reports, these include pilots, crews, and technicians for up to two fighter-bomber squadrons, although the actual Pakistani cadres in the air force seem to be significantly smaller.[15]

There has since been some discussion of replacing the Pakistanis with up to 10,000 Egyptian, Turkish, and Jordanian troops or making contingency arrangements for the rapid deployment of such forces.[16] The UAE also may be considering hiring a 10,000 man Egyptian contingent, and Kuwait is reported to be making contingency arrangements for Egyptian reinforcements. Any such forces would lack air power and armored strength, but would have the advantage of being Muslim and, in the case of Egypt and Jordan, of speaking Arabic. They also would ease the problems the West would face in committing ground and air troops to any complex contingency where the case for intervention was politically ambiguous.

The prospects for such aid, however, are uncertain. Egypt has stated on several occasions that it will not provide such forces, although the validity of such public statements is uncertain. Turkish support presents the problem that there is still lingering resentment of Turkey's former domination of the Arab world.

The history of efforts to provide Jordanian support to Saudi Arabia has been filled with problems. U.S. efforts to help Jordan build up its contingency capabilities to intervene in the Gulf were leaked to the press in October 1983. These reports indicated that Jordan was developing a two-brigade force of 8,500 men. The Jordanian forces were to receive training assistance from the U.S., to airlift from 8 C-130 transports, and to be armed with TOW anti-tank weapons, Stinger surface-to-air missiles, and C^3I equipment. The U.S. was to sell Jordan 3 additional C-130s, to give it a total of 8, but this would provide lift for only 70 paratroops and 100 troops per aircraft. The U.S. was to provide large-scale lift in an emergency.

The initial cost of the program, which included the land force equipment, was $220–230 million. This was to come from U.S. foreign military sales loans, and Saudi Arabia was to provide Jordan with the money. Further sales of up to $470 million of FMS equipment were to include F-16 fighters and other major equipment necessary to allow Jordan to resist a major air or armored threat. This Jordanian force, however, seems to have been tailored largely to provide a rapid reaction force that could support the smaller GCC states. The U.S. air package also seems to have been tailored to help Jordan develop its own defense capabilities in the face of a massive Syrian military build-up.

In any case, the U.S. did not fund the package and told Jordan in January, 1986, that it would be unable to meet its request for modern fighters and other major combat equipment. The key questions now are whether Saudi Arabia will be able to afford to fund Jordan's military modernization, whether Saudi Arabia and Jordan can come to some political arrangement as to the proper role of Jordanian forces, and how this would impact on each nation's complex strategic relations with Syria.

As for the internal effects of Saudi dependence on foreign personnel, they seem unlikely to have any significant destabilizing effects. The situation in Saudi Arabia differs radically from that in Iran under the Shah in that the foreign military advisors have a relatively low profile and are not dominated by one Western nation. Many such advisors are Muslim, and the living standards of most advisors do not contrast unfavorably with those of the Saudis. Saudi Arabia also does not have large numbers of recently educated students or workers who feel displaced and discriminated against by the hiring of a foreign labor pool. In any case, this dependence on foreign manpower is not excessive or destabilizing, and much of the foreign manpower supporting Saudi forces will leave in the late 1980s, as Saudi Arabia completes its massive military construction program.

MILITARY INDUSTRY AND OFFSETS

Saudi Arabia is taking several steps which should ease its defense expenditure and foreign manpower problems. It has coupled most of its recent major arms sales either to arrangements that force the seller to buy Saudi oil, to arrangements ranging from barter to market share agreements, or to offset to arrangements that force the seller to buy Saudi goods and services from the Kingdom's civil sector or to help expand Saudi Arabia's small but growing military industries. These sales arrangements affect the Kingdom's recent aircraft purchases from Britain and the Peace Shield air defense C^3I system it has bought from the U.S.

Saudi Arabia already assembles some of its small arms, including the Heckler and Koch G3 assault rifle. It is expanding its industries to produce newer types of small arms, military spares, and electronic components, and has developed plans to produce naval equipment and to assemble light armor. One new major project may involve a deal worth up to $3 billion with a German consortium led by Rheinmetall G.m.b.H and Thysenn AG to create the plants necessary to produce most Saudi munitions and to provide an export capability to sell to the rest of the Arab world.[17]

There are also reports that the Kingdom signed an agreement in October, 1984, with Brazil to train Saudi workers and help Saudi Arabia use the same manufacturing techniques that have proved successful in Brazil. Reports also surfaced in early 1986 that Saudi Arabia had signed an arms for oil and technical cooperation deal which would have included the Tucano T-27 trainer, 1,000 Osoro battle tanks, Urutu and Cascaval armored vehicles, and Avibas Astros II multiple rocket launchers. These reports have been denied, but Saudi Arabia continues to show an active interest in industrial cooperation with Brazil and has short-listed the Osoro tank as a prime candidate for a new tank purchase.[18]

What is more significant is that Saudi Arabia has made major efforts since 1985 to create a GCC-wide effort at military production. It is creating training facilities for the skilled manpower it needs, and set up a General Establishment for Military Industries for the production of small arms and ammunition in 1986. This establishment is reported to have 4,000 workers and to meet nearly 50% of Saudi needs for these systems. Saudi Arabia also carried out talks with Egypt about reviving the Arab Industrial Organization shortly after the Arab League summit meeting in November, 1987.

Saudi Arabia's $5 billion Peace Shield air defense C^3I system involves a contract calling for offset arrangements affecting $3.4 billion worth of the contract. These offsets call for each element of the Boeing-led consortium to invest in economically viable projects where Saudi firms will provide matching funds. The investments are to go into Saudi manufacturing industries, preferably in high-technology areas. Projects include telecommunications, aircraft engine overhaul, and electrical equipment. Saudi Arabia is also attempting to get GCC support for the creation of a joint GCC arms industry, where Saudi Arabia would be a major producer.[19]

Britain and Saudi Arabia also seem to have renegotiated their 5 million pound Tornado package of 1985 in both 1986 and 1987, in order to revise the repayments scheme to allow Saudi Arabia to pay the entire contract price in oil over an extended period of time. Britain also seems to have agreed to up to 35% of the contract's technical element in Saudi Arabia under an offset arrangement similar to that used in the U.S. Peace Shield program.[20]

NEAR-TERM FORCE TRENDS

These factors help explain the force trends shown in Table 7.5. Saudi military modernization cannot be based on a conventional approach to military spending or to the use of military manpower. It must be based on slowly evolving the ability to operate a limited number of high-technology forces, on a high degree of dependence on other states, and on giving priority to the services and equipment which can most rapidly strengthen Saudi Arabia's deterrent and defense capabilities.

THE DEVELOPMENT OF SAUDI GROUND FORCES AND THE AIR DEFENSE CORPS

The broad thrusts in Saudi Arabia's military development, which are summarized in Table 7.5, help explain the emphasis Saudi Arabia now places on the development of its air force, its search for improved combat aircraft and dual capability in the attack role, and its emphasis on close military relations with the West. This emphasis on high

Table 7.5: Trends in the Modernization of Saudi Forces, 1987–1995

Army:

- The army will grow from about 45,000 to 53,000 men.

- The Army will fully man and equip 4 mechanized brigades, 2 armored brigades, 1 infantry brigade, and 1 paratroop brigade.

- The airborne brigade will slowly expand from a strength of 1 battalion and several companies to 2 full paratroop battalions and 1 special forces company, and become fully operational. The Royal Guard brigade will become a light mechanized unit.

- Armor will rise from 550 tanks to over 650, with Saudi Arabia concentrating on modernization rather than forced expansion. OAFVs will rise from 550 AFVs to about 700 and APCs will rise from about 930 to 1,000. Major self-propelled artillery weapons will rise from 330 to about 400.

- The army will acquire large numbers of modern anti-tank weapons and SHORADS.

- Some prepositioned unit equipment will exist for use by French and Jordanian or other external combat units and special forces.

- Improved deployment of the Improved Hawk missiles, acquisition of large numbers of Shahine mobile surface-to-air missiles (as part of a $4.5 billion deal in January, 1984) to provide oil field and ground installation protection, and integration into the Peace Shield C^3I/BM system will significantly improve the new Saudi Air Defense Corps and Saudi Arabia's ground-based air defenses. The Air Defense Corps will still have readiness and proficiency problems in the early 1990s, however, and ground-based air defense of maneuvering army units will only be poor to fair.

- Major procurements will give Saudi Arabia relatively large numbers of top quality weapons but leave it with an extremely diverse mix of equipment to support, and make it even more heavily dependent on external technical and service support.

- The Kingdom will acquire Blackhawk helicopters with an option to buy 12 more. It is studying plans to acquire a regiment-equivalent strength in combat helicopters, with emphasis on attack helicopters, in the 1990s.

- Training and readiness will improve steadily, but not keep pace with the impact of the Iran-Iraq War in improving the quality of Iranian and Iraqi forces, or the impact of the Arab-Israeli War in improving Israeli and Syrian forces. Saudi Arabia will continue to rely on U.S., French, and Pakistani (Bangladesh) assistance in training.

Table 7.5 (Continued)

- The Saudi Army must modernize its tank fleet. It will convert its 150 M-60A1s to M-60A3, and possibly adopt a West German proposal to update its AMX-30s with new engines, fire control systems, and other modifications. It is examining candidates for the purchase of a new main battle tanks, including the British Challenger, Leopard II, M-1, AMX-40, and two variants of a Brazilian tank. The M-1 and Osoro have been short-listed.

- The Saudi Army is examining coproduction arrangements with Turkey to build the tank and other German armored equipment including the Marder family of MICVs. It is also examining a major buy of French OAFVs, and French artillery both for military reasons and because of the political problems in buying arms from the U.S. This will reduce U.S. military ties to Saudi Arabia and reduce interoperability and standardization with USCENTCOM.

- The Saudi Army will probably standardize on U.S. equipment at Hafr al-Batin and build up to a two brigade level there. It will standardize on French equipment at Sharurah and Tabuk with a mix of equipment at Khamis Mushayt. It is unclear how Saudi Arabia will distribute any new purchases of German armor or its next generation of tanks, but it is most likely to deploy any new French armor at Tabuk. The same uncertainties exist regarding reports of Saudi arms deals with Brazil and/or the FRG that would lead to substantial deliveries of new and non-standardized equipment.

- By the early 1990s, Saudi Arabia will have an army capable of fighting reasonably well in prepared defensive positions, and of dealing with a relatively low-level threat like the PDRY, but lack offensive and maneuver warfare capability. The Saudi Army will, however, remain dependent on air cover and support and be unable to defend or fight without it.

National Guard and Security Forces:

- The future of the National Guard will depend upon the political influence of Crown Prince Abdullah. He and several collateral princes would like to create an independent force of at least two fully armored brigades equipped with M-1, Leopard II, or AMX-32/34 tanks, and suitable OAFVs and artillery, although the upgrading of the Guard's present V-150s with lighter weapons now seems far more likely.

- The Guard has evolved from a light infantry force to a brigade headquarters with 8 all-arms, 16 active infantry, and 24 irregular battalions. It also has one ceremonial cavalry squadron. It now has fully upgraded 4 of its battalions to light mechanized forces with 4 more in the upgrade process, but it only formally inaugurated its first full "arms" unit early in 1985. It should, however, still have 3 full light mechanized brigades by the early 1990s.

(continued)

Table 7.5 (Continued)

- The Guard is equipped with 240 V-150s, 50 M-102 105 mm artillery weapons, 81 mm mortars, TOW ATGMs, 106 mm recoilless rifles, and 30 M-40 Vulcan 20 mm AA guns. The Guard will acquire at least 489 more OAFVs, including V-300 APCs and V-150 self-propelled 20 mm AA guns, TOW-equipped vehicles, and vehicles equipped with 90 mm gun turrets. It will acquire substantial additional TOW and self-propelled TOW weapons, some heavy artillery, and improved air defense.

- The Guard will improve in readiness and proficiency, but will remain a highly politicized internal security force dependent on foreign technical and service support. It will not be able to operate effectively at a full brigade level until the early 1990s.

- The Ministry of the Interior anti-terrorist, Frontier Force, and Coast Guard will expand from 8,500 men to about 11,000. They will acquire some highly trained and equipped cadre units supported by foreign technical advisors with combat capability. Overall internal security standards will, however, depend more on political activity than on effective internal security forces.

Navy:

- The navy will increase from a token strength of about 3,500 men to 4,500 by 1995. It will, however, be at less than half its minimum required strength, and will depend on foreign technical, service, and basing support.

- Both the Red Sea (Western) and Gulf (Eastern) flotilla headquarters are already fully operational. Major active force deployments to the Western Fleet should take place over the next two years.

- Both fleets will acquire significant combat strength by local standards. Saudi Arabia has taken delivery of a total of 4 F-2000 French frigates in the late 1980s, and at least 2–4 more major ships will be delivered by 1990.

- Guided missile patrol boat strength will increase. The Kingdom will acquire 100 Improved Harpoons for both its ships and aircraft.

- The Saudi Air Force started extensive joint training with the navy in the late 1980s, using its E-3As in the maritime patrol mode. F-5E-IIs, F-15s, and the Tornado now practice air support.

- The navy would like to buy at least 2 MPAs of its own by 1990 and a total of 5 by 1995. According to some reports, it has 2 French Atlantiques on order.

- The navy will slowly build up its helicopter strength (now 4 SAR and 20 ASM Dauphine 2) and transition from AS-15TT ASMs to a more advanced variant during the early 1990s.

Table 7.5 (Continued)

- The Kingdom is seeking a more advanced mine warfare capability than its 4 MSC-322s now provide, and is studying purchase of mine warfare helicopters, but probably cannot develop a meaningful mine warfare capability against advanced mines until after the mid-1990s. It will remain heavily dependent on U.S. or French support.

- The navy will acquire Otomat shore missile defenses for the point defense of key areas by the late 1980s.

- Amphibious capability is being built up. There are 16 U.S. landing craft and vessels, and 2 logistic support ships. At least 4 more large amphibious vessels may be purchased by 1993.

- A small marine force has been created with 1,500 men and 140 BMR-600Ps.

- Overall naval readiness and proficiency will remain low to moderate well into the early 1990s. The Kingdom will become steadily more dependent on French advice and service, and on other European suppliers, and U.S. influence and role will decline. The navy will not begin to emerge as a fully effective force even by regional standards until the mid-1990s.

Air Force :

- Manpower will gradually increase from 15,000 to about 21,000 men.

- Combat aircraft strength will increase from about 226 aircraft to 260.

- First-line combat aircraft will increase from 60 F-15 C/D, optimized for the air defense mission, to 126 aircraft. This will include 42 attack and 24 air defense variants of the Tornado. By late 1989, Saudi Arabia will have 48 Tornado attack fighters, 24 Tornado air defense fighters, and 60 F-15C/Ds.

- Saudi Arabia will transition from advanced IR missiles to a full BVR capability by the late 1990s. It will transition from a reliance on Maverick and LGBs to extensive use of smart area and hard target submunitions during the early 1990s.

- All Lightnings will be phased out during 1986–1988, and Saudi Arabia may expand its second-line combat strength from 62 F-5E IIs to a total of 80 aircraft including 40 of a new type by the early 1990s. It will then slowly shift its F-5Es to a third echelon role.

- The Kingdom will phase out its F-5Bs and most of its F-5F and 38 BAC-167 trainers. It will deploy 36 new Hawk light attack trainers by 1988.

- Saudi Arabia has completed the initial deployment of its 10 RF-5E recce fighters, and will complete deployment of its 5 E-3A AWACS by the late 1980s. It may buy at least 3 additional E-3As, or other heavy recce/AC&W/MPA aircrafts to give it a two front coverage, and a more advanced recce fighter during the early 1990s.

(continued)

Table 7.5 (Continued)

- The new ground-based Peace Shield C^3I system should start to become effective by the late 1990s and greatly improve land/air coordination and the Kingdom's air battle management capabilities.

- Saudi Arabia will expand its tanker force from 8 KC-130H to include 6 KE-3A heavy tankers by the late 1980s. It will buy additional tankers in the early 1990s. It will use refueling to increase effective sortie rates and deployment capability and as a substitute for aircraft numbers.

- The Saudi transport fleet will increase from 35 C-130E/Hs, 2 VC-130Hs, 9 L-100-30HS hospital aircraft, 35 C-212s, 2 CN-235s, and 2 C-140s. It will absorb 1 B-747, 10 CN-235s, 5 C-212-200s, and 2 Jetstream 31s by the late 1980s and early 1990s. Saudi air lift will be used to redeploy Saudi ground forces and to provide a strategic lift capability for GCC use and to bring in ground forces from outside the area.

- Helicopter strength will rise from 65 (15 AB-206A, 15 AB-205, 15 AB-212, 15 KV-107, 5 Sea King) to at least 22 more AB-212 by 1990. Effective strength is likely to exceed 80 helicopters by the early 1990s, and the air force may acquire some combat helicopters by 1992.

- All of the main Saudi bases should be fully sheltered by 1990, and be able to take major reinforcements. The key combat bases will be at Dhahran, Taif, Khamis Mushayt, Riyadh, and Tabuk, although some discussion is being given to creating an advanced base at Hail rather than Tabuk. The air strips and facilities at Hafr al-Batin and Sharurah may be expanded to contingency bases with shelters and MOB support capabilities by the mid-1990s.

- The Saudi Air Force should be the most effective air force in the Gulf and Red Sea areas by the late 1990s, although it will not be able to compete with an air force like Israel's. It will remain fully dependent, however, on foreign technical and support services.

- The Saudi Air Force will seek interoperability or cooperability with the GCC air forces during the next decade, but the result will be little more than a facade. Although the SAF will provide advanced training facilities, the lack of standardization in combat aircraft and C^3I/BM capability will prevent full integration, and most other GCC air forces will buy "glitter factor" aircraft in larger numbers than they can support or make effective.

technology in air power is reinforced by the problems Saudi Arabia faces in modernizing its army, National Guard, and navy, and by the limits it faces in expanding its air strength.

The current structure of the Saudi Army is shown in Table 7.6. This table shows that Saudi Arabia is now in a state of transition towards a mix of French and U.S. Army equipment, with a total of about 550 main battle tanks. The Saudi Army must now concentrate on filling out its present paper strength of two armored, four mechanized, and one air borne brigade. It would, however, like to build up substantially by the late 1990s or early 2000s.

In theory, this expansion would mean an expansion from a current paper strength of eight brigades to eleven brigades, but the Saudi Army is now undermanned by about 30–50%, and has significant problems in retaining skilled technicians and NCOs. Even by Gulf standards, an eleven-brigade force would require a minimum of 110,000-150,000 men, and the Saudi Army will be hard pressed to build up above 60,000 before the mid-1990s.

THE PROBLEMS OF EQUIPMENT DIVERSIFICATION

The manpower problems in the Saudi Army will be compounded by its need to operate a complex mix of different equipment from the many different nations shown in Table 7.6. The end result of the diversification of its sources of army equipment has been to effectively double its life cycle costs and training and support burden. These problems have been increased by a combination of politically oriented purchases from its major oil customers and the inability to obtain a consistent supply of equipment from the U.S. because of internal U.S. domestic politics.

Much of the equipment the Saudi Army has purchased has required modification of its original technical and logistic support plan before it could be operated in large numbers, and some key items still present major servicing problems. These problems have been compounded by the need to disperse most of the army's combat forces to three distant corners of the Kingdom, by the erratic quality of contractor support, and by an over-ambitious effort to create a modern logistical system that has lacked proper Saudi and U.S. advisory management.

The Saudi Army's mix of different types of armor has been a particular cause of such problems. The U.S. M-60s have proved reasonably reliable and effective, but the crew compartment cannot be

Table 7.6: The Current Structure of the Saudi Army

Total Manning	
Active:	45,000
Conscript:	0
Reserve:	0
Major Formations	
Armored Brigades (French equipped) (3 tanks, 1 Mechanized Infantry Battalion) [a]	3
Mechanized Brigades (U.S. equipped) (1 tank, 3 Mechanized Infantry Battalions) [b]	3
Airborne Brigade (2 paratroop battalions, 3 special forces companies)	1
Royal Guard Regiment (3 battalions)	1
Artillery Battalions[b]	5
Anti-Aircraft Battalions[b]	18
Surface-to-Air Missile Batteries[b] (12 Hawk with 216 missiles, and 2 with 12 Shahine launchers and 48 missiles and AMX-30SA 30mm SP AA guns)	14

Key Equipment

- Main Battle Tanks: 550: 300 AMX-30, 100 M-60A3, 150 M-60A1 (with 150 M-60A3 conversion kits on order), 100 M-60A3.
- OAFVs/APCs: 1,510: 200 AML60/90, 350 AMX-10P (some with HOT), 800 M-113, 30 EE-11, and 130 Panhard M-3. (60 AMX-10P, and Urutu APCs on order.)
- Artillery: 505: 24 Model 56 105mm pack howitzers, 100 M-101/2 105mm howitzers, 72 FH-70 and 34 M-198 towed 155 mm howitzers, 224 M-109 and 51 GCT 155 mm self-propelled howitzers, 81 mm M-30 and 107 mm mortars. Some ASTROS II multiple rocket launchers. (8 M-198 and up to 400 JPz SK-105 105 mm anti-tank guns on order.)
- Anti-Tank Weapons: 2,000+ TOW anti-tank guided missiles, including 200 launchers on VCC-1 self-propelled vehicles, 4,292 M-47 Dragon, HOT (including some on AMX-10P AFVs). Saudi Arabia also has up to 2,000 obsolete SS-11 and 300 obsolete Vigilant)anti-tank guided missiles. It has 75, 90, and 106 mm recoilless rocket launchers; and 5,548 Improved TOW and Jpz SK-105 mm anti-tank guns on order or in the process of delivery)
- SHORADS: M-42 Vulcan 20 mm self-propelled and 15 M-117 90 mm AA guns. AMX-30SA 30 mm self propelled AA guns, 200 35 mm towed AA guns. Some 500 FIM-43 Redeye manportable and 200 FIM-92A Stinger surface-to-air missiles. (200 Stinger launchers and 600 reload missiles on order.)

Key Bases and Facilities:

• Hafr al Batin (now one brigade, building up to division size). Tabuk (building to one division, plus one brigade). Khamis Mushayt (building to one division). Sharurah (one brigade plus one brigade set of equipment prepositioned). Riyadh (Royal Guard Regiment plus some elements of Airborne Brigade).

a. Another unit is being formed. Given the actual strength of these units, they are missing sufficient elements so that they are only equivalent to 3 brigades.
b. Some elements are significantly short of full complement of Saudi troops.

Source: Adapted from IISS, JCSS, and DMS data bases by the author.

cooled effectively and the M-60 can develop internal temperatures of well over 120 degrees. Saudi Arabia's 300 French AMX-30s have presented more substantive problems. They lack the armor, firepower, and operational availability to be kept in service much past the 1980s.

The AMX-30 has relatively light armor and is not competitive with any of the newer Soviet and Western-made tanks now being deployed in the region (e.g., T-62/72/80, M-60, Khalid, Merkava, Chieftain, and Challenger). While the adoption of newer anti-armor round technology has made up for the lack of penetrating power in the Obus G rounds that France originally sold the Saudi Army, the AMX-30s fire control and range-finding capability is inadequate to help Saudi tank crews make up for their lack of experience, and the AMX-30 lacks the power, cooling, and filtration for desert combat.

This dependence on the AMX-30 compounds Saudi Arabia's modernization problems, and has created a source of tension with the United States that serves as a good illustration of the need for stable military relations between the two countries.

THE M-1 TANK ISSUE

The U.S. had a natural interest in testing its new M-1 tank and M-2 Bradley fighting vehicle in the Gulf, as well as in seeing Saudi Arabia buy heavy armor that would be directly compatible with the equipment used by USCENTCOM forces. The U.S. began preliminary discussions of an M-1 tank test and evaluation effort in Saudi Arabia in 1982. By early 1983, these discussions produced a detailed test plan, and Saudi tank and maintenance crews had begun training on the M-1 tank at Fort Knox, Kentucky. The Saudi National Guard requested the sale of 57 M-1 tanks, and the Saudi Army made it clear that it was actively interested in procuring the M-1 and in standardizing on U.S. equipment.

These Saudi-U.S. discussions began at a time when shifts in the politics of the Arab-Israeli conflict and the impact of President Reagan's 1982 peace initiative made an M-1/M-2 sale seem possible without serious Israeli and Congressional opposition. Such a sale offered both the U.S. and Saudi Arabia major advantages. The net effect of such a Saudi purchase would be (a) to lower the unit cost of the M-1 and M-2 to both the U.S. and Saudi Arabia, and (b) to allow both armies to standardize on a mix of armor that could be used by both Saudi forces and U.S. reinforcements.

Unfortunately, these discussions were leaked to the press, and in a form that stressed a goal of 1,200 tanks, which neither the U.S. nor Saudi Arabia had ever seriously discussed.[21] The senior Pentagon official who had provided the data on a background basis had failed to consider Saudi, Israeli, or Congressional sensitivities. As a result, the story turned preliminary talks into a "plan" and triggered immediate U.S. domestic political opposition. Worse, the leak was accompanied by discussion of Saudi Arabia's problems with reduced oil revenues and with paying for M-1 tanks at a reported cost of nearly $2 million each. This immediately led to debates within Saudi Arabia regarding the "waste" of Saudi funds on a highly controversial tank that was still having teething problems in the U.S. forces.

This embarrassment was compounded when the U.S. tried to issue a denial of the story.[22] By then, however, a web of false rumors surrounded the Saudi-U.S. discussions, including reports that 800–1,000 of the tanks would be prepositioned solely for the use of USCENTCOM forces. By March 26, the BBC was picking up major propaganda attacks on the sale from groups like the Soviet-backed Radio Peace and Progress, which attacked both the Saudi royal family and U.S. "imperialism".[23]

These problems grew still more serious when the first U.S. demonstration team sent to Saudi Arabia attempted to demonstrate the M-1. The team did not speak adequate Arabic, and during the resulting trials several problems occurred simply because the two sides could not communicate. While the M-1 tank did present some legitimate problems in terms of service and fuel needs, and interior heat and crew fatigue, the demonstrations and trials failed largely because of communication problems. As later chapters will show, this mishandling of U.S. and Saudi military relations is all too typical of what has happened in a U.S. domestic political climate where there is little practical understanding of the strategic importance of U.S.-Saudi military relations, and where virtually every sale is debated almost solely in terms of its theoretical impact on the security of Israel.

This mishandling also came during a period when the U.S. was actively trying to persuade Saudi Arabia to put pressure on King Hussein of Jordan to support President Reagan's peace initiative and on Syria and Lebanon. And it came at a time when the U.S. was seeking Saudi support in persuading the other Gulf states to standardize on U.S. air defense systems. Needless to say, it did little to enhance the prospects for peace or to improve Saudi-U.S. military cooperation.

The political embarrassment that occurred over the M-1 issue was all too typical of U.S. treatment of Saudi arms requests in other respects. It publicly exposed the kind of informal relations that can be of great strategic value to both the U.S. and Saudi Arabia. It made Saudi military ties to the U.S. an embarrassment in the Gulf and other Arab states; it was a source of weakness rather than of strength. It undercut relations at a time when the U.S. badly needed Saudi political and diplomatic support. Like the F-15 sale debate, the political debate in the U.S. also did nothing to protect Israel. Saudi Arabia immediately proved it could turn to Western Europe for equally effective arms. If anything, the controversy weakened a Saudi dependence on U.S. arms and technical support that acts as a powerful factor to persuade Saudi Arabia to avoid any confrontation with Israel.

In fact, Saudi Arabia was virtually forced to respond by seeking other sources of supply and it renewed previous contacts with West Germany. In late May 1986, Saudi Arabia announced that Prince Sultan would visit Germany for health reasons. In fact, he met with West German Chancellor Helmut Kohl at the latter's home in Ludwigshafen to discuss possible arms purchases. This was announced officially in Bonn on June 13, and the announcement was followed by reports that Saudi Arabia might buy up to 800 Leopard II tanks, 800 Marder armored fighting vehicles, and 200 Gepard air defense vehicles.

In practice, however, a combination of politics and funding problems has delayed any Saudi action and the FRG has never been able to decide whether it can make such sales to a state that is nominally at war with Israel. Saudi Army experts have since visited France again to examine France's new AMX-32 and AMX-40 tanks. They also have examined the Brazilian Osoro P-1 (105 mm gun), P-2 (120 mm gun), and British Challenger, and General Dynamics sent additional M-1 evaluation teams into Saudi Arabia in 1985–1987 that were far more successful than the first.

It is still unclear, therefore, where Saudi Arabia will buy its new armor from, although this is now as likely to come from Brazil as from the U.S. Saudi Arabia announced in February, 1988, that it had short-listed the M-1A1 and EE-T1 Osoro for some form of coproduction in a purchase that might involve some 315 vehicles and a $1 billion contract. One main issue was U.S. willingness to release the M-1A1 with a 120 mm gun. Another was the fact that the Brazilian Osoro existed only in prototype form and that production could not begin until 1990 at the earliest.[24]

It also is unclear whether the Saudi Army will buy the Bradley M-2 Armored Fighting Vehicles. It has shown strong interest in this vehicle because its speed, protection, and firepower allow it to outmatch the Soviet armored fighting vehicles in most potential threat armies, all of which have far better protection and firepower than the U.S. M-113 armored personnel carrier. Saudi Arabia has asked for a U.S. offer for some 220 M-2 vehicles at a cost of $500 million.[25]

OTHER MODERNIZATION ISSUES

To return to the broader problems in Saudi Army and Air Defense Corps modernization, the Saudi Army needs to implement its plans to improve its air defense, artillery, and helicopter strength as quickly as possible. Unfortunately, an initial U.S. contractor effort to improve the integration of the Saudi Air Defense Corps' Improved Hawks, Shahines (Improved Crotale), anti-aircraft guns, and land-based radars and C^3I systems has not been successful.

Even if the Saudi Army and Air Defense Corps can rapidly develop the ability to use the air defense equipment it is buying, its ability to fight in defensive positions against superior forces will depend heavily on the quality of its air cover, the ability of the Saudi Air Force to link its operations with those of the army, and its ability to provide close air and interdiction support.

It seems unlikely that the Saudi Air Defense Corps can hope to do more than properly integrate its Hawk defenses before the mid-1990s, and create a few effective mobile Shahine units.[26] Saudi forces will have to be far more dependent on air power than the strength of their land-based air defense forces indicates, and will need systems like Stinger which do not require sophisticated training or full integration into the new Saudi "Peace Shield" air defense system.

The Saudi Army also has serious problems in making its artillery properly effective. It has now acquired suitable numbers of modern types, and better mobile fire-control and ammunition supply equipment. It seems likely, however, that Saudi Army artillery capabilities will suffer from major manpower quality, and some mobility and support, problems well into the 1990s.

The Saudi Army's search for helicopter forces raises different issues. The Saudi Army is now deployed nearly 600 miles from the Kingdom's main oil facilities in the Eastern Province, and even though the combat elements of a brigade are now deploying to the new Saudi Army base at King Khalid City near Hafr al-Batin in 1984, the Saudi Army will still be dispersed so that roughly one-third of its strength is deployed at the ends of a triangle reaching to Saudi Arabia's borders with the angles located at Tabuk, Hafr al-Batin, and Sharurah-Khamis Mushayt.

Helicopters offer a limited solution to this problem. They can both provide rapid concentration of force and allow Saudi Arabia to make up for its lack of experience in large-scale maneuvers. It is far from clear, however, how Saudi Arabia can absorb or support the large numbers of attack and troop lift helicopters it needs, or the kind of advisory and technical support required. The Saudi Army is seeking to buy 12 Blackhawk helicopters, with an option to buy 12 more, from the U.S. It is also studying the long-term option of developing a helicopter force using a total of 60–100 U.S. AH-64 attack, Blackhawk utility and support, and Chinook CH-47 transport helicopters by the mid-1990s.

Saudi Arabia is also examining the purchase of attack and support helicopters from Italy, France, and a Franco-German consortium. The U.S. Army is probably the only force that could support such a purchase with the mix of conversion, training, and service capabilities the Saudis need, but such a purchase again opens up the problem of military relations with the U.S. and U.S. domestic politics. If these political barriers again block such a sale, the U.S. Army could lose precisely the kind of forward interoperable weapons and support capabilities it needs to make USCENTCOM effective, while Saudi Arabia will be forced to turn to France or the FRG for similar weapons.

Regardless of these future purchases, the Saudi Army will not be large enough to concentrate significant forces on a given front unless it can move forces from another major military city, and all the way across Saudi Arabia. This would take a minimum of a week to ten days. Even then, Saudi Arabia would lack the massive armored forces of its stronger neighbors.

Training has been a problem in the past, and will continue to be so. Many of the Saudi Army's training plans have not been executed, and maneuver training has been poor. The army's mix of U.S., French, German, Spanish, and British equipment presents major conversion problems, and the army has been much slower in providing the trained manpower necessary to absorb such equipment than the air force. Once again, this highlights the fact that the Saudi Army must depend on air force support to help make up for its own deficiencies.

THE IMPACT OF THE SAUDI NATIONAL GUARD

All of these problems are compounded by the fact that Saudi Arabia must divide its manpower between the army and the National Guard. Although the National Guard's future may depend upon the complex politics within the Saudi royal family following King Khalid's death, the Saudi National Guard seems likely to remain a lightly armed internal security force whose main mission is to ensure the loyalty of

Saudi Arabia's traditional tribes.[27] At the same time, the National Guard will use at least 10,000 men, or about 30–40% of the Kingdom's active trained military manpower, in a paramilitary force that is far more suited to internal political and security needs than to creating an effective deterrent or defense against outside threats.

The current force structure of the National Guard is shown in Table 7.7. While the National Guard's current purchases do not seem over-ambitious, and the Guard is now better trained and deployed, it cannot absorb large numbers of heavy arms. Even if it is given such arms for political reasons, the National Guard will continue to have little value as a regular combat force. In fact, the greatest single uncertainty in the Saudi military modernization process is whether the National Guard can be effectively trained and equipped to deal with terrorism and paramilitary threats, and what role the army, air force, and navy should play in aiding it in this mission.

The National Guard began to hold significant training exercises for its first 6,500 man mechanized brigade, the Imam bin Mohammed al-Saud Brigade, during the early 1980s. It has established a brigade-sized presence, and a limited oil field security force in the Eastern Province. The Mohammed al-Saud Brigade held a 10-day exercise in the desert about 250 miles west of Riyadh in early 1983. While it experienced its habitual problems in translating tribal discipline into regular military discipline, and the force was well below its authorized manning level, the maneuvers were relatively successful. Units moved from as far away as the Eastern Province, and the key mechanized elements performed relatively well.

The National Guard formally inaugurated its second mechanized brigade in a ceremony on March 14, 1985. This new unit was called the King Abdul Aziz Brigade, and was formed after another relatively successful round of set piece exercises called "Al Areen" near Bisha. Prince Abdullah then spoke of expanding the Guard to 35,000 men, and of building up to three mechanized brigades by 1989. While each of the present Saudi brigades have a strength closer to two reinforced motorized infantry battalions by Western standards, rather than the four shown in the Saudi order of battle, they have modern infantry support and anti-tank weapons.

Nevertheless, the bulk of the Guard remains a traditional tribal force. It is dominated by the 11,000–15,000 men in its *Firqa* (full-time tribal) and 25,000 men in is *Liwa* (part-time irregular tribal levy) units. Many of its "troops" are actually retired members of the military, descendants of the troops who fought with King Abd al-Aziz, or the sons or relatives of tribal leaders.

Table 7.7: The Current Structure of the Saudi National Guard

Total Manning

Active:	10,000
Conscript:	0
Reserve (part-time or reserve):	15,000
	25,000
Tribal levies	25,000

Major Formations

Mechanized "Brigades"	1-2
All Arms Battalions[a] (Are part of mechanized brigades)	8
Regular Infantry Battalions (Closer to Companies by Western definition)	16
Irregular Infantry Battalions (Largely part-time tribal forces)	24
Cavalry Squadron (Ceremonial)	1
Support units	(?)

Key Equipment:

- 240 V-150 Commando Cadillac Gage armored cars, with 206 equipped with British Racal communication systems
- 50 M-102 105 mm towed howitzers
- TOW anti-tank guided missiles, and 106 mm recoilless rifles
- 81 mm mortars
- 30 M-40 20 mm Vulcan AA guns and towed 90 mm AA guns
- 489 additional Commandos are on order, including V-300 APCs and V-150 AFVs with a mix of heavy machine gun turrets, TOW, 90 mm guns, and 20 mm AA guns.

Key Bases and Facilities: Eastern Province (now one brigade). Riyadh (gradually forming brigade). Other tribal areas: Battalion-sized elements are scattered throughout the country, and serve as a major source of subsidies to the traditional tribal leaders.

Additional Internal Security Forces:

- Ministry of the Interior: A counterterrorist unit equipped with helicopters, some guided missiles, and non-lethal gas. A Frontier Force and Coast Guard with 8,500 men. The Coast Guard has a large inventory of small craft with several hundred small ships, 16 SRN-6 Hovercraft, and 8 BH-7 Hovercraft. Readiness, training, and combat effectiveness is generally low. There is a General Civil Defense Administration unit with 10 Kawasaki helicopters.
- European: French: A small battalion-sized equivalent of advisors. Pakistani: Some. British: Ex-SAS provide security outside Saudi Arabia for the royal family and some other key officials.

[a]Up to 8 units are shown in order of battle. Effective strength seems to be 4.

Source: Adapted by the author from IISS, DMS, and JCSS data bases.

The Guard's manpower still serves to the age of 60–65, much of it is directly recruited and paid by tribal or regional chiefs, and many positions have a quasi-hereditary status. It is heavily recruited from the Otaiba and Mutair tribes (of the areas between Makkah and Riyadh and the northeast). It has only gradually recruited new and educated personnel, who are loyal to a service rather than to a given leader or subleader.

The National Guard is also more a means through which the royal family allocates funds to tribal and Bedouin leaders than a modern combat or internal security force. The Guard helps key princes maintain close relations with the tribes in each region. It has not evolved into a force that can deal with urban disorder, oil field security problems, or border security problems, although it can do a good job of dealing with ethnic and tribal divisions.

This makes the Guard politically vital to ensuring the integration of Saudi Arabia's tribes into its society, but it does not mean the Guard has found a clear military mission. As Table 7.6 shows, the Guard's current force structure and equipment also fail to provide air mobility and the specialized units necessary to deal with urban warfare and terrorist activities. Such specialized forces might come from the army and air force, but there seem to be no clear plans for this. If anything, creating new internal security forces under the Ministry of Interior means that there is yet another force competing for a role and for manpower—particularly since this force is under Prince Naif, who is King Fahd's full brother and a member of a competing branch of the royal family.

This lack of a clear thrust behind the Guard's modernization also means that Saudi Arabia is not doing an adequate job of preparing for the low-level military threats that may be more dangerous on a day-to-day basis than the major military threats building up on its borders. French and other external aid can help in the interim, as can the small security units being built up under the Ministry of Interior, but the Guard does more to weaken the army's manpower pool than provide an added source of military capability. This again increases the importance of the air force in providing the reach, reaction capability, and firepower missing in Saudi ground forces.

THE SAUDI NAVY

The Saudi Navy has limited capability to absorb its modern equipment. It has made significant progress in recent years, but it faces a decade of expansion before it can become a true "two sea" force capable of covering both Saudi Arabia's Gulf and Red Sea coasts. Even then, it will depend heavily on air support, and ultimately on reinforcement by USCENTCOM and the British, French, and U.S. navies.

The current size of the Saudi Navy is shown in Table 7.8. It is now completing the construction of two major, fully modern naval bases at Jiddah and Jubail. When it deploys fully to the Red Sea, it will be divided into a Western Fleet with its main facilities at Jiddah and an Eastern Fleet with its main facilities at Al Qatif/Jubail. The navy will also have facilities at Ras Tanura, Dammam, Yanbu, and Ras al-Mishab.[28]

By the end of 1987, the Saudi Navy had taken delivery on all of the major frigates and support craft it had ordered earlier in the 1980s. It had a 34-ship force, plus 24 missile-equipped helicopters. It was also seeking to expand its manpower from 3,500 to 4,500, was examining the possible purchase of mine-hunting helicopters, and had begun to conduct low-level joint exercises with the U.S., French, and Omani fleets in the Indian Ocean.

The major deliveries under the U.S. phase of the Saudi naval expansion effort had been completed for several years. The U.S. had delivered nine patrol gunboat/missile (PGG) craft, and four patrol chaser missile (PCG) craft. It had also delivered four coastal minesweepers from the U.S., two large harbor tugs, two utility landing craft, and eight LCM-6 mechanized landing craft. Other deliveries included Harpoon missiles, Mark 46 torpedos, and ammunition for the Saudi Navy's 76 mm guns and other weapons.

These deliveries were all part of a U.S.-managed Saudi Naval Expansion Plan (SNEP) that experienced severe political and quality problems and led the Saudis to turn to France for naval aid and equipment. Nevertheless, the U.S. Navy Expansion Program team in Saudi Arabia still has about 25 military and civilian employees. Saudi Arabia has also regularly renewed its contracts for U.S. naval support.

The four minesweepers the U.S. has delivered have also had an unanticipated importance. They began to see active service in 1987, when Iran expanded its attacks on tankers moving through the Gulf. Saudi Arabia has the only GCC Navy with specialized mine warfare vessels, and its success in using the U.S. vessels has led it to consider a $800 million program to buy eight more mine vessels. It has invited bids from a Franco-Belgian-Dutch consortium, from France, and from the U.K. The key candidates are the Alkmaar-class minehunter, Lerici-class minehunter, and Vosper Thorneycroft Sandown. The first of these vessels could enter service as early as 1989.

Saudi Arabia has turned to France as the major source of its naval ships and weapons. The Saudi Navy signed its first major contract with

Table 7.8: The Current Structure of the Saudi Navy

Total Manning

Active:	3,500
Conscript:	0
Reserve:	0

Major Formations

Western Flotilla
HQ Jiddah, base at Al Wajh
Eastern Flotilla
HQ Jubail, bases at Al Qatif, Ras Tanura, Al Bammam, Ras al Mishab

Major Ships and Equipment

- 4 X F-2000 French-made frigates with 8 Otomat surface-to-surface missile launchers, Crotale launch unit and 26 surface-to-air missiles, one 100 mm surface gun and 4 40 mm twin AA guns, 6 torpedo tubes (4 X F-17p and 2 X Mk 46), Sea Tiger fire control system, and 1 SA-365 missile equipped helicopters. Some are not fully operational.
- 4 X 732 ton Tacoma-class guided missile corvettes with two Harpoon RGM-84 launchers with 4 missiles each, 1 76 mm L62 Oto Melara gun, 1 20 mm Phalanx, 2 40 mm grenade launchers and 2X3 Mark 32 torpedo launchers.
- 9 X 384 ton Peterson guided missile patrol boats with 2 Harpoon RGM-84 missile launchers with 2 missiles each, 1 76 mm Oto Melera L-62 gun, 1 20mm Phalanx, 2 20 mm guns, and 6 Mark 32 torpedo tubes.
- 1 X 100 ton large ex-U.S. coast guard cutter with a 40 mm gun.
- 3 X 160 ton Lurssen Jaguar patrol boats with 2 X 40 mm guns and 4 X 21"/53 mm torpedo tubes.
- 4 X 320 ton MSC-322 coastal minesweepers.
- 45 light patrol craft: 8 X 90 tons, 12 X 26 tons, and 15 X 25 tons, many now with Coast Guard.
- 16 landing craft: 3 LCU, 8 U.S. LCM-6, and 4 LCVP.
- 24 AS-365N (Dauphine II) helicopters: 4 for the sea air rescue mission and 20 equipped with AS-15TT air-to-surface missiles.
- 2 Mod Durance 10,500 ton logistic support ships with 4 X 40mm guns and 2 Dauphine helicopters each.

Major Orders: 200 AS 15TT missiles, 100 Harpoon ship-to-ship missiles, and possibly two Atlantic maritime patrol aircraft.

Key Bases and Facilities: Western Flotilla is headquartered at Jubail. Eastern Flotilla is headquartered at Al Qatif/Jubail. Main command facility is at Riyadh. Other bases include Ras Tanura, Dammam, Yanbu, and Ras al Mishab.

Source: Adapted by the author from IISS, DMS, and JCSS data bases

France in 1980, in an effort to accelerate its modernization, obtain better support, and obtain more advanced ships than it could get from the U.S. It signed a modernization package costing $3.4 billion, and then signed another contract that effectively made the French the primary source of support and modernization for future Saudi orders. This follow-on French program, which began in 1982, is called Sawari (Mast) I. It has reached a minimum value of 14 billion French francs, or $1.9 billion, and may have escalated in cost to $3.2 billion.

France delivered the last of four missile-equipped 2,000-ton frigates in August, 1986. It has also delivered 2 modified Durance-class fuel supply/replenishment vessels, Otomat missiles for the frigates, 24 SA-365F Dauphin 2 helicopters (20 missile-equipped and 4 SAR-equipped), AS-15 missiles for the helicopters, and additional training services. The Otomat is the longest range anti-ship missile in Gulf service, with a range of 160 kilometers. Saudi crews trained in France to operate the vessels and helicopters.

Four U.S.-supplied PCGs constructed by Tacoma Boatbuilding are in active service with the Saudi Navy. These ships displace 732 tons and are armed with twin RGM-84A Harpoon launchers with four missiles, plus the Mark 309 anti-submarine warfare system, two triple Mark 32 324 mm torpedo launchers, a 76 mm gun, one 20 mm Phalanx air defense gun, and two 40 mm grenade launchers.

The Saudi Navy's nine Al Siddiq-class PCGs/Corvettes were built by Peterson Builders, and were operational at Jubail by mid-1984. These PCGs have computerized fire-control with twin Harpoon surface-to-surface missiles. They have both gas turbine and diesel engines and can go from 16 to 30 knots. They displace 385 tons and are 190 feet long. Each is also armed with Oto Melera 76 mm guns, Mark 67 20 mm cannon, Mark 19 40 mm grenade launchers, Mark 2 81 mm mortars, and the Vulcan/Phalanx close-in defense system. They have Sperry Mark 92 fire-control systems and full communications, radar, navigation, and IFF equipment.

The Saudi Navy also has three Jaguar-class fast torpedo boats in service. There are roughly 70 other patrol vessels in service, many of French or West German design. The bulk of these are in service with the coast guard or other security forces.

Saudi Arabia has been happier with the support it has obtained from the French Navy than with the support it has received from the French Army. This is reflected in its plans for the new Sawari II program, which could cost an additional $1.6–2.12 billion. Prince Sultan first met with France's President Francois Mitterrand and Defense Minister Charles Hernu to discuss this program in May, 1983. Only Saudi Arabia's reduced oil revenues seem to have prevented agreement on a new program called Sawari II.

The program would provide at least two more 2,000-ton frigates and possibly 4,000-ton frigates as well. It also may include mine–sweeping helicopters and maritime patrol aircraft as the first step in the procurement of a much larger force. Other equipment may include lift and troop-carrying helicopters, surveillance and intelligence equipment, and special warfare equipment.

Saudi Arabia's plans to expand its naval forces now center on its mine warfare units. Saudi Arabia feels this program is so important that it may defer plans to buy coastal submarines. Nevertheless, Saudi Arabia has sought to buy six to eight submarines, and has discussed program coasts of up to $1.5–$3 billion. Saudi Navy representatives visited several European manufacturers in 1986 and 1987—including the builders of the Walrus-class boats in the Netherlands, Vickers Type-2400 in the U.K., and ILK 209/2000 and Kockums 471 in West Germany.

Saudi Arabia may already have ordered two AMD-BA Atlantique 2 (ANG) maritime patrol aircraft, and is discussing the order of two more Atlantique 2, Fokker F-27 Maritime Enforcers, or Lockheed P-3 Orions as part of a GCC maritime surveillance force.[29]

Saudi naval facilities are also excellent. The Saudi Navy's bases are exceptionally capable and well stocked. The main bases will eventually have up to five years of stocks on hand, and will have initial deliveries of two years' worth of inventory. The Jubail base is now the second largest naval base in the Gulf and stretches nearly eight miles along the coast. It already has its own desalinization facility, and is designed to be expandable up to 100% above its present capacity.

The Saudi Navy is procuring an automated logistic system similar to that in the other services, and with extensive modern command and control facilities. It had made this system operational, and had hardened command centers at Riyadh, Jubail, and Jiddah, by the end of 1985. It is acquiring automated data links to the E-3A and the ability to obtain data from the E-3A AWACS as it operates in the ocean surveillance mode. Other U.S.-designed facilities include a meteorology laboratory, a Harpoon missile and Mark 46 torpedo maintenance facility, an advanced technical training school, and a Royal Naval Academy.

Given these trends, the Saudi Navy will be a very powerful force in terms of equipment. It will create a two-fleet force with ocean surveillance, coastal defense, anti-air, anti-surface, and anti-submarine capabilities, and some of the most modern equipment in the world.

The Saudi Navy, however, is unlikely to meet its goal of 4,500 men by the mid-1990s. Further, its equipment mix requires a force of at least 8,000 men, and probably closer to 15,000. It is already having problems operating its new French frigates, although it has gradually become fairly effective in operating its U.S.-supplied vessels. Even with automation and foreign support, the Saudi Navy will not be able to operate much of its equipment effectively before the mid-1990s.

The Saudi Navy will certainly be in for serious "indigestion" problems during 1989–1995. It should be able to use some of its major combat ships effectively, and counterbalance the limited surface capabilities of regional powers like Iran, Iraq, South Yemen, and Ethiopia—all of which have severe naval readiness and modernization problems of their own. At the same time, the Saudi Navy will not be able to absorb what it already has on hand or in delivery, and new orders will simply increase the overload.

The role of advisor to the Saudi Navy may also become more difficult. It is unclear that the French Navy is fully ready for this challenge. It now has to make a transition to the role of training and supporting a major naval force that uses equipment that is not standard in the French Navy. This is a far more difficult task than selling ships and training crew cadres. As for the U.S. Navy, its advisory role will continue at a low level. The Saudi Navy added $31 million more worth of technical services contracts with the U.S. to its previous total of $49 million on March 1,1983. The U.S. once informally proposed the sale of FFG-7 class frigates, but it is more likely that the U.S. Navy will be limited to providing support and training for its past deliveries.

This indicates that both the Saudi Navy and its principal advisors are in for a troubled decade. Certainly, the Saudi Navy's mission capability will be heavily dependent on the C^3I/BM data it receives from the E-3A AWACS, and on the degree of air cover and attack support it can get from the air force. This will be particularly true of any missions in support of other GCC states. The navy is making progress, but not enough to operate as an independent force. It will be heavily dependent on support from the Saudi Air Force in low- to medium-level contingencies and will require support from the U.S. Navy in any major conflict.

THE SAUDI AIR FORCE

All of these trends in defense spending, manpower, and the development of Saudi Arabia's other services help explain why Saudi Arabia has put so much effort into the expansion of the Saudi Air Force. The air force offers the fastest increase in deterrent capability per dollar and unit of skilled manpower. It is the only service that can cover Saudi Arabia's 2.3 million square kilometers of territory. It represents the investment most capable of cross-reinforcement of the other services. It also has the most impact in terms of regional prestige, and the most credibility in terms of being able to support other GCC states or to operate with USCENTCOM forces in a major crisis.

It is not surprising, therefore, that Saudi Arabia has achieved its greatest success in this aspect of its military modernization program. The Saudi Air Force's current strength is shown in Table 7.9, and it is backed by excellent foreign support. Saudi Arabia has been able to draw on U.S. Air Force and contractor support to create some of the most modern air facilities in the world. No U.S. or NATO base has sheltering or hardening equal to the Saudi bases at Dhahran and Khamis Mushayt, and similar facilities will be built at all of Saudi Arabia's main operating bases.

As has been mentioned earlier, Saudi Arabia now performs most of the support and service for its Lockheed C-130s, and its F-5E/F units have also reached proficiency levels approaching those of many Western squadrons. Saudi Arabia has so far been remarkably successful in converting to its new F-15C/Ds. In fact, the Saudi Air Force's transition to the F-5E and F-15 has been smooth enough to indicate that the Saudis can absorb its new Tornados and effectively operate a modern air force of 250 combat aircraft by the early 1990s.

While the role of the Saudi Air Force will be discussed in depth in Chapter 8, it is important to note that Saudi Arabia has done a good job of operating today's most advanced fighters. The first of its 60 F-15C/Ds were operational in Dhahran by early 1984. A second squadron was formed at Taif by the end of 1983, and a third became operational at Khamis Mushayt in July 1983. By late 1984 and early 1985, the Saudi Air Force was conducting major joint exercises in both the Gulf and Red Sea areas, and conducting Red-Blue or aggressor exercises similar to those employed by the U.S. Air Force. Saudi aircraft attrition levels are significantly higher than those of the U.S., but overall training levels are good.

While Saudi Arabia lacks the C^3I/BM systems, advanced avionics and electronics, munitions, and attack capabilities to match USAF proficiency levels, it has also demonstrated a high level of squadron readiness, has begun to perform much of its own major support on the F-5, and provides Saudi support of the F-15 at its bases in Dhahran and Khamis Mushayt.

This experience is important because the Saudi Air Force badly needs to replace many of its existing aircraft. Its 38 BAC-167 trainers are armed only with 7.62 mm machine guns. They no longer can be used in any combat function and soon will be too old to use as trainers. The Saudi Air Force bought its now obsolete Lightning fighters from the U.K. under

Table 7.9: The Current Structure of the Saudi Air Force

Total Manning	
Active:	15,000
Conscript:	0
Reserve:	0
Major Formations (Squadrons/Aircraft)	
• Total Combat Aircraft	222
• Fighter Ground Attack: 20 Tornado GR.1, 3/67 F-5E	87
• Air Defense: F-15 C/D[b]	56–59
• Reconnaissance: 10 RF-5E (2 more kept in U.S. as attrition reserve).	10
• AWACS: E-3A	5
• OCU: 13 F-5B[a]	13
• Training: 2/37 BAC-167, 12 PC-9, 8 Hawk, 16 Cessna	73
• Helicopter: 2 squadrons with mix of 15 AB-206, 15 AB-205, and 15 AB-212, 15 KV-107, 5 Sea King	65
• Transport: 3 squadrons with mix of 38 C-130E/H, 7 KC-130H, 2 VC-130 H, 9 L-100-30HS (hospital), 9-10 KE-3A (B-707-320C), 2 CN-235, 35 C-212, 2 KE-3A, and 2 C-140 Jetstar.	105
Air Defense Command:	
New element of RSAF to control missile, gun, and radar elements.	
• Vulcan M-163 20 mm AA guns	100
• AMX-30SA 30 mm AA guns	
• 35 mm AA guns	180
• Shahine surface-to-air missiles (2 Shahine batteries with 12 fire units and 148 missiles.)	60
• MIM-23B Improved Hawk surface-to-air missiles (12 MIM-23B Improved Hawk Batteries with 216 missiles).	70

Major Orders:

24 Tornado ADV air defense fighters; 48 Tornado IDS/GR.1 ground attack fighters; 18 Pilatus PC-9 trainers; 22 BAe Hawk trainers; 101 shipsets of F-15 conformal fuel tanks; 909 AIM-7F and AIM-9P/L; 100 Harpoon ASM; and 1,600 Maverick missiles, JP-233 and BL-755 bombs and munitions.

Key Weapons and Munitions:

- AIM-7F, 1,041 AIM-9L and 2,552 AIM-9P4 air-to-air missiles.
- 2,566 Maverick AGM-65 air-to-surface missiles; CBU laser guided bombs.
- British JP-233 airfield attack munitions and BL-755 cluster bombs.
- AN-TPS 43, 43E, and 63 radars; Ferranti surveillance radars.
- Chukar II and two other types of target drone aircraft.

Key Bases and Facilities:

Headquarters at Riyadh, plus transportation, training, and E-3A/K-7 base. Major operating bases at Dhahran, Taif, Khamis Mushayt, and Tabuk. Dispersal bases at Sharurah and Hafr al-Batin. Support base at Jiddah. Normal deployments are 22 F-15C/D, 8 Hawk, 20 Tornado, and 5 AB-212 at Dhahran; 25 F-5E/F 17 F-15C/D and 3 AB-212 at Khamis Mushayt; 42 F-5E/F, 20 F-15 C/D, 10 RF-5E, 20 AB-206, and 3 SH-3D at Taif; 22 BAC 167, 26 C-130, 16 Cessna 172, 5 E-3A, 8 KE-3, 2 G-III, 2 Lear Jet, 5 C-=130 medical, 3 SH-3D, 4 Casa CN-235, 12 PC-9, and 3 AB-212 at Riyadh; 27 C-130 and 2 AB-212s at Jiddah, and 23 F-5E/F and 3 AB-212 at Tabuk.

a. effective strength may be less than 13.
b. 2 more in U.S. as attrition reserve.
Source: Adapted by the author from IISS, DMS, and JCSS data bases.

pressure from former Secretary of Defense Robert S. McNamara—as part of a then covert three-cornered deal designed to allow the U.K. to buy the F-111.[30] The Lightning never had the range, dual capability, avionics, and performance Saudi Arabia needs. The remaining 23 Lightnings are now at the end of their useful life, and have been phased out of the force structure.

The Saudi F-5E-IIs and F-5Fs are advanced models equipped with INS, refueling probes, and the ability to fire Maverick (the F-5F can also fire laser guided bombs). They have proved to be excellent fighter aircraft. The oldest aircraft, however, are now thirteen years old and nearing the end of their useful life, and the F-5 production line is closed. The F-5s also are too short-ranged and too limited in avionics and payload to cope with the kind of advanced threat aircraft being introduced into the region, or to deploy from one air base in support of another. They will have to be phased gradually into a training and light support role, and 20–30% of Saudi Arabia's F-5 strength is already devoted to full-time training missions.

It is also important to note that Saudi Arabia badly needs the attack capabilities of its new Tornados. The Saudi F-15C/Ds do an excellent job of meeting Saudi air defense requirements—particularly since Saudi Arabia obtained the conformal fuel tanks necessary to extend their range, tankers for refueling, and advanced air-to-air missiles as part of the U.S. Air Defense Enhancement Package sold to Saudi Arabia in 1982. The Saudi F-15C/Ds, however, are currently virtually one-mission aircraft.

The Saudi Air Force cannot use the F-15 effectively in the air-to-ground role, although the F-15C/D has the potential ability to carry up to 10,960 lbs of air-to-ground ordnance without reducing its ability to carry ECM gear and its normal load of Sidewinder air-to-air missiles if it is equipped with modern attack munitions racks and dispensers.[31] While Saudi Arabia has modified the bomb racks for its F-15s so that they can carry a five-bomb rack under each wing pylon, they constitute a very high drag system and force the F-15 to fly at subsonic speeds until it releases its bombs.

The Saudi F-15C/Ds also lack the radar and bomb delivery computers and software necessary to use the F-15 in low-altitude penetrations and to continuously calculate weapon release and impact point. This means they cannot use the automatic air-to-ground, electro-optically (EO) guided missile, or continuously displayed impact point (CDIP) modes on their heads-up displays, or fly the kind of attack sortie that minimize exposure to SAMs and enemy fighters and have relatively poor accuracy.

The Saudi Air Force does not have the MER-200 racks necessary to carry air-to-ground ordnance efficiently under the wings of the F-15s (each MER can carry six 500 lb bombs and one MER can be carried under each wing and another under the body in addition to the conformal fuel tanks), nor does it have the tangential racks that allow each of the two conformal fuel tanks under the body of the F-15 to carry 4,400 lbs of air-to-ground ordnance even when the F-15 is maneuvering at 5.5 G. [32]

Although the U.S. Air Force recommended that the Saudi Air Force be given a dual capable advanced fighter back in 1977, when it conducted the original studies leading to the U.S. sale of the F-15, U.S. domestic politics have precluded any sale of the bomb racks and the attack systems necessary to make the F-15C/D effective in this role. This means that approximately half of Saudi Arabia's total first-line fighter strength has been unable to perform effective attack missions, or provide attack support to Saudi land and naval forces. This problem is also compounded by the fact that Saudi Arabia will need most of the advanced avionics being developed for the U.S. Air Force F-15C/D MSIP program to allow it to provide optimal air cover at long ranges.

The success of Saudi Air Force modernization has, therefore, depended on the Saudi Air Force and on its eventually acquiring either a modern dual capable fighter or full dual capability for its F-15s. It is this requirement which has triggered Saudi Arabia's original arms request to the U.S., and which explains why it turned so quickly to Britain for Tornado attack fighters when the U.S. rejected the sale.

The Tornado sale will give the Saudi Air Force the aircraft it needs. The Saudi Air Force is, however, near manpower saturation. It has had to delay taking over operation of its E-3As because of shortages in trained crews, and is having problems in finding all the combat pilots necessary to keep its existing aircraft flying and to convert to its new Tornado fighters.

Saudi Arabia will also face deployment problems in replacing the Lightning, because it has informally stated that it does not plan to base its F-15s and Tornados in Tabuk—the only active air base near Israel. This means the Saudi Air Force must rely on its F-5Es, or on some successor fighter, once the Lightnings phase out at one of its most important main combat bases. While Tabuk is the one active Saudi base near Israel, it is also a key base in defending Saudi air space near the upper Red Sea and Syria.

NOTES

1. Estimate based on CIA; *World Factbook, 1986*, pp. 215–217.
2. *Jane's Defense Weekly*, January 16, 1988, p. 59.

3. The FY1988 budget has a $10 billion deficit, and involves $8 billion in foreign borrowing. It involves the first foreign borrowing in 25 years and the first increase in taxes in eight years—all on foreign businesses. *Economist*, January 16, 1988, p. 59; *Defense News*, January 18, 1988, p. 4.

4. *Wall Street Journal*, December 31, 1987, p. 4; January 5, 1988, p. 21, January 6, 1988, p. 12, January 7, 1988, p. 16, January 12, 1988, p. 2; *Washington Post*, December 31, 1987, p. E-3, January 12, 1988, p. C-3; *Economist*, Janury 16–22, 1988, p. 59; *New York Times*, January 6, 1988, p. A-1; *Chicago Tribune*, January 27, 1988, pp. 3–7.

5. These data are based on excerpts from the new Saudi national budget, and on reporting by the IISS. CIA, ACDA, SIPRI, and other estimates often differ significantly. The reader should also be aware that many major arms transactions in all Middle East countries are handled privately by their defense ministers, often on a multi-year basis. Many of these transactions are not reported to or through the central bank. Saudi Arabia, like other oil exporting states, complicates this situation further by using oil barter arrangements and offset arrangements, and by constantly renegotiating major arms deals while deliveries are in progress. The data published in IISS, *Military Balance, 1987–1988*, indicate that the total manpower fit for service could be about 50–70% of the CIA estimate.

6. *Defense News*, January 18, 1988, p. 4.

7. Unless otherwise specified, the military data quoted here are taken from the relevant country sections of IISS, *Military Balance, 1987–1988*; CIA, *World Factbook, 1986*; and Zeev Eytan, *The Middle East Military Balance, 1986*, Jaffe Center for Strategic Studies, Tel Aviv University, Tel Aviv, 1987.

8. As is discussed later, these Pakistani forces may leave the Kingdom in 1988.

9. Estimates of active manning in the National Guard differ sharply. The most recent IISS estimate is 10,000 full-time actives, 15,000 semi-active reserves, and 25,000 tribal levies.

10. These figures are based on estimates taken from CIA, *World Factbook, 1986*, pp. 199–200.

11. *Jane's Defense Weekly*, May 15, 1985; *New York Times*, April 28, 1985.

12. These figures do not include the 500-plus men in the ELF-1 AWACS detachment.

13. C3I/BM means *c*ommand, *c*ontrol, *c*ommunication, and *i*ntelligence/*b*attle *m*anagement system.

14. This unit will be sent to one of the less defended positions on the Indian border. Ironically, it has no equipment of its own. The PRC, however, has agreed to provide it, including some 100 T-59 tanks.

15. These data are based on interviews with Pakistani and Saudi officials. See also *Economist*, December 12–18, 1987, pp. 48–49; *Jane's Defense Weekly*, January 9, 1988, p. 13. Pakistan is reported to be about 15% Shi'ite and about 10% of the Pakistani troops in Saudi Arabia were Shi'ite. The contract was worth about $300 million annually to Pakistan.

16. According to one report, the Egyptian government would provide a 10,000-man contingent by allowing the officers and men to take early retirement from

the Egyptian Army. Saudi Arabia would pay Egypt some $1.8 billion annually in aid. *Jane's Defense Weekly*, January 9, 1988, p. 13.

17. *Wall Street Journal*, October 8, 1985, p. 35.

18. *Los Angeles Times*, March 7, 1986, pp. 2–14.

19. *Economist*, January 21, 1984, p. 34.

20. *Jane's Defense Weekly*, June, 14 1986, p. 1075.

21. These discussions were published in the February 1983 issue of the *International Defense Review*. They received broader exposure in the *New York Times* on March 4, 1983, in an article by Richard Halloran.

22. On March, 15 the Associated Press carried both a report on the sale and a categorical denial by Henry Catto, the Assistant Secretary of Defense (Public Affairs). When Halloran repeated the story with more details on April 4, the Pentagon issued another denial, this time on a nonattributable basis.

23. Ironically, the Congress was notified of an actual Saudi tank purchase on August 2, 1983. The purchase, however, was of 100 M-60A3 tanks and not 1,200 M-1s. The M-60A3 sale had long been in train, and had no significant impact on the regional balance in the Gulf, much less on the Arab-Israeli conflict.

24. *Jane's Defense Weekly*, February 6, 1988, p. 191.

25. *Defense News*, February 22, 1988, p. 3.

26. The Saudi Air Defense Corps renewed its contract for technical assistance support from Raytheon for its Hawk surface-to-air missiles in May, 1986. This contract has been running since 1976, and was renewed for three years at a cost of $518 million. *Jane's Defense Weekly*, June 7, 1986, p. 1019.

27. For an interesting Israeli view of the role of the National Guard, see Mordechai Abir, "Saudi Security and Military Endeavor", *The Jerusalem Quarterly*, No. 33, Fall 1984, pp. 79–94.

28. The sources for the analysis of the Saudi Navy also include James Bruce and Paul Bear, "Latest Arab Force Levels Operating in the Gulf", *Jane's Defense Weekly*, December 12, 1987, pp. 1360–1361; and Michael Vlahos, "Middle Eastern, North African, and South Asian Navies", *Proceedings*, March, 1987, pp. 54–55.

29. *Jane's Defense Weekly*, December 12, 1987, pp. 1360–1361.

30. See the author's *The Gulf and the Search for Strategic Stability*, pp. 122–126.

31. The aircraft cannot carry its normal load of Sparrows or AIM-7s. The missiles are too heavy. R.D.M. Furlong, "Operational Aspects of the F-15 Eagle, *International Defense Review*, Vol. 3, 1975, pp. 129–139.

32. Ibid,.;and *Aviation Week and Space Technology*, April 26, 1982, p. 27.

8

The Regional Role of the Saudi Air Force

It is clear from the preceding two chapters that the military forces of the Southern Gulf states are not sufficient to provide a regional deterrent to meet more than low-level threats. The Saudi Air Force does, however, have the potential to play a major regional role. In fact, it is important for several reasons:

- It is capable of performing medium-intensity combat missions on a regional basis and is perhaps the only Southern Gulf military force capable of doing so.
- Air power is not a fully adequate substitute for land power or sea power. The Saudi Air Force can, however, take advantage of geography, its ability to win local air superiority, its tank killing capability, and its combination of maritime surveillance and ship killing capability to act as a partial substitute for Saudi Arabia's lack of land and naval power.
- The two kinds of "over-the-horizon" reinforcements that the West can deliver most quickly, and with the least political problems for both Western and Southern Gulf states, are sea power and air power. Only the U.S., however, has large carrier strike forces and these cannot operate in the Gulf. Saudi air bases and facilities provide outstanding basing capability with excellent C^3I facilities and strategic depth.
- In any major regional contingency, the pro-Western side of the strategic equation will ultimately be determined by two states: the U.S. and Saudi Arabia. The cooperation of other Western and Gulf states will be vital; they can only help shape the outcome.

As this and the following chapters will show, U.S. ability to maintain close strategic relations with Saudi Arabia and the other Southern Gulf states has been heavily influenced by U.S. ability to sell the Southern Gulf states weapons and military equipment. This has been particularly true in the case of the Saudi Air Force, and such sales provide an important case study of the strengths and weaknesses of U.S. policy in the region.

THE IMPORTANCE OF THE SAUDI AIR FORCE TO SAUDI ARABIA

Saudi Arabia can afford delays in the modernization of its other services, but it cannot afford delays in the modernization of its air force. The Kingdom is forced to make air power the pivotal element of its efforts to create a national and regional deterrent. The Saudi Air Force is the only way Saudi Arabia can create a defense capability that can deter its larger neighbors, and make up for its limited supplies of skilled manpower.

At the same time, the Saudi Air Force must rely on small numbers of high-performance aircraft. Saudi Arabia can at best support a first-line combat force of about 200–220 aircraft. It has no hope of operating an air force with the air strength of its larger neighbors. Iraq, for example, has over 530 combat aircraft rather than any effort to match threat air strength. Syria has 650, and Iran had over 400 fighters—plus 150 combat helicopters—before the Shah's fall.

Saudi Arabia must use its air force to defend a large territory with widely scattered population centers and oil facilities. While the Kingdom is most vulnerable on its Gulf coast, it faces threats on all its borders. The air force is also the only Saudi military force that can rapidly deploy in support of its smaller neighbors, particularly the oil-rich states of Kuwait and the UAE.

Saudi access to first-line force structure aircraft, sensors, C^3I assets, and air munitions is essential if the Kingdom is to create a convincing deterrent to attacks on its own territory and the prestige and capability to build the military unity of the Gulf Cooperation Council. In fact, the real issue shaping Saudi Arabia's military modernization is not whether it needs advanced fighter aircraft, but whether it can obtain the right mix of fighter aircraft and foreign support to serve its own interests, those of regional stability, and indirectly those of the West.

The answer to these question is dependent on the extent to which such Saudi request are a proper extension of its new C^3I/BM systems, of its existing force structure, and of its need to combine improved air defense coverage with an improved attack capability to compensate for Saudi and GCC weakness in land forces.

SAUDI ARABIA'S NEED FOR AIR DEFENSE

In order to make this evaluation, it is necessary to understand that the current thrust of Saudi Air Force modernization is the product of joint Saudi and Western planning, which began shortly after the fall of the Shah of Iran, and which resulted in detailed force plans during the last days of the Carter Administration.

The Iran-Iraq War was the major catalyst affecting this joint planning process. The need to give Saudi Arabia effective air power took on new meaning when Iran retaliated against Iraq's invasion by launching its first major strikes against Iraqi oil facilities in December, 1980. These Iranian air and naval attacks demonstrated that a few sea and air strikes could destroy all of Iraq's oil-loading facilities in the Gulf and halt up to 3.2 MMBD of oil exports to the West. They transformed the climate of U.S. and Saudi planning from planning for worst case scenarios to one of dealing with immediate threats.

The Saudi Air Force and U.S. Air Force had already begun a joint feasibility study to examine the options for modernizing the Saudi fighter force and air defense system as the result of Saudi Arabia's desire for a new first-line fighter and the fall of the Shah of Iran. The U.S. agreed to provide Saudi Arabia with a detailed plan for the modernization of its air force in April, 1980, although the study was given relatively low priority because of the fear of domestic political problems with the supporters of Israel in an election year. The study was initiated in September, 1980, in the midst of President Carter's campaign for re-election, and was completed in December 1980—after the Iran-Iraq War had begun, and Ronald Reagan had defeated President Carter.

The U.S. study was not formally released to the Saudis until March, 1981, in order to give the new Reagan Administration time to take a policy position on its recommendations. By this time, however, the Saudis had been fully aware of the study's contents for nearly four months, and both the Saudis and the USAF had already used it to plan a more comprehensive approach to Saudi air defense. This plan called for major improvements in Saudi Arabia's ability to use its F-15s for air defense, for a new Saudi air control and warning (AC&W) system, and for improved Saudi air attack capabilities.

The study focused on basic problems that still shape Saudi Arabian defense planning and are likely to do so for the foreseeable future. Saudi Arabia must concentrate on defending key population centers and economic facilities. Saudi territory is too large for comprehensive air defense of all its air space or borders. Saudi Arabia's land borders are 4,537 kilometers long and it has an additional coastline of 2,510 kilometers. Its total territory is 2.3 million square kilometers. To put this in perspective, Iraqi territory totals only 445,000 square kilometers, and Iranian territory totals only 1.6 million kilometers. The total air space of all the other member states of the GCC is only about 321,000 kilometers of air space, and both Yemens cover 518,000 square kilometers. To put it differently, Saudi Arabia must defend an air space roughly 110 times that of Israel.[1]

Saudi Arabia is also forced to disperse its air assets to a relatively few air bases, and these bases are too far apart to support each other without the use of long-range fighters and refueling. Saudi Arabia can afford to provide only one major air base to cover each threatened area of the country. Saudi Arabia now bases its first-line combat aircraft in four key air bases: in *Dharhan,* which covers its main oil facilities and key Gulf cities; in *Taif,* which covers Saudi Arabia's main port, its holy cities, and key agricultural and population areas in the lower Red Sea area; in *Khamis Mushayt,* which covers the Yemens and which can reinforce the air strip at Sharurah; and at *Tabuk,* which must cover the upper Red Sea, Jordan, Syria, and Israel, and potentially help reinforce the air field at *Hafr al-Batin,* which is the Saudi Army's key defense point to meet any threat from the Northern Gulf.

These few bases can provide only limited cross-reinforcement capability, even if they refuel and rearm at Riyadh, unless Saudi squadrons change their deployments from one base to another. It is 650 nautical miles (NM) from Khamis Mushayt to Dhahran. It is 443 NM from Khamis Mushayt to Riyadh, 736 NM from Dhahran to Tabuk, 590 NM from Riyadh to Tabuk, 394 NM from Riyadh to Jiddah, and 465 NM from Tabuk to Taif. In practical terms, this means that refueled F-5 fighters cannot safely reinforce any main Saudi combat operating base from any other base. Even F-15s and Tornados cannot provide the loiter time necessary to provide air defense coverage from any other base without the added fuel provided by conformal fuel tanks and additional in-flight refueling.

By reducing its air coverage of one front to strengthen the defense of another, Saudi Arabia can develop a limited concentration of force. Even so, Riyadh is 250 NM from Dhahran and 394 NM from Jiddah. This means that the only additional major Saudi air base that can cover the oil fields at Dhahran is outside the range where an F-5 can provide effective air defense without refueling, and that an F-15 without conformal fuel tanks will still lack the loiter time to fly effective CAP missions. Even Khamis Mushayt and Taif are 230 NM apart—too far for effective fighter cross-reinforcement much beyond the other base.

Saudi Arabia also faces the problem that it may have to move combat aircraft forward from their normal bases to military air strips in the same general area. Saudi Arabia does not have the air strength to permanently base combat aircraft at its key dispersal strips at Hafr al-Batin or Sharurah.

These size and coverage problems mean the Saudi Air Force will be strained to defend its vital areas even with the most advanced fighters, support system, and C^3I/BM capabilities. Saudi Arabia cannot, however, rely simply on point defense. It must be able to maintain a perimeter defense against the Yemens, Iraq, Jordan, Syria, and Israel, and be able to concentrate its air forces to defend its "strategic axis". This strategic axis extends south from the Saudi oil facilities and cities on the Gulf (Jubail, Juaymah, Ras Tanura, and Dammam) to cover the capital at Riyadh, Makkah, Taif, and the key coastal ports of Jiddah and Yanbu. It adds still another requirement for high-performance fighters.

It is particularly important that the Saudi Air Force should be able to defend key population centers. Saudi Arabia's modernization has made Saudi cities acutely dependent on a limited number of water, desalinization, gas, and electric power facilities and on the use of a limited number of airports for food supplies. As is the case with the Saudi oil fields, past economies of scale have led Saudi Arabia to buy extremely large one-of-a-kind facilities. An air attack on such facilities could deprive a given Saudi city in the "core area" of essential human services, and this vulnerability is growing steadily as more and more of the Saudi population become dependent on modern urban utilities.

The key interest that unites both the West and Saudi Arabia, however, is the need to protect Saudi Arabia's oil facilities. The total oil area covers about 850,000 square miles, but Saudi oil facilities are concentrated largely in a 300-by-100-mile core area. Although Saudi Arabia has 47 oil fields scattered from west to east in the Gulf and near the Gulf coast, its core area ranges along a 250-mile axis that extends from the offshore field at Safaniya (the world's largest offshore field) to Ras Tanura and Berri on the coast to the southern tip of the Ghawar oil field (the world's largest onshore field). If various peripheral territories are included, the core area covers 10,000 square miles, or twice the area of the state of Connecticut. Depending on world demand, the facilities in this area must produce, distribute, and ship up to 9.6 million barrels a day and 3,500 million barrels a year.

Aside from the offshore portion, almost all of this core area is located on dry exposed plains with no natural cover or protection. During most of the year, oil targets are easy to locate by air, and there are long periods during the day when such Saudi oil facilities produce enough image contrast for even first-generation air-to-surface missiles to be effective. Many such facilities are vulnerable to area bombing or to heliborne raids, and many key facilities are vulnerable to a single bomb, and a few are even vulnerable to light, shoulder-fired anti-tank rockets like the RPG-9 or to light anti-tank guided missiles like Milan or Dragon.

The terrain favors terrorist raids and air and sea operations. The coastal area is covered by shallow sand flats and small hummocks that severely limit rapid movement by armored vehicles except along a limited number of roads. The shore line is generally too shallow for amphibious landings, and the waterfront can shift back and forth by several miles in some places, depending upon the wind and the tide. Salt flats are common and can become impassable with even limited rain. Coastal marshes often merge with the sea and create major mobility problems. Away from the coast, the dunes near the shore give way to the Jafura sand desert.

This desert terrain, a lack of water and facilities, and temperatures well over 100 degrees limit the ability to rapidly seize the Saudi oil fields. In contrast, a successful air attack or amphibious assault on the oil facilities within 40 miles of the Saudi coast—or even a raid on Abqaiq, Ras Tanura, and their sea islands—could destroy or seize most of Saudi Arabia's oil-loading and/or processing capabilities.

Weather conditions complicate the problem of providing air defense for the core area. In the summer, prevailing northwest winds, called "shamal", create summer sandstorms. Line squalls are common, and visibility at the ocean surface can drop to zero. Thermal currents create a ducting phenomenon that can "blind" or obscure most radars. By taking advantage of these weather conditions, an airborne or seaborne attacker could hit Saudi oil facilities at a time when radars on the ground would be virtually "blind" and have difficulty operating surface-based sensors and missile defenses.

Critical targets for air or sea attack are scattered throughout the core area. Most of the roughly 650 onshore wells in Saudi Arabia require water injection to maintain pressurization. While the individual wells are scattered targets, a more limited series of air attacks on key high-output wells would have a considerable effect on Saudi oil production, as would destruction of key components of the water-injection system.

Similarly, much of the Berri field is under water and feeds into eleven offshore platforms that regulate four to six wells each. Eight more offshore platforms maintain the water well and injection system. These are connected by 25 miles of undersea pipeline ranging from 10 to 24 inches in diameter. Much of this piping is vulnerable to sabotage by ordinary scuba divers. Each of these platforms is an attractive target, and carefully planned air strikes against a few of them could halt production from the entire Berri field. An attack on Saudi Arabia's twelve offshore drilling and workover rigs could also affect production, in the long run. So could an attack on the small island of Abu Ali, which is 20 miles off the Saudi coast, where several main well and water-injection facilities are centralized, in addition to those for gas-oil separation.

The Saudi oil and gas collection system is vulnerable to air attack. Crude petroleum from all the Saudi fields, and a considerable amount of gas, converge on Ras Tanura through well over 1,200 miles of pipe. Pipe diameters vary between 12 and 48 inches, and many increase dramatically in size as lines pass gathering points. The overall security problems in protecting such facilities are less difficult, however, than the total pipeline mileage indicates. The length of the major pipelines in the Kingdom exceeds 6,160 miles, and many redundant or back-up links exist. Large pipes are difficult to destroy and small ones are easy to repair or replace. More than eight pipes share the same route from Abqaiq to the Qatif junction, five continue to the coast, and four pipes parallel the seaboard south of Khursaniyah. This redundancy makes them resistant to air attack and sabotage.

There are, however, many vulnerable choke points. Saudi production depends on up to 12.5 MMBD of processed, nonpotable saline water. A single plant, the Quarayyah Seawater Plant, processes more than 4.2 MMBD of filtered and conditioned sea water. This is a highly vulnerable "one-of-a-kind" facility that could take up to two years to replace.

The Saudi power plants that service the fields are extremely large, and the loss of key plants could shut down large parts of Saudi oil operations. Destruction of the 400-plus megawatt facility at Ras Tanura might severely limit shipments from the port, as would hitting key power distribution points. The electric power available in the Eastern Province increased from 1,200 megawatts in 1976 to 2,928 megawatts in 1981, and four new 400 megawatt plants have been built at Ghazlan; but there are still only a limited number of high-payoff targets.

The evolving Saudi power system is also highly dependent on a single central computer facility located in Dammam. A successful attack on this facility might paralyze key aspects of Saudi oil production for months. A wide range of other computer centers controlling key oil facilities are equally vulnerable, and their destruction or seizure could have serious effects. These include key computerized facilities such as the EXPEC Computer Center and the Supervisory Control and Data Acquisition (SCADA) installation, which provides remote management of more than 90 wells and platforms for the Berri oil field.

The Saudi oil pipeline system is now over 19,000 kilometers long and depends on less than 50 major pumping stations, dotted anywhere from 5 to 50 miles apart. One new pumping station at Abqaiq alone processes 1.7 MMBD. One natural gas generator activates three pumping stations with three pumps each east of Abqaiq, and all nine pumps would be inoperative if the turbine failed. Other sensitive facilities are even more critical, and a loss of power to these facilities, or their sabotage, would virtually halt Saudi oil production.

These power plants, which provide electrical current for a comparatively limited number of gas-oil separators (GOSPs), desalting facilities, pumps, and other purposes, are controlled from a single dispatch center at Dhahran. If this center were destroyed, the ability to distribute power in the Eastern Province would be severely limited. Ain Dar is the gathering place for all petroleum produced in northern Ghawar, plus Khurais on the west flank. Lines from Ain Dar and Haradh (in southern Ghawar) funnel into Abqaiq. Even more critical junctions lie farther east at Dhahran, Qatif, and Ras Tanura, and each of these centers has special vulnerabilities.

A new major "target complex" has been created at Shedgum at the north end of the Ghawar field, the center of the new transpeninsular gas and oil pipelines. This center has a new gas plant which processes 1.5 billion cubic feet a day, and a twin plant which was completed at Uthmaniyah in 1983. Each of the four gas modules at Shedgum is a target that processes 375 million cubic feet a day, and 237 kilometers of vulnerable high-pressure gas piping feeds the facility. The combination of gas, oil, and pipeline facilities—many of which use one-of-a-kind equipment with long replacement lead times—will eventually make Shedgum a target almost as attractive as Ras Tanura or Juaymah.

The new petrocity at Jubail, at the southern end of the Berri field, is also an important target complex. Jubail combines a new coastal population center with plants using gas fuel and feedstock. It can process over 270,000 barrels per day (BPD) in gas products, as can Yanbu on the Red Sea. Jubail and Yanbu will not be as attractive as the previous targets, but they will still require Saudi Arabia to extend its air defense perimeter.

Saudi oil terminals are vulnerable, as Iraq's recent attacks on Iran's much better dispersed facilities at Kharg Island have demonstrated. About 2,700 to 4,000 ships a year call at the marine terminals at Ras Tanura, Juaymah, and Yanbu. The main terminal facilities that serve all Saudi oil fields occupy a 50-mile arc along Tarut Bay between Ras Tanura and Al Khobar. They shipped nearly 80% of Saudi production in 1984, with about 5% going to the Ras Tanura refinery and the rest going to the crude oil export terminals.

The Ras Tanura refinery and loading facility plays a critical role in ensuring the smooth flow of Saudi oil production. The refinery has a capacity of 450,000 BPD and processed 416,421 BPD of oil and 254,300 BPD of natural gas liquids (NGL) during the peak export period in 1981.

Equally critical are the stabilization plants on adjoining land, particularly those at Abqaiq, which remove poisonous and corrosive hydrogen sulfide from crude petroleum before it is piped aboard tankers.

Crude oil and refined products awaiting shipment are stored in four great tank farms at Abqaiq, Dhahran, Ras Tanura, and the new port complex at Juaymah. The latter location includes 5 containers holding 1.5 million barrels each, and 14 that hold 1.25 million barrels each. Including pipelines and smaller storage facilities, the location stores up to 30 million barrels. The major storage facilities are extremely vulnerable to air or other attack.

Ras Tanura is the world's foremost oil port. It shipped over 175 million barrels of oil and gas product in 1984, and has shipped peak levels of 256 million barrels. It can berth 18 tankers. Its T-shaped northern loading pier can berth 6 tankers simultaneously, and its south pier can handle 4 more. An artificial "sea island", a mile farther out, can accommodate five supertankers up to 200,000 deadweight tons (dwt). As propane gas (LPG) facilities have "layered" tanks and pipelines over the existing crude oil facilities, and as they have placed LPG tanks near new 1.5 MMBD oil storage tanks, there is an acute risk that a major air attack could cause a massive "chain reaction" of explosions. Ras Tanura's loading facility may be the most attractive conventional bombing target in the world.

ARAMCO off-loads Arabian light crude oil at Juaymah, and Juaymah's capacity is steadily expanding. In addition to a new 10 kilometer offshore trestle and an LPG twin-berth loading facility, Juaymah has single-point moorings (SPM) in deep water about 11 kilometers offshore. These can handle 500,000 deadweight ton tankers with 95 foot drafts (currently the world's largest). The Ras Tanura terminal is limited to 65 foot drafts.

Juaymah can simultaneously provide several grades of oil to six vessels. A pair of 56 inch pipes now connects the oil storage tanks on shore with a giant metering platform in 45 feet of water, and this platform can shunt oil to waiting ships at rates of over 140,000 barrels per hour. Juaymah started loading LPG in 1980, and its 30,000-barrel-per-hour loading facilities can handle 200,000-cubic-meter tankers. The key facilities at Juaymah are concentrated, exposed, and might also "chain react" as the result of a large-scale air attack. The throughput capacity at these two ports serviced 4,067 tankers in the peak year of 1981, and loaded 3.2 billion barrels of crude oil product and natural gas liquids.

These Saudi defense problems are compounded by several other factors. Many of the facilities in the Saudi core area are pressurized systems. Unless Saudi Arabia can protect its oil systems with confidence, it must respond to a major threat of air attack by shutting in—or depressurizing—a wide range of facilities. Saudi Arabia also cannot produce more oil or gas than any given choke point allows, and most of these choke points depend on a few items of tightly grouped and specially fabricated equipment that were designed for maximum economy of scale and were located without any regard to vulnerability to military weapons. There is little or no redundancy or in-country repair capability for such equipment, and most such key items are not readily available on the world market. They must be specially fabricated to order, which can take up to two years.

This Saudi vulnerability extends beyond Saudi territory. Saudi Arabia is dependent on the free movement of tankers to ship its oil. Virtually all of these tankers now move in and out of the Gulf through its narrow eastern opening at the Strait of Hormuz. Approximately 50 to 65 ships move through the Strait each day. About 25 to 40 of these ships are loaded outbound tankers. During peak production periods in 1981, an average of 11 tankers, carrying 8.9 MMBD of gas, oil, and petroleum products, came from Saudi Arabia. Another 0.2 MMBD, or 2.2% of total Saudi production, was shipped from the refinery at Bahrain.

Saudi Arabia now has only two possible alternatives to the Gulf route—both of which have vulnerabilities to air and sea attack of their own. The first such alternative has little practical value. It is the "Tapline" pipeline, with a 470,000 BPD capacity, which runs from Dhahran and Qaisumah through Jordan, Syria, and Lebanon to a port and a 17,500 BPD refinery south of Sidon.

The Tapline was built in 1950, and it requires exceptionally large amounts of manpower and service by modern standards. It is not economic in comparison with tanker shipment through the Gulf, and it has been made even less economic through the lack of modernization and service that has resulted from the long series of disputes over transit fees.

The Tapline's operation has also been highly erratic. It intermittently shipped about 50,000–60,000 BPD from 1979 to 1982, a total of 20–29 million barrels—only about 0.62–0.83% of Saudi crude production. The main impact of the pipeline is that roughly 36,000 BPD has been shipped to support a refinery in Jordan. Even when the rest of the line has operated, it has done little more than support a refinery in Sidon. There have been no significant exports to the West. In fact, the Tapline is virtually useless. The link to Lebanon was cut during the fighting in 1981, and by the Israeli invasion in 1982, and has never flowed since. While the line could be restored to 470,000 BPD, major investment would be needed to upgrade the line, repair port facilities, improve efficiency, and improve security against sabotage. ARAMCO officials privately regard such an effort as a waste of money.

The second—and far more meaningful—alternative is a pair of 1,170 kilometer "East-West" pipelines from Shedgum in eastern Saudi Arabia to the new petrocity at Yanbu on the Red Sea. One of these lines, the "Petroline", is a 48 inch pipeline with a 1.85 MMBD normal capacity and a peak capacity of 2.35 MMBD. The other is a 26–30 inch gas line with an initial capacity of 270,000 BPD of gas liquids. A second oil line will be added, with a capacity of between 1.15 and 1.35 MMBD. This line runs parallel to the existing line, and will be completed in March, 1987. It was rushed ahead at a cost of $600 million because of the threat of an escalation of the Iran-Iraq War.

There is also a strong possibility that Iraq will build a full pipeline to the Red Sea, rather than simply connect to the Saudi line. This, however, would shift Saudi vulnerability to an increasingly unstable Red Sea area, and strain the limited Saudi air defense forces based at Tabuk.

Saudi Arabia is also becoming heavily dependent on the free movement of tankers through the Gulf and on its ability to persuade tanker owners and captains that they can safely move through Gulf waters. The new pipelines will make the Kingdom dependent on convincing such owners and captains that they can safely move through the Red Sea.

The new pipelines and pumping stations along the route to Yanbu also open up new opportunities for sabotage and long-range air strikes that would bypass Saudi Arabia's C^3I/BM and E-3As and other air defenses along the Gulf coast. The large 130,000 barrel an hour shipping pumps at each of Yanbu's three berths would be good targets. So would the polyethylene plant. The new complex at Yanbu forces the small Saudi Air Force to defend two coastal areas more than 1,000 miles apart, and this defense burden will increase when Iraq connects to the Saudi Petroline in 1986. It will be further heightened if Iraq goes on to construct its own Red Sea pipeline to a new facility near Yanbu. Current plans indicate this line will have a capacity of over 1 MMBD, and it could link Saudi Arabia to any future attacks on Iraq.

Finally, Saudi Arabia also plans to provide air and maritime defense for the other GCC states. Iran's air attacks on Kuwait, and air raids on tankers east of Qatar, have shown that an attacker might well choose to put pressure on Saudi Arabia by attacking urban facilities in each of the smaller conservative Gulf states, or on the wider range of oil facilities located in a broad arc along the southern Gulf from Kuwait to Oman. As a result, Saudi Arabia publicly declared in June, 1986, that it would help defend Kuwait against an Iranian attack.

Like the Saudi facilities, the other major GCC oil facilities have generally been designed as one-of-a-kind installations on the basis of purely economic criteria. They tend to use extremely large individual equipment items, without redundancy, to achieve economies of scale. The destruction of a few targets in the form of pumping stations, gas-oil separators, or other key equipment could lead to major cuts in production and in the loss of equipment that can take up to two years to replace fully.

THE IMPACT OF SAUDI VULNERABILITY ON SAUDI MILITARY MODERNIZATION

These vulnerabilities explain why the U.S. and Saudi Arabia paid so much attention to improving Saudi air defenses after the Shah's fall in 1979, why Saudi Arabia pressed the U.S. so hard for the sale of advanced air combat equipment, and why the Kingdom was so quick to turn to Britain when the sale fell through. Saudi Arabia's mix of vulnerabilities created the need for an exceptionally efficient and credible Saudi air defense system. To be effective as a deterrent or defense, this system had to be able to guard against sudden raids or saturation attacks on a 24 hour basis and to offer broad coverage over a wide area.

While the most critical part of the Saudi core area in the northern oil fields occupies only 10,000 square miles, the total critical air space to be defended in northern Saudi Arabia must include its cities in the area, and this creates a critical air space of over 30,000 square miles. If the offshore fields and Neutral Zone are included, the area to be defended becomes at least 70,000 square miles, and the relevant air space of friendly Gulf states adds another 35,000 square miles. Moreover, the Saudi defense problem is not simply one of detecting and intercepting against offshore and shore line facilities, which means that ships have to be tracked and characterized.

The Saudi Air Force faces major resource constraints in creating an effective air defense. It must rely on limited numbers of fighter aircraft, sensors, and surface-to-air missiles to cover this territory in a way that does not lock its forces into a "one-front" defense that can cover only the Gulf. It must deal with special terrain problems. No mountains or high points exist near the Gulf coast. The terrain rises an average of only 5 feet for every mile from the high-water line until it reaches the Summan Plateau, which is too far inland to be useful in covering the Gulf. As a result, groundbased Saudi radars in the Gulf area can provide only about 30–50 nautical miles of low-altitude coverage or 2–4 minutes of warning.

Saudi Arabia's resource constraints mean its air defense system must rely on the Kingdom's five existing major air bases, and their related sectoral radar and ground-based air defense centers, to cover the entire country. The air base at Dhahran is especially critical in defending against Iran. It must cover all the major oil facilities in the Gulf, the Saudi cities in the area, and those of the other conservative Gulf states.

These five main Saudi air bases do have Improved Hawk surface-to-air missiles and 35 mm anti-aircraft (AA) guns. They are also being sheltered with aircraft shelters and underground command facilities superior to those in NATO. The first such shelters were completed at Dhahran in late 1981. Similar facilities have now been completed at Khamis Mushayt, and will eventually be completed at Tabuk and the other major Saudi bases.

The Saudi shelters allow four aircraft to be worked on in each shelter, with a fifth on quick-reaction alert (QRA). They are twin-door shelters with shielded, camouflaged entrances and exits and four taxiways each, and they can house air and service crews for 30 days without resupply. Dispersion is excellent, and the taxiways from each shelter are long enough to serve as runways. The Saudi munitions and fuel facilities are equally well sheltered, as are certain classified service facilities as well as the battle management or sector coordinating center (SCC).

Saudi bases can fully shelter two squadrons in this manner, and this capability is being expanded and supplemented by dummy shelters. The base construction teams also have a rapid runway repair capability, and Saudi Arabia is constructing dummy shelters and expanding its taxiways and alternative air strips to help deter the use of penetrators or air field suppression munitions.

This base design reduces Saudi Arabia's problems in relying on a single base per sector, and the air force can deploy to civil air fields as a dispersal measure. Saudi Arabia has 55 additional airfields with paved runway and taxiway facilities, 5 with runways over 3,659 meters, 22 with runways over 2,440–3,659 meters, and 28 with runways less than 2,440 meters.[2]

Passive and active base defenses cannot, however, protect Saudi air fields against new air base suppression techniques, such as earth penetrators, cluster or modular glide bombs with minelets, or sensor weapons. They cannot protect such bases against updated versions of the hard-point munitions that Israel first used against Arab targets in 1973, and which are now becoming commercially available in Western Europe. They also cannot fully protect all the key facilities on these bases against conventional munitions delivered with the accuracy Israel demonstrated in its attack on the Iraqi Osirak reactor.

All these factors shaped Saudi perceptions in 1979 and 1980. Saudi planners had to consider the risk that Saudi Arabia might lose a forward air base like Dhahran, and most of its fixed ground radars in a given area, by the late 1980s or early 1990s. This situation imposed requirements almost as demanding as the sensor coverage problem discussed earlier. The Saudis also faced the problem of deploying and operating a reasonable number of individual radars. Even if their air bases could survive, their effectiveness depended on land-based radars. Virtually all of these had to be sited in fixed and exposed positions dictated by the need to maximize low-altitude coverage. Both U.S. and Saudi planners concluded that no expansion of Saudi Arabia's land-based radar system could eliminate the problems of vulnerability and the possible "blinding" of a large part of each base's defense zone by anti-radiation missile (ARM) attacks on its radars.

The main radar at Dhahran, for example, was permanently fixed on an artificial hill, and many other main radars were similarly fixed in a position to allow rapid data transmission to each of the five sector operating centers serving the local air base and army air defense control center (ADC). Even the base at Khamis Mushayt, whose main radar is located on a sawed-off mountain top, had limited low-altitude surveillance capabilities.

Further, Saudi planners had to consider that if they lost the use of the base at Dhahran, they would have to use remote sensors and the air bases at Taif and Khamis Mushayt to cover the Gulf, which would mean flying fighter missions of 600–800 miles. This is a longer operational radius than Israeli fighters had to fly in attacking the Osirak reactor, and is roughly equivalent to attempting to defend Chicago from Wichita or Dallas, or London from Athens. Any such Saudi sorties would also be meaningless without local radar coverage, since the reinforcements would be too blind to be effective.

THE NEED FOR THE AIR DEFENSE ENHANCEMENT PACKAGE

These factors shaped the Air Defense Enhancement Package that the USAF and Saudi Arabia developed in the late 1970s and early 1980s. The details of this package are shown in Table 8.1, and it is clear that the package represented a logical way of making Saudi Arabia's fighters effective enough to meet its complex and demanding air defense system requirements.

Table 8.1: The Saudi Air Defense Enhancement Package

- 5 E-3A AWACS Aircraft: The aircraft were to be delivered within 4 years. The procurement cost of $3.7 billion included 3 years of spares and support equipment, plus logistical, maintenance, training, and technical support.[a] The E-3A is highly sophisticated and requires 170 aircrew and 350 maintenance personnel for 7 days of 24 hour flight. The initial requirement for U.S. support manning was 480 contractors and 30 U.S. government personnel.
- Ground Defense Environment: 22 major system elements of hardened command and control facilities, data processing, and display equipment, new radars, and ground entry stations were to be procured as the result of a 2 year USAF study over a period of 6 years at a cost of $2.1 billion. This was the first major modernization of the Saudi air control and warning system since the 1960s. The RSAF was to operate 10 sites, and the Civil Aviation was to operate 12.
- 6–8 KC-707 or KE-3A Tankers: The tankers were to be delivered over 40–44 months at a total cost of $2.4 billion or $120 million each.[b] The price included 2 years of spares plus training, maintenance, and support. Saudi Arabia also obtained the option to buy 2 more tankers. The aircraft required 96 aircrew and 320 contractor support personnel during the initial phase of delivery. They had parts, engine, and air frame commonality with the E-3A.
- 101 Sets of Conformal Fuel Tanks or "Fast Kits": These were to be delivered over a 27 month period at a total cost of $110 million, or $900,000 per unit. The cost included related spares; support for the F-15; and equipment, training, and publications. Saudi Arabia informally agreed that the F-15s using the tanks were to be deployed at Dhahran, Taif, and Khamis Mushayt.
- 1,177 AIM-9L Missiles: These new air-to-air missiles supplemented Saudi Arabia's existing stocks of AIM-9F-3 and AIM-7F missiles, and were to be delivered over 30 months as the USAF replaced them with stocks of AIM-9M missiles. The total cost was $200 million, or $98,000 per missile. The price included 42 months of contractor training, maintenance, and logistic support. No new Saudi personnel were required. Nine U.S. contractor personnel were required during the first 3 years of conversion.

Source: Department of Defense

[a]The total price of the AWACS portion of the program was $5.8 billion. The price for 5 aircraft was $1.2 billion (including software modifications, engineering change orders, and an avionics integration laboratory). The price of spares was $536 million, support equipment was $44 million, contractor support and training were $1.3 billion, and miscellaneous program expenses were $664 billion. The price for upgrading the ground radar network was $2.0 billion.

[b]The total price for the tankers was $2.4 billion. This included $962 million for 8 aircraft ($120 million each, includeding extensive engineering, design, and test work to make the aircraft into a tanker), $316 million for spares, $34 million for support, $700 million for contractor support and training, and $387 million for miscellaneous program expenses.

Ambitious as the Air Defense Enhancement Package was, it represented a vital step in dealing with the problems posed by the threat from Iran. The collapse of the Shah's regime had added 432 fighters to the threat against Saudi Arabia, including 188 F-4D/Es, 166 F-5E/Fs, and 77 F-14As. While many Iranian fighters ceased to be fully operational once U.S. support was withdrawn or they were lost in the Iran-Iraq War, the U.S. and Saudi Arabia had to plan for Iranian re-equipment and for the possibility that Iraq might come under hostile rule.

The Carter Administration had never considered this possibility when it pledged not to deploy additional advanced air defense weapons to Saudi Arabia in 1978, and far more was involved than force numbers. As a staff report which the Senate Foreign Relations Committee issued on the E-3A sale pointed out, the Iranian threat posed an incredibly demanding problem in defending the Saudi core area.[3]

Defending the marine terminals at Ras Tanura and Juaymah was particularly difficult. These terminals transshipped over 90% of Saudi crude oil exports, and were situated at points on Saudi Arabia's coast that faced the Iranian air bases at Bushehr and Shiraz. The key processing and distribution facilities at Abqaiq and Saudi Arabia's multibillion dollar natural gas development project, the Master Gas System, were only minutes farther inland.

This combination of geography and threat created an extremely difficult time and distance problem for Saudi and U.S. military planners, and remains a driving factor behind Saudi Arabia's effort to upgrade its F-15C/Ds, its purchase of the Tornado, and its request for additional F-15s.

An Iranian F-4, flying at 200 feet and 480 knots, can bomb the Sea Island at Ras Tanura or the offshore mooring stations at Juaymah with a payload of over 5,000 lbs of munitions only 16 minutes after taking off from Bushehr. If such an enemy fighter loiters peacefully within Iranian airspace over the Gulf and then suddenly veers directly for Ras Tanura, the Saudi reaction time is reduced to 8–9 minutes.

An F-4, however, is a relatively slow-flying and limited-range attack fighter. It was clear even in 1981 that the threat aircraft of the mid-1980s were likely to be faster, have much more lethal avionics and munitions, and have greater range and endurance. This has since been proven all too clearly by Soviet deployment of the new Su-24, Su-27, and MiG-29 fighters discussed in Chapter 4, which have the ability to carry conformal or internal attack munitions at supersonic speeds and can substantially cut the warning and reaction time available to Saudi forces.

In addressing this problem, both Saudi and U.S. Air Force planners had to take into account the fact that intercepting an enemy air attack involved more than the intercept itself. The Saudi Air Force had to be able to perform several other tasks to provide an effective air defense for its cities and oil fields. These tasks are summarized in Table 8.2, and they still describe the basic tasks the Saudi air defense system must perform.

These requirements made it plain that the Saudi Air Defense Enhancement Package had (1) to improve the Saudi sensor and air control and warning systems, (2) to improve Saudi head-on intercept capability, and (3) to improve Saudi air endurance and deployment capability. Without an airborne sensor platform as capable as the E-3A, the Saudi Air Force would have to rely on fixed, ground-based radars in the Dhahran-Juaymah area.

Even once an Iranian F-4 was detected, it would still take the Saudi Air Force 12–15 minutes to intercept it east of Ras Tanura, and again, U.S. and Saudi planners had to consider that the fighters of the mid-1980s would be much faster and be able to penetrate at lower altitudes using low conformal munitions.

This meant Iranian fighters could be over the target within 3 minutes of the time Saudi ground radars could detect them. Even under optimal assumptions, a Saudi intercept based on a warning from ground radars would then occur 10 minutes too late. This then would force the Saudis to defend the target with only surface-to-air missiles and anti-aircraft guns.

While the Ras Tanura–Juaymah–Dhahran area was defended by army air defense battalions made up of Hawk and Crotale missile batteries and 35 mm Oerlikon guns, the Hawk batteries normally need about 10 minutes warning to fire effectively, and the Crotale and Oerlikon weapons systems would probably be able to fire only as the attacker left the target. Even with extensive warning available with the AWACS, and additional short-range systems like the Shahine, the Saudis could not afford to rely on ground-based defenses.

The theoretical sensor range of Saudi Arabia's land-based air defenses had no practical meaning in the face of low-altitude threats. The Hawk could theoretically detect fighter-sized targets at a range of 120 km, the Shahines and Crotales could detect such targets at 70 km, and the 35 mm Oerlikon guns could do so at ranges of up to 40 km. At low altitudes, however, detection ranges rarely exceed 30 km, and the ground-based radars associated with each air defense weapon must work perfectly in detecting small, low-altitude targets to provide even 3 minutes of warning.

In practice, however, ground-based radars operating in the Gulf experience serious performance degradations about 75% of the time due to the temperature gradient between the air mass above the hot desert

Table 8.2: The Military Tasks Necessary for Effective Saudi Air Defense

- Detection: Recognize that a potentially hostile target is airborne. This task must be accomplished using radar. Radar detection, however, is a function of target altitude, size, and range. The higher the target, the greater the distance at which it can be "seen" by the radar; the larger the target, the better the chances become of receiving a usable radar return. Radar operates on a "line of sight" principle, and a low-flying fighter aircraft is screened by the curvature of the earth at ranges beyond 25–30 miles. Only by elevating the antenna or sensor system could Saudi Arabia extend the horizons of its radars significantly beyond this distance.

- Identification: Identify all aircraft. Once a radar operator "sees" a target on his scope, he must attempt to determine its identity or at least to categorize it as "potentially" or "probably" hostile. In wartime, one can assume that any target detected coming from a certain area is probably hostile. In peacetime, however, a specified sector of airspace is likely to be filled with numerous commercial, private, and military aircraft. By using identification of friend or foe and by cross-referencing filed flight plans and pilots' radioed position reports, a certain number of targets will remain unknown, however, and the determination of possible threats must rest on a more subjective reading of likely intent.

- Decision. Decide whether to intercept. A decision to order an intercept of an unidentified target can be pegged to an arbitrary rule: Any "unknown" aircraft entering a certain zone will be intercepted. This is the approach used by the United States, which has established an Air Defense Intercept Zone, or ADIZ, which extends up to 300 miles beyond its east and west coasts. In the Saudi case, geography does not permit the luxury of such an extensive air defense zone. It had to be kept within its borders or the limits of territorial waters in the Gulf and the Red Sea. The Gulf is just over 100 miles wide opposite Dhahran, and major air routes—connecting Baghdad, Bahrain, Abu Dhabi, Tehran, and other major cities to the northwest and southeast—lie just miles off the Saudi Gulf coast. To be sure of intercepting any unidentified aircraft flying as close as 100 miles from the oil fields, the RSAF has to be ready to fly around the clock.

- Combat Air Patrol and Scramble: Have sufficient forces to maintain combat air patrol and to scramble. Saudi F-15 sorties take 3–5 minutes to launch. Fighter reaction time could be cut by several minutes by keeping the pilots in the fighter cockpit with the aircraft engines running, or by using new quick-reaction laser gyros, but Saudi F-15s normally must fly combat air patrol and loiter in the air to be sure of being able to react in time.

- Intercept: Make successful intercepts. Once fighters are airborne, they must be able to engage the enemy, which means both closing on the target (Dhahran air base is about 30 miles from Ras Tanura) and maneuvering into a position from which short-range missiles can be fired. As a general rule, it takes much longer to maneuver behind an attacking aircraft and this means that the defending fighter must expose itself to hostile fighters. While an RSAF F-15 could also launch radar-guided AIM-7 Sparrow missiles from beyond visual range, it must then track the enemy until the missile hits. This method would mean tying up the Saudi fighter while other hostile fighters could fly through or attack the Saudi aircraft.

in Saudi Arabia's Eastern Province and the cooler air over the Gulf. This phenomenon is known as "ducting" and often impedes target detection until the enemy fighter is virtually on top of a ground-based radar. While Ferranti has developed a technology that can detect its existence, only an airborne platform can take advantage of the fact that radar coverage will continue at other altitudes even when ducting shortens radar range at sea level.

The potential threats to Saudi Arabia's other key border areas are only marginally less demanding. For example, the South Yemeni Air Force, which attacked in the Sharurah area as recently as 1973, has since been equipped with Soviet MiG-21s and Su-20/22s, and has more advanced aircraft on order. Its fighters have the combat radius to strike at most targets in southwest Saudi Arabia, although Makkah, Riyadh, and the oil fields in the east are too distant for current Yemeni fighters.

The key Saudi base defending against the Yemens at Khamis Mushayt is supported by ground-based radars, as is the base at Dhahran. Ground-based radar coverage at Khamis Mushayt and Sharurah is limited, however, because of the mountainous terrain. This potentially allows low-flying Yemeni fighters to penetrate deeply into Saudi Arabia without detection. Saudi officials were already highly impressed with the U.S. AWACS' performance during a deployment in 1979, and with its superior capability to look over the mountains into South Yemen and provide advance warning of aerial activity during a burst of fighting between North and South Yemen.

SELECTING THE E-3A

All of these factors contributed to the selection of the E-3A. Work done by USAF Studies and Analysis in the "Air Feasibility Study"of July, 1980, concluded that it would take a minimum of 48 large, fixed, and vulnerable radars to cover all of Saudi Arabia's borders without an airborne warning and control system. These radars would still leave major penetration gaps, provide only 20–30 miles of low-altitude coverage, and give only token maritime surveillance capability.

The USAF also found it would take 11–15 E-2C Hawkeyes and 34 ground radars to provide minimal coverage of the Gulf. Even this force could not provide 24-hour surveillance of Saudi Arabia's oil facilities for a 7-day period. Moreover, the E-2C could not effectively control Saudi fighters in the face of a large-scale attack over Saudi airspace, which is over 100 times the airspace of Israel, and the E-2C had electronic intelligence (ELINT) capabilities that would be useful in Saudi land and air offensives against Israel.

In contrast, the USAF estimated that 5 E-3A and 18 ground radar installations could provide single-front coverage of the oil facilities and cities in the Gulf—even under the assumption that 1 E-3A aircraft would be in a constant overhaul and refit status, another would be in regular maintenance, and 3 would be fully operational. Each E-3A could provide a minimum of 175 nautical miles of low-altitude and maritime coverage and extend warning of an air attack on the Gulf front to a minimum of 7 minutes.

This warning did not solve all of Saudi Arabia's C^3I problems, but it was sufficient to allow the E-3A to vector in Saudi fighters in time to make at least one pass on an attacker before it hit a critical oil facility. It was sufficient to allow the Saudi Army to vector the acquisition radars of its Hawk batteries to enable them to engage effectively and would allow the F-5E to play a back-up role. Further, 5 AWACS could cover the entire front over the Gulf.

USAF planners estimated that even this small number of E-3As would also be survivable in defending the forward area over the Gulf front because there would be virtually none of the terrain masking at low altitudes that exists on Saudi Arabia's western border with Israel and Jordan, and the E-3As could operate far enough to the rear to withdraw if attacked. Moreover, the long flight times and limited air density of the Gulf would allow the E-3As to know they were under attack in time to react. Saudi fighters could then cover the AWACS; the E-3A could retreat or it could "pop down" below the coverage of most fighter radars.

An individual E-3A could also fly for 11 hours before refueling. This long flight time allowed a crew of 17 men per aircraft (13 in the actual electronics role) to carry out the air control and warning mission more efficiently. USAF studies showed that the E-3A had more than three times the endurance of a force using the E-2C plus added ground radars, and would require far fewer technicians.

Most important, the E-3A had the radar power and side lobe suppression to be highly jam-resistant, to track large, medium-altitude targets at 360 mile ranges, and to identify even small cross-section fighters flying at altitudes above 200 feet at ranges of up to 175 miles ("cross-section" refers to the minimal radar profile of the aircraft, usually expressed in square meters).

These features allowed the AWACS to fly about 50–100 miles inland from the Gulf coast and still provide a minimum of 5 minutes warning before the required moment of intercept. Each AWACS could independently track up to 300 targets and accept data on up to 200 more from another AWACS or ground radar. Under ideal circumstances an

E-3A could engage and characterize up to 240 targets in terms of size, altitude, identity, speed, and direction. It could also direct 6 to 8 closely spaced intercepts and 3 to 6 simultaneous intercepts.

THE ISSUE OF TECHNOLOGY TRANSFER

The USAF also concluded that the configuration of the AWACS sold to Saudi Arabia could omit the frequency-agile, secure voice, and jam-resistant joint tactical information distribution system (JTIDS) necessary to manage an air battle of the density that would be common in a NATO–Warsaw Pact or Arab-Israeli environment. The USAF was able to tailor the E-3A to meet Saudi needs, concluding that such a configuration would be able to avoid creating a potential threat of technology loss to the U.S.—or military threat to Israel—but still be able to handle most Gulf threats through the late 1980s.

This "tailoring" sets an important precedent for current and future U.S. and European efforts to meet Saudi Arabia's current arms needs from the West, and it is interesting to note the details:

- No advanced secure digital data links such as the JTIDS were to be provided.
- The computer software was tailored to limit capability against U.S.-made fighters of the kind flown by the USAF and Israel: Air Force.
- Training and software support were optimized for defensive missions in the Gulf and Red Sea areas only.
- Major overhauls, annual service, and all modifications were to be performed in the U.S.

At the same time, Saudi Arabia agreed not to fly the E-3A outside its own borders without U.S. consent, to protect all classified systems using U.S. procedures, to U.S. approval of its security plan, to the computer software remaining in U.S. possession, to a new information security agreement, and to limiting access to all sensitive documentation. It agreed to share all E-3A data and to refuse to allow third country personnel to fly on, modify, or maintain the E-3A—or to have access to E-3A data without U.S. consent.

Equally important, the U.S. estimated that the first E-3A configured to Saudi needs could be delivered in 1986, and the other four at two-month intervals thereafter. The Saudis would thus be able to operate without U.S. crew support as early as 1988–1989, and such Saudi operation of the E-3A was absolutely critical. Any other long-term arrangement for deployment of the AWACS or basing of U.S. combat forces in Saudi Arabia would have presented impossible political problems in terms of Saudi sovereignty, internal political stability, ability to lead the other conservative Gulf states, and provocation of hostile reactions from the more radical Arab states.

The U.S. also concluded that it could sell the AWACS to Saudi Arabia with little fear that it would be hijacked or operated in ways hostile to U.S. interests. The E-3A aircraft could not be flown for more than a month without the full cooperation of the U.S. support personnel in Saudi Arabia, and all depot-level maintenance would have to be done in the U.S. Because of sensitive radar design and operating requirements, the key equipment on the AWACS could not be operated without U.S. support for more than a few days.

The E-3A's mean time before service required to deal with major system failures is 27.4 hours. The variant of AWACS sold to Saudi Arabia could meet all Saudi requirements and still be left vulnerable to sophisticated U.S. communications jamming and to jamming by Israel. The E-3A could also be sold without the sophisticated ECM gear or any of the ELINT capability of the four Grumman E-2C Hawkeyes now in Israel.

The sale of the E-3A presented little risk of technical compromise because of its age and the special character of its technology. The AWACS was first flown in 1971. Only 10 of the more than 1,000 AWACS technical manuals were classified, none was classified higher than "secret," and all had already had broad release in NATO. The basic radar design of the E-3A was not sensitive—only the equipment used to produce it. This design could not be compromised by reverse engineering through acquisition of the aircraft. The production equipment involved vast amounts of software that would not leave U.S. hands and that represented an investment in special manufacturing capabilities of well over $1 billion.

In fact, since the first E-3A would not be delivered to Saudi Arabia until 1986, it seemed likely that the Soviets would be in the process of deploying their new AWACS variant of the Il-76 Candid by the time U.S. aircraft deliveries ended. The rest of the electronics in the E-3A dated back to the late 1960s and would also be well within Soviet technical capabilities at the time the aircraft was transferred to Saudi Arabia. The E-3A's basic ECM defense and its key characteristics—operating frequencies, pulse repetition rates, pulse widths, side lobe characteristics, effective radiated power, antenna gain, transmitter power, scan rates, and receiver dynamic range—had already been acquired by Soviet ELINT systems.

THE IMPACT OF THE AIM-9L MISSILE

The sale of 1,177 AIM-9L missiles contributed to the Air Defense Enhancement Package by giving Saudi Arabia an all-aspect air-to-air missile with the shoot-down capability that it could use in "head-on" intercepts against low-flying attackers, without having to sacrifice the time and probability of intercept necessary to maneuver into a long, stern chase, or "dogfight" position. Such a capability was essential given the limited warning time the AWACS could provide and the need for each Saudi F-15 to be able to engage more than one attacking fighter per encounter.

Although the Saudis already had 2,000 second-generation AIM-9J air-to-air missiles, 600 fourth-generation AIM-9-P3s, and large stocks of "product-improved" AIM-7F radar-guided Sparrows, these missiles lacked the multi-aspect capability to enable the F-15 to avoid time consuming and complex fighter maneuvers in meeting an attacking aircraft and lacked the energy of maneuver to be lethal at very low altitudes. While the sale of the AIM-9L did create some risk of technology transfer, no other missile in the U.S. inventory had the requisite capabilities, and the U.S. gained compensating advantages.

PROVIDING ENHANCED RANGE AND REFUELING CAPABILITY

The sale of 101 sets of conformal fuel tanks or "FAST" kits (two tanks per aircraft) for Saudi Arabia's F-15 fighters solved another problem. It gave individual Saudi fighters the endurance and range to maintain combat air patrol with a comparatively limited number of fighters and the ability to mass in the Gulf area for short periods even if Saudi Arabia should lose most of the facilities on its air base at Dhahran. The FAST kits gave the Saudi F-15 about 1,500 gallons more fuel per set. They extended the F-15s, radius by 79% in the air-superiority mode and 93% in the interdiction mode and increased endurance by 65% in the CAP mode. They also allowed the F-15 to retain its capability to carry four AIM-7F radar-guided missiles.

This increase in range was essential if Saudi Arabia was to use the F-15 in anything other than a relatively short-range point defense mode. Without the FAST tanks, the F-15 remained a relatively short-legged and low-endurance fighter. It was for this reason that the U.S. Air Force studies in 1978, which had led the U.S. to encourage Saudi Arabia to buy the F-15 rather than the F-14 or F-18, had recommended that Saudi Arabia be given these tanks. The kits also had the advantage that they could be provided in a form that would not allow Saudi F-15s to carry offensive attack munitions without U.S. technical support, since Saudi Arabia could not modify them unilaterally in ways that would allow a loaded F-15 to sustain its speed and maneuverability.

THE NEED FOR A FULL C^{3}I/BM SYSTEM: THE GENESIS OF "PEACE SHIELD"

The precise interface among the Saudi version of the E-3A, its air defense guns, and its Shahine and Improved Hawk missiles was not decided upon when the USAF study was completed, and was still under study when the U.S. Congress debated the Air Defense Enhancement Package. In fact, the actual award of the first phase of the Peace Shield contract was not made until late February, 1985. Nevertheless, it was planned from the outset that Saudi Arabia would acquire data terminals at each major air defense site with digital processing capability and commercial encryption gear similar to the U.S. TADIL C.

The radar problems discussed earlier made it essential that Saudi Arabia's limited fighter strength be able to "net" its fighters and Hawk surface-to-air missiles through the six ground-based sector coordinating centers and army air defense control centers (ADCs), so that the data collected by AWACS could be interpreted and transformed to provide the Hawk units with the search data they needed to back up the fighter screen. This combination of air and ground defenses could protect its Hawk against much of the ECM gear likely to be common in the Gulf in the mid-1980s. Similar terminals planned for selected Saudi, short-range air defense systems were too limited in range, C^3 flexibility, and electronics sophistication to benefit greatly from AWACS data.

As the Saudi Air Defense Enhancement Package progressed through the Pentagon, the U.S. added further features to improve the Saudi C^3 system and make optimal use of the improvements planned in Saudi Arabia's sector operating centers and General Operating Center (GOC) near Riyadh. These efforts drew on the Peace Hawk VII study and other studies begun after the sale of the F-15 to Saudi Arabia in 1978.

Such an upgrading was vital. Many elements of the existing Saudi C^3 system dated back to the abortive Anglo-U.S. air defense system of the 1960s, and major improvements were required to make effective use of the AWACS. These improvements included new hardened command and control facilities, new data-processing and display equipment, links to the new al-Thaqib package of Shahine missiles and other land-based defenses that France had sold to the Saudi Army, and improvements to the ground radar surveillance network through replacement of existing radars and the addition of new radar sites to extend coverage. Special ground entry stations also had to be provided to allow optimal communications with the AWACS.

In any case, a contract award went to a consortium headed by Boeing Corporation in March, 1985, for the first phase of the Peace Shield command, control, communications, and intelligence (C^3I) system. This contract was worth about $1.18 billion, and the entire system was priced at roughly $3.7 billion. The deal also illustrated the commercial as well as the strategic importance of Western military ties to Saudi Arabia.

About 75% of the Peace Shield contract went to subcontractors, including Westinghouse (displays and software), ITT (communications engineering and long -aul communications), Standard Electric of the FRG (communications switching and mobile telephones), BDM (engineering support), Frank E. Basil (personnel support services), General Electric (radar installation services), and the Saudi Amoudi Group (in-country support).

The award involved three contracts to be managed by different elements of the USAF. One was for $848 million to be managed by the USAF Electronic Systems Division for equipment and system integration, the second was for $331 million to be managed by the Air Force Logistics Command for organizational and intermediate maintenance, personnel support services, and on-the-job training, and the third was for $3 million to be managed by Air Force Training Command for initial operations and maintenance training. Separate contracts had already gone to General Electric for radars ($330 million), and to other firms for design, training support, and so on. The Boeing awards did not include an estimated $900 million in additional construction contracts.

Interestingly enough, the award involved a new feature in U.S. and Saudi relations. The bidding companies had to offer a 35% offset for programs to create a high-technology industrial infrastructure in Saudi Arabia. These programs will fund increased Saudization of the Kingdom's defense support as well as such activities as jet engine overhaul and long-distance telecommunications. It seems likely that most future U.S.-Saudi military contracts will have similar features.[4]

The main features of the final Peace Shield system have already been shown in Figure 8.1. They include:

- A fully integrated network of control centers.
- 17 modified "minimally attended" AN/FPS-117 ground-based surveillance radars based on the "Seek Igloo" radars that the USAF used in Alaska as part of the NORAD defense system.

- Upgrading of the centralized Command Operations Center (COC) that already exists in Riyadh.
- Building of 2 Base Operations Centers to coordinate country-wide air activities such as aerial refueling and search and rescue.
- Creation of 5 Sector Command and Operations Centers with fully automated computer capabilities. Each center will be located at a major Saudi air base and is capable of surveillance, control, and management of air operations for a given sector and is linked to the COC in Riyadh.
- Eventual growth to provide improved data links and C^3I links to the Air Defense Corps Improved Hawk missiles and to the Shahines in the al-Thaqib air defense package of 1984.

Peace Shield is specifically designed for defense against attacks from across the Gulf, attacks by Yemen, and attacks from across the Red Sea. Like the AWACS, it will be heavily dependent on continued U.S. technical support throughout the 1980s and uses U.S. communications systems and operating concepts. This will help ensure that Saudi air units will be directly interoperable with USCENTCOM forces in spite of the Saudi purchase of Tornado. It also reduces the risk that Saudi air units can be used unilaterally against Israel.

The Peace Shield air defense system has the further advantage that it can be modified to interface with the ships in the Saudi Navy. It can link their radars and surface-to-air missiles to Saudi Arabia's other air defenses and conduct joint sea-air operations against any seaborne attacking force. The E-3A has 35–108 nautical miles of range in the maritime surveillance mode against metal ships 40 feet or more in length, operating in a normal sea. Given the probable proficiency of the Saudi Navy in the 1980s, such a battle-management system may be essential if Saudi Arabia is to use its new vessels with any effectiveness.

Although it is not part of the present Peace Shield, Saudi Arabia will also have the option of using the AWACS to monitor transponders on Saudi Army vehicles and helicopters and potentially those of the National Guard. This capability would the Saudi Army and Air Force provide with offers air support for the land forces defending the Saudi oil fields and other critical facilities. Finally, Peace Shield a significant potential improvement in the ability to coordinate Saudi defenses against guerrilla or larger-scale terrorist attacks. Such combinations of ground, helicopter, and fighter operations are exceedingly difficult to coordinate even for Western armies, and the AWACS provides a potential battle-management and communications center for the kind of small-scale but delicate operation that would be involved in actions against guerrillas and terrorists.

THE ADVANTAGES TO THE WEST

Each of these elements of the 1981 Saudi Air Defense Enhancement Package contributed to a system that would upgrade Saudi air defense capabilities and lay the cornerstone of the deterrent that Saudi Arabia had sought since the early 1960s. The West had a great deal more to gain from the sale, however, than simply strengthening Saudi ability to protect the oil fields. In the course of the negotiations, the U.S. acquired potential over-the-horizon reinforcement capability which was far more important than the full-time bases it had sought at the end of the Carter Administration, and which was far better suited to the region.

In fact, the advantages to the West of the U.S. sale of the Air Enhancement Package closely parallel those that were later provided by the British sale of the Tornado and advanced air munitions. They are listed in Table 8.3, and provide a detailed illustration of the ways in which Western and Saudi objectives can coincide.

SETTING THE STAGE FOR THE ARMS SALE CRISIS OF 1985–1986

These close relations with the U.S. shaped the pattern of Saudi Air Force modernization up to the time of Saudi Arabia's request to the U.S. for additional F-15 aircraft, air munitions, and other force improvements. They represented major progress. U.S.-flown E-3As worked with Saudi fighters to shoot down Iranian F-4s that intruded into Saudi space on June 5, 1984. The Saudi F-15s were vectored to their target by USAF E-3As, and were refueled by USAF tankers, and the incident unquestionably played a major role in deterring Iranian attacks on tankers moving through the Southern Gulf.[5]

Still, the Air Defense Enhancement Package and Peace Shield failed to deal with several key issues. First, they did not provide Saudi Arabia with a balanced mix of air defense and air attack capabilities. Second, they did not provide the combat aircraft numbers that Saudi Arabia would need in the future. Finally, they made a number of serious compromises in the C^3I equipment the U.S. made available to the Kingdom.

The limits in C3I capability were least important. The Saudi configuration of the E-3A lacked the JTIDS, secure digital data links, the Have Quick and Seek Talk frequency agile voice links, and the five extra data-display consoles in the USAF versions of the E-3A. This made the Saudi E-3As potentially vulnerable to the level of ECM-

Table 8.3: The Advantages to the U.S. of the Saudi Air Defense Enhancement Package of 1982

- The Air Defense Enhancement Package gave Saudi Arabia a credible local deterrent and defense capability against the most probable regional threats that could affect its security or the West's supplies of oil.
- The package gave Saudi Arabia the military power and status it needed to act as a counterweight to Iran and Iraq. It meant that the other smaller Gulf states could turn to Saudi Arabia for a credible strengthening of their air defense. It was an ideal form of "linkage" between Saudi Arabia and the smaller conservative Gulf states, which had something like 50% of Saudi Arabia's oil production capacity. It gave them the defensive coverage they need without threatening them and was an ideal means of uniting them behind the Gulf Cooperation Council and its nascent National Security Council.
- The package was a partial solution to the long-standing problem of convincing Saudi Arabia, particularly the Saudi military, that the country's military expenditures would be effective and would not jeopardize Saudi sovereignty, and that the Saudi government had not miscalculated in maintaining its ties to the U.S. No single factor had been more important in destabilizing Third World military forces than the perception that their governments have failed to provide a credible basis for national defense and for maintaining national strength and independence.
- The package avoided formal military ties between the West and the conservative Gulf states, which half a century of Arab nationalism, the Arab-Israeli conflict, and the political history of Saudi Arabia had made untenable. It provided the U.S. with a means of strengthening Saudi Arabia while simultaneously preserving Saudi Arabian sovereignty and stability.
- The resulting Saudi-U.S. agreement gave the U.S. continued direct access to the information gathered by the AWACS while retaining control over transfers to third nations. Direct digital look down links were to be available to the U.S.; these included data not only from the AWACS, but from the full range of sensors that make up the Saudi air defense system, many of which were not otherwise available.
- The resulting improvement in Saudi capabilities greatly reduced the risk of having to send U.S. or other Western forces to deal with local or low-level threats. It also decreased the destabilizing effects of any such U.S. deployment without a threat great enough to unite the conservative Gulf states and moderate Arab opinion in support of such U.S. intervention.
- At the same time, the Air Defense Enhancement Package gave the U.S. the ability to deploy up to two wings, or roughly 140 USAF fighters, to support Saudi Arabia and the conservative Gulf states in an emergency. In fact, each main Saudi air base had the basic support equipment for 70 U.S. F-15 fighters in addition to supplies for its own F-15s. The package meant that Saudi Arabia would have all the necessary basing, service facilities, refueling capability, parts, and key munitions in place to accept over-the-horizon reinforcement from USAF F-15 fighters. No conceivable improvement in U.S. air lift or USAF rapid deployment and "bare basing" capability could come close to giving the U.S. this rapid and effective reinforcement capability.

(continued)

Table 8.3 (Continued)

- The package allowed the U.S. Navy to use its air power effectively in the Gulf. For example, U.S. carriers were too vulnerable to deploy into the Gulf during the start of the Iran-Iraq War in 1980. The TF-70 Task Force the U.S. sent to the region had to stay in the Gulf of Oman because of the threat posed by Iranian forces, and carrier-based fighters lacked the range and AC&W capability to overcome the problem of having to deploy from outside the Gulf.
- None of the key elements of the system would be operable without U.S. support throughout their useful life.
- The facilities that would become part of the Saudi system would also help to strengthen U.S. ability to deploy forces from the eastern Mediterranean and project them as far east as Pakistan in those contingencies that threatened both U.S. and Saudi interests. No conceivable build-up of U.S. strategic mobility, or of U.S. staging bases in Egypt, Turkey, Oman, Somalia, or Kenya, could act as a substitute for such facilities in Saudi Arabia. The closest other U.S. facility in the region was at Diego Garcia, which is as far from the western Gulf as is Dublin, Ireland.
- The systems' maritime mode provided a potential method of solving many of the problems in Saudi Arabia's naval defense and the modernization of its naval forces and in improving the coordination of its diverse U.S. and French vessels and equipment.
- The package solved the critical problem of financing the defense of the Gulf by having Saudi Arabia assume the full cost of the system. The Air Defense Enhancement Package was projected to cost $8.5 billion in current dollars, with only three years of spares and support. Even this cost estimate depended on exceptionally efficient U.S. assistance in managing the construction and training effort. Allowing for a 10-year life-cycle, systems growth, and normal cost growth, the true cost was projected to be at least $15 billion. The package also built on past and future U.S. military assistance expenditures, which then totaled $33–38 billion, some $20–25 billion of which was still to be delivered, and it meant the dollars were to be spent in the U.S. rather than in France or Britain. At least 50% of this expenditure was to be on construction, and some 80% of it will contribute directly to the defense of the Gulf's oil exports. The U.S. lacked the financial resources and the forces to create any credible alternative.
- The package was inherently defensive in character and did not threaten other regional powers. At the same time, the Saudi ground stations, SAM defenses, E-3A, and other elements of the package could be upgraded in the future in direct proportion to any growth in the threat.

capability electronic warfare that neighboring powers could acquire in the late 1980s.

This risk, however, has largely been addressed in the years that have followed, and the U.S. has found ways of tailoring its technology transfer to meet Saudi needs without compromising key Western military technology or increasing the threat to Israel:[6]

- Secure radio teletype equipment and software for the E-3A were ordered in May, 1983, that were superior to commercially available software, and compatible with the NATO data links, but did not present a risk of technology loss to the U.S.

- UHF relay equipment was ordered at the same time. This will greatly improve data transfer over the use of the E-3A's radios, but it is oriented to the ground links in the Peace Shield system and will not threaten Israel.
- Color display monitors were added in September, 1984, when the USAF ordered such equipment. These help compensate for the limited number of stations on the Saudi AWACS.
- Infra-red countermeasure sets were added similar to those installed on other Saudi aircraft in August, 1984. These were provided in the "tailored technology" form releasable to other foreign countries.

These compromises present some military risks to Saudi Arabia in a high intensity conflict, although Britain may now be able to supply some of the missing capabilities as part of its follow-on sales to the Tornado package. The lack of JTIDS digital data link and Have Quick secure voice link reduces the capability of the Saudi E-3A to handle high-density attacks because of the need to rely on slower communications with less security. The Saudis have been sold sanitized Modes 1, 2, and 3 of identification of friend and foe (IFF) capability, rather than U.S. military secure Mode 4 IFF cryptographic equipment. The Saudi E-3As are not to be provided with the computer software needed for TADIL C data links with interceptors or U.S. Navy F-14 fighters; and the Saudi F-15s still lack the advanced air-intercept program support package to be provided on U.S. F-15C/Ds.

Even with these limitations, however, the revised Saudi air defense system offers the Western and Saudi air forces a substantial degree of interoperability. For example, the Saudi E-3As transmit or receive digital data using TADIL A high-frequency links with the U.S. AWACS, U.S. Navy ships, or air force ground-based tactical reporting and control posts (provided the formatted data is not passed through the incompatible Saudi and U.S. cryptography equipment).

The Saudi E-3As will also be able to communicate by voice with U.S. E-3As and fighters using UHF radio (again, provided that neither side passes transmissions through incompatible secure voice equipment). The Saudi E-3A can interrogate U.S. E-3As, E-2Cs, and U.S. and British fighter IFF transponders using Mode 3 (but not the secure Mode 4). Further, the communications equipment on the Saudi E-3As is designed to have "form, fit, and function" interchangeability with the USAF E-3A, which means that Saudi interoperability with U.S. forces can be significantly upgraded with modifications that can be quickly and easily performed in the field.

In an emergency, the Saudi Air Force's secure voice and IFF equipment can be removed and replaced with U.S. "black boxes" in about four hours, giving both air forces fully compatible secure voice data and IFF capabilities. Alternately, the U.S. AWACS could replace the gear with Saudi "black boxes". U.S. Air Force mission tapes and TADIL C software can quickly be programmed into the RSAF AWACS computers. Only the installation of JTIDS and Have Quick requires depot-level installation in the U.S.

It is also worth noting that one follow-on modification to the Saudi E-3A has been of direct benefit to the U.S. The Saudi purchase of the CFM International CFM56 engines for the E-3A, instead of the Pratt and Whitney TF33-PW-100A, in October, 1983, gave Saudi Arabia improved fuel efficiency and thrust. This has resulted in longer unrefueled time on station, a slightly higher optimal cruising altitude, and faster or shorter takeoffs. The R&D and design costs for this substitution were paid for solely by the Saudi government, but have had the side effect of saving the U.S. Navy $50–60 million on its Boeing E-6 program.[7]

Saudi Arabia's new Tornados should be able to integrate smoothly into this system. British and U.S. air control and warning concepts differ slightly in detail, but they are broadly compatible. The air defense variant of the Tornado can interface effectively with the E-3A, and its basic IFF and secure communications systems should be compatible with Peace Shield. The uncertainties lie more in the efficiency of the interface between Peace Shield and the Tornado IDS attack fighters, which may require some software and training changes both in the Saudi use of the E-3A and ground-based C3I system, and in determining the best way to provide the Tornado with full intercept data in beyond-visual range or complex intercept modes.

NOTES

1. These data are taken from CIA, *World Factbook, 1983,* CR 83-11300.
2. CIA, *World Fact book, 1984,* Washington, April, 1984, p. 199.
3. Senate Foreign Relations Committee, *The Proposed AWACS/F-15 Sale to Saudi Arabia,* GPO; Washington, 84-557-0, September, 1981.
4. For a good detailed description, see *Defense and Foreign Affairs Daily,* May 21, 1984, June 11, 1984, and March 4, 1985; and the *Wall Street Journal, Washington Post,* and *New York Times,* February 26, 1985.
5. *Washington Post, New York Times,* and *Christian Science Monitor,* June 6 and 7, 1984.
6. *Aviation Week,* July 22, 1985, p. 21.
7. *Aviation Week,* July 22, 1985, p. 22.

9
The Saudi Arms Sale Crisis of 1985–1986

Much of this patient progress by Saudi Arabia and the U.S. was halted by the Saudi-U.S. arms sale crisis of 1985–1986. In spite of considerable patience in trying to preserve its military relations with the U.S., the Kingdom was forced to turn to Europe. The resulting events which shaped the original Saudi request to the U.S., the refusal of the U.S. Congress to grant the request, and the eventual Saudi arms purchase from Britain make a classic case study in both the strengths and weaknesses in the strategic ties between the West and the Gulf states, and between the U.S. and Saudi Arabia.

THE ORIGINAL SAUDI ARMS REQUEST: THE "F-15 PACKAGE"

Saudi Arabia originally sought to continue the modernization and expansion of its air force by obtaining arms from the U.S. Even at the time of the E-3A debate, it was clear that Saudi Arabia would need to expand its F-15 force, acquire a more advanced attack mission capability, and begin to replace its aging F-5s at some point in the late 1980s to early 1990s. This led to a series of low-level planning efforts by the Saudi Air Force and U.S. Air Force. After long consultation with the U.S. State Department and Department of Defense, the Kingdom developed an arms package which called for the further expansion and modernization of Saudi air defense capabilities, but which deferred the full modernization of Saudi attack mission capabilities, and involved a complex series of technical constraints to help protect Israel.

Both sides agreed that any new Saudi arms had to be deferred until after the presidential election campaign of 1984. Shortly after President Reagan's sweeping re-election victory, therefore, both Saudi Arabia and the Administration began to move towards a formal Saudi request for additional arms and toward dealing with the Congressional debate that was certain to follow.[1] The Reagan Administration initially felt its new mandate was sufficiently strong to rush the sale through, and planned to announce the sale during King Fahd's visit to Washington on February 11, 1985. [2]

The details of this new Saudi arms request are shown in Table 9.1. The request involved $3,612 million in immediate U.S. FMS sales. It also involved a carefully crafted compromise in which Saudi Arabia gave up the acquisition of an advanced attack capability and access to some advanced air combat technology for an increase in the strength of the Saudi Air Force improved air defense capability, and a capability to deal with the naval threat from Iran and the growing naval threat in the Red Sea. The new "F-15 package" was a logical next step in eliminating the remaining gaps in the ability of the Saudi Air Force to provide a strong deterrent in the Gulf and Red Sea areas.

The key developments in the new arms package were the gradual conversion of Saudi Arabia's existing 60 F-15 C/Ds to more advanced versions of the aircraft, the purchase of additional F-15C/Ds, and several major arms acquisitions. These new acquisitions included 40 more F-15 C/D with MSIP plus 8 additional aircraft to be held in the U.S. as an attrition reserve, 1,620 more AIM-9L/P missiles, and the remaining 800 Stingers out of a 1984 order for 1,200. Saudi Arabia also sought 100 Harpoon air-to-ship missiles for use on its F-15s, and 12 unarmed Blackhawk UH-60 helicopters, with an option to buy 12 more, to provide a limited amount of heliborne lift for its army.[3]

The upgrading of Saudi Arabia's F-15 force, however, was the key element of the new arms request and the Reagan Administration felt it could obtain sufficient support in the Congress to obtain the votes of one-third of the Senate, which is required to prevent a Senate resolution against an arms sale from going into force.

The key was the timing of the modernization and expansion of the Saudi F-15 force. While the U.S. could begin deliveries and conversions of the F-15C/D MSIP in 1988, it could complete them only by 1991–1992. This would not have allowed Saudi Arabia to fully absorb its new purchases and bring them on line in combat ready forces until 1993–1994. Many within the Reagan Administration felt this would defuse opposition by Israel's supporters, and Saudi Arabia agreed because both Saudi Arabia and the U.S. estimated that this would be in time to counter Iranian rearming and deal with the fact that radical Soviet-supplied states would have large numbers of Soviet-made aircraft roughly equivalent to Saudi Arabia's current version of the F-15 C/D.

Table 9.1 The Saudi Arms Request of 1985

- Conversion of Saudi Arabia's existing F-15C/D to the Multi-Stage Improvement Program (MSIP).

  — Total cost: $250 million.
  — Conversion to be accomplished in batches of 3 in Saudi Arabia. Would require a team of 80 contract personnel.
  — Conversion would begin 46 months after signing letter of offer (1988), and be completed in 3 years (1991).

- Purchase of 40 additional F-15C/D MSIP with an additional 8 fighters to be kept in the U.S. as an attrition reserve.

  — Total cost: $2.8 billion.
  — Delivery would begin 40–46 months after signing of letter of offer (1989), and be completed in 1990. Full operational training would begin in the summer of 1990 and be completed in 1992.

- Purchase of 980 AIM-9L air-to-air missiles for the F-15C/D, and 630 AIM-9P4 for the F-5E/F to round out previous purchases of 1,177 AIM-9L and 2,500 AIM-9P4.

  — Total cost: $100 million for AIM-9L and $45 million for AIM-9P4.
  — Delivery time is 30 months after LOA for AIM-9L and 34 months for AIM-9-P4.

- Purchase of 800 manportable Stinger surface-to-air missiles for the Saudi Air Defense Corps to complete a 1984 authorized program of 1,200, and provide defense for Saudi Arabia's ports, oil, and other key facilities in its Eastern Province.

  — Total cost: $89 million.
  — Delivery time is 29 months after signing of LOA.

- Purchase of additional 100 Harpoon missiles with air-to-ship capability for Saudi Arabia's F-15C/Ds to provide an improved maritime deterrent and defense capability once these complete MSIP conversion.

  — Total cost: $106 million.
  — Delivery time is 30 months after LOA, but cannot be used until F-15 C/D MSIP is completed.
  — Cannot be used against ground targets.

- Purchase of 12 unarmed Blackhawk UH-60 helicopters with an option to buy 12 more to provide the Saudi Army with limited heliborne lift.

  — Total cost: $267 million.
  — Maximum lift is 11–14 troops or 2,640 pounds of cargo. Maximum speed is 145 knots and range is 324 NM.

THE STRATEGIC RATIONALE FOR THE NEW F-15 PACKAGE

It is important to understand the full strategic rationale for the Saudi arms request because it illustrates not only why virtually every major foreign policy and defense official in the Reagan Administration at least privately agreed to the military need for the sale, but also the reasoning behind Saudi Arabia's decision to turn to Britain once it became clear that the U.S. could not go ahead with the sale. The previous chapters have made it clear why Saudi Arabia feels it must invest in the most advanced fighter it can obtain, and in one which is directly compatible with its existing force structure.

In broad terms, the new arms request offered Saudi Arabia the following advantages:

- Superior air combat technology: This is the only way to impress its radical neighbors with its ability to use force quality to make up for its inferior force numbers;
- The future option of upgrading its F-15s to provide superior air attack capability: This is the only way Saudi Arabia can compensate for its acute weaknesses in land force numbers relative to threats from the Northern Gulf and the Yemens;
- A demonstration of U.S. support for Saudi Arabia: This could give "over-the-horizon" reinforcement from USCENTCOM more credibility and military effectiveness;
- Enhanced Saudi prestige and ability to lead the GCC states;
- A demonstration of "balance" in the U.S. treatment of Israel and Saudi Arabia: This could help defuse the political and strategic backlash in the Arab world resulting from Saudi ties to the U.S.;
- Adequate provision for the necessary lead time in force modernization: U.S. fighter deliveries lag at least 2–4 years behind U.S. approval of an order. This lead time—combined with the 2–3 years it takes to fully convert to new aircraft and advanced technologies—means much of the advantage of advanced technology is lost by the time the equipment actually becomes fully operational.
- Cost-effective force improvement: The marginal cost of buying the most advanced air combat technology is negligible in "life cycle" cost relative to the overall price of conversion and support, and buying the most advanced systems usually adds five years more life to a major weapons system relative to the improvement in potential threat systems. Coupled with investments in infrastructure and munitions, advanced technology purchases can save Saudi Arabia 30–50% in real life cycle costs over mid-level technologies.

THE MAJOR TECHNICAL ISSUES

The rationale for the new request becomes even clearer when one examines the technical issues involved in each major part of the new arms package.

The Upgrading of Saudi Arabia's Existing 60 F-15C/Ds

Saudi Arabia was virtually forced to obtain some form of the multistage improvement program for its F-15C/Ds. The existing versions of the F-15C/Ds were going out of production, and Saudi Arabia faced major eventual increases in its training, munitions, and operating costs if it could not improve its F-15C/Ds to U.S. standards because the U.S. Air Force would cease to support some of the systems in the existing configuration of the F-15C/D as standard equipment.

The Saudi Air Force also had an obvious interest in upgrading the performance of its F-15C/Ds, and these upgrades were of considerable value in meeting its mission requirements. Table 9.2 shows what the MSIP program could add to Saudi Arabia's existing F-15C/Ds.

The Saudi F-15 MSIP was not, however, to include all the features in the USAF MSIP-2 program. The omissions from the Saudi MSIP package included the new dual mode threat warning system, secure voice system, full TEWS, new electronic warfare system (EWWS), Mode 4 identification of friend and foe (IFF), MER-200 multiple ejection racks, and Joint Tactical Information Distribution System (JTIDS). The omission of JTIDS was particularly important, and illustrates the extent to which the proposed arms package was tailored to protect Israel.

JTIDS is a key part of the USAF MSIP program. It provides the F-15 C/D MSIP with more than a secure communications system. On the improved F-15C/D, it replaces today's limited voice cueing, and sector scan radar, with far more advanced fighter control and communications. It provides digital communications and conversion of voice information; grid referenced TACAN and navigation information; automatically updated position, speed, and vector data on unknown and unfriendly aircraft; and coordination data for local flights and air-to-ground missions. It also allows the new tactical scope to display the equivalent of a 360 degree picture of the current combat situation including target and mission data. Expanded modes may also provide data on hostile surface-to-air missiles, allow automated coordination with other air defense fighters (an automated mini AWACS capability), and other features.

Even without JTIDS, however, the new 5X5 inch color cockpit display

Table 9.2: Multi-Stage Improvement Program (MSIP) for the Saudi F-15C/Ds

- A Programmable Armament Control Set (PACS) with:
  —Improved signal data processing.
  —Improved data processing.
  —Multicolor displays.
  —Conversion programs to allow more effective use of modern air armament.
- An improved central computer with four times the memory and three times the processing capability of the existing computer.
- A modified APG-63 radar with wider band width, frequency agility, and increased dynamic range.
- Incorporation of MIL-STD 1760 wiring.
- A modified Tactical Electronic Warfare System (TEWS) allowing for the use of advanced weapons such as the AIM-9M, Improved AIM-7, and AMRAAM—although none of these weapons are to be approved for release to Saudi Arabia.
- Modifications to allow use of the Harpoon air-to-ship missile.
- HF radio.
- ALR-56 radar warning receiver (RWR).
- ICS in Variant B with group A provisions.
- Internal provision for JTIDS upgrading without JTIDS.

would still have provided Saudi Arabia with several important advantages. These would have included built-in test displays, graphic displays of armament stores to replace the existing armament control panel, and a video display of electro-optical sensor systems and weaponry.

Further, the Saudi MSIP did not include any of the features of the far more ambitious MSIP-3 program. The MSIP-3 is a full dual-role fighter conversion of the F-15C/D which includes many of the features of the F-15E. These include the APG-70 high-resolution radar, a much more advanced HUD digital flight control, automatic terrain-following modes, expanded weapons capabilities, and a 9 G flight envelope. It did not include the full integrated flight/fire control system (IFCCS-1) which uses the Firefly III electro-optical sensor tracker to both improve air combat capability and allow far more survivable maneuvering attacks instead of standard pop-up attacks.[4]

The Saudi MSIP program could have made the Saudi Air Force compatible with the USAF on a support and block improvement basis, but it scarcely represented a threatening transfer of technology either in terms of potential loss to the Soviet Union or in terms of giving Saudi Arabia an edge over Israel.

This was particularly true because of the timing involved in the

the sale as early as October 1985, and the letter of offer (LOA) had been signed the next day, the first conversions could not begin in Saudi Arabia until 46 months later because of contracting (6 months) and manufacturing (40 months) lead times. The 80 man contractor team to be sent to Saudi Arabia would then be able to process only three aircraft at a time, and will take three years to complete the conversions. The full conversion process would not have been completed until 1992.

The Purchase of 40 More F-15C/D MSIP

The most expensive and controversial aspect of the Saudi arms request, and the one that did the most to provoke Congressional resistance, was Saudi Arabia's desire to buy 40 more F-15C/D MSIP fighters, plus 8 more aircraft to be kept in the U.S. as attrition reserves. The issues affecting this controversy, however, were more apparent than real. They involved F-15s with the same performance, and technology transfer limitations, as the F-15 C/D MSIP conversions discussed above. They also involved similar lead times.

Even if the Congress had approved the sale in October, 1985, and Saudi Arabia had signed the LOA the next day, work on the new aircraft could not have started until April, 1986. The first aircraft could not have been delivered until February, 1989, and delivery could not have been completed until mid-1990. Even with preliminary training on the existing F-15C/Ds, or in the U.S., full squadron-level Saudi training could not have begun until the summer of 1990, and could not have been completed until 1991–1992. By this time, the new Saudi aircraft would face far larger regional hostile forces equipped with Soviet and European fighter types which would be far more comparable to the F-15C/D MSIP than today's hostile fighters are to the F-5EII and F-15C/D.

Further, Saudi Arabia did not have the option of buying more existing F-15C/Ds because the production line of these aircraft had already closed. Further, by 1992, many Saudi F-5s would be twenty years old. Nearly half will be at the end of their useful life, and restricted to a training role. As a result, the additional F-15s would not have represented a major increase in Saudi air strength, but rather would have served as replacement aircraft. In fact, Saudi Arabia could not have maintained its existing first line combat strength unless it had ordered 40–60 other fighters during this time frame.

The Purchase of Additional AIM-9L/P Missiles

The 980 additional AIM-9L and 630 AIM-9P4 air-to-air missiles in the new Saudi request were designed to create a total stockpile of about 1,900 AIM-9Ls and 2,800 AIM-9P4s. While Saudi Arabia already had ordered 1,177 AIM-9Ls and 2,500 AIM-9P4s, deliveries of the new order would have been staggered over 30–34 months after the approval of the sale and signing of the letter of offer, and training would have used up a significant number of existing missiles in the interim. The timing involved is illustrated by the fact that only 250 AIM-9Ls have so far been delivered out of the 1981 order for 1,177.

These total future inventories were approximately half the original force goal that Saudi Arabia had set based on USAF planning factors. These planning factors were based on target density, and the U.S. has since persuaded Saudi Arabia to accept goals based on (a) the need to disperse munitions throughout the Kingdom, (b) a 5–10 year life cycle and training attrition, and (c) probable maximum Saudi sortie rates and attrition against probable enemies. The resulting goals were conservative in that they make an informal allowance for U.S. ability to provide over-the-horizon reinforcement and compensate for Saudi losses in combat or lack of force numbers.

The new weapons were to be stored under U.S. security procedures, and could not increase the risk to Israel or the risk of technology loss. Saudi Arabia cannot hope to generate with Israel the additional sorties in air combat which it can achieve against a Gulf or Red Sea threat, and would not have enough surviving aircraft to use the additional missiles in a war with the Israeli Air Force.

The Request for 800 Stinger Missiles

The Saudi request for 800 manportable Stinger surface-to-air missiles would have supplemented earlier deliveries of 400 missiles out of an order of 1,200, and they would have been delivered 29 months after the signing of the letter of offer, or no earlier than 1989. The Reagan Administration first delivered the existing 400 Stingers under emergency conditions in May, 1984, at a time when Iran was constantly issuing threats to broaden the war into the Gulf. The delivery consisted of 200 shoulder-fired launchers and 200 missiles with a follow-on delivery of 200 improved Stinger missiles.

The Stingers would have filled a critical gap in Saudi Arabia's ability to provide close-in protection against air and helicopter attacks, and U.S. studies validated a total Saudi requirement for up to 5,000 missiles.

The Saudi Air Defense Corps' land-based air defense units do not match the Saudi Air Force in effectiveness, and have only a limited capability to make effective use of their Crotale and Shahine systems. This situation could not be corrected until the deployment of the Shahine II and the Saudi E-3As, and until Saudi Arabia's new Peace Shield system could integrate the Improved Hawks, French-supplied short- to medium-range air defense missiles, and radar controlled AA guns into a true Saudi air defense system.

In addition, Saudi Arabia needed a quick reaction weapon for its ports, the oil facilities in the Eastern Province and Red Sea area, its ships, and its seven army brigades. Saudi Arabia cannot predict when Iran might launch air attacks, or use helicopters against Saudi Arabia. The Stingers could provide a highly mobile means of point defense, and one that can be used with only limited training. They have only moderate effectiveness against the kind of first-line attack fighters the USSR is beginning to deploy, but they are effective at line-of-sight ranges against helicopters, the kind of aircraft now flown by Iran, and most of the hostile aircraft likely to be in inventory through the mid-1990s.

Once again, Saudi Arabia was already using U.S. security procedures to protect its existing Stinger missiles. The additional deliveries could have provided nation-wide coverage over a three to five year period without significantly increasing the risk of technology compromise or the number that would be usable against Israel.

While the Stingers would have commercial identification of friend and foe (IFF) capability, it is important to note that they did not have the passive optical seeker technique (POST) version of Stinger. This sensitive system had a dual infra-red and ultraviolet (sometimes called "multicolor") counter-countermeasure capability against flares and IR decoys, new mirror alignment technology, and improved electronic signal processing. This ensured that such systems cannot be used with high effectiveness against U.S. aircraft with advanced countermeasures, and gives the U.S. the option of transferring such countermeasure technology to Israel.

It was also clear that Saudi Arabia could buy roughly the same arms elsewhere. France, Britain, Italy, and the USSR were all developing or selling roughly equivalent systems to Middle Eastern countries. According to U.S. officials, the heavier SA-13 missiles that the USSR had sold to Syria already had roughly equivalent counter-countermeasure and lethality characteristics to the Stinger, and a classified variant of a Soviet manportable weapon is available which also has similar capability.

At the same time, U.S. experts estimated that the USSR would not allow any other nation to have the countermeasure technology that Stinger POST is designed to be used against, and that the Stinger would fully meet Saudi needs over a 5 to 10 year period. The U.S. Army will only slowly convert to the POST variant—which is not yet in production—and it will be deployed only where first-line Soviet threats seem likely.

The Sale of 12 UH-60 Blackhawk Helicopters

The sale of 12 UH-60 Blackhawk helicopters, with the option to buy 12 more, was the least controversial aspect of the Saudi arms request. The sale met the urgent Saudi Army need for heliborne mobility discussed in Chapter 6. Each UH-60, however, can carry only 11–14 troops, or 2,640 lbs of payload, under typical Saudi mission conditions. Maximum speed is 145 knots and maximum range (not radius) is 345 nautical miles. The total purchase would have provided a total lift of only 132–168 men for all 12 helicopters, and the UH-60s were to be unarmed. This buy was adequate for urgent missions in support of other GCC states, to protect key facilities, and to deal with terrorism or infiltrators, but it scarcely presents a threat of a major Saudi airborne assaults.

The Sale of 100 Harpoon Air-to-Ship Missiles

The sale of 100 AGM-84A Harpoon air-to-ship missiles was designed to supplement Saudi Arabia's existing stocks of ship-to-ship missiles with the ability to deliver air strikes on threats like the Iranian Navy, the navies of the radical Red Sea states, and the other threats described in Chapter 4. It would also have helped give the GCC a rapid reaction capability against infiltrating ships, arms smugglers, seaborne terrorists, and small landing parties or raids.

The Harpoon missiles are highly effective against ships at ranges up to 90 km, depending on the altitude and radar resolution/horizon of the launching aircraft; but they cannot be used against land targets, and the U.S. would have retained all classified data on countermeasure technology. While the Harpoons could be delivered within 30 months of the letter of offer, the Saudi Air Force could not begin training to use the new missiles until the F-15C/D MSIP conversions were completed in significant numbers, and this would not have taken place until 1990 at the earliest. The Saudi Air Force could not have been fully effective in the anti-ship mission role until 1991–1992 at the earliest. By this time, it had to be prepared to face far stronger threat forces from both Northern Gulf and Red Sea states.[5]

THE MISSING LINK: DUAL CAPABILITY FOR THE SAUDI F-15

The one key need which the new arms request did not meet, and which was a powerful factor leading Saudi Arabia to buy 48 Tornado IDS attack aircraft, was strengthening the air-to-ground capabilities of the Saudi Air Force. In the long run, Saudi Arabia could not afford to try to defend solely by matching air force against air force. It had to have enhanced air-to-ground capabilities to make up for the weakness of Saudi ground and naval forces, and to use air power against the Northern Gulf states' overwhelming superiority in land force numbers.

Saudi Arabia's F-15C/Ds now use a relatively low-quality attack munitions delivery device called the MER-10, which Saudi Arabia obtained from a third country with U.S. permission. The MER-10 does not allow accurate or high-capacity attack munitions delivery, and forces the plane to fly at subsonic speeds, which make it extremely vulnerable to air attack in a high-density air environment.

The upgrading of Saudi Arabia's F-15C/Ds to full attack mission capability would not have given the Saudi Air Force attack aircraft equal to the Tornado IDS, but it would have provided considerable air-to-ground capability. The improved F-15C/D MSIP could be equipped to carry up to 17,800 lbs of munitions to typical attack sorties. Conformal fuel tanks could be provided to carry ordnance tangentially to reduce drag, and the multipurpose color video screen and upgrades to the computer would support low-altitude penetration and accurate weapons delivery against the probable threats in the Gulf and Red Sea areas.

The Saudi F-15s could carry up to three 30 mm gun pods, and electro-optical missiles such as the new imaging infra-red version of the Maverick AGM-65D. The AGM-65D has already been cleared for technical release to Saudi Arabia. While it is still experiencing some development troubles, it should be much more reliable and effective than Saudi Arabia's 1,600 existing AGM-65A/Ds. The new Tri-service IIR seeker would have allowed the missile to lock on in mist or poor contrast conditions, and the F-15s could be converted to use a FLIR pod or radar warning receiver to strike at standoff ranges against most of the systems that will be in the Gulf and Red Sea forces.

The modifications to the F-15C/D MSIP would also have allowed them to deliver such advanced munitions as HARM, Harpoon, runway suppression systems, modular glide bombs, and "smart submunitions". Like the Tornado, the F-15 C/D had the heavy munitions "lift" and delivery technology to provide an advanced attack mission capability well into the 1990s.

Further, the F-15 C/D MSIP could have been equipped with an improved version of the existing lightweight X-Band APG-63 Pulse Doppler Radar. The Improved APG-63 will be a significant upgrade of the original version. It has programmable signal processors with high-density storage elements in the radar which make it possible for the APG-63 to reject ground clutter. While the precise success of these upgrades is uncertain, it seems likely that the Improved APG-63 should be able to provide a considerable increase in ground resolution under all weather conditions, and at slant ranges of up to 20 miles. The upgraded radar has been tested as accurate to 3 meters at ranges of 18 kilometers [10 feet at 11.5 miles] and had a radar mapping navigation accuracy of 127 feet at a possible range of 173 miles.

Such an upgrading would not, however, have made the F-15C/D MSIP competitive in the air-to-ground mode with the F-15s that will be in the USAF inventory during this period. Even fully equipped air-to-ground versions of the Saudi F-15 MSIP would not have approached the lethality of the USAF F-15C/D MSIP-3, full F-15E, or Tornado IDS. The key differences between the U.S. and Saudi versions would have been the lack of the F-15's additional weapons officer position, new radar, and tactical weapons system (see Table 9.3).

Combined with an expanded Tactical Electronic Warfare System (TEWS), the new APG-70 radar and computer system will have up to 1,000 times the memory of the present APG-63 and about three times the processing speed. The APG-70 can store up to four times as much data and will have far greater growth capability to handle improved munitions, more demanding mission profiles, and more sophisticated countermeasure requirements.

The second seat and dedicated weapons officer in the full F-15E will be able to use an integrated LANTIRN capability to provide a full all-weather/night-attack capability, and the ability to simultaneously process both radar and FLIR data for attack missions. The imagery from the APG-63 can also be integrated with the FLIR system of Pave Tack, but the success of such integration is far more uncertain, and the "growth" capability to handle more complex munitions and mission profiles of the 1990s will also be uncertain.

The weapons officer will be able to use the APG-70 and its associated computer and data links to allow the F-15E to (a) coordinate with other fighters, (b) use the F-15 radar to vector other fighters, (c) make optimal use of data from an AWACS or ground sensor system, (d) conduct

Table 9.3 Features of the F-15E Which Would NOT Have Been on the Saudi F-15C/D MSIP

- Second seat and four multipurpose cathode ray tube displays, plus two hand controllers, for optimal weapons officer role in air-to-air and air-to-ground combat.
- APG-70 radar and computer with a high resolution Synthetic Aperture Radar (SAR) which is capable of all-weather detection of tank sized targets at 30 NM, of moving target detection, and of discriminating targets only 10 feet apart.
- Tactical Situation Display with electronic moving map bearing orientation, threat status, navigation data, and sensor management information.
- Advanced terrain following/terrain avoidance system.
- LANTIRN target forward looking infra-red (FLIR) for high resolution imaging and targeting.
- Advanced ECM and other countermeasure gear and Mode 4 IFF.
- Combined gun/internal countermeasure set.
- Increase to five air-to-ground store stations.
- Optimal avionics and platform capability to use AMRAAM, the final form of the Advanced Joint Tactical Missile (JTACM), boosted/conventional glide bombs, and smart munitions dispensers and glide bombs.
- Increase to 81,000 lbs maximum gross take-off weight, and add internal fuel, permitting payloads to 24,500 lbs.
- 9 G maneuver capability with payload.

anti-ship surveillance and attacks, (e) fight in beyond-visual-range (BVR) and non-dogfight intercepts, and (f) make optimal use of the AIM-9M, advanced AIM-7, AMRAAM, and successor air-to-air missiles.

IMPACT ON WESTERN CONTINGENCY CAPABILITIES AND NEEDS

The U.S. would have gained significant strategic advantages from meeting the Saudi arms request, as well as over $10 billion in total export earnings. The sale would have created larger interoperable stocks, facilities, and forces in an area 12,000 miles away from the U.S., and in the country best positioned to defend the West's oil assets. It would effectively have "prepositioned" more support facilities, maintenance and repair capabilities, and C^3I assets of the kind that ease the deployment problems a U.S. force faces in moving to Saudi Arabia, and in covering the upper Gulf or lower Red Sea areas. The arms sale package would have allowed the U.S. to simultaneously build up regional deterrent capabilities while improving its own contingency capabilities, and allowed the U.S. to do so at Saudi expense.

The fact that the Saudi Air Force would have operated a force of 100 F-15C/D/E MSIPs might have meant that the U.S. would have been able to fight with regional allies and not in isolation. Saudi Arabia will continue to equip the largest, most modern, and best-sheltered air bases in the Middle East. It will also steadily improve the quality of its "Peace Shield" command, control, communications, and intelligence (C^3I), and air control and warning (AC&W) systems in ways which will aid USCENTCOM deployment capabilities.

As has been touched on earlier, these Saudi air facilities can base up to two wings of U.S. Air Force fighters, and give them full munitions and service support. The bases could allow large amounts of U.S. air power to deploy to the most threatened areas in the Gulf in 48 to 72 hours. While Oman, Turkey, and Egypt provide useful contingency facilities on the periphery of the Gulf and lower Red Sea, they could not compensate nearly as well for the range and reinforcement problems USCENTCOM would face in defending the critical oil facilities in the upper and central Gulf.

U.S. air power also has the greatest contingency impact of all the forces the U.S. can deploy. As Table 9.4 shows, the U.S. estimate of the balance of U.S. and Soviet capabilities in the Gulf at that time led it to believe it could not hope to match the USSR or Iran in terms of land forces, and although it is steadily improving its ability to speed the deployment of USCENTCOM forces, it could still take more than a month to deploy its full strength. It can, however, deploy massive amounts of tactical air power within days. This air power may well be enough to offset the probable threat advantage in land forces, provided that it can deploy to compatible bases, use the same munitions and support facilities, use an advanced C^3I system, and operate with Saudi forces using the same tactics and doctrine.

In short, the Saudi F-15 package would have supported precisely the kind of broad strategic relationship with Saudi Arabia and the other moderate Gulf states that the West needs. It would have raised the threshold at which U.S. forces must be committed and allowed them to be far more effective when they were committed. It would have allowed U.S. forces both to stay "over the horizon" and to be effective. This would have minimized the political problems of maintaining a U.S. military presence in the Gulf area, yet would still have helped to ensure that such U.S. reinforcements had deterrent credibility.

WHY THE U.S. DENIED THE SALE

U.S. domestic politics, however, drove Saudi Arabia in a totally different direction. Even before King Fahd's visit to the U.S. in February, 1985, the Reagan Administration had rapidly learned that it had seriously miscalculated the effort it would have to put into getting Congressional support for the sale. The problem was not so much one of winning the resulting debate in the House and Senate; it was one of being willing to pay the political price of a long and grueling political battle.

The domestic policy advisors in the White House, and senior Republicans in the Congress, were worried that any major arms sale to an Arab state would lead to a major fight with every supporter of Israel. They were concerned about the slow rate of progress in making new Arab-Israeli peace initiatives, about the impact on the President's ability to achieve tax and budget deficit reform, and about the potential cost in Jewish-American votes and support at a time when this part of the ethnic vote was turning towards the Republican Party and could be critical to maintaining Republican control of the Senate in the 1986 elections. A growing gap developed between the advocates of U.S. strategic interests and advocates of the President's effort to strengthen the conservative political movement within the U.S.[6]

The Reagan Administration never firmly came to grips with this division between the advocates of strategy and the advocates of domestic politics. By mid-January, 1985, it had to warn King Fahd that it could not send the new arms request forward during his visit. This almost led the King to cancel his trip, until he was quietly promised that the President would make a personal commitment to him to send the sale forward at a future date.[7]

At the same time, however, the Administration was running into serious problems with a similar arms sale to Jordan. The Congress was increasingly demanding a clear picture of the Administration's arms policy towards the Middle East, and "linkage" between any sale to Saudi Arabia or Jordan and major progress on a Middle East peace settlement. For many supporters of Israel, this progress meant direct talks among Israel, Saudi Arabia, and Jordan that excluded the PLO—a requirement that neither Saudi Arabia nor Jordan could meet without isolating themselves in the Arab world and fundamentally undermining their security.

The result was that the Administration bought itself time by stating it would defer all new arms requests for the Middle East until it had prepared a comprehensive report to the Congress on such sales. This announcement was made on January 31, 1985, by Richard W. Murphy, the

Table 9.4 Comparison of U.S. and Soviet Forces Available for A Gulf Contingency: 1986-1987

U.S. Combat Forces: Service / Units	Personnel	Soviet Combat Forces: Service / Units	Personnel
1. Army			
82nd Airborne	16,200*	1 Tank Division	11,000
101st Air Assault	17,000*	22 Motorized Rifle Divisions	152,000
24th (Lt. Mech) Division	12,300*	1 Airborne Division	9,000
Ranger Battalions (2)	1,200	3 Artillery Divisions	15,000-18,000
Air Cavalry Brigade/ 5th Special Forces Group	2,500		
6th Air Combat Brigade	?		
7th Light Division	?*		
9th Light Division	?*		
2. Amphibious Assault			
1 Marine Amphibious Force (Division/Air Wing)	47,500	2 Naval Infantry Regiments	5,000
3. Air Force		Frontal Aviation	
9th Air Force		2 TAAs with 6 fighter and 6 fighter-bomber regiments:	
7 Wings Plus 3 Wings Reserve		6th TAA (Turkestan MD)	
Including:			
F-15 (1 Wing)	72	with MiG-21, 23, 21R, Mi-8	175
F-4E (1 Wing)	72	34th TAA (Transcaucasus MD)	
A-10 (1 Wing)	85	with MiG-21, 23,27, Su-24,25,27	300
F-111 (1 Group)	144		
	373		

B-52H (2 Squadrons)	35	Long-Range Aviation Tu-95, Tu-16, Tu-22, Tu-26 Bombers	100+
		Naval Aviation Naval Bombers	100+
4. Naval Forces			
2-3 Carrier Air Task Forces		1–2 ASW Carrier Task Groups	
Combat Aircraft	258	Combat Aircraft (VSTOL)	28
5 ASW Aircraft Squadrons		2–3 Guided Missile Task Groups	
Prepositioned Ships	17-21		
1 MAB Amphibious Lift		1 Naval Infantry Regiment plus Amphibious Lift	
? Nuclear Missile Submarines		? Nuclear Missile Submarines	

*To be reorganized during 1986–1995 as part of army's new light division concept. Decisions as to assignment of 7th and 9th Divisions are being re-evaluated.

Source: Adapted from Anthony H. Cordesman, *The Gulf and the Search for Strategic Stability*, Westview, Boulder, 1984, p. 818; and IISS, *Military Balance, 1985–1986.*

Assistant Secretary of State for Mideast Affairs, in testimony before the Congress. It was the result of considerable debate at the staff level, but Secretary of State George Shultz, Secretary of Defense Caspar Weinberger, and the President's National Security Advisor, William C. "Bud" McFarlane, finally agreed they had no other choice.[8]

Senator Alan Cranston of California had already gathered some 51 signatures of U.S. Senators on a letter opposing the sale before the Administration had announced its terms or begun to defend it. Senator Richard Lugar, the Chairman of the Senate Foreign Relations Committee, had warned the President he might lose a political battle over the sale, and McFarlane was particularly concerned about the impact on Republican control of the Senate after the 1986 elections. Further, it had become clear during Israeli Defense Minister Yitzhak Rabin's visit to Washington the week before that Israel would strongly oppose a sale to Saudi Arabia, in spite of the Administration's willingness to increase military aid to Israel to $1.8 billion in FY1986, a $400 million increase over the aid provided in FY1985.[9]

The timing of the Administration's announcement of a delay in arms sales to Arab states was acutely embarrassing because it came just before the King's visit, and was so clearly linked to Rabin's visit. Nevertheless, the Reagan Administration felt it had found a way to deal with Israeli and Congressional pressures. Several senior Administration sources agree that President Reagan felt confident enough to make a direct verbal promise to King Fahd that he would send the sale forward when the King visited Washington in February, 1985. Exactly what the President said is still unclear, but the Saudis certainly were confident enough in this promise that they slowed efforts to find a European alternative to the U.S. arms sale.

The Reagan Administration did not, however, use the time it had bought to push the sale forward. While the "Middle East arms sale study" continued to go on within the Administration at the staff level, regurgitating material and work that had actually been fully completed at least six months earlier, the Administration found itself in more and more trouble on domestic political issues. The various lobbying groups supporting Israel, and opposing not only the arms sale but U.S. strategic relations with Saudi Arabia, continued to work on the Congress without any serious opposition. A well-orchestrated press campaign was launched to oppose the sale, and the heavily Democratic House of Representatives became more and more vehement in opposing the sale.[10]

The Reagan Administration had originally planned to issue its arms sale report to the Congress in both classified and unclassified form in April or May, 1985, and then send it forward as the opening gun in a major political effort to support the sale. As time went on, however, domestic politics began to take more and more precedence over foreign policy and strategic interests, and the impact on Republican chances for control of the Senate after November, 1986, became the primary issue. While even a number of very senior Administration officials still pushed hard for the sale, including the Secretary of Defense, some of the White House and National Security Council staff began to push for killing it. Others simply stalled, hoping for some dramatic catalyst from the Iran-Iraq War or the Middle East peace process.

The Middle East arms sale report was distinctly put on a the backburner. When it was finally issued, it was issued only in classified form and became one of the most unread documents in Washington. The U.S. Air Force was not allowed to conduct any meaningful briefing effort on the need for the sale, U.S. foreign policy and defense experts were not allowed to brief the Hill, and the U.S. defense industry was put on hold. The only lobbying effort for the sale consisted of small and badly organized pro-Arab groups which spent so much time defending the PLO and Libya that they did at least as much harm as good.

The Administration simply dithered. The State Department and Defense Department both prepared actively to present the new arms package to Congress until virtually the day President Reagan finally decided not to go ahead. Even officials at the Assistant Secretary level were so convinced of the President's continuing support for the Saudi sale that the State Department notified Saudi Arabia during Ramadan that the Administration was ready to give Congress informal notification of the sale.

When this notification was delayed by the White House, a whole series of false notification announcements followed. Indeed, senior State Department officials again gave Saudi Arabia a "green light" that the sale would go forward less than a week before the President's final decision not to proceed with the sale. The whole impression on the Saudis was one of an Administration in chaos, captive to its own fears and lobbying efforts, and unable to make a decision. Worse, even those Saudis who understood the U.S. best could never fully understand that getting the President's promise was not the same as getting a King's promise. The impression that the President was too weak or indecisive to act was reinforced by the Administration's inability to send clear and consistent signals to the Saudis even from the senior policy level.

The Administration did make one final attempt to help the sale go through in May, 1986, but this only made things worse. A senior Administration official called Prince Bandar bin Sultan, the Saudi Ambassador to the U.S., when he was already in London discussing options to buy the Tornado with the Thatcher government. The U.S. official stated that the U.S. would send the F-15 package forward to the Congress *if* the Saudi government agreed to support the Jordanian peace initiative, *if* the Saudis agreed to support a parallel U.S. arms sale to Jordan, and *if* the Saudis agreed to pay for the arms sold to Jordan.

After urgent consultation with King Fahd, the Saudis tentatively accepted the last two conditions. The first condition, however, threatened to drag them into the middle of a political battle among Jordan, the PLO, and Syria at a time when the Iran-Iraq War was escalating and Saudi Arabia was trying to promote Arab solidarity in dealing with Iran. Further, the whole offer rapidly proved to lack the President's support. When Bandar returned to Washington, he found that Shultz and McFarlane did not really support the plan, nor did President Reagan. The U.S. had sent another high-level false signal to the Kingdom.[11]

By July, 1985, the situation had gotten so bad, and U.S. officials had sent so many conflicting messages and signals, that the Saudis finally asked President Reagan for a formal letter stating that the sale could not go through and that the U.S. would understand if Saudi Arabia turned elsewhere for arms. The President sent the letter, and this triggered a mad scramble between Britain and France to pick up the Saudi market.[12]

When President Reagan's letter finally came, however, it arrived under conditions which were fully transparent and deeply embarrassing to the Saudis and most other Arab embassies in Washington,. It was clear that the President decided against the sale because of pressure from senior Republican members of the Senate, including Senator Richard Lugar, the Chairman of the Senate Foreign Relations Committee.

It was also clear to the Saudis and other Arab states that the Senators only gave this advice for one reason: the fear that the sale would lead to an additional punishing political battle in the midst of domestic political battles over the budget and national debt, and that the backlash from a public debate might produce enough hostile reaction from Israel's supporters to weaken Republican chances for control of the Senate in the 1986 elections.

Domestic political pressure had obviously taken clear precedence over Secretary of State George Shultz and Secretary of Defense Caspar Weinberger's support for the sale and their repeated reassurances to the Saudis. Senior officials in the National Security Council and several of the President's senior domestic political advisors opposed the sale only because of their concern for its impact on the 1986 elections. In fact, no major actor in the Administration's decision-making process seriously challenged the merit of the sale from the viewpoint of U.S. national security. The only reason for blocking the sale was domestic politics and the desire to avoid a political fight with pro-Israel lobbyists.[13]

This political context added insult to injury. It virtually forced Saudi Arabia to turn to another arms supplier. It was a further step in a long U.S. history of rejecting or delaying Saudi arms requests that had already helped push Saudi Arabia to turn to Europe for equipment and advice for its army and navy. It capped the similar mismanagement of U.S. arms sales efforts to the other Southern Gulf states, and represented the virtual paralysis of efforts to improve U.S. military ties to the Gulf states and contingency capabilities in the Southern Gulf.

It is hard to blame Israel or its supporters, however, for what happened. The problem lay in the Reagan Administration, in its initial failure to properly check Congressional opinion and then its failure to even try to actively prepare the Congress for the sale. The White House failed when it decided it could wait until the last minute and then use the President's prestige to rush the sale through Congress. While the Administration did nothing, strong and well organized pro-Israel lobbying groups did what special interest groups are supposed to do in the U.S. and made every possible effort to block the sale. By the time the White House finally did decide to act, it was simply too late. The Congress had been thoroughly briefed and mobilized by supporters of Israel.

THE SAUDI DECISION TO BUY FROM BRITAIN

This long series of uncertain U.S. signals, of "green lights" followed by slow indicators and then rebuffs, helped keep Saudi Arabia's interest in buying from Europe alive, and helped shape the Saudi decision to turn to Britain. The Saudi Air Force had had other options to the U.S. sale since at least early 1983 and Saudi Arabia had begun to lay the groundwork for buying substitute weapons and aircraft from Britain in early 1984.[14] Both the British and French realized they were unlikely to make any major fighter sales to Saudi Arabia as long as the U.S. was able to overcome its domestic political problems in providing such arms, but by late 1983, they had concluded that there was a good chance that the U.S. could not make further fighter sales. By early 1984, Britain and France had massive lobbying efforts under way both in Saudi Arabia and in their respective capitals.

In January, 1984, the British Minister of Defense, Michael Heseltine, went to Riyadh to meet with the Saudi Defense Minister, Prince Sultan ibn Abdul Aziz. Prince Sultan was normally considered strongly pro-

French, but the British felt they had two good arguments for an aircraft sale. The first was that the Tornado was a much more sophisticated and survivable aircraft than the competing Mirage 2000, had two engines, and could fly much further and deliver a heavier payload. The second was that France had just won a major arms sale in the form of a $4 billion short-range air defense system for the Saudi Air Defense Corps. The British had also done an excellent job of briefing some of the younger Saudi princes with experience in the Saudi Air Force, virtually all of whom felt the Tornado was a more capable fighter. The result was that Prince Sultan signed a non-binding letter of intent to buy 40–50 Tornados.

This did not stop the French from competing, however, and France felt it had three major advantages: better contacts with the royal family, a better sales team in France and in the Kingdom, and a cheaper aircraft that would allow the Saudis to buy the Mirage 2000 as a replacement for their U.S.-made F-5s and then to buy more F-15s when the U.S. political climate shifted.

The French made real headway when King Fahd visited the Cote d'Azur in April, 1984. The French Minister of Defense, Charles Hernu, flew down to see the King. He capitalized on Saudi dissatisfaction with the U.S. and the British withdrawal from Lebanon, and persuaded the King to come to Paris to visit President Mitterrand. The King's talks went so well that he left the French with the clear impression that he had agreed to the sale, although the Saudis feel he did no more than express his interest.

The French position seemed further strengthened between April, 1984, and March, 1985, as the U.S. sent new signals that it might sell Saudi Arabia the F-15 package. Saudi oil revenues had begun to decline, and the Tornado was too expensive and too complex to be bought in addition to the F-15. The Mirage 2000, however, was a good replacement for the F-5 as a second-line fighter, and it seemed unlikely that the U.S. could both sell the F-15 package and make a timely later sale of an F-5 replacement. The French also were successful in entertaining a number of senior Saudi officers and officials, and in using Akram Ojjeh—a Saudi arms agent living in France—to reach Khalid bin Abdul Aziz al Ibrahim. Ibrahim was the brother of the King's favorite and most recent wife.

The British, however, launched their own entertainment and briefing battle during this period, and they used Wafiq Said, a Syrian-born Saudi national, with good contacts within the Saudi Ministry of Defense. Said normally worked with the French, but had helped arrange Sultan's signature on the British letter of offer, and was in an excellent position to brief the Saudis on every problem in the French offer.

Prince Sultan's son, Prince Bandar, wanted the Tornado for both technical and strategic reasons. While some Saudi Air Force officers saw the sale only in terms of an F-5 replacement and strongly supported the Mirage 2000, the French fighter lacked military impact and political prestige. Both Sultan and Bandar did not like the French use of agents to pressure the King and tried to win influence with the Saudi officer corps. Further, Prince Bandar, who was not only the Saudi Ambassador to the U.S. but an ex-British-trained fighter pilot, was increasingly worried that the U.S. F-15 sale might fall through.

Prince Bandar arranged for Prime Minister Thatcher to visit King Fahd in Riyadh in April, 1985, on her way back from a trip to the Far East. The Prime Minister brought the prestige of having won in the Falklands and strongly impressed the King. This gave Prince Sultan the influence to cancel a meeting on the Mirage sale with Hernu in May, 1985, and led Prince Bandar to begin to spend almost as much time trying to close the Tornado sale in May and June, 1985, as he did trying to salvage the F-15 package. His efforts were aided after June both by the U.S. inability to make the F-15 sale and by the fall in oil prices. Britain proved ready to deal in oil, and even to offer a limited "buy back" of Saudi Arabia's obsolete British Lightnings. The French did not prove to be good oil traders, and already had problems disposing of Gulf oil because of their previous refinancing deals with Iraq.[15]

The groundwork for the sale was well prepared, therefore, when Saudi Arabia began the final negotiations with Britain in September, 1985. The a surprise was the extent to which Saudi Arabia expanded the original offer to buy 48 Tornados into a 132 aircraft package that raised the cost from around $4 billion to $7 billion and shifted the entire structure of the Saudi Air Force from dependence on the U.S. to dependence on Britain.

The Saudis seem to have concluded that they had no real hope of a future major U.S. sale until after the 1988 elections, and then only after a new President had had some time in office. They also concluded that the Reagan Administration would not see a major purchase from Britain as in any way hostile, and was actually encouraging the Saudis to distance themselves from the problems of U.S. domestic politics.[16]

The Saudi negotiating team had been in London for about eight weeks before the public announcement of the sale on September 26, 1985. It was only on September 24, however, that the chief of the Saudi Air Force received instructions from Prince Sultan to expand the sale from 48 attack versions of the Tornado to include 24 air defense versions of the Tornado, and 60 trainer aircraft, including 30 British-made Hawks and 30 Swiss-built Pilatus PC-9s. The catalyst seems to have been a new package deal that offered more rapid delivery of the Tornados at the expense of the RAF and was linked to a deal in which Shell Transport and Trading Company and British Petroleum were to work with ARAMCO over the next three years to market the necessary oil to pay for the all 132 aircraft.[17]

The final letters of offer and acceptance were signed on February 17, 1986. The British also agreed to invest 35% of the technical content of the contract in Saudi Arabia under the same kind of offset arrangements the U.S. had provided as part of the Peace Shield sale. Further, the contract allowed renegotiation of the timing and volume of the oil to be sold to protect Saudi Arabia against a further drop in oil prices. This became important in April, 1986, when oil had fallen from $26 per barrel at the time of the sale to $12.25 a barrel and the Saudis were threatening to reduce their aircraft purchases. In May, 1986, the British Secretary of State for Defense, George Younger, rushed to Riyadh, and in June, Chancellor Nigel Lawson renegotiated the deal to increase the volume of oil and stretch the repayment period out over six years to allow Saudi Arabia to remain within its OPEC quota. By that time, the Kingdom had already paid the U.K. some $300 million worth of oil and the British had already delivered six Tornado GR.1 attack fighters.[18]

U.S. EFFORTS TO MAINTAIN STRATEGIC RELATIONS

The final act in the arms sale crisis of 1985–1986 was a U.S. effort to rebuild its strategic relations with Saudi Arabia after the failure of the F-15 package and the Saudi decision to buy from Britain. By mid-1985, the Reagan Administration had begun to consider a smaller arms package that would at least modernize the Kingdom's F-15s, and provide most of the munitions the Kingdom had requested in the original F-15 Package. It became clear during the course of the rest of 1985, however, that the Reagan Administration could not obtain solid political support for any major new arms sale to Saudi Arabia, including the modernization of Saudi Arabia's existing 60 F-15C/Ds, without having to sacrifice some of its domestic political support. The Administration was also forced to let a major arms sale to Jordan die without full Congressional debate or action.[19]

The Congress kept putting pressure on the Administration, to limit sales to the Arab world. In July, 1985, the Senate introduced legislation that forced the Administration to report every major technical upgrade in the equipment it sold to foreign nations. This was clearly targeted towards Saudi Arabia since there had been some minor changes to the E-3A AWACS being sold to the Kingdom. These included better secure teletype equipment and software, because commercial equipment had proved unable to communicate with NATO teletype data links, better UHF equipment, the same color data displays that had become standard on all U.S. E-3As, and infra-red countermeasure equipment of a kind already released for use on all other Saudi aircraft.[20]

Several members of the Congress, led by Senator Cranston, then began a low-level effort to try to block the actual transfer of the E-3As that Saudi Arabia had bought on October 28, 1981. A section was inserted into the International Security and Cooperation Act of 1985 that required the President to make fully public all the details of his certification and of the precise ways in which the Saudis would meet U.S. conditions for buying and using the E-3As.[21] Some of these provisions were embarrassing to the Saudis, including "continuously and completely" sharing all data, prohibitions on the transfer of data to third countries, and limitations on its use in flights over Saudi territory. The intent of the opposition, however, was not to embarrass the Saudis with more publicity, but to lay the groundwork for challenging the President's certification, and to try to get enough votes in the Senate to block the sale.[22]

The Reagan Administration then began to fight back. The certification problem threatened to shatter every aspect of U.S. policy and military relations in the Gulf. It was also clear that a token arms package had to be pushed through the Congress to ease the growing internal pressure on the Senior Saudi officials that supported strong ties to the U.S., including Prince Sultan.[23]

On March 11, 1986, the Administration decided to send forward a token arms package of missile systems. This new arms package included 995 AIM-9L air-to-air missiles ($98 million), 671 AIM-9P4 air-to-air missiles ($60 million), 100 Harpoon air-to-sea missiles ($107 million), and 200 Stinger man-portable surface-to-air missile launchers and 600 reload missiles ($89 million). The total value of the new arms package was $354 million. The Administration used the excuse that the Iranian victory in Faw in early 1986 had led Saudi Arabia to urgently request additional missiles to improve its defenses against Iran. In fact, however, this was simply a Reagan Administration political cover to justify announcing the sale. [24]

The new "mini"-package was carefully designed to minimize the political cost of sending it forward. The Reagan Administration could defend the package on the technical and military grounds described earlier for these portions of the F-15 package, and on the basis that Saudi Arabia already had most weapons systems in inventory and had demonstrated its ability to provide security from espionage and terrorists. It could quietly make it clear that the reason for the high number of munitions was as much to provide stockpiles to enhance U.S.

contingency capability in the Gulf as to meet Saudi needs.[25] The only systems that were even slightly different were the AIM-9P4 and the air-to-surface version of the Harpoon. The AIM-P4 was only a slight upgrade over the AIM-9P3 already in Saudi Arabia, and Saudi Arabia already possessed a surface-launched version of the Harpoon.

The Reagan Administration also adopted a strategy based on a very low-profile lobbying effort for the sale, and on trying to make a tacit bargain with pro-Israel lobbying groups to support a high level of aid for Israel and to strengthen U.S. and Israeli strategic relations in return for limited opposition to the sale. This strategy, however, backfired. The lobbying effort was so low level, so badly organized, and so confused by the efforts of various pro-Arab lobbying groups whose real goal was to defend the PLO, that it almost collapsed. The various pro-Israel lobbying groups saw no reason in an election year to make any bargains and became more active than ever.

As a result, the Reagan Administration met so much Congressional resistance to the sale that it was forced to drop the Stinger man-portable surface-to-air missile from the package in spite of the facts that (a) the Saudis already had the Stinger and had proved they could preserve its security, (b) an independent study by the Congress General Accounting Office conducted in Saudi Arabia indicated the arms were safe in Saudi hands, and (c) Soviet SA-14 missiles of similar capability were being delivered to Syria and other Arab countries. IThe Administration faced a situation where 64 Senators had already joined Senator Alan Cranston in signing a resolution against the sale, and about 290 members of the House were ready to vote against the existing package. This was enough votes in the Senate to block the sale without a Presidential veto and enough votes in the House to override the veto if the President made it.[26]

The Reagan Administration was forced to recognize that it could never get the support of the House of Representatives, and that the President could never get enough members of the Senate to support the sale on his first try. Under U.S. law, his only way of making the sale was to accept the fact that the Senate would pass a resolution against the sale, but to use his influence to prevent the Senate from overriding a Presidential veto of the Senate bill. This meant the President would need the support of only one-third of the Senate.

When the first real key vote came on May 6, 1986, four-fifths of the Senate voted for a bill opposing the sale. This forced the President to rely on his ability to sustain a veto of the Congressional legislation blocking the sale. Ironically, the vote against the sale also illustrated the clumsiness of the Administration's lobbying effort even at the highest levels. The President's new National Security Advisor,

Admiral John A. Poindexter, wrote an editorial in the *Washington Post* designed to influence the Senate vote on the sale, but Administration lobbyists knew so little about the opposition that the editorial only appeared a day *after* the Senate vote.[27] The signal to both the Congress and the Saudis was that the Administration simply did not know what it was doing.

This again raised the specter of the Congress voting against the transfer of the E-3As as well and catalyzed a more intensive effort by the President.[28] The problem became one of whether the President could now salvage the sale even if he did commit all his personal prestige. After weeks of political arm twisting, however, President Reagan finally did win enough support to sustain his veto. When the vote came on June 5, 1986, President Reagan won by only a single vote. This victory was so minor, so uncertain, and involved so many Congressional attacks on Saudi Arabia, that it did almost as much to undermine U.S. and Saudi relations as to strengthen them.[29]

The only positive event in two years of U.S. and Saudi strategic relations was that the battle over the "mini"–arms package was sufficiently bitter and exhausting that the pro-Israel lobbyists chose to give priority to increased U.S. aid to Israel over yet another battle over the AWACS. Saudi Arabia formally agreed to comply with all the conditions the U.S. had set for using the E-3As in a signed agreement with the U.S. in mid-June, 1986.[30] As a result, the President was able to transfer the first of the five E-3A AWACS that Saudi Arabia had bought three years earlier with a simple letter of certification later that month. The President's certification, however, was scarcely a triumph. The U.S. had clearly reached the point where it was unable to maintain full-scale strategic relations with Saudi Arabia, and with any other moderate Arab state in the region except Egypt.

THE STRATEGIC IMPACT OF THE ARMS SALE CRISIS OF 1985–1986

The strategic impact of the arms sale crisis of 1985–1986, and of the growing shift in Saudi military relations towards dependence on Europe, becomes clear when the results are examined for each of the major actors involved. These impacts and trends are scarcely disastrous for the West, and can still be transformed into an effective structure of military relations, but they have unquestionably weakened Western military contingency capabilities in the Gulf. Ironically, they have also weakened Israel.

Impact on Britain

Britain will get major benefits from the sale, and faces only indirect and long-term risks. It will get a minimum of some $4–6 billion in additional exports. The sale will probably provide at least another $2–4 billion in support costs and other benefits, and could lead to substantial additional exports. This is of special importance to Britain because its arms exports have dropped from 2.7% of all exports in 1981 to less than 1.5% in 1984, and the value of its arms exports has dropped from $2.8 billion in constant 1982 dollars to under $1.5 billion.[31] The sale to Saudi Arabia re-establishes Britain as a serious competitor in the world sales of military electronics and aviation.

The first of the risks to Britain is that it must now demonstrate that it can actually provide the support and training necessary for Saudi Arabia to operate one of the world's most advanced fighter aircraft. Britain has not performed this role recently, although it has long provided much of Saudi Arabia's air training and services and has supported Saudi operation of the Lightning and BAC-167 fighters.

More significantly, Britain faces the risk that any erosion of U.S. military influence in Saudi Arabia weakens the U.S. ability to deter threats against the Gulf and Europe's key future source of oil imports. This risk, however, is a long-term and uncertain one. It must be balanced against the fact that Britain's aviation and defense electronics industries desperately needed the sale to Saudi Arabia to avoid massive layoffs in the late 1980s.

Impact on Saudi Arabia

The risks to Saudi Arabia are also limited. Saudi Arabia must trust in Britain's ability to provide the support it will need, although Britain lacks America's recent experience in managing large-scale foreign military sales programs. More significantly, the Kingdom runs the risk that the new turn of events will diminish the military assistance the U.S. can provide in an emergency.

U.S. ability to provide immediate "over-the-horizon" reinforcements is heavily dependent on U.S. air power, and the effectiveness of U.S. air reinforcements is heavily dependent on the extent to which Saudi Arabia can already provide interoperable bases and forces. Nevertheless, Saudi Arabia will not experience any such loss in U.S. reinforcement capabilities as the result of buying the Tornado until the late 1980s and early 1990s, when the 40 additional U.S. F-15s will have been delivered. By this time, oil prices are likely to have recovered and U.S. policy may have changed.

The Saudi Air Force will also get major offsetting benefits in return for any risks the Kingdom will run. It will get a total of 72 first-line Tornado fighters, rather than 40 F-15C/Ds—and the option of buying more whenever the Kingdom needs them and without future political restraints. It will get 30 Hawk light attack trainers as part of the package, and a wide range of technologies that were not available from the U.S.

The U.S. had not planned to provide advanced attack mission capability, or dual capability in the attack role on the variant of the F-15 C/D MSIP, all of which were to be heavily optimized for the air defense mission. The Saudi Air Force now will get 48 Tornado GR.1 or IDS (Interdictor-Strike) fighter-bombers. These Tornados can fly missions at ranges over 500 nautical miles (NM) in LO-LO-LO profiles, over 700 NM in close air support missions with one hour of loiter near the FEBA, and over 850 NM in HI-LO-HI interdiction missions.

The Tornados have one of the most advanced terrain-following systems in the world, full fly-by-wire capability, an excellent combined radar and map display for precision bombing, an advanced radar warning receiver, and the avionics, data display, and laser to make effective use of virtually any air-to-ground munitions including virtually all of NATO's "emerging technology systems" like the MBB MW-1 munitions dispenser and the GBU-15 laser guided bomb. They will be sold with full ECM gear, and have an excellent ability to penetrate below ground-based radar coverage and to use terrain masking to minimize the "look down" capability of advanced fighters.[32]

The Tornado IDS will give the Saudi Air Force the advanced attack mission capability it has sought from the U.S. for a decade. Britain will also sell Saudi Arabia the Sea Eagle anti-ship missile (the Tornado has a 700 nautical mile HI-LO-LO-HI range capability to deliver this missile including a terminal supersonic dash), the Dynamics Alarm anti-radiation missile, and far more advanced attack munitions than the U.S. was willing to provide. These will include many advanced- or emerging-technology weapons such as the new JP-233 air base suppression system and the smart submunition weapons Britain has under development.[33]

Moreover, the Saudi Air Force gets 24 Tornado ADV air defense fighters with more advanced air defense avionics and munitions than the U.S. was willing to release for the F-15. These will include full all-weather capability, more advanced tactical air displays, a multiple target track with scan radar and more advanced computer with up to 100 mile look down–shoot down and BVR missile capability, excellent

ECCM capability, and the ability to carry and fire four state of the art air-to-air guided missiles such as Skyflash and to fire them in a near "ripple" mode.[34]

The comparative performance specifications of the Tornado, the F-15, and the F-5E are shown in Table 9.5, and it should be clear even to the layperson that the Tornado has excellent performance specifications.

While the British Aerospace Hawk P.1182 T.1 is not an advanced combat aircraft, it is a relatively long-range light fighter with transsonic speed capability. It carries a 30 mm gun and two AIM-9Ls and has effective enough avionics to act as a light day interceptor. It also can carry up to 6,800 lbs (3,085 kgs) worth of attack munitions, and can fly patrol missions for up to 3 hours or fly over 200 NM HI-LO-HI attack profiles. While it lacks an advanced radar, its avionics are adequate for locating nearby fighters at medium to high altitudes.[35]

Finally, the Saudi Air Force will get its new British aircraft two to four years earlier than if it had ordered 72 F-15s from the U.S. While it has stated that it will not initially base its Tornados in Tabuk, the Saudi Air Force will have more long-term flexibility in basing the Tornados anywhere in Saudi Arabia—which the U.S. would have prohibited in the case of the F-15 because of opposition by Israel. It will be free of U.S. restraints on the employment of its aircraft and on the release of emerging technology that will allow the Saudi Air Force to equip itself with force multipliers.

Impact on the United States

It is the U.S. and its military capabilities which will suffer most. In the case of the U.S., the risks and costs will be strategic and financial. The main strategic risks are the erosion of U.S. military relations with Saudi Arabia, and the loss of interoperable bases, forces, and equipment. The main financial risks are the loss of a major arms sale worth an immediate $2.8 billion in new F-15 sales and the probable loss of $3–6 billion more in follow-on support sales, plus some $7–14 billion for replacing Saudi Arabia's F-5E-IIs. In addition, the U.S. will inevitably lose more civil sales in what has recently been a $4–5 billion dollar annual market for U.S. commercial exports.[36]

It is hard to say whether the damage to U.S. and Saudi military relations or that to U.S. contingency capabilities in the Gulf will be more significant. There is no question that the manner in which the U.S. failed to support the F-15 sale has eroded a great deal of Saudi and moderate Arab confidence in the Reagan Administration, and in its ability to maintain military relations with the Gulf states. The U.S.

decision also came at a uniquely bad time. The Saudis not only had at least some reason to believe that the U.S. had broken a Presidential promise to the King, but the U.S. rejection of the F-15 sale occurred within days of *New York Times* and *Washington Post* leaks of a classified U.S. study that mentioned that Saudi Arabia had made secret agreements for U.S. contingency use of Saudi bases.

The Saudi Defense Minister, Prince Sultan, did announce immediately after the signing of the agreement with Britain that the Tornado sale did not mean the Kingdom was turning away from the U.S., and stated he still had an interest in buying up to 50 more F-15s. He also visited Washington in early October, 1985, to negotiate the modernization of Saudi Arabia's existing F-15s and some additional arms purchases. Nevertheless, the Kingdom cancelled several other arms requests, and it was clear that the U.S. denial had left a distinctly bitter political aftertaste.[37]

Further, the fact that the sale went to Britain will affect U.S. contingency capabilities in the region. There has never been any question that Saudi bases could be vital in a U.S. effort to defend the West's oil supplies. The issue has always been their availability and their interoperability with U.S. forces. Until now, the U.S. has made steady progress in both areas. A Saudi shift to British aircraft will reduce U.S. ability to use Saudi facilities in both political and military terms. It makes it more politically difficult for Saudi Arabia to plan with the U.S. or to accept U.S. reinforcements, and it will undercut the interoperability between U.S. and Saudi forces.

Impact on Israel

While it is tempting to blame many of these problems on Israel or its supporters, the blame must rest with the U.S. No political group in the American political system can be blamed for having an effective lobby or for advancing its own self-interest. The blame lies with those in the Reagan Administration and the Congress who are charged with serving the U.S. national interest—and that of the free world—even at the cost of domestic political advantage.

At the same time, it is clear that the efforts to serve Israel's interests have ended in damaging Israel's strategic interests as well as those of the U.S. As has been mentioned earlier, this damage has taken three forms: It has increased the barriers to Saudi participation in the peace process, it has reduced U.S. ability to influence Saudi policy and military actions, and it has increased the theoretical military threat that Saudi forces could present to Israel in some future war.

Table 9.5: Comparative Performance Characteristics of Key Fighter Aircraft Affecting the Saudi Air Force Modernization Program

Part One: Aircraft Performance

			Mirage	Mirage		
	F-5E	F-15C/D	F-15E	F-1C	F-2000	Tornado
Mission Distance (NM)						
Air-to-Air Radius	360	720	720	400+	454	500+
Ferry Range	1,340	3,450	3,450	2,050	1,754	2,400
Attack Radius (NM)						
Hi-Lo-Hi	480	750	750	400	465	863
Hi-Lo-Lo	-	580	580	340	390	750
Lo-Lo-Lo	325	450	450	205	260	
400Payload (Lbs)						
Max.	8,000	-	-	8,820	16,758	-
W/ Full Internal Fuel	8,000	16,000	16,000	-	13,225	16,000
Air-to-Air Armament						
Gun	2 X M39	M61A1	M61A1	DEFA 30mm	DEFA2X 30mm	27mm
IR Missile	2XAIM-9	4XAIM-9	4XAIM-9	S-550	S-550	6XAIM-9
Radar Missile	No	Yes	Yes	S-530	S-530	Yes
		AMRAAM	AMRAAM			
AMRAAM						
Maximum Endurance/Loiter in CAP Missions (Hrs.)	-	-	-	3.7	-	3+
Absolute Ceiling (Ft)	51,500	100,000	100,000	60,695	65,000	51,000+
Maximum Mach						
Optimum Altitude	1.64	2.5+	2.5+	2.12	2.4	2.2
Sea Level	1.04	-	-	1.2	1.2	1.1
Maximum Turn Rate						
8 M at 15,000'	18.5	-	-	-	15.81	14
Wing Loading (lbs/ft.)	72	-	-	76.1	83.7	-
Thrust to Weight Ratio	0.75	1.4	1.4	0.73	0.77	-
Maximum Load Factor	7.33	9.0	9.0	-	8.0	7.5
Sea Level Rate of Climb (Ft/Min)	34,300	-	-	41,930	59,050	46,500
Maximum Climb to 40,000' (Min)	3.4	1.0	1.0	-	1.5	3.1
Subsonic Turn Radius (Ft)	4,700	-	-	-	-	-
Loiter Time (Min at 100 NM)	-	-	-	-	-	170
Landing Distance (Ft)	2,450	2,500	2,500	1,700	-	1,200

Table 9.5 (Continued)

Part Two: Avionics and Maintenance Performance

			Mirage	Mirage		
	F-5E	F-15C/D	F-15E	F-1C	F-2000	Tornado
Radar	APQ-159	APG-64/5	APG-70	Cryano	CSF	British
Detection Range 5M²						
Up (NM)	14	80-120	80-120	30	50	50
Down (NM)	No	47	47	15	20	30
Shoot Down	No	Yes	Yes	Lim	Yes	Yes
Ground Mapping	Lim	Lim	Yes	Lim	Yes	Yes
Doppler Beam Sharpening	No	Yes	Yes	No	Yes	Yes
BVR Missile Capability	No	Yes	Yes	No	Yes	Yes
Terrain Avoidance	No	No	Yes	No	Yes	Yes
Moving Map Display	No	No	Yes	No	Growth	Yes
Track While Scan	No	Yes	Yes	No	Yes	Yes
Raid Assessment	No	Yes	Yes	No	No	Yes
Radar Ground Track	No	No	Yes	No	Yes	Yes
Radar Altitude	No	Yes	Yes	No	Yes	Yes
FLIR	No	No	External	No	External	Optional
Laser Designator	No	No	External	No	External	Yes
Fire Control						
Memory/Speed	No	64K/500K	–	–	64K	–
HUD	No	Yes	Yes/TV	Yes	Yes	Yes
Displays	Conv.	2 CRT	3 CRT	Conv	3CRTMulti	Multi
INS	Litton	Litton	Laser	No?	Yes	Yes
Radar Warning	No	ALR-67	ALR-67	–	Yes	Yes
ECM	No	Conformal	ASPJ	Pod	Pod	Internal
Data Transfer	Voice	Automatic	Advanced	Voice	Manual	Advanced
Built-in Test	No	Yes	Yes	Yes	Yes	Yes

The first form of damage is unquestionably the most important. Saudi Arabia is not, and has never been, a serious military threat to Israel. It faces immediate threats in the Gulf, in the Red Sea, and in the Yemens, and its role in the Arab-Israeli conflict is limited to political and economic support of Arab causes. The problem the U.S. faces in persuading Saudi Arabia to make peace with Israel is essentially one of making the political risks of such a peace acceptable. King Fahd long ago advanced a peace plan of his own, and no key Saudi official privately objects to King Hussein's peace initiatives.

Saudi Arabia is not, however, going to readily take the risk of joining a peace process which will alienate Syria, a radical state and a powerful military presence near its northwestern border. It is not easily going to take the risk of having its political and religious legitimacy thrown into question by Iran or the other radical states that would capitalize on its support of what remains a highly uncertain peace process. Above all, it is not going to take the risk of seeming to support the U.S. at a time when the U.S. conspicuously refuses to provide it with the arms it needs and does so under the political influence of Israel.

This lack of Saudi support for the peace process and of improved Arab relations with Israel is a far more real problem than the theoretical capability of Saudi forces to join in an Arab attack on Israel. At the same time, the previous analysis has shown in detail that the U.S. refusal to sell Saudi Arabia the F-15 and the Saudi purchase of the Tornado have unquestionably increased Saudi Arabia's theoretical capability to engage Israel.[38]

BROADER IMPACT ON WESTERN STRATEGIC RELATIONS WITH THE GULF

The U.S. refusal to sell Saudi Arabia the arms package it requested has an impact which goes far beyond U.S. interests or even the damage done to Western military relations with Saudi Arabia. In combination with the U.S. refusal to sell arms to Jordan, it represents the culmination of similar problems in dealing with the other Southern Gulf states. Although none of these states is a threat to Israel, the Reagan Administration's fear of domestic political problems has led it to delay, block, or mishandle every major opportunity to sell U.S. equipment or air defense fighters. The U.S. has lost every major market in the Gulf to France and Britain except Bahrain's small air force.

This has not only cost the U.S. billions of dollars in exports; it has also prevented the creation of effective U.S. military relations with most of the smaller Southern Gulf states. The resulting diversification has helped ensure that many of their forces are not interoperable with each other, and the general Western failure to ensure that they are interoperable with U.S. forces has severely reduced the potential effectiveness of any over-the-horizon reinforcements from the U.S. Navy and USCENTCOM. This is illustrated all too clearly in Table 9.6, which shows just how chaotic the efforts of the Gulf Cooperation Council states to create an effective air deterrent have become.

This is not the kind of military chaos that either the U.S. or Western Europe can afford to encourage. The West faces very real threats in Southwest Asia. It cannot predict the nature of the threat Iran poses to the other Gulf states and the flow of Gulf oil, or what will happen to

Iraq as a result of the Iran-Iraq War, or the threat posed by the growing radicalism in several Red Sea states.

The USSR may not be a pressing threat in the Gulf today, but it will face major oil problems in the 1990s. It has already established a permanent naval base in Dahlak off the coast of Ethiopia and major naval and air facilities in South Yemen. It will certainly capitalize on any opportunity that may come out of the political turmoil in Iran or the Iran-Iraq War, and its freedom of action will greatly increase if it can win in Afghanistan or if Ethiopia can finally suppress its various rebel movements.

This makes it vitally important to Western strategic interests that the conservative and moderate Gulf oil exporting states—Kuwait, Bahrain, Qatar, Saudi Arabia, the UAE, and Oman—should create an effective deterrent. They are now linked together in the Gulf Cooperation Council, but they cannot develop strong enough naval and air forces to build a meaningful regional deterrent without U.S. aid. It is equally important that the U.S. Navy and USCENTCOM forces should have a contingency access to friendly bases in as many Gulf states as possible, and have interoperable facilities, stocks, and major combat equipment.

LESSONS OF THE ARMS SALE CRISIS

The lessons to the West from this arms sale crisis are all too obvious, and virtually all are lessons that come from American mistakes. Britain's willingness to sell Saudi Arabia the Tornado has minimized the overall impact on the West and friendly Gulf states of U.S. political decisions which failed to benefit either Israel or the U.S. Western Europe, however, cannot make up for the collapse of U.S. military relations with key Arab states like Saudi Arabia.

It is one thing for the U.S. to make hard choices between aiding one friend at the cost of threatening another. It is another for the U.S. to make choices that damage a friend like Saudi Arabia without aiding a friend like Israel. Worse, it is absurd to damage the strategic interests of the U.S. at the cost of damaging the security of Israel when Israel is the very friend that the U.S. actions were intended to protect.

Both U.S. politicians and Israel's friends and supporters need to look beyond narrow and short-sighted definitions of Israel's security, and avoid the kind of thoughtless "taking sides" that blocked the U.S. F-15 sale to Saudi Arabia. There also is a clear need for more leadership from the Reagan Administration. It is far from clear that the Administration

Table 9.6: The Lack of Standardization in the Air Forces of the GCC States

Air Defense Systems		Aircraft		SAM Missiles	
Type	Supplier	Type	Supplier	Type	Supplier
Kuwait					
Radars* ($12M)	Thomson CS	32 Mirage F-1	Dassault (France)	1 IHawk Bn	Raytheon (US)
AN/TSQ-73* C^3 System**	Litton, ITT (US) ($8.2 M) U.S.	12 Hawk 34 A-4KU	UK US	SA-7 Missiles SA-8 Missiles SA-6 Missiles	USSR USSR USSR
Saudi Arabia					
5 E-3A AWACS and air, ground C^3I links	Boeing (US)	48 IDS & 24 ADV Tornado*	UK	2 Shahine Btes.	Thomson CSF (France)
		30 Hawk Lt. Attack*	UK		
	30 Short	UK PC-9 Trainer*			
C^3 System	Litton Data (US) ($1.6B)	62 F-15A/B AIM-9L	US	16 IHawk Btes. ($270M)	Raytheon (US)
Peace Shield C^3	Boeing consortia (U.S) ($3.9B)	80 F-5E/F 10 RF-5E	US US	400 Stinger	GD (US) ($30M)
17 Seek Igloo Radars	General Electric US ($330M)	35 F-5B/F	US	500 Redeye 30mm SP AA guns AMX-30SA	US France
5 Underground Centers ($184M)	CRS-Sirrien, Metcalf & Eddy (US)	15 Lightning	UK		
Al Thaqib System with 100 radars	Thomson CSF (France) ($4B)			12 Shahine Btes. and Crotale point and naval defense and 100 missiles	

Bahrain					
Teleprinter data Link to Saudi Arabia		12 F-5E/F w/AIM-9P3*	Northrop (US)	6 RSB-70 Btes. 50+ Missiles	Sweden
				IHawk***	Raytheon (US)
				Stinger	GD (US)
Qatar					
Negotiations on C^3I System	France, U.S.	14 Mirage F-1	Dassault (France)	18 Rapier	British Aerospace (UK)
				5 Tigercat	
		6 Alphajet	FRG	Blowpipe	
		3 Hunter	BA (UK)		
UAE					
Project Lambda*** air defense C^3I and Electronic Warfare with C-130s ($422M)	U.S.	18 Mirage 2000	Dassault (France)	Rapiers	BA (UK)
				Crotale	Thompson CSF (France)
		18 Mirage 2000*		24-42 I Hawk	Raytheon (US)
		30 Mirage 5	Dassault	RSB-70	Saab (Sweden)
		6 Alphajet	FRG	Skyguard 35mm	Contraves (Italy-Swiss)
Oman					
28 Blindfire Radars	Racal (UK)	8 Tornado	Europe ($280M)*	28 Rapiers	BA (UK)
				Blowpipe	
		20 Jaguar	BA (UK)	4 ZU-23-2	USSR
		12 Hunter	BA (UK)	12 40mm	Bofors L/60
		13 BAC-167	BA (UK)		
		300 AIM-9P	US		

*Ordered **Planned ***Discussion only.

Sources: IISS, *The Middle East*; SIPRI, *Defense and Foreign Affairs.*

ever chose to make the true nature of the U.S. sale clear to Israel's supporters or the Congress, and it is equally far from clear that the Administration ever honestly faced the consequences of what its area experts were telling it about the prospects for a British or French sale if the U.S. did not sell the F-15.

European arms sales to Saudi Arabia or other friendly states in the Middle East generally serve the West's strategic interests and reduce the impact of the Arab-Israeli conflict. If Saudi Arabia should turn to Europe for *all* its fighters because of its political problems with the U.S., however, this would have strategic implications which go far beyond those of an aircraft sale. The net impact of full Saudi reliance on European aircraft would be to weaken or break the one major remaining military relationship between the U.S. and Saudi Arabia, and the one most critical to USCENTCOM's deployment capability in terms of interoperable stocks, bases, and combat service facilities. Since no other Western state can assume the U.S. role of power projection in the Gulf, the result would be far more serious than the present Saudi shift to Britain.

NOTES

1. *New York Times,* November 18, 1984, p. A-12.

2. *Washington Post,* January 24, 1985, p. A-18; *Chicago Tribune,* January 28, 1985, p. I-1; *Washington Times,* January 29 ,1985, p. 1A, and January 30, 1985, p. 6A.

3. The original sale of the F-15 had been approved on May 15, 1978, by a Senate vote of 54 to 44, after a long political battle.

4. See "Integrated Systems Evaluated on F-15", *Aviation Week,* April 11, 1985, pp. 47–57; and "F-15C/D Display Nears Test Stage", *Aviation Week,* February 20, 1985, pp. 77–81.

5. For a good description of the role of the current version of Harpoon, see John F. Judge, "Harpoon Missile Targets Ships and Cost", *Defense Electronics,* April, 1985, pp. 92–98. Also see Bill Gunston, *Modern Airborne Missiles,* ARCO, New York, 1983, pp. 104–107.

6. *Washington Post,* January 24, 1985, p. A-18; *Chicago Tribune,* January 28, 1985, p. I-1; *Washington Times,* January 29 ,1985, p. 1A, and January 30, 1985, p. 6A.

7. *Washington Times,* January 29, 1985, p. 1A, and January 30, 1985, p. 6A; *Chicago Tribune,* January 28, 1985, P. I-1, January 31, 1985, p. I-5, February 1, 1985, p. I-2; *Baltimore Sun,* January 31, 1986, p. 2A; *New York Times,* January 31, 1985, p. A-1; *Washington Post,* February 1, 1985, p. A-25.

8. *Chicago Tribune,* January 31, 1985, p. I-5, *Washington Post,* February 1, 1985, p. A-25.

9. *New York Times*, January 31, 1985, p. A-1; *Boston Globe*, February 3, 1985, p. 1.

10. For typical reporting see *Washington Times*, February 5, 1985, p. 7B, February 11, 1985, p. 1D; *Los Angeles Times*, February 10, 1985, p. IV-5, February 12, 1985, p. II-5, February 13, 1985, p. I-1; *New York Daily News*, February 13, 1985, p. 24; *Economist*, February 9, 1985, pp. 25–26; *Aviation_Week*, July 22, 1985, p. 21.

11. John Newhouse, "Diplomatic Round", *New Yorker*, June 9, 1986, pp. 52–54.

12. Ibid.

13. Ibid.

14. See Anthony H. Cordesman, "The Saudi Arms Sale: The True Risks, Benefits, and Costs", *Middle East Insight*, Vol. 4, Nos. 4 and 5, pp. 40–54, and "U.S. Middle East Aid: Some Questions", *Defense and Foreign Affairs*, June, 1986, pp. 15–18; and John Newhouse, "The Diplomatic Round, Politics and Weapons Sales", *New Yorker*, June 9, 1986, pp. 46–69.

15. *International Defense Intelligence*, Vol. 6, No. 53, December 31, 1984, pp. 1–2; *Sunday Times*, September 15, 1985; *New York Times*, September 16, 1985; *Financial Times*, December 21, 1984, and September 16, 1985; *Aviation Week*, September 30, 1985, p. 29, October 21, 1985, pp. 73–74, March 24, 1986, p. 59; *Defense and Foreign Affairs Daily*, October 22, 1985, p. 1.

16. See Newhouse, "Diplomatic Round", *New Yorker*, June 9, 1986, pp. 60–61.

17. *Sunday Times*, September 15, 1985; *New York Times*, September 16, 1985; *Financial Times*, December 21, 1984, and September 16, 1985; *Aviation Week*, September 30, 1985, p. 29, October 21, 1985, pp. 73–74, March 24, 1986, p. 59; *Defense and Foreign Affairs Daily*, October 22, 1985, p. 1.

18. *Jane's Defense Weekly*, June 14, 1986, p. 1075.

19. *Washington Post*, February 13, 1986, p. A-1, February 27, 1986, p. A-14, May 6, 1986; *Wall Street Journal*, February 23, 1986, p. 34, February 28, 1986; *New York Times*, March 1, 1986, p. 3, March 5, 1986, p. B-6; *Aerospace Daily*, April 22, 1986, p. 123.

20. *Aviation Week*, July 22, 1985, p. 21.

21. Section 131 of U.S. Public Law 99-83, August 8, 1985.

22. *Wall Street Journal*, February 13, 1986, p. 34; *Washington Post*, February 13, 1986, p. A1.

23. The decision was not totally divorced from trying to preserve a market of immense value to the U.S. As of September 30, 1984, the Saudis had signed foreign military sales agreements worth a total of $47.7 billion, including $19.9 billion in construction and design services. U.S. FMS sales were averaging $4 billion annually, with some $20.2 billion in sales still to be delivered.

24. *Washington Post*, February 27, 1986, p. A-14.

25. The sale raised the total Saudi inventory of U.S. AIM-9P and AIM-9L air-to-air missiles to 5,407. It raised the ratio of AIM-9Ls to Saudi F-15s (the only Saudi plane aside from the Tornado that could fire the AIM-9L) to 37:1 versus 6:1 for Israel and less than 10:1 for the U.S. The Saudis realized as well as the U.S. that Saudi fighters would never survive long enough in combat intensive enough to require such stock levels. The rationale was enhancing U.S. over-the-horizon capabilities.

26. *New York Times*, March 1, 1986, p. 3; *Wall Street Journal*, March 21, 1986, p. 28; *Philadelphia Inquirer*, May 5, 1986, p. 11A; *Washington Times*, May 5, 1986, p. 3A; *Washington Post*, May 5, 1986, p. A.5.

27. *Washington Post*, May 7, 1986, pp. A-1 and A-23.

28. *Aviation Week*, May 12, 1986, pp. 30–31.

29. *Wall Street Journal*, June 6, 1986, p. 29; *New York Times*, June 6, 1986, p. 1; *Washington Post*, June 5, 1986, p. A-35, June 6, 1986, p. A-1; *Chicago Tribune*, June 5, 1986, p. I-4; *Baltimore Sun*, June 5, 1986, p. 17A.

30. *Wall Street Journal*, June 16, 1986, p. 37; *New York Times*, June 15, 1986, p. 1.

31. Estimate based on ACDA, *World Military Expenditures and Arms Transfers, 1973–1983*, GPO, Washington, 1985, p. 127.

32. See D. Fiecke, B. Kroqully, and D. Reich, "The Tornado Weapons System and Its Contemporaries", *International Defense Review*, No. 2/1977; and current Panavia brochures.

33. *Aviation Week*, September, 30 1985, p. 29.

34. D. Fiecke, B .Kroqully, and D. Reich, "The Tornado Weapons System and Its Contemporaries", *International Defense Review*, No. 2/1977, and current Panavia brochures.

35. Bill Gunston, *NATO Fighters and Attack Aircraft*, ARCO, New York, 1983, p. 40 .

36. For fuller details see U.S. Department of Commerce, *Saudi Arabia: Foreign Economic Trends and Their Implications for the U.S.*, FET 84-80, July, 1984. Recent British reports have raised the total value of the sale to Britain to $7 billion, although the offset provisions involving oil make any analysis of its value uncertain.

37. *Washington Post*, September 30, 1985.

38. For further technical details, see "Integrated Systems Evaluated on F-15", *Aviation Week*, April 11, 1985, pp. 47–57; "F-15C/D Display Nears Test Stage", *Aviation Week*, February 20, 1985, pp. 77–81; and "Saudis, British Define Terms of Tornado Sale", *Aviation Week*, September 30, 1985, p. 29.

10
The West Enters the Gulf: 1985–1987

The issues surrounding Western strategic relations with the Gulf states acquired a far more immediate importance during the course of 1986 and 1987. The fighting in the Iran-Iraq War shifted to the point where an Iranian victory seemed possible for the first time, and where the Iranian threat to Gulf shipping and the Southern Gulf states shifted from theory to practice. These pressures gradually thrust the U.S., and then several of its principal European allies, into a high-profile military intervention in the Gulf. Almost immediately, this intervention spotlighted Western dependence on the support of Saudi Arabia and other friendly Southern Gulf states, and their dependence on Western military reinforcements.

At the same time, events also exposed virtually all of the problems in Western strategic relations with the Gulf states that have been discussed in the previous two chapters. The U.S. was deeply embarrassed by the disclosure of covert arms for hostages deals with Iran. It found itself involved in new political disputes over arms sales to the region, and over its ability to obtain support from its European allies and the Gulf states. The Southern Gulf states, in turn, found that their vulnerabilities, divisions, and limited military capabilities could pose a major threat to their security.

Both the West and the Southern Gulf states had ample reason to realize their dependence on each other. This was particularly true of Saudi Arabia and the U.S. The U.S. rapidly found in the course of intervening in the Gulf that it had to depend on its military relations with Saudi Arabia, Bahrain, and Oman. Saudi Arabia and the smaller Gulf states found, in turn, that a strong U.S. presence in the Gulf was essential in checkmating a growing threat from Iran.

THE SEEDS OF WESTERN INTERVENTION: COVERT U.S. ARMS SALES TO IRAN

In order to understand the events involved, it is necessary to understand two key developments that occurred during 1985 and early 1986: The covert U.S. arms sales to the Khomeini regime in Iran, and the impact of Iran's seizure of Iraq's Faw Peninsula. The covert U.S. arms sales created a crisis in U.S. relations with other Gulf states that was a key factor in leading the U.S. to rush into an arrangement to reflag and convoy Kuwaiti tankers through the Gulf. Iraq's conquest of Faw suddenly gave it renewed credibility as a real threat to both Iraq and Kuwait, and helped trigger a major expansion of the tanker war that increasingly involved Iranian strikes against neutral shipping and the Southern Gulf states.

The history of the U.S decision to trade arms for hostages is the complex history of a process in which an attempt by the Reagan Administration to improve U.S. relations with Iran drifted into becoming an arms for hostages deal, and then became a deal that used the funds paid by Iran to finance the Nicaraguan Contras. In the process, the management of this covert operation shifted from the regular interagency process of decision making within the U.S. government, which largely opposed the covert arms transfers, to a small and inexperienced cadre of "true believers" in the Reagan White House.[1]

The arms deals began in 1985, when the Reagan Administration made a series of decisions that eventually led to a massive scandal in 1987. In a series of exchanges which have still not been fully explained, several Israeli officials and U.S. consultants outside the government helped persuade the senior officials of the U.S., and the Director of the Central Intelligence Agency, that it might be possible to rebuild U.S. political ties with "moderates" in the Iranian government, and to free Americans held hostage by pro-Iranian Shi'ite factions in Lebanon.

In spite of serious reservations throughout the U.S. national security system, the U.S. was driven to make contact with the Khomeini government by its inability to find military solutions to the problems of Iranian-sponsored hostage taking and terrorism.

The seizure, torture, and murder of William Buckley, the CIA station chief, in Beirut on March 16, 1984, was a key catalyst in this process. It provided what some NSC staff members regarded as further evidence that U.S. counterterrorist efforts could not suppress the activities of pro-Iranian groups in Lebanon or win the release of the U.S. citizens they had captured. It led them to believe that the only way the U.S. could recover the hostages, and generally strengthen its position in the Gulf, was to improve relations with Iran.

Several senior Israeli officials had long encouraged such contacts, and on May 4–5, 1985, Michael Ledeen, a consultant to the NSC, visited Israel and talked to senior Israeli officials about Iran. These officials included Prime Minister Shimon Peres and senior members of Israeli intelligence. It is still unclear, however, whether Ledeen did this as the result of prior contacts with Israeli officials, or of contacts with European intelligence officers. It is also unclear whether he acted at the formal request of National Security Advisor Robert McFarlane.

In any case, Ledeen briefed McFarlane in mid-May, 1985. McFarlane then tasked the intelligence community with producing a Special National Intelligence Estimate (SNIE) on Iran on May 20, 1985. The Ledeen talks scarcely had a broad mandate. Secretary of State George Shultz formally objected to Ledeen's activities on June 5, 1985—and McFarlane advised Shultz on June 14 that he had instructed him to discourage any arms sale initiative. Nevertheless, the NSC proceeded to develop plans for contacts with Iran.

These efforts were accelerated when a TWA airliner, Flight 847, was hijacked by pro-Iranian groups in Lebanon on June 14, 1985. While President Reagan gave a speech calling Iran part of a "confederation of terrorist states" and "a new international version of Murder Incorporated", the NSC staff concluded that there was little punitive action the U.S. could take.[2] The resulting discussions of what Israel could do to help also brought the Iranian issue to the attention of the President's Chief of Staff, Donald Regan. The whole issue of how Iran should be treated then steadily became one which was handled at the political level, as distinguished from the foreign policy level.

McFarlane transmitted a draft National Security Decision Directive (NSDD) recommending the option of both arms sales and of intelligence to Secretary of State George Shultz and Secretary of Defense Caspar Weinberger on June 17, 1987.[3] Both Secretaries rejected the NSC's arguments in writing. Shultz stated on June 29 that it was "perverse" and contrary to American interests, and Weinberger stated on June 16 that it was "almost too ridiculous to comment upon."[4]

At this point, however, the problem of American hostages began to take steadily greater precedence over strategic considerations, a trend that was steadily reinforced by each hostage incident—particularly when Syria informed the U.S. in late 1985 that Buckley had been tortured and killed.

Further, several Israeli officials continued to push the U.S. towards opening contacts with Iran. David Kimche, Director General of Israel's Foreign Ministry, had indirectly been pushing the idea of a U.S. initiative towards Iran since at least the early fall of 1984, and on July 3, 1985, he visited the White House and asked McFarlane to take up the proposal again, stating that his request was on the instructions of Shimon Peres.

Another "private emissary" from Israel visited McFarlane on July 13. The next day McFarlane sent Shultz a cable supporting the Iran initiative, and indicating that it might help the seven U.S. hostages in Lebanon. This led Shultz to reply that the U.S. should show interest without formally supporting any action and that McFarlane should take the initiative.

While there are conflicts in the testimony as to what happened next, McFarlane visited President Reagan who had just had an operation, in the hospital at some point during July 13–17. McFarlane interpreted this conversation as giving him the President's approval of covert contacts with Iran. In initiating these contacts, he used two private arms dealers, a Saudi named Adnan Khashoggi and an Iranian named Manucher Ghorbanifar—sometimes described as an advisor to Iran's Prime Minister. Further, Ledeen introduced an Israeli arms dealer, Al Schwimmer, to McFarlane at the suggestion of David Kimche.

Ledeen met in Israel in late July with Kimche, Ghorbanifar, Schwimmer, and another Israeli arms dealer, Yaacov Nimrodi—an ex-Israeli military attache to Iran.[5] The outcome was a proposal for new contacts with Iran and for trading the sale of U.S. TOW anti-tank and Hawk anti-air missiles for the U.S. hostages. This led to a meeting in the White House on August 6, 1985, where President Reagan presided. Although the details of the meeting are unclear, it is clear that the key issue was whether to trade arms for hostages, that CIA Director William Casey, Regan, and Vice President Bush supported the arms initiative, and that Secretaries Shultz and Weinberger opposed it.[6]

The meeting does not seem to have had a formal outcome, but the practical result was that Israel felt it obtained authority from the NSC on August 30 to sell 508 TOW missiles to Iran, which the U.S. would replace. The first 100 missiles arrived in Iran on August 30, 1985, and the remaining 400 arrived on September 14, 1985.[7]

This arms transfer helped lead to the release of one U.S. hostage—Reverend Benjamin Weir—on September 14, but the sale did not lead to the broader release of U.S. hostages that the White House officials had expected. In fact, Iran established a pattern that it was to repeat in 1986, when the U.S. shipped arms directly to Iran. Iran released one hostage, but kept the others to retain its leverage over the United States. Iran repeated this pattern when it released Reverend Lawrence Jenco on July 26, 1986, and David P. Jacobsen on November 2, 1986.

On each of these subsequent occasions, the pro-Iranian groups in Lebanon carried out only one-third to one-half the releases that the U.S. had expected, and Iran seems to have encouraged such groups to take new hostages in compensation. By the time the covert U.S. arms deals became public in early November, 1986, pro-Iranian groups had as many U.S. hostages as they did before the arms deal.[8]

Even so, the hostage problem led to President Reagan's decision to initiate covert contacts with Iran, and these contacts were aided by Khashoggi and Ghorbanifar. Further, an Israeli arms dealer, Al Schwimmer, was introduced to McFarlane by Ledeen at the suggestion of David Kimche.

The covert U.S. sales to Iran also did a great deal to reduce the impact of "Operation Staunch", the U.S. effort to restrict arms deliveries to Iran which the U.S. had begun in 1983. They also gave Iran good reason to believe that U.S. policy was uncertain and opportunistic. Iran found it could improve its access to other suppliers as they gradually became aware that U.S. controls were weakening. While the exact pattern in arms flows is uncertain, it is clear that Israel stepped up its other arms sales to Iran to the point where they totalled around $500 million in value by mid-1986. It is also possible that other Western nations such as Greece, Spain, and Portugal became more willing to expand their arms shipments.[9]

IRAN LAUNCHES ITS ATTACK ON FAW

It is somewhat ironic that the Reagan Administration initiated many of these contacts with a hostile Iran at precisely the time it was unable to carry out its arms deals with Saudi Arabia or expand its military relations with any of the other friendly Southern Gulf states. It is equally ironic that the Administration tried to deal with "moderates" in the Iranian government at a time when Iran was actively attempting to conquer Iraq and dominated the Gulf.

Another shipment of Hawk missiles and Hawk parts went by plane from Israel to Iran in November, 1985. The flight was authorized by Lt. Colonel Oliver North, of the NSC. Virtually all of the shipment seems to have been returned, however, because Israeli middlemen substituted older parts and missiles, and sent equipment for the regular Hawk missile, rather than the improved Hawks in Iranian forces.[10]

Iran's attention, however, was occupied elsewhere. In early 1986, Iran built up a massive concentration of forces in the southern part of its border area with Iraq. The Iranian forces then divided into two major concentrations. One smaller concentration was to the north of Basra, opposite the flooded Hawizah marshes. The main concentration was to the south opposite the Faw Peninsula.

On February 9, 1986, Iran launched the first phase of its three-pronged attack (the Al Dawa or Dawn 8 offensives) against the Iraqi 3rd Corps and the Iraqi 7th Corps. The new offensive involved over 150,000 men. The Iranian forces in the north launched two thrusts across the marshes. These attacks were repulsed, although it is unclear whether they were ever intended to be more than diversionary efforts. The story was different in the south. On the night of February 9, 1986, Iran carried out a major amphibious landing near Iraq's largely abandoned oil port of Al Faw.

The Iranian landings were successful at all six points and Iranian troops began to move forward over a 40 mile front along the coast of the Faw Peninsula. Iran was also able to rapidly reinforce its original bridgeheads during the next few days. Iran then continued to build up its forces with only limited opposition, and succeeded in virtually routing the Iraqi forces in Faw by the night of February 10.

It is doubtful that the covert U.S. arms shipments to Iran before the attack had a critical impact on the fighting, but it is important to note that the U.S. was now attempting to trade some 4,000 TOW anti-tank guided missiles for the release of hostages and better relations with Iran.[11] The first 500 TOWs were shipped to Bandar Abbas in Iran on February 18, 1986. Another 500 TOW missiles, and possibly Hawk and F-4 parts as well, were shipped to Bandar Abbas on February 27, 1986.

While the TOW was of potential value to Iran's infantry-dominated assault forces as a means of killing Iraqi tanks and hardened defense points, these deliveries probably arrived too late to influence the battle at Faw, and at this point the U.S. had still not made any effective transfer of Hawk missiles or parts. While there were three Hawk sites covering Iranian forces in the area, only one was operational. Further, the new arms transfers did not produce the release of the hostages that the U.S. had expected.[12]

In any case, Iraq was not able to halt Iran's invasion of Faw. By February 16, Iran occupied over 300 square miles of the Faw Peninsula, although much of this was marsh land. At the height of Iran's success, it threatened to break out of the Faw Peninsula. Iranian troops reached the Khaur Abdallah waterway opposite Kuwait, and there were even reports that Iranian forces had surrounded the Iraqi Navy base at Umm Qasr. Iran also captured Iraq's main air control and warning center covering the Gulf, which was located north of Al Faw.[13]

After February 17–18, however, Iraq was able to fully deploy its superiority in air power, firepower, and armor. By early March, a pattern of fighting was established in Faw that lasted for the rest of 1986. Iran retained about 120 square miles of the Faw Peninsula, and was able to keep its supply lines open across the Shatt al-Arab. It now, however, could not break out of Faw or expand its bridgehead.

Nevertheless, the conquest of Faw gave Iran a new and important strategic position. While Faw was on a long marsh-ringed peninsula, and was relatively isolated from Basra and the main roads between Iraq and Kuwait, it was still of special strategic importance because of its impact on the Southern Gulf. The Faw Peninsula juts out between the Shatt al-Arab and the Island of Bubiyan in Kuwait. The successful Iranian thrust thus offered Iran three major benefits: (a) It cut Iraq off even more firmly from the Gulf; (b) it positioned Iran to attack Basra from the South, and (c) it positioned Iran to cut off Iraq's main lines of communication to Kuwait or even to launch amphibious attacks on Kuwait. Equally important, it created conditions that encouraged Iran to put pressure on all the Southern Gulf states and to threaten Western oil supplies.[14]

THE COVERT ARMS DEALS CONTINUE

Iran responded to the halt of its Faw offensive by launching occasional raids against tankers in Saudi and UAE waters, and pressured the GCC states to end their ties to Iraq and the U.S. It also, however, expanded the scope of its talks with France and the USSR—both of which were impressed by Iran's victory at Faw—and carried on with its secret contacts with the U.S. The Iranian talks with the U.S. were now conducted directly by Iran's Prime Minister, Ali Akbar Hashemi Rafsanjani, and he used a relative as a "second channel" to deal with the U.S. government.

Iran made steadily more ambitious demands. It was unsuccessful in demands for Harpoon anti-ship missiles, reconditioned Hawk radars, and 200 advanced Phoenix air-to-air missiles for its F-14As. It did, however, obtain U.S. agreement to provide more TOWs and to provide critically needed Hawk parts—partly in compensation for the defective parts Israel had shipped in November, 1985.

Two more Southern Air Boeing 707s flew from Texas to Tel Aviv in May, 1986, carrying the Hawk parts and TOWs. Then, on May 28, McFarlane, Lt. Colonel North, George Cave, a CIA official, and Amiran Nir flew to Iran in a plane carrying an initial load of spare parts for the Hawk missiles Iran needed to help protect its oil facilities. The rest of the Hawk parts and the 500 more TOW missiles were kept back in Israel awaiting the release of the U.S. hostages.

The U.S. officials were supposed to see President Khamenei, Prime Minister Mousavi, and Majlis Speaker Rafsanjani, but McFarlane and the rest of his party got few results from their arrival in Tehran. They spent several days waiting in the Tehran Hilton, and talking to low-level officials, before being asked to leave.

Although the circumstances are still unclear, Rafsanjani may have faced a major problem in dealing with the Americans because Montazeri—and/or one of his most senior assistants and chief of staff, Mehdi Hashemi—found out about the U.S. visit and strongly opposed it. Rumors surfaced later that more radical factions within the Iranian government threatened to arrest the Americans, and it became all too clear in the fall that the Iranian government was as divided about the wisdom of covert dealings with the U.S. as American officials were divided about the merits of dealing with Iran.

Rafsanjani, however, did make a guarded speech on June 10, 1986, discussing improved Iranian relations with the U.S., and one hostage—Father Jenco—was released on July 26, 1986. Although this hostage release fell far short of U.S. expectations, the U.S. sent Iran more Hawk parts on August 3.

Iran then pressed the U.S. for more Hawk missiles, Hawk radars, Hawk electron tubes, 1,000 more TOWs, and intelligence on Iraq. By late August, after discussions of returning the body of William Buckley for burial, the U.S. began preparation for the shipment of 500 more TOWs. After some complex delays, the 500 TOWS were delivered to Iran on October 28. Another hostage—Peter Jacobsen—was released on November 2, 1986.[15]

THE COVERT U.S. ARMS SALES BECOME A GLOBAL SCANDAL

One of the many risks inherent in the U.S. covert arms sales was that one of the divided factions ruling Iran would expose the sales to the world, knowing that these sales would immediately force Iran to break off its ties to the U.S. and seriously embarrass the U.S. throughout the Arab world.

In November, 1986, this risk became reality. A member of the staff of Khomeini's designated heir, the Ayatollah Montazeri, reacted to an ongoing power struggle between Montazeri and Iran's Prime Minister Rafsanjani, by giving the details of the covert arms deals to *As Shira*, a pro-Syrian magazine in Lebanon. The magazine announced McFarlane's visit to Tehran in its November 4, 1986, issue.[16] *As Shira's* editor, Hassan Sabra, later stated that the source of his story was Montazeri's office, and it is clear that the leak was designed to undercut Rafsanjani, as well as end Iranian dealings with the U.S.[17]

Rafsanjani officially confirmed the story on November 5, 1986, and again on November 24 and December 5.[18] The broad outlines of the rest of the U.S. transactions with Iran became public by the end of the first week of November. One leak after another led to the disclosure that the U.S. had shipped at least 2,008 TOW missiles and 235 Hawk assemblies. It also became clear that Israel had provided much larger shipments of U.S. parts and arms than had previously been believed.[19]

These disclosures inevitably triggered a series of investigations by the U.S. Congress and months of media attention. The Reagan Administration found itself under a siege which was not to diminish until the late summer of 1987, and which constantly threatened to undermine U.S. relations with the Arab world. Rather than strengthening Iranian "moderates", the Reagan Administration ended in strengthening Iran's extremism and commitment to the war in the Gulf. It had added another symbol of American incompetence and irresolution to the memories of the U.S. withdrawal from Lebanon.

It is important to note, however, that the Reagan Administration's failures did not occur because the U.S. national security planning system did not work. They occurred because it was not allowed to work, and because the Administration bypassed it when it gave the correct advice. The Secretary of State and the Secretary of Defense both advised against what they were allowed to know of the covert arms sales. So did virtually every professional area expert in the State Department, the Department of Defense, and the U.S. intelligence community. If the lack of depth and expertise in the U.S. Congress had shaped the Saudi arms sale crisis of 1985–1986, it was the Reagan Administration's bypassing of its own professionals, and their expertise, that led to the Iran-Contra arms scandal.

THE IRANIAN ATTACK ON BASRA

In early 1987, Iran demonstrated just how committed to war it really was. Iran prepared for two new thrusts: one directly against Basra with support from Iran's forces in Faw, and another thrust by its regular forces in the area north of Baghdad between Qasr-e-Shirin and Sumar. Iran's precise strategic and political objectives in launching its new offensive are unclear, but its possible objectives include the following: (a) Seize Basra and create an alternative pro-Khomeini "capital" in the south; (b) destroy the Iraqi Army in the south and bring down the Ba'ath government; (c) weaken the Iraqi Army through sheer attrition; and/or (d) lay siege to Basra while providing a further political lesson to the southern Gulf states.

Regardless of such uncertainties, Iran's forces launched a massive attack on the Iraqi positions near Basra in January, 1987. This offensive scored significant initial gains before coming up against Iraq's main defense positions, which included massive earth berms, bunkers, and concentrations of weapons. The Iranian troops lacked the weight of firepower and the mobility they needed and may have begun to have problems with supplies and in moving sufficient ammunition forward. By some reports, Iran suffered nearly 50,000 dead, and nearly half of the infantry committed to the attack became casualties. Even so, intense fighting continued well into the spring of 1987. Once again, the Gulf and the West were confronted by the risk that Iran might seize southern Iraq and pose a direct threat to Kuwait.

IRAN REBUILDS ITS SEA POWER

Iran, however, did not stop at land offensives. During the first three years of the Iran-Iraq War, sea power played only a limited role in the conflict. Iran achieved a decisive advantage in the first few months of the war when it cut the Iraqi Navy off from the Gulf and destroyed Iraq's ability to export oil from its Gulf terminals. The remaining oil facilities on both sides also proved to be sufficiently vulnerable so that neither side chose to attack the other's oil export facilities after the first months of the war, except for some largely demonstrative strikes during key land battles.

Beginning in 1984, however, the nature of the naval war changed. Iraq had been forced back on the defensive in the land war and saw no clear hope of ending the war on favorable terms as long as Iran could take advantage of its revolutionary fervor and strategic depth to attack at will. At the same time, the Iranian Air Force had decayed to the point where it could no longer effectively defend its air space and oil facilities, or attack Iraq's new pipeline system through Turkey and Saudi Arabia. This gave Iraq a potential edge in attacking Iran's oil exports and main source of income, and it was an edge that Iraq immediately chose to exploit.

As Table 10.1 shows, Iraq began what quickly came to be called the "tanker war". During 1984–1986, it also steadily expanded the range of its air strike capability—eventually reaching virtually all Iranian naval, oil, and tanker targets in the Gulf area. While Iraq never was able to cut off Iran's oil exports to any significant degree, it was able to threaten them and greatly increase the cost of such exports and Iran's need to sell oil at a discount. Further, when Iraq launched successful land

Table 10.1: Patterns in Iraqi and Iranian Attacks on Gulf Shipping: 1984 to End June, 1987

Date	Iraqi Attacks	Iranian Attacks	Total Attacks	Deaths	Ships Lost
1984	36	18	54	49	32
1985	33	14	47	16	16
1986					
October	1	3	4	–	–
November	9	2	11	–	–
December	5	0	5	–	–
Total 1986	66	41	107	88	30
1987					
January	7	6	13	–	–
February	6	3	9	–	–
March	3	3	6	–	–
April	2	3	5	–	–
January–June	29	29	58	10	4

Source: Adapted from the *Economist*, April 25, 1987, p. 34; and Lloyd's Maritime Information Service.

offensives against Iraq's Faw Peninsula in 1986, and against the eastern defenses of Basra in 1987, Iraq was able to retaliate by striking at Iran's oil export facilities at Kharg Island, the tanker shuttle it used to ferry oil to tankers waiting in the more secure waters of the lower Gulf, and Iran's inland and offshore oil facilities.

These strikes increased the number of Iraqi attacks on shipping to Iran from 27 in 1984, and 40 in 1985, to 65 in 1986. They put powerful pressure on Iran to limit its offensives, and countered some of the political advantages Iran had gained from its highly visible land victories. Iran, however, soon showed that it could retaliate. It had launched only 16 attacks on shipping to Iraq in 1984, and 13 attacks in 1985. It launched 41 in 1987.[20]

In January 1987, for example, Iraq's use of the "tanker war" to retaliate against Iran's attacks on Basra led to more attacks on shipping by Iran and Iraq than in any previous month in the war. Iraq struck at Kharg Island, Iran's transloading facilities at Sirri, and Iran's shuttle tankers and oil facilities. This did not make major cuts in Iran's oil exports, but Iran was forced to send a purchasing mission to Greece, London, and Norway to buy 15 more tankers.

Iran was not willing, however, to allow Iraq to continue to exploit its

its air power to strike at Iraq's inland oil facilities or its air defenses to protect its oil facilities and tankers, but it could find new ways to use sea power that could threaten all international traffic through the Gulf, and particularly shipping to and from Kuwait and Saudi Arabia—the states that were giving Iraq the most aid in terms of financial payments and transshipments.

While attacks on such traffic did not directly hurt Iraq, and forced Iran to strike at ships moving in international waters, they did offer Iran a reasonable hope that Kuwait and Saudi Arabia would threaten Iran with the halt of financial aid. Further, Iran almost certainly hoped that such attacks would be successful enough to force Kuwait and Saudi Arabia to sharply reduce their aid and transshipment of goods to Iraq.

This strategy also led Iran to create new naval capabilities. The regular Iranian Navy was not an adequate means to carry out such a mission. It was far too visible a force in terms of international politics to use openly in attacking international shipping, and it already had suffered serious damage to at least two of its destroyers, two Saam-class frigates, and one US PF-103 corvette. It had lost two minesweepers and two Kaman-class patrol boats, and two more were seriously damaged. Iran was experiencing serious difficulties in maintaining its sensors and weapons systems, and shortages in both anti-ship and anti-air missiles.

Most of the larger Iranian warships had serious problems in maintaining their combat readiness, and many key radar and electronic systems were no longer operational. This included most of the Contraves Sea Hunter, SPG-34, and Mark 37, 51, and 61 fire control systems; the WM-28 tactical and fire control radars; the Plessey AWS 1 and SPS6 search radars; and the SPS-37 air surveillance radars. Stocks of most major missiles were very low, including the RIM-66 Standard (anti-aircraft), Sea Cat (anti-aircraft), and RGM-84 Harpoon (anti-ship). Further, most of the missiles delivered during the time of the Shah now exceeded their maximum reliable storage life, and Iran's stocks of Harpoon may have been limited to seven missiles for its Kaman-class fast-attack craft.[21]

Iran, however, still had the largest local naval force in the Gulf. Its operational ships included the U.K.-built air defense destroyer *Artemiz,* and the U.S.-built destroyers *Babr* and *Palang,* with FRAM II conversions and improved air defenses. While it was not clear how many were operational, Iran had four British-built frigates with Sea Killer anti-ship missiles: *Saam, Zamm, Rostam,* and *Faramarz*. The navy had at least four of its original eight Combattante II–class patrol boats operational, equipped with 76 mm and 40 mm guns and Harpoon launchers, although supplies of missiles had evidently been exhausted.[22] These 154 foot ships could reach speeds of up to 33.7 knots, and had a range of 700 to 2,000 miles, depending upon speed. One 320 ton minesweeper, the *Shahrokh*, survived, but this was deployed in the Caspian.

There were four LSTs: The *Hengam, Larak, Lavan,* and *Tonb*. The latter two had been delivered in 1974 by Britain, under the pretext that they were hospital ships. They carried Agusta-Bell AB-212 helicopters with AS-12 missiles, smaller landing craft, TACAN beacons, and minelaying equipment. Three more 67.5 meter, 2,024 ton landing ships had just entered service, and four 654 meter Dutch-built freight barges were being used as landing ship tankers. Two small FRG-built 5,000 ton replenishment tankers—the *Bandar Abbas* and *Booshehr*—were modified in 1984 to carry 40 mm guns and a telescopic helicopter hanger.

The navy still had some BH-7 and SRN-6 Hovercraft and up to 150 patrol craft operational—although many of the latter might have been sunk, damaged, or turned over to the naval Guards. Iran had the repair ship *Shah Bahar* and the floating docks "400" (300 ton lift) and *Dolphin* (28,000 tons lift), which it had bought from the FRG in 1985.

The regular Iranian Navy also did not not have to rely on direct naval engagements. It had extensive stocks of mines. These may have included some U.S. Mk 65 and Soviet AMD 500, AMAG-1, and KRAB anti-ship mines, as well as supplies of modern Italian influence mines, although most were PRC- and North Korean–made versions of older Soviet contact mines.[23] Iran was beginning to manufacture non-magnetic, acoustic, free-floating, and remote controlled mines. While Iran only had one surviving minesweeper, which was deployed in the Caspian, it could lay mines with virtually any small ship, including dhows and small cargo vessels.

The Iranian Navy retained up to seven AB-212 helicopters, each with two Sea Killer Mk II or AS-12s. Both systems could be used for sudden attacks against tankers or U.S. warships. Iranian naval air capability consisted of a maximum of two PF-3 Orion maritime patrol aircraft, which lacked operational sensors, and could be used only for visual reconnaissance missions, seven Sikorsky SH-3D ASW helicopters, and two RH-53D minelaying helicopters. It had Hercules C-130 and four F-27 Mark 400 Fokker Friendship aircraft for minelaying and patrol missions.

This mix of forces would have been largely useless, however, if Iran had not felt it could find a solution to several key political and military problems. The most serious political problem was that Iraq could accomplish all its goals in the tanker war by striking at ships and oil targets that were clearly in a war zone, and without interfering with shipping in international waters. Iran, however, had to strike at shipping in international waters and strike at shipping moving to and from non-belligerents. It could not use its regular navy directly without a highly visible violation of international law, and without triggering the almost certain intervention of the Western powers. If it did trigger such intervention, it then faced the problem that its ships had virtually no remaining air defenses.

Iran's solution was to create a new naval branch of its Guards force under the command of Mohsen Rezai. This force took nearly two years to build, and began operations in the Gulf in October, 1986. By that time, it had roughly 20,000 men—more men than Iran's regular navy. They were equipped with a mix of hundreds of small craft, including roughly 80 Swedish-built fast interceptor craft, and Zodiak rubber dinghies. Such boats were very difficult to detect by radar, and could carry out high-speed rocket, machine gun, small arms, and 107 mm recoilless rifle attacks. They also had the political advantage that it was far easier to disclaim attacks by Guards forces in small craft than attacks by regular Iranian Navy vessels.

The 29 to 50 Swedish boats were built by Boghammer, and were 42 feet long. They were purchased in 1985, and could carry a 6 man crew and up to 1,000 pounds of weapons for ranges of about 500 nautical miles. They could cruise at 46 knots, and could reach speeds up to 69 knots. The Guards were reported to have ordered at least 40 to 50 and to have deployed at least 29.[24] The Guards were given dhows equipped with cranes. These boats were hard to identify from the small commercial ships in the area, and could carry up to 350 tons worth of mines. The Guards also had at least some landing craft and small craft borrowed from the navy, and a North Korean miniature submarine of 6–9 meters in length.

The new naval Guards forces gradually built up bases at a number of offshore islands and oil platforms, with key concentrations at Al Farisiyah, Sirri, Halu Island, Abu Musa, the Greater and Lesser Tunbs, and Larak.[25] They trained in "suicide boats" designed to ram ships with high explosives, and claimed to have other boats it could fill with fast-drying concrete and could use to block key ports or shipping channels. They acquired extensive stocks of Scuba equipment and an underwater combat center at Bandar Abbas.

The Guards also obtained air and missile capabilities. They were given the 35 to 46 Pilatus PC-7 light training/attack aircraft that Iran had bought from Switzerland, and were trained by North Korean instructors. They were rumored to be receiving Chinese-supplied F-6 and F-7 fighters and to be seeking more advanced fighters from the Soviet bloc. They had a facility at Nowshahr Naval Academy on the Caspian

Sea, where some Guards were reported to have had training in suicide attacks using light aircraft as well as small craft.

The Guards were also given control of Iran's new HY-2 Silkworm anti-ship missiles.[26] In late 1986, the naval Guards began work on hardened sites for 35–50 Chinese Silkworm missiles, and created hardened concrete bunkers for the radar sites for the missiles.[27] These missiles were initially located on Quesham Island and the mainland. The Guards also reinforced their naval forces on Farsi Island, Abu Musa, and the Greater (Sughra) and Lesser (Kubra) Tunbs to levels of about 1,000 men, fortified their positions, and deployed helicopters. The Shah had seized these latter islands from Sharjah in 1971: Abu Musa is about 60 miles west of the Straits and the Tunbs are about 40 miles west.[28] They also began to use more of Iran's offshore oil platforms as military bases.

Iran's leaders must certainly have realized that such forces could scarcely win a direct battle with the U.S. or Soviet Navy. This combination of forces did, however, give Iran considerable ability to conduct hit and run attacks using mines and small craft, while retaining the ability to escalate using its Silkworm missiles and suicide attacks by boats and aircraft.

Iran was well postured to fight the naval equivalent of guerrilla war, and to exploit any political weakness in the willingness of U.S. or other outside naval forces to engage in a long low-level conflict, or take casualties. Equally importantl Iran had considerable capability to conduct mining and sabotage operations that would be very difficult to trade directly to the Iranian government, and could use the Guards under conditions that made it difficult to tell whether they were small units acting on their own or units directly under the control of the government.

IRAN BEGINS TO USE ITS NEW NAVAL CAPABILITIES

While Iran began to use the naval elements of its Guards in the early 1980s, its first major use of their new naval capabilities in the Gulf began in January, 1987. Iran started by using its smaller missiles to strike at the cargo ships and tankers in the Gulf, often at night. Confirmed attacks included missile strikes on the *World Dawn* on January 8, the *Atlantic Dignity* on January 12, the *Saudiah* on January 14, the *Isomeri* on January 23, the *Ambia Fortuna* on February 4, and the *Sea Empress* and *Wu Jiang* later in the month.[29] In each case, the missile used was identified as the Italian-made Sea Killer. On at least one occasion in early March, the Sea Killer also showed that it could do catastrophic damage. It hit the small 998 ton tanker *Sedra* in a vulnerable area and turned the ship into an inferno, killing at least seven crewmen.[30]

Few of the cargo ships that Iran and Iraq hit during this period were sunk or seriously damaged, but 16 ships were damaged between January 1 and early February—raising the total number of hits since the war began to 284.[31] As a result, the USSR and the West began to take preventive action. The USSR reacted in mid-January by sending a Krivak-class missile frigate to escort four Soviet ships carrying arms to Iraq from the Straits to Kuwait. This was the second Soviet warship to enter the Gulf since 1982—the first had been sent when Iran detained two Soviet ships in September, 1986—and was clearly intended as a signal to Iran, Iraq, and the southern Gulf states that the USSR would protect its ships.

The U.S. did not increase its six-ship force in the Gulf, but it did send patrols further north into the Gulf, and it increased its force in the Indian Ocean to a full carrier group—including the 85,000 ton carrier *Kitty Hawk* and eleven escort ships. The U.S. deployed the carrier task force just east of Masirah, off the coast of Oman. Further, the U.S. sent F-111s to a conspicuous "exercise" in Turkey. Britain and France increased their ship activity, and the British Armilla Squadron in the Indian Ocean began to spend roughly 50% of its time in the Gulf. The squadron then consisted of the air defense destroyer *Nottingham*, the frigate *Andromeda* with the Sea Wolf anti-missile defense system, and the support ship *Orangeleaf*.[32]

The movement of the U.S. carrier group to the Gulf of Oman was in reaction to more than Iranian use of the Sea Killer. The U.S. had found that Iran was deploying its much heavier land-based Silkworm anti-ship missile near the Straits of Hormuz, and detected at least one test firing of the missile at the Qeshim Island site in late February.[33]

While the resulting mix of anti-ship missile capabilities scarcely gave Iran a powerful strike force by Western standards, the combination of the Sea Killer and Silkworm missiles did give Iran a significant capability to strike at tankers and cargo vessels in the Gulf, a growing capability to sink ships rather than damage them, and even a limited ability to close the Straits of Hormuz. Further, Iran continued to buy more of the outboard motor-powered aluminum small craft it had begun to use in the summer of 1986 to mine the waters off Umm Qasr. Iran had used these boats to sow coastal mines in waters up to 60 feet deep, and had the capability to launch free-floating mines that could at least ha[illegible] traffic in the Gulf.[34]

THE GROWING INTERNATIONALIZATION OF THE TANKER WAR

This rise in the Iranian threat to Gulf shipping created a very different problem for the West from the one created by Iraq's strikes on tankers in Iranian waters. Iran was not arming to strike at ships in Iraqi waters. There had not been any traffic through these waters since the first weeks of the war, and all Iraqi ports were closed. It was arming to strike at third country ships moving through international waters to ports outside the war zone. This meant Iran planned to threaten or attack international shipping to the Southern Gulf states.

It was also apparent that the build-up in the tanker war that had taken place in 1986 was continuing in 1987. Iraqi attacks on Gulf shipping were far exceeding the number in 1986, and Iranian attacks were doubling in frequency.

This helped catalyze a U.S. movement towards more active intervention in the Gulf, but it was only one of the factors involved. The Reagan Administration had to deal with the political backlash from the full Congressional exposure of its covert arms sales to Iran, and it badly needed to reassert the political and military credibility of its willingness to support friendly Arab states in the region. Further, the Iranian advances at Faw created a growing risk that Iran might seize southern Iraq and even invade Kuwait.

In any case, the key problem American planners faced was not the current level of the tanker war as much as the broader trends and risks that were affecting the security of Western oil supplies. The statistics involved are summarized in Figure 6.3. Even in a period of comparative "oil glut", the U.S. faced the problem that the Gulf region was a $7 billion market for U.S. goods, provided 63% of the free world's oil reserves, and supplied over 25% of all oil moving in world trade, 30% of the oil used by the OECD European states, and 60% of Japan's oil. Even with the expansion of pipelines through Turkey and Saudi Arabia, 17% of the West's oil still moved through the Straits of Hormuz.[35] (See Table 10.2.)

The Iranian threat to Kuwait was particularly important. Kuwait had nearly as many proven oil reserves as Iraq and Iran combined. It had more than 100 billion barrels or oil reserves, or 13% of the world's proven oil reserves (versus roughly 3.9% in the U.S). It had overseas investments of over $100 billion (80% government and 20% private), and about 40% of these investments were placed in the U.S.[36]

The U.S. could not ignore the risks inherent in a scenario where Iran might gain direct control of Kuwait. It also could not ignore the fact that Kuwait and Saudi Arabia would virtually have to become more

Table 10.2: Western Dependence on Gulf Oil in 1986 and Early 1987

Category	U.S.	Japan	FRG	France	Italy	U.K.
Total Oil Consumption in 1986 (MMBD)	16.1	4.3	2.5	1.9	1.7	1.6
Total Oil Imported in 1986						
MMBD	5.3	4.4	2.3	1.7	1.7	(a)
Percent	33	100	92	90	100	(a)
Total Oil Imported in 1986 from the Gulf by Pipeline or Tanker						
MMBD	.9	2.6	.2	.6	.8	.2
Percent	6	61	8	32	47	13
Percentage of Total Oil Consumption passing through the Straits of Hormuz in First Quarter of 1987	4	48	(11% for all of Western Europe)			

(a) U.K. is a net exporter because of North Sea Oil.

Source: U.S. Energy Information Agency, June, 1987

accommodating towards Iranian political demands if the West did not offer them more support.

It is hardly surprising, therefore, that the Reagan Administration became increasingly receptive to a series of Kuwaiti requests to provide U.S. protection and reflagging for the tankers Kuwait was using to shuttle oil to loading points outside the Gulf. Kuwait seems to have openly raised its concern with the vulnerability of its shipping in informal discussions with the U.S. in September, 1986. It raised the reflagging idea again after the GCC meeting on November 1, 1986. The Kuwaiti Oil Tanker Company (KOTC) tasked the U.S. Coast Guard for information on reflagging on December 10, and raised the idea of obtaining U.S. flag protection with the U.S. Embassy in Kuwait on December 22, 1986.[37]

On January 13, 1987, Kuwait asked the U.S. Embassy if reflagged tankers would receive U.S. Navy protection and informed the U.S. that it had an offer to provide such protection from the USSR. Kuwait made it clear that it had begun these negotiations in Moscow in November, 1986, shortly after it had first approached the U.S.

In making these initiatives to the U.S., Kuwait was as concerned with the need to find outside protection against all forms of Iranian threats and attacks as it was with attacks on Kuwaiti flag ships.

Although the "tanker war" was serious—and ships moving to Kuwait had been the target of 15 of the 19 attacks Iran had conducted on Gulf shipping after September 15, 1986—the tanker war still affected only a small portion of the 168–196 tankers a month that moved through the Gulf. The total tanker movements to Kuwait averaged at least one ship per day, and often reached a total of as many as 70–80 tankers per month.

By early 1987, only seven Kuwaiti flag ships had been hit, out of some 284 attacks on shipping since the beginning of the tanker war. No Kuwait flag ship had been included in any of the 34 vessels Iran had attacked between Christmas day on 1985, and September 17, 1986, when the 290,085 ton tanker *Al Funtas* was struck. The most recent Kuwaiti target had been the 263,679 ton tanker *Al Faiha* on October 22.[38] This was enough to virtually double the war insurance premium for tankers moving towards Kuwait, but the rate was still only 0.5% of the value of the ship versus 0.25% for the ships in the rest of the Gulf.[39]

Kuwait, however, could not ignore the broader strategic implications of Iranian gains at Faw and Basra, the build-up of the naval elements of the Iranian Guards, and the fact that it was coming under a steadily growing series of Iranian political attacks for allowing Iraq to use its port of Shuaiba to ship Soviet and other arms and war material. Kuwait had also increased the risk of Iranian attacks on its own ships or other ships moving towards Kuwait by allowing Iraqi planes to overfly Kuwait so that they could fly down the southern coast of the Gulf and attack Iranian shipping without warning. Kuwait also seems to have allowed the Iraqi Navy to send small ships down the Sebiyeh waterway between Kuwait and Bubiyan Island, and may have allowed Iraqi helicopters to stage out of Kuwaiti territory.[40]

In short, the Reagan Administration and Kuwait both felt they had good reasons to move forward with the reflagging option. It was the U.S., however, that now began to take the lead for reasons that had more to do with the political backlash from U.S. arms sales to Iran than with military developments in the Gulf. The White House tasked the National Security Planning Group (NSPG) with determining how the U.S. could rebuild its position in the Gulf. This group was formally chaired by the President and included Secretary Weinberger, Secretary Shultz, National Security Advisor Frank C. Carlucci, and a representative from the Joint Chiefs.

The NSPG concluded that the U.S. had to take major new initiatives to restore its position in the Gulf, reduce any increase in Soviet influence, and reduce the risk of an Iranian victory.[41] These initiatives included a recommendation that the U.S. support the reflagging plan, make further efforts to halt arms sales to Iran, and make a new effort to use the key members of the UN Security Council to support a peace initiative that included sanctions on such arms sales.

Three factors had catalyzed the NSPG into giving this major U.S. policy impetus to the reflagging plan. The first was the U.S. discovery that the Soviet Union had agreed to a similar reflagging request, and that Kuwait was now proposing that the U.S. and USSR share protection of eleven of its tankers—with the USSR protecting five and the U.S. protecting six. The second factor was the fact that Iran had deployed the Silkworm missile. The third was that the NSPG concluded that Iran had not attacked the flag carriers of the major powers earlier in the war, was unlikely to do so now, and lacked the conventional naval and air strength to challenge the U.S. Navy. As a result, the NSPG reacted by supporting a plan for the U.S. to escort all eleven ships.

It is important to note, however, that in reaching this decision, the NSPG repeated one of the major mistakes that had contributed to the Iran-Contra arms scandal. It did not go through the full interagency process normally used in the U.S. government to shape its decision to reflag the Kuwaiti tankers. In fact, the Chairman of the Joint Chiefs, the Secretary of the Navy, most of the experts in the U.S. intelligence community, and most of the Reagan Administration's leading officials on Gulf affairs were not consulted in detail.[42] No effort was made to develop a formal National Security Decision Directive (NSDD). In fact, by that time, the most recent NSDDs relating to the Gulf were already several years old. As a result, the NSPG acted without any formal analysis of the political and military risks involved and without any clear picture of the potential military requirements necessary to escort the tankers.[43]

Further, the U.S. repeated a common mistake in low-level wars. It focused on the potential for success and underestimated the full range of risks. The Department of Defense did prepare a contingency plan to strike at the Silkworm missiles, but it did not prepare plans to deploy a force that could deal with all of the threats the U.S. would face in conducting a convoy operation that had to extend from the Gulf of Oman to ports in Kuwait. As later became clear, these threats were very real. Further, the NSPG failed to consider both the risk of an extended military test of wills with Iran and the risk of military escalation. There was little, if any, consideration of what would happen if the U.S. became involved in a long military test of wills with Iran, or in active and prolonged fighting.

Because the Reagan Administration underestimated the risks involved, it also did not pay proper attention to the probable reaction of the U.S. Congress if the reflagging effort led to actual fighting, and it ignored many of the risks inherent in a rising Iranian threat to the southern Gulf states. It saw the Soviet problem almost solely in terms of the risk that the Soviets might gain influence with the GCC states, and not in terms of the risk that it might force Iran to improve its relations with the USSR. It ignored Iraq's incentive to try to exploit the reflagging to bring the U.S. into the war against Iran, and it largely ignored Western Europe in its desire to reassert U.S. leadership and demonstrate the importance of a U.S. military presence to the Gulf states.

Finally, the Reagan Administration ignored the uncertainties in Kuwait's political position. In spite of the Administration's relatively quick acceptance of the Kuwaiti request, Kuwait did not immediately accept the U.S. proposal. It delayed for several reasons. It wanted either to obtain U.S. protection with a low political profile, or to obtain it on some kind of international basis. It even indicated that it might refuse the offer of a U.S. tanker escort when the U.S. made it clear that it would not operate as part of a joint task force with the USSR.

This helps explain why the Reagan Administration's rush to support the reflagging effort did not win immediate gratitude from Kuwait. On January 29, the Reagan Administration notified Kuwait that it would permit reflagging of its tankers. President Reagan declared on January 23, 1987, that the U.S. would use armed force to secure the passage of cargo ships through the Gulf, and on February 6, the U.S. notified Kuwait that such reflagging would involve protection by the U.S. Navy. On February 25, President Reagan repeated his guarantee that the U.S. would protect the flow of oil.

The Reagan Administration then, however, proceeded to give its offer to provide U.S. Navy escorts for Kuwait's reflagged ships a very high profile. It did so prematurely, and without fully consulting Kuwait. While this U.S. publicity was partly intended to have a deterrent effect on Iran that would protect Gulf shipping, it was also clearly part of the Reagan Administration's efforts to restore the political credibility it had lost because of its covert arms sales to Iran.

This publicity confronted Kuwait with a Hobson's choice. It was clear that it might need protection from Iran. The Khomeini regime had already shown that it could strike at Kuwait with both military force and terrorism. It had attacked Kuwait at least twice by air early in the war. In 1983, Iran had supported the bombing of the French and U.S. embassies. In 1984, it had supported the hijacking of an airliner from Kuwait to Tehran. In May 1985, Iran was associated with an attempt to assassinate the Emir. In 1986, Iran seems to have supported Kuwaiti

Shi'ites in an effort to start a fire at Kuwait's largest refinery at Ahmadi in June, 1986. Similar groups carried out car bombings and damaged its main loading terminal.

Kuwait also, however, had to deal with the fact that Iraq was still highly suspicious of U.S. intentions in the Gulf, and senior Iraqi leaders like Tariq Aziz made it loudly clear that they felt the U.S. was offering help only in an attempt to exploit the war to increase its influence in the Gulf. Further, Kuwait still resented the past U.S. failure in 1984 to provide the arms Kuwait requested to improve its air defense, and many members of Kuwait's royal family regarded the USSR as more reliable in its dealings with Kuwait than the U.S.[44]

This led Kuwait to delay its acceptance of the U.S. offer. To put this delay in perspective, Kuwait faced serious internal problems as well as external ones. It had been forced to dissolve its parliament in July, 1986, and Kuwaiti police officials had announced in November, 1986, that some 26,898 people had been deported during the previous year. Informed sources put the true figure at 40,000–50,000, including many Iranian Shi'ites, large numbers of Palestinians, and many of the top foreign journalists in the country. While Kuwait still claimed that most of the Shi'ites involved were tied into Al Dawaa, the pro-Iranian radical underground in Iraq, many were actually native Shi'ites.

Even with these deportations, only about 40% of Kuwait's population had citizenship, only 22% (720,000 out of 1.1 million) was native born, and nearly 33% of its total population was Shi'ite. Many of the Shi'ites with citizenship were of Persian origin and many of the resident Shi'ites without citizenship had emigrated from Iran.[45] Some of Kuwait's oldest and most important Shi'ite families had members who were implicated in pro-Iranian groups—some of which were involved in guerrilla or terrorist incidents.[46]

Kuwait also had to consider the fact that it still had some 400,000 resident Palestinians, and had to be sensitive to the desire of the UAE to avoid any major increase in superpower military activity in the Gulf because Kuwait was quietly exporting much of its oil by a chartered tanker fleet that sailed between Khor Fakken in the UAE (near the Straits of Hormuz) and Kuwait. This fleet had built up to 22 Kuwaiti flag ships, plus up to 40 more on charter, and was critical to keeping Kuwait's insurance costs low and ensuring the flow of oil.[47]

All of these factors led Kuwait to react to the potential foreign and domestic backlash from the U.S. reflagging offer—and the growing Iranian threat—by making the Soviet agreement to reflag five Kuwait tankers equally public. Kuwait then waited until March 2 to ask to put six of its tankers under the U.S. flag.

Kuwait's implied threat of expanding the role of the Soviet Navy in the Gulf led the U.S. to react in a rush. On March 7, the U.S. informed Kuwait that it would protect all eleven of the tankers that Kuwait had originally discussed with the U.S., and Kuwait tentatively accepted this offer on March 10. The Reagan Administration informed the Congress of this plan on March 12, and began secret briefings of the House and Senate Foreign Affairs Committee staffs on March 19. It also had Admiral Crowe, the Chairman of the Joint Chiefs of Staff, reiterate the U.S. military commitment to protect the tankers on March 17, 1987.[48]

As a result, Kuwaiti personnel from Kuwait's tanker company, and naval officers from the Middle East Task Force in the Gulf, began talks on protection procedures and emergency communications on March 22. These talks were successful enough for Richard W. Murphy, the Assistant Secretary of State for Near Eastern and Asian Affairs, to begin classified briefings of the House Foreign Affairs Committee on March 30, and of the Senate Armed Services Committee on March 31.

Kuwait signed a formal reflagging agreement with the U.S. on April 2, 1987. Kuwait still, however, was careful to avoid giving its arrangements with the U.S. any unnecessary publicity. It continued to emphasize the fact that it had chartered Soviet tankers.[49] This gave the entire tanker escort activity an "international character" and further defused some of Kuwait's domestic and regional problems in accepting U.S. support.

Reflagging under both the U.S. and Soviet flags had other advantages from Kuwait's perspective: It gave Iran a clear message that Kuwait's tanker fleet was getting both Soviet and U.S. naval protection; it avoided an immediate commitment to obtaining direct military support from the superpowers; it was more acceptable to Iraq, which was then far more willing to trust the USSR than the U.S.; and it put growing pressure on both the U.S. and the USSR to find some way of ending the war.[50]

Like the U.S., however, Kuwait seems to have underestimated the risks involved in the reflagging effort. It almost certainly felt that Iran would not challenge a combination of U.S. and Soviet guarantees, and that it now had an implied U.S. and Soviet guarantee to defend it from Iranian attacks or invasion. It does not seem to have considered the risk that Iran would try to reestablish some kind of relationship with the USSR, or that it would challenge the U.S. in much the same way it did during the U.S. hostage crisis and intervention in Lebanon.

IRAN ADDS POLITICAL TO MILITARY INTIMIDATION

Iran continued to show it would conduct its own political struggle to influence other states. As its attack on Basra proceeded, Iran launched new efforts at regional political intimidation and carefully timed these efforts to put indirect pressure on the Islamic Conference that was to meet in Kuwait at the end of January. A series of small bombing incidents occurred in Baghdad, artillery shells hit the Kuwaiti Island of Failaka, and on January 20, three fires were set in oil installations in Kuwait. There were also indications that Iran had covertly attempted to mine Kuwait's harbors.[51]

It is unclear that this Iranian pressure succeeded in influencing the behavior of its neighbors, but such intimidation was probably unnecessary. The Arab world was already divided, and states like the UAE and Qatar were reluctant to take any steps that would increase their involvement in the war. The Islamic Foreign Ministers' meeting in Kuwait in February, 1987, produced nothing more substantive regarding the Gulf than rhetoric and new peace initiatives. Algeria and Libya blocked any substantive debate of the issues relating to the war, and Syria concentrated on the Arab-Israeli issue.

Iran also continued to increase its pressure on Kuwait to reduce its support of Iran, and on Iraq to reduce its attacks on Iranian oil facilities. Pro-Iranian Shi'ites carried out a new series of protests in April. These reached the point where it was clear that Kuwait faced problems with its own Shi'ite citizens and was forced to stop attributing all such incidents to foreign elements. The government admitted that some 25 native Shi'ites were under arrest for terrorism. It also removed at least 1 Shi'ite Minister from his position and some 200 Shi'ites from sensitive jobs in the oil industry.

The risks Kuwait faced in terms of internal and external sabotage were made even more clear on May 22, when a major fire was set at a propane storage tank at the Ahmadi refinery, near the city of Fahaheel. The fire burned for three days (this was the second attempt to burn the refinery in two years). It was designed to explode six 450,000 barrel propane and butane tanks, and it threatened to spread to the entire oil installation, and even to part of Fahaheel city.

Iran, however, seems to have badly miscalculated the Kuwaiti reaction to such pressure. The Kuwaiti royal family clearly felt that the political, ideological, and military threat from Iran was far too serious to attempt to appease. Kuwait pressed ahead with obtaining U.S. and Soviet escort of its tankers. It reached a formal charter agreement with the USSR to charter three tankers, with the option of chartering two more "at short notice," on April 1, 1987.[52]

Kuwait also put 16 Kuwaitis on trial on April 4, 1987, for creating a sabotage network called the Forces of the Prophet Mohammed in Kuwait to bomb the Kuwaiti oil fields. In spite of efforts to halt the trial and some small protest demonstrations, it went on until June 6. The State Security Court of Kuwait then sentenced 6 of the defendants to death for sabotage and subversion.[53] Kuwait kept in jail some 17 Shi'ite bombers who had been convicted of attacks on the U.S. and French embassies, and on facilities in Kuwait in March, 1984. It executed 3 of these terrorists in July, 1987.

At the same time, Iran continued to develop a threat to the Southern Gulf states that went far beyond terrorism. It completed the siting of two to four full batteries of Silkworm missiles near the Straits of Hormuz. By mid-April, 1987, it had at least 12 Silkworm launchers and 20 missiles on line, with up to 28 more in reserve or on order. The only problem with delaying active deployment of the missiles seemed to be a shortage of critical parts, and even this was controversial.

One of the batteries detected early in the year was relocated from the Island of Queshim (Onqeshim) on the edge of the Straits to the Iranian naval base at Bandar Abbas on the northwest shore of the Straits of Hormuz, to provide surface-to-air missile protection. The other battery was located at Kuhestak to the east. While the missiles could just barely cover the 64-kilometer-wide Straits with reasonable accuracy and with their normal payload, they could reach up to 75 kilometers with less accuracy and a lower payload.[54]

Equally important, Iran began to build another battery site on the southern tip of the Faw Peninsula. This site gave Iran the ability to site the missile within firing range of Kuwait city and its port, and gave Iran increasing ability to put pressure on Kuwait and its tanker shuttle.[55] Iran also quietly built-up a considerable force from the naval branch of its Guards units on four islands in the Gulf.

FURTHER INTERNATIONALIZATION OF THE TANKER WAR

Iran's efforts to deprive Iraq of its ability to dominate the "tanker war", and the Kuwaiti reflagging effort, led to the growing internationalization of the naval conflict in the Gulf. This began with a confrontation between Iran and the USSR, but it soon became dominated by a far more serious process of confrontation between Iran and the U.S.

During April, 1987, Iran began to respond more actively to the Soviet agreement to escort the three Soviet tankers which Kuwait had leased from the USSR and had made part of its tanker shuttle. Iran warned on April 15 that Soviet leasing of oil tankers to Kuwait could create "a very dangerous situation".[56] Four days later, on April 19, Iran responded to a visit to Kuwait by Soviet Deputy Foreign Minister, Vladimir Petrovsky, by warning the USSR that the Gulf could become a "second Afghanistan".[57] The Iranian statement went on to warn the Soviets that they were acting, "little knowing that the people of Iran consider this immaturity and naivete to be similar to the presence of the U.S. fleet in the Persian Gulf".

Rafsanjani issued a similar warning to the U.S. on April 20.[58] The next day, U.S. intelligence experts announced that Iran had deployed its Silkworm missiles on Faw.[59] The U.S. government began to actively publicize each new development in the "tanker war", and U.S. government sources reported that 300 ships had been fired on by both sides since the war began.

On April 27, Hussein a'Lael, the Revolutionary Guards' naval commander, announced that Iran now had "full control" of the northern Gulf, and had established the "first IRGC naval zone". On April 28, the commander of the Iranian Navy, Commodore Mohammed Hoseyn Malekzadegan, claimed in a speech over Iranian radio that Iran had boarded 1,200 ships since the war had begun and had seized 30 cargos. He also warned that Iran would close the Straits of Hormuz if there was any disruption of Iranian exports and imports.[60]

These claims took on a special meaning because of the geography involved. Iraq had been relatively modest in defining its naval war zone, which it set at 29 degrees, 30 minutes, North latitude. This included Kharg Island, but did not include a large part of the Gulf's navigable waters. The Iranians, however, claimed both a 12 mile limit from the shore and an "exclusion zone" that ran along the Gulf at a point roughly 40 miles from the shore.

This position was arguably legal under the law of the sea, which permits a nation to declare a "zone of danger" if it does not directly interfere with international trade. In practice, however, the Iranians declared so large a zone that it created points at which traffic to Kuwait was confined to a very narrow channel outside the Iranian zone. This was particularly true of the waters south of Farsi Island, where the exclusion zone permitted a main navigation channel only two miles wide, although the actual depth of the Gulf permitted passage over a 60 mile area. The Iranian exclusion zone acted to force tankers into a vulnerable area and created routes that made tanker passage relatively predictable if Iran chose to carry out mining and other attacks outside its declared zone.

By early May, Iran had further increased its naval activity. On

May 2, 1986, Iranian naval units boarded a total of 14 tankers. The naval Guards attacked the 89,450 ton Indian tanker *B R Ambedkar* off of Sharjah. On May 4, Guards units attacked the 31,120 ton *Petrobulk Regent*, and on May 5, the Guards attacked the 258,000 ton Japanese tanker *Shuho Maru* about 48 kilometers off the Saudi coast.

These naval Guards units launched many of their attack craft from the Guards base closest to Bahrain and Saudi Arabia. This was the small island of Farsi (Al-Farisiyah), midway between the Iranian coast and northern Saudi Arabia.[61] The Guards' naval units also attacked from at least four other bases: an oil platform near Halul Island, Sirri Island, Abu Musa, and Larak. Their normal pattern of attack was to challenge the tanker at night, establish its identity and destination, and then return several hours later. Such attacks normally used machine guns and rockets, and while these did not do serious damage it was obvious that they were often aimed at the crew's quarters.

THE U.S. TRIES TO ADOPT A "TWO TRACK" POLICY IN THE GULF

These shifts in Iranian capability led the U.S. to establish a "two track" policy towards Iran. The first track was to use the UN to force a general cease-fire in the war than would lead to a peace settlement based on the 1974 Algiers Accord, and a return to pre-war borders. This led the U.S. to conduct a series of unannounced meetings with the USSR, PRC, Britain, and France at the United Nations regarding a new peace settlement initiative.[62]

The second track was to take military steps to contain the war in the Gulf, limit the growth of Soviet influence, and help push Iran towards accepting a cease-fire. The means the U.S. chose was to speed up the implementation of the agreement with Kuwait over registering 11 Kuwaiti tankers under the U.S. flag. Ambassador Robert Murphy made the fact public that negotiations were nearing completion in testimony to the House Foreign Affairs Committee on April 21, and technical talks between the KOTC and the U.S. Coast Guard followed on April 22.

These U.S. actions, and the initial operation of Kuwait's charters from the USSR, led to an Iranian reaction during the first week in May. Iran launched a new series of political attacks on Kuwait. On May 6, Iranian Prime Minister Mir Hossein Mousavi warned Kuwait to stop its search for the "protection of the superpowers". Then, on May 8, 1987, Iran's verbal attacks turned into military ones. Unmarked Iranian small craft using machine guns and rocket launchers attacked a 6,459 ton

Soviet freighter, the *Ivan Korotoyev*, while it was sailing near Iran's Sassan offshore oil field, about 30 miles from Iran's Rostam oil platform.

While it was unclear whether the attack was a deliberate effort by the Iranian government or the effort of some more extreme faction of the Revolutionary Guards, later evidence pointed strongly towards a deliberate attack. During the months that followed, there was no evidence that the naval Guards units acted independent of government policy—or even of the regular armed forces.

The Iranian Navy used its helicopters, small craft, and P-3F maritime patrol aircraft to track ship movements in the Gulf and help target such Guard activity. The Guards' bases on Al-Farisiyah and Abu Musa seemed to be firmly under central government control. Further, the *Ivan Korotoyev* was a cargo vessel which regularly visited the Gulf, and was exiting Kuwait en route to the Saudi port of Dammam via Dubai along a route precisely known to Iran. This made it somewhat unlikely that the attack was accidental.[63]

Iran's actions did little to deter either Soviet or U.S. support of Kuwait. While the USSR was careful to avoid any break with Iran, it responded with a strong warning. It then quietly demonstrated the strength of its air power to the Iranian government. It sent 50 combat aircraft across parts of Iran's northern border. The U.K. announced that it was increasing its Armilla patrol to three combat vessels so that two British ships could be in the Gulf at all times, and the U.S. Secretary of Energy formally announced the details of the U.S.-Kuwaiti registration deal on May 10.[64]

Ambassador Richard Murphy—who was visiting the Gulf and later talked to Sadam Hussein—issued a further warning to Iran the next day, when it attacked another Kuwaiti tanker. While the damage involved in this attack was comparatively limited, Japanese tanker owners responded by temporarily suspending traffic to Kuwait.[65] This was the third Iranian tanker attack in eight days, and the twentieth ship attack since the beginning of the year. Sixteen of these attacks had been against ships moving to and from Kuwait.[66]

The U.S. responded by completing its efforts to reflag Kuwait's tankers. The U.S. Coast Guard sent a team to Kuwait to begin inspecting the eleven tankers to be reflagged on May 12, and Kuwait defense officials met with the Commander of the U.S. Middle East Force on May 13. On May 14, the Department of Defense issued a one year waiver to Kuwait to free the eleven tankers from compliance with some U.S. safety requirements, and a two year waiver from the requirement to be drydocked in the U.S. U.S. law already exempted the ships from the requirement to use 75% U.S. crews imposed on U.S. ships using U.S.

ports. The only major provision of U.S. law that was left was that the ships have a U.S. captain.[67]

Meanwhile, Iraq stepped up its attacks on Iranian ships and, on May 13, launched its first major air strikes on Iranian refineries since the start of 1987. The Iraqi fighters struck at the refineries at Isfahan and Tabriz. Iraq sought U.S. and British air charters to help carry new purchases of small craft and outboard motors to Baghdad, for use in its rivers and water barriers and the Gulf.[68] While the land war remained relatively quiet, Iraq also responded to a step-up in PUK activity by launching a number of raids on Kurdish camps and villages. It seems to have begun to use mustard gas on Kurdish villages and PUK camps near the border to raze villages and displace their populations.[69]

Iran responded by continuing to escalate. Its next step, on May 16, was to mine one of three Soviet tankers that had been leased to Iran. The *Marshal Chuykov* was damaged by a mine in the upper Gulf not far from the Neutral Zone. While this damage might have been coincidental, and could have come from a free-floating mine, the combination of location and timing made an accident seem highly unlikely—particularly because Iran rapidly demonstrated it could carry out such attacks with excellent precision.[70] Further, that same day, Iranian radio quoted Chief Justice Abdulkarim Mousavi as saying that Iranian forces might have hesitated in attacking Kuwaiti tankers, but would never do so in attacking foreign ships.[71]

THE ATTACK ON THE *U.S.S. STARK*

These events might well have led to a confrontation between Iran and the USSR, if external events had not acted to change the U.S. role in the Gulf. The attack on the *Marshal Chuykov* was rapidly overshadowed by the events of the next day. At approximately 9:12 on the evening of May 17, an Iraqi Mirage F-1EQ attacked the U.S. radar frigate, the *U.S.S. Stark*, about 85 miles northeast of Bahrain and 60 miles south of the Iranian exclusion zone. It fired two Exocet missiles, both of which hit the ship, and one of which exploded. This was perhaps the last event that anyone had expected, and while the Iraqi attack was unintentional, it sparked a whole series of debates over the U.S. role in the Gulf, U.S. defense capabilities, and U.S. relations with Saudi Arabia.

While Iraq promptly apologized for the attack, and agreed to both reparations and a joint investigation of the incident, the attack on the *U.S.S. Stark* had several major policy and military impacts, and changed the entire political context in which the U.S. was to begin reflagging the Kuwaiti tankers.

The first policy impact was that the loss of 37 American lives suddenly transformed the growing U.S. commitment in the Gulf from a low-level political-military activity—which received little public attention in the U.S., and which was handled at the expert level—into a major American political crisis. The attack on the *Stark* immediately led to a heated domestic debate over whether the U.S. presence in the Gulf needed approval under the War Powers Act, and to a debate over whether the U.S. should even have military forces in the area.

Many senior members of the U.S. Congress challenged the Reagan Administration's commitment of forces to the Gulf at precisely the time the Reagan Administration was trying to demonstrate its resolution to all the states in the region. These Congressional critics challenged the need for any role in defending transit through the Gulf, raised the specter of Vietnam and of escalation to a major war, and challenged the Administration on the grounds that the reflagging plan could lead to another pointless loss of American lives like the loss of Marines in Lebanon.[72]

The second major policy impact of the attack on the *Stark* was to expose the domestic political weaknesses in the U.S. military position towards Iran, and the West's lack of unity in supporting the U.S. reflagging effort. It is impossible to know just how much this encouraged Iranian weakness to try a military test of wills with the U.S., but it seems likely that it at least encouraged Iran to believe that a "peripheral" strategy of indirect attacks on U.S. ships and forces, the tankers under convoy, or those nations supporting the reflagging effort might lead the Congress and the American people to demand that the U.S halt its reflagging effort, or even withdraw from the Gulf.

Iran had, after all, already achieved two significant limited victories over the U.S. under somewhat similar circumstances. The first was the hostage crisis after Iranian "students" seized the U.S. Embassy in Tehran. The second was U.S. withdrawal from Lebanon after the car bombing of the U.S. Embassy and then of the U.S. Marine Corps barracks in Beirut.

The Attack on the U.S.S. Stark *and U.S. and Saudi Relations*

The third major policy impact of the attack was to do serious harm to U.S. and Saudi relations at a time when all the events in the Gulf highlighted the potential U.S. dependence on support from friendly

Gulf states. The U.S. faced the practical problem that its carrier-borne fighters needed refueling to reach Kuwait safely and that this could be done efficiently only by tankers based in Saudi Arabia or some other Gulf state. Equally, it faced the problem that the most efficient solution would be to base USAF F-15s in Saudi Arabia. Nevertheless, the third major policy-level impact of the attack on the *Stark* was to damage U.S. and Saudi relations at a time when cooperation was of critical importance. A series of false reports right after the attack indicated that the Saudi fighters escorting the AWACS had refused to come to the aid of the ship before it was attacked.[73]

These reports ignored both the role the Saudi fighters were supposed to play in supporting the USAF E-3As and the actual chronology of events. The USAF E-3A AWACS routinely flew 24 hour patrols of the Gulf. They had a Saudi controller on board who was in constant touch with the Saudi operations centers on the ground. As part of the rules of engagement established by the U.S. long before the attack on the *U.S.S. Stark*, Saudi Arabia agreed to provide protection for these U.S. AWACS aircraft.

During the day, the Saudi Air Force provided a constant fighter escort to protect the AWACS and Saudi air space south of the so-called Fahd line which runs down the center of the Gulf. Because Iran has so few remaining operational fighters, and as most are limited in capability, the Saudi fighters do not fly escorts at night. They instead remain on the runway at Daharan in combat ready status. The U.S., however, had told Saudi Arabia that no air cover was necessary for U.S. Navy ships in the Gulf because they could defend themselves. The rules of engagement agreed to between Saudi Arabia and the U.S. thus specifically excluded Saudi planes from protection of U.S. ships.

The Iraqi attack on the *Stark* was the third Iraqi attack on Gulf shipping on that day. Both of the first two attacks occurred during daylight. Both followed roughly the same flight pattern of going south over international waters on the edge of the Saudi air defense zone and then turning north when they reach positions above Jubail. Neither provoked any reaction from either U.S. or Saudi military forces since such attacks have become a virtually routine part of the Iran-Iraq War.

The tracking of the Iraqi fighter that eventually attacked the *Stark* began when the E-3A was flying without Saudi fighter escort. The Saudi fighters scrambled when it became clear that the Iraqi fighter was flying unusually far to the southwest, and might come near the E-3A. When the Iraqi fighter suddenly turned to the northeast, the Saudi fighters stayed near the E-3A.[74]

Neither the E-3A nor the Saudi fighters had reason to think the Iraqi pilot would fire on the *U.S.S. Stark*. The Iraqi pilot used his radar to illuminate the *U.S.S. Stark,* but did not lock his radar onto the ship for several minutes before firing as was normal in previous Iraqi attacks. In any case, there was never any chance that the Saudi fighters could have intercepted the Iraqi fighter before the attack. The Saudi fighters were not in the air near the *Stark,* and they were never requested to take any action to protect the *Stark* before the Iraqi attack.

While some early press reports were published in the U.S. that Saudi fighters refused U.S. requests to come to the aid of the *Stark,* these reports were totally false. The issue never arose. In fact, it was one of the Saudi fighter pilots who asked after the attack on the *Stark* and after the Iraqi fighter had turned back towards Iraq whether the Saudi fighters should intercept the Iraqi fighter. Neither the U.S. officers nor the Saudi controller on the USAF AWACS had the authority to authorize this intercept because it was a violation of the U.S-Saudi rules of engagement.

The two Saudi fighter pilots then waited while the Saudi controller on the AWACS sent the request back to the Sectoral Operating Center in Dhaharan. Since such action involved firing on an aircraft from a friendly state, the request was forwarded up the chain of command. Saudi Arabia was not at war, however, and no senior policy-level officials could be contacted in the minutes that were available before the Saudi fighters ran low on fuel and the Iraqi pilot was too far along his return to Iraq.

It is also important to note that such an intercept could have done nothing to protect the *Stark,* and that neither the U.S. nor Saudi Arabia had anything to gain from such an intercept. If the Saudi pilots had intercepted, they would almost certainly have made a tragic situation worse. Trying to force the Iraqi fighter to land at night might well have led to a further military incident under conditions that would have sharply raised the tensions between Iraq and the U.S. and Saudi Arabia. The last thing on earth anyone needed—after the damage was done—was more political complications.

All of the facts affecting the role of Saudi Arabia during the attack on the *U.S.S. Stark* were made public by the White House, the State Department, the Office of the Secretary of Defense, and the Saudi government. Unfortunately, facts were not enough. The misleading first accounts of Saudi failure to aid the *U.S.S. Stark* were used by pro-Israeli lobbying efforts to gain strong support in Congress in halting the latest round of U.S. arms sales to Bahrain and Saudi Arabia and—more important—acted to limit U.S. military cooperation with friendly Arab states in the Gulf.

This was true even though the arms the Reagan Administration was proposing to sell to Bahrain and Saudi Arabia did not involve any serious risk to Israel. The Administration proposed to sell 16 Stinger launchers and 70 missiles to Bahrain. The "package" the Administration was then proposing to sell to Saudi Arabia involved[75]

- Sale of 1,600 AGM-65D Maverick missiles,
- Sale of 12 more F-15C/D aircraft to be kept in the U.S. as an attrition reserve to ensure that Saudi Arabia would have an operational strength of 60 aircraft,
- Conversion of all Saudi F-15C/D fighters to include a modified Multi-Stage Improvement program (MSIP),
- Upgrade of 150 Saudi M-60A3 tanks to the M-60A3 configuration, and
- Sale of 93 M-992 field artillery support vehicles.

The most important part of the sale to Saudi Arabia was 1,600 U.S. AGM-65D Maverick air-to-surface missiles. The Saudi Air Force had first acquired the TV-guided version of the Maverick in 1976, when it began to take delivery on 2,500 AGM-65A/B, for use with its F-5E aircraft. The Saudi Air Force had ordered 1,600 more AGM-65Bs in 1984, and had notified Congress, but had delayed the order when it became apparent that a new and more effective model with imaging infra-red guidance would be manufactured.

The new AGM-66D allowed the use of the missile at night and in low-visibility conditions. More important, it was far less vulnerable to the lack of visual contrast and the shadow effects common in desert warfare, and was far more effective against small maritime targets like Iran's motor boats and patrol boats. The IR seeker produces a TV-like picture of a ship, and the contrast between the cold ocean water and the heat of the ship meant it could easily track targets moving at up to 50 miles per hour and in virtually any sea state and weather.[76]

The AGM-65D had the same warhead, rocket motor, and logistical support as the previous model, so the only issue involved in the new Maverick sale was whether the Saudis should be allowed to buy a Maverick with the infra-red guidance system instead of the TV sensor originally authorized in 1984. The U.S. also stood to benefit because the Saudi purchase meant longer production runs for the AGM-65D and a lower unit cost to the U.S. Air Force, and the price of the sale climbed from $119 million to $365 million.

The second-most important part of the sale involved giving Saudi Arabia 12 more F-15C/Ds. This sale did not involve any build-up of the Saudi Air Force. The U.S. had sold Saudi Arabia 60 F-15C/Ds in 1978. It then sold 2 additional aircraft in 1981 to make up for the loss of 2 aircraft because of operational accidents. By 1987, the Saudis had lost 4 fighters. Although the Saudi Air Force was meeting most USAF maintenance and training standards, its attrition levels continued to be much higher than forecast, with a loss of 8 aircraft per 100,000 flying hours, versus a predicted level of 3. This led the USAF to calculate that the Saudi Air Force would lose 12 more aircraft during the useful life of the F-15 in Saudi forces.

The Reagan Administration had informed the Congress earlier in 1987, in its "Javits Report" on Middle East arms sales, that it would propose the sale of 3 more F-15C/Ds to make up for attrition. It now faced a new problem because the production line for the F-15C/D was scheduled to close in May, 1988, and the new F-15E involved advanced technology the Congress had already refused to sell Saudi Arabia in 1986. As a result, the Saudis now had to buy all of the 12 F-15C/Ds they needed for an attrition reserve immediately, and had to do so at a cost of $502 million.

The Saudis attempted to ease the Reagan Administration's problems in getting Congressional assent to the sale by agreeing to keep the 12 attrition aircraft in the U.S., and to take delivery only as they lost one of their fighters in active service. They also agreed to a delivery schedule that meant the first aircraft would not be delivered until 1991, although U.S. projections of Saudi attrition meant Saudi Arabia would then have lost 8 aircraft, even if it did not suffer combat losses.

These projections ignored the fact that the Saudi F-15s had already had one air combat exchange with Iran and were acting as the escort for the E-3As in the Gulf. Further, they made no allowance for the fact that the USAF had originally recommended that Saudi Arabia receive two more squadrons of advanced fighters, or 120 aircraft. The Administration's inability to get Congress to approve the sale of advanced F-15E fighters in 1986 had led the Saudis to buy 72 Tornados, but only 24 of the Tornadoes were air defense versions and none of them would be fully operational for several years. This made it likely that Saudi losses would probably be substantially larger and earlier than planned, and a large percentage of the additional losses would be incurred in support of U.S. forces.

The Multi-Stage Improvement Program (MSIP) part of the sale has been described in earlier chapters. It was part of a U.S. Air Force upgrade program for all F-15C/Ds which provided improved cockpit displays, electronics, and computers. It improved reliability and allowed the F-15C/D to operate the most recent forms of air-to-air missiles. The $500 million program was important both to allow Saudi Arabia to make maximum use of its fighters and to stay compatible with the USAF logistic and technical support system—which was going to phase out all support of the earlier electronics.[77]

The MSIP sale was timed to begin the conversion of the remaining Saudi F-15s and attrition reserve in 1992, when the USAF would have had the MSIP conversion for four years and after Israel would have converted its aircraft to a more advanced version of the MSIP. It did not include the attack mission capabilities of the F-15, and was tailored to eliminate some sensitive technologies and advanced systems like the APG-70 radar. While the MSIP potentially allowed the Saudis to operate more modern air-to-air missiles like the AMRAAM, these were not part of the deal, and they required the transfer of advanced U.S. software.

The 150 M-60A3 tank conversions were designed to bring all 250 of Saudi Arabia's M-60s to the same standard. Saudi Arabia already had 100 M-60A3s and the conversion meant it could standardize on the same variant of the M-60. The conversion kits added three new functions to the earlier M-60A1s:

- Tank thermal sights, allowing detection of targets in darkness, smoke, and dust.
- Laser range finders, allowing quicker engagements.
- M-21 ballistics computers, which—in combination with the thermal sight and range finder—allowed more rapid target engagement and an increased probability of a first round hit.

This $120 million conversion package was urgent because the U.S. production line for M-60 upgrades was due to close in 1987 or 1988, and Saudi Arabia might have to go to the aid of Kuwait. Its small armored forces needed any help they could get.

Finally, the request for the sale of 93 M-992 field artillery ammunition support vehicles involved a tracked, armored, and automated ammunition transporter to support the 117 M-109 self-propelled howitzers in the Saudi Army. The M-992 is a tracked, automated, armored ammunition system that can carry 93 155 mm artillery shells with propellant and fuses. It can transfer ammunition under combat conditions, and uses the same chassis and engine, and most of the same parts and logistic support systems, as the M-109. The sale would ease Saudi Arabia's manpower and maintenance problems, and improve its rate of fire, but it did not make a substantial difference in effectiveness. Saudi Arabia was already using trucks and M-113s to perform the same mission. It required Congressional approval because its cost was $62.7 million—well above the Congressional notification threshold of $14 million.

The Stinger sale to Bahrain involved only 16 launchers and 70 missiles, at a total cost of $7 million. It was designed to give

Bahrain the ability to provide some air defense for its patrol boats and key economic and political targets, including the U.S. Navy's Administrative Support Unit. Bahrain's only other air defenses were a limited number of F-5s and French-supplied Crotale missiles in fixed emplacements, and Bahrain had agreed to adopt all U.S. security procedures. The sale was still sensitive, however, because the Congress feared the transfer of the missiles to the enemies of Israel.

Most of the Congressional opponents of the arms sale package, however, did not care about its technical details or whether it posed a real threat to Israel. They were acting under pressure from pro-Israeli lobbyists who wanted to block any expansion of the U.S. strategic and military relationship with any Arab state that had not reached a peace settlement with Israel. They also had an extremely powerful domestic political base in the U.S. and were more than willing to capitalize on the false reports about the *Stark* incident. The resulting propaganda barrage created enough Congressional pressure to make the Reagan Administration defer any mention of the F-15 sale to the Congress and then defer the Maverick sale indefinitely on June 11. The Administration did so, although Richard Murphy called the action a "slap across the face" to Saudi Arabia.[78]

These actions embarrassed Saudi Arabia, and divided it from the U.S. at a time when the U.S. was actively negotiating to establish new contingency agreements to use Saudi air bases and facilities to provide air cover over U.S. ships in the Gulf, and to use the five E-3A AWACS aircraft it had sold Saudi Arabia to extend USAF Elf-1 AWACS patrol in the Gulf to cover the southern Gulf. The Saudis were experiencing major problems in training their own AWACS crews as rapidly as originally planned, and the use of Saudi aircraft gave the U.S. the option of increasing its operational fleet in the Gulf from four to nine aircraft without drawing down on its own limited total force of E-3As.[79]

Coupled with the Congressional efforts to invoke the War Powers Act, the Reagan Administration's inability to sell arms to the Saudis raised serious questions in both the Gulf and Europe as to whether the U.S. could stay the course. It increased the fear that the U.S. might respond to Iranian behavior by withdrawing—just as it had done in Lebanon. This not only encouraged Iran to take a tougher line than it might otherwise have done; it also indicated that any nation whose aid to the U.S. became too visible might simply end up as the target of Iranian hostility and associated with a U.S. defeat.

Iran's "War of Nerves" with the West

Once the Iraqi attack on the *Stark* took place, Iran shifted its tactics in dealing with the West from attempts to capitalize on any U.S.-Iraqi tensions and attempts to embarrass the Reagan Administration with a slow war of nerves, to a war of nerves designed to push the U.S. Congress into canceling U.S. plans to escort Kuwaiti tankers and even into withdrawing U.S. naval forces from the Gulf.

Iran's public reaction to the attack on the *Stark* began with comments by Iran's Prime Minister, Hussein Mousavi, to the effect that "The Great Satan has been trapped" and "We urge the Soviet Union and the United States not to be further embarrassed". Iran then charged that the entire incident was a plot against the U.S. by Iraq and the USSR. This was followed by claims that the attack was a plot among the U.S., Iraq, and Saudi Arabia to justify a U.S. presence in the Gulf, and then by claims that Iran had decoyed the Iraqi fighter into attacking the *Stark*.[80]

This initial round of verbal fencing was followed by more serious Iranian actions. These began with an attack by Iranian Revolutionary Guard boats on a Norwegian tanker, the 219,387 ton *Golar Robin*, which hit the crew's quarters and set the ship afire.[81] Guard units then attempted to seize the offshore terminal near Faw, which involved some 40 small craft armed with 106 mm recoilless guns, machine guns, and rocket launchers. A 272,000 ton VLCC, the *Primrose*, was hit by a mine in roughly the same area as the *Chuykov* on May 16. Guard units attacked a small 2.6 ton cargo vessel, the *Rashidah* on May 22, and the LPG carrier *Nyhammer*, on May 24. A speed boat attacked the *Nyhammer* without warning while it was en route to Ras Tanura. The Iranian boat fired two rockets which missed the vessel, and then left.

Iran claimed on May 31 to have seized seven Kuwaiti fishing boats near the Khour Abdullah estuary at the head of the Gulf which it said were spying for Iraq. The same day, Iranian Foreign Minister Ali Akbar Velayati used his visit to Abu Dhabi as a platform to warn both the U.S. and USSR that Iran was the "most powerful in the Gulf" and would not tolerate superpower interference.[82] It became increasingly clear that Iran saw the attack on the *U.S.S. Stark* as an incentive to increase its pressure on Kuwait and shipping through the Gulf, and felt it could do so without escalating into major military involvements with U.S., European, and Soviet naval forces.

These Iranian actions, however, did little to deter the Reagan Administration, which simply viewed them as an additional reason to reassert the U.S. commitment in the Gulf. They helped lead the Reagan Administration to reiterate its commitment to reflag the 11 Kuwaiti tankers, and its determination to protect U.S. ships, in spite of the growing Congressional debate over the War Powers Act and a Senate resolution requiring the President to tell the Congress how the U.S. Navy ships protecting the Kuwaiti tankers would protect the tankers and themselves before any escort arrangement started.[83]

President Reagan made it clear that the U.S. would take every step necessary to protect U.S. combat ships and cargo ships, as did the Chairman of the U.S. Joint Chiefs of Staff.[84] By late May, the Reagan Administration had also begun to use carefully planted leaks to the press to threaten Iran with attacks on its Silkworm missiles. The U.S. first announced that the missiles were not operational, and that Iran was waiting for the PRC to deliver key parts. It then leaked the fact that it was considering preemptive attacks if Iran deployed them to its pre-surveyed sites at Faw and near the Straits of Hormuz.[85]

IRAN ORGANIZES FOR CONFRONTATION WITH THE U.S.

Both Iran and Iraq temporarily halted their attacks on Gulf shipping after May 21. The most Iran did militarily until mid-June was to challenge a U.S. merchant ship, the *Patriot,* to identify itself when it entered the Gulf on its way to Bahrain. The Iranian destroyer turned away immediately when the merchant ship's escort, the U.S. Destroyer *Coynigham*, came to its aid. No major attack took place on Gulf shipping until June 11, when an unidentified warplane attacked the 126,000 ton Greek tanker *Ethnic* near Kuwait.[86] Iran continued to avoid incidents with any ships carrying the U.S. flag, and U.S. analysis of the Iranian strikes on Kuwaiti ships showed that no Kuwaiti flagged ship had been deliberately hit since the *al Faiha* was struck on October 22. Further, only 8 of the 93 ships that Iran had hit since the heightening of the tanker war in 1984 had Kuwaiti flags.[87]

Nevertheless, there were few signs of Iranian moderation. Khomeini publicly took the position that the war must continue, and Iran continued to build up its naval Guards fleet and to increase its armament. Rafsanjani gave the risk of suicide attacks more credibility when he threatened to use Iranian fighters in such attacks in a speech to Iranian Air Force technicians. He warned in several of his speeches that Iran must prepare for war with the U.S. and USSR—a theme that was promptly echoed by Iran's Prime Minister Hussein Mousavi.[88]

Iranian diplomats visited Bahrain and the UAE, and issued thinly veiled warnings against any support of the growing U.S. role in the Gulf. Iran's Deputy Foreign Minister, Mohammed Javad Larijanti, visited Europe to warn against European support of either the U.S. military role in the Gulf or its peace initiative in the UN.[89] Khomeini also denounced any remaining "lovers of America".

JOCKEYING FOR POSITION BEFORE U.S. CONVOY ACTIVITY BEGAN

The events between the Iraqi attack on the *Stark* and the first U.S. convoy of the reflagged Kuwait tankers did, however, lead to some important jockeying for position. They also exposed serious divisions within the West as to how it should deal with intervention in the Gulf.

The USSR began to shift away from trying to exploit its ties to Iraq and its relationship with Kuwait to trying to exploit the weaknesses in the U.S. position. It began to quietly court Iran, while publicly reiterating its support for Iraq and Kuwait. Soviet officials indicated that the USSR would not increase its small three ship force in the Gulf and that there was no reason for the U.S. to do so. At the same time, they leaked the fact that the USSR had sent 50 combat aircraft into Iranian airspace after the attack on a Soviet ship on May 6. When Alexander Ivanov, the head of the Gulf section of the Soviet Ministry of Foreign Affairs, visited Kuwait in early June, he announced that the USSR would take every means available under international law to respond to any attack on a Soviet flag vessel, particularly if it occurred in international waters, but he was careful to avoid any endorsement of the U.S. reflagging effort.[90]

The U.S., in turn, emphasized its two track approach and continued to press forward with the initiatives in the Security Council it had begun early in the year. It tried to unify British, French, and Soviet support around a U.S.-sponsored cease-fire which would demand that Iran and Iraq join together in a cease-fire, and which would enforce sanctions against either side if it refused. This U.S. strategy called for such sanctions to take the form of an embargo on all oil imports.[91]

The U.S. had mixed success in the UN. It found it difficult to get firm French, British, or Soviet support for sanctions on arms sales, and Chinese agreement to anything other than the call for a cease-fire. This led the Reagan Administration to take an unusually hard line in pressuring the Chinese. When the PRC denied it was shipping arms to Iran, Frank Carlucci, the new Director of the U.S. National Security Council, publicly stated that the Chinese denial was false. The U.S. then circulated reports that the PRC leadership was divided over the issue of arms sales to Iran.

The U.S. encouraged many of the Arab states to begin to step up their pressure on the PRC to halt its arms shipments, and Kuwait went so far as to ask the PRC to provide a further source of tanker escorts. It leaked the fact that China had signed a new $560 million arms agreement with Iran—including the sale of MiG-19 and MiG-21 fighters, SA-7 and CSA-1 SAMs, and 200 more T-59 tanks—to both the U.S. press and Arab capitals, and encouraged the Arab states to protest this to the PRC.[92] This pressure seems to have worked. The U.S. soon claimed it had PRC support for at least the first part of the UN cease-fire initiative.[93] These claims proved to be founded in fact, and the PRC formally announced its support for a cease-fire to the UN Security Council on June 22, 1987.

The Reaction of the Southern Gulf States

In spite of past Western concerns about the political courage and resolve of the Southern Gulf states, they continued to show considerable resistance to Iranian pressure. They ignored Iranian threats to attack any bases on Arab soil which were used by U.S. forces. While the Gulf Cooperation Council meeting on June 6–8, 1987, did not lead to any dramatic new steps towards sanctions, the GCC did endorse Kuwait's effort to obtain escorts for its tankers. While the GCC states continued to be reluctant to do anything they could to avoid alienating Iran, they also were careful not to embarrass the U.S.[94]

Oman steadily expanded its in-shore traffic zone at the mouth of the Gulf, which offered shipping that hugged the Omani coast at least some protection against Iranian action. Further, Saudi Arabia moved towards agreement with the U.S. that it would provide emergency tanker bases and extend the Elf-1 patrol to cover the Southern or lower Gulf by allowing USAF crews to make use of the five E-3As that Saudi Arabia had bought from the U.S. By June 22, 1986, the Saudis reached agreement with the U.S. on this activity in spite of the continuing problems arising from the fact that the Congress was blocking U.S. arms sales to the Kingdom.[95] Bahrain did state that it would not serve as a U.S. military base, but this was a routine denial of Bahrain's actual roles which was based on the technicality that U.S. no longer had formal basing rights on the island.

Only the UAE seemed to avoid taking any action that reflected Iran's efforts to counter the U.S. presence. It barred U.S. warships from its waters on June 11, 1987. This UAE action coincided with more Iranian warnings of suicide attacks on U.S. and Soviet ships in the Gulf, and of attacks on any Arab state that provided bases for such ships.[96]

The U.S. Search for Support from Its European Allies

This mix of U.S. and Southern Gulf resolve led to new Iranian threats and military actions. On June 12, Iran's President Khamenei warned that Iran could attack U.S. ships with weapons not known to Washington. At virtually the same time, Foreign Minister Velayati denied that the PRC had sold Iran Silkworm missiles. Iran then mined Kuwaiti waters in an effort to show it had an alternative to direct attacks on U.S. flag carriers. The discovery of these mines led Iraq to announce on June 15 that it would resume attacking ships going to Iran.[97]

During this time, the internal U.S. political debate over the U.S. role in the Gulf continued. Senator Claiborne Pell, the Chairman of the Senate Foreign Relations Committee, made it clear that he opposed the reflagging effort. The Democrats of both houses also continued to show considerable opposition to the plan, and a negative vote in the House of Representatives was avoided only by delaying action on the issue until July.

President Reagan responded to these actions by the Congress by giving another speech supporting the escort plan on June 15, and claimed it was essential to limit the growth of Soviet influence. Soviet Ambassador Dubrinin reacted by claiming that the USSR was not escorting tankers and had only a minor military presence in the Gulf, and Soviet Deputy Foreign Minister Ului Vorontsov called for U.S. forces to leave the Gulf.[98]

This mix of pressures led the Reagan Administration to make more serious efforts to obtain allied support for its military operations in the Gulf. It had initially been somewhat lukewarm about the need for allied support, and had hoped to unilaterally reassert U.S. power in the Gulf. It became increasingly clear, however, that a cooperative effort would help defuse the impact of superpower rivalry and opposition to the U.S. presence in the Congress, reassure the Southern Gulf, put added pressure on Iran, and possibly provide some of the specialized military forces the U.S. lacked.

On June 1, President Reagan announced before leaving for the Western Summit Conference in Venice that he would seek support for the reflagging effort from the major European nations and Japan.[99] Senior U.S. officials continued to back this theme during the period before the Venice Summit and then actively sought some kind of formal financial and military support behind the scenes as the heads of state continued their open sessions. In practice, however, the U.S. got little more than a vague pledge to consult and a declaration that the "free flow of oil and other traffic through the Straits of Hormuz must continue unimpeded".[100]

France was particularly cautious. Although it had four frigates in the

Indian Ocean, it flatly rejected a Kuwaiti request that it join the tanker escort effort and U.S. efforts to obtain French support in a combined force or in providing mine clearing services. Italy took the position that it was still facing a threat from Libya, and the FRG and Japan claimed that their constitutions and laws prevented even indirect support of military action—although Japan's Finance Minister indicated that Japan might be able to provide some financial support. Britain provided some political support, but refused to increase its escort activity.[101]

Britain hardened its attitude towards Iran, but it did so for reasons that had nothing to do with U.S. diplomatic initiatives. In late May, Revolutionary Guards abducted Edward Chaplin, a British diplomat in Tehran, and charged him with "drugs, corruption, theft, and undermining the economy in time of war". This abduction seems to have been the result of a prior British arrest of Ahmed Gassemi, an Iranian counselor in Manchester, for shoplifting.

Iran picked the wrong country to try to intimidate. When Iran refused to apologize for its actions, Britain responded by forcing Iran to close its consulate in Manchester. Iran responded by expelling five British diplomats, and Britain expelled two Iranian diplomats from London and then all Iranian diplomats but one. While this exchange had a somewhat farcical character, Britain also took a harder line towards British arms exports to Iran. Iran effectively succeeded in cutting off a source of military equipment and high-technology gear that had been worth nearly $600 million in 1986.[102]

On June 19–20, Iraq made its first major air strikes in the tanker war since the attack on the *Stark*. It hit one of the Iranian shuttle tankers, the *Tenacity*, with an Exocet missile. The new Iraqi attacks came after a hiatus of nearly one month and at a point where Iran had reached its OPEC oil quota of 2.3 million barrels a day for the first time. The resumption of the tanker war also came at a time when insurance premiums on Gulf tanker traffic had just been slightly reduced for all shipping except the tanker shipments to Iran. This announcement was followed by an Iraqi air raid on one of Iran's shuttle tankers, and on the "western jetty" at Kharg Island.

The next day, it was reported that Iran had begun active minelaying in the northern Gulf, and had laid mines in Kuwait's port of Ahmadi. These reports led to a new round of concern over mine warfare, and to an announcement by U.S. officials that Saudi Arabia had agreed to allow its four minesweepers to help clear mines, and to use its minesweeping helicopters and their sleds to help clear the mines in Kuwait harbor. Further, it was announced that Saudi Arabia would allow the USAF to use Saudi Arabia's E-3A AWACS to expand surveillance coverage over the Gulf. The planes were to have Saudi pilots with U.S. crews.

As June ended, the U.S. also rejected an Iranian proposal for a partial cease-fire that would have affected only the Gulf, while preventing Iraq from shipping through the Gulf and allowing the land war to go on. Iran replied by staging naval maneuvers, and both Iran's Prime Minister and Hassan Ali, a commander of the naval branch of the Guards, used the occasion to warn that the U.S. would get a " bitter and unforgettable lesson" if it challenged Iran in the Gulf.[103]

All of this jockeying for position made it increasingly clear that Iran was unlikely to back down in the face of the U.S. escort effort, and that the U.S. faced a serious threat from mines, terrorism, suicide attacks, and small craft raids. As for the overall intensity of the tanker war, some 53 vessels were damaged in the first five and a half months of 1987. This total compared with 107 attacks in 1986—the highest number of attacks in any year since the beginning of the war. The tanker war had produced a total of 226 deaths between 1981 and 1986. It had killed 47 people so far in 1987, some 37 of them aboard the *U.S.S. Stark*. The total insurance claims paid since the beginning of the war totalled roughly $1.5 billion. That figure, however, included 110 vessels paid out as total losses, and all but 17 of these ships had been trapped in the Shatt al-Arab at the start of the war.[104]

THE LESSONS OF U.S. ENTRY INTO THE GULF

It is difficult to fault the Reagan Administration for becoming involved in the tanker war in the Gulf, for demonstrating that the U.S. would pursue its strategic interests, and for acting to preserve freedom of navigation. In fact, the Administration showed considerable courage in acting in the face of substantial Congressional opposition.

At the same time, the Reagan Administration can be criticized for acting in a fashion that ignored so many of the hard-fought lessons that had shaped the U.S. national security process.

In far too many ways, American actions during 1986 served as a case study in how not to shape strategic relations with the Gulf. The U.S. repeated virtually all of its mistakes of 1985, and added new ones. Above all, the Reagan Administration failed to trust its own senior policy appointees and failed to use the experts in the State Department, Department of Defense, and intelligence community. It is striking that after all the hearings on the Iran-Contra scandal, it is still impossible to determine why senior White House officials chose to ignore the warnings of virtually every expert in the area, and sustained their contacts with Iran in the face of so many obvious problems and failures.

It is equally difficult to understand why the new set of officials in the White House repeated their predecessors' failure to use the interagency system as they rushed to totally reverse U.S. policy in the Gulf. They not only failed to fully consider the risks; they failed to define a clear objective. Reflagging Kuwaiti tankers and escorting them through the Gulf was at best a means to an end. It also marked an awkward halfway point between limited reassurance of the Southern Gulf states and a commitment to defend all neutral shipping outside the war zone. The U.S. effectively took sides in the Iran-Iraq War without making a commitment to try to force an end to the war based on the cease-fire policies the U.S. was supporting. In fact, one of the strangest aspects of both the covert arms sales to Iran and the reflagging effort is that virtually all of those concerned in shaping U.S. policy do not seem to have looked beyond their immediate tactical goal. They paid lip service to strategic objectives, but they did not define them or calculate what would really be necessary in terms of U.S. forces and actions in order to achieve them.

It is always easy to second-guess policy decisions and military actions. Reality, however, rarely provides ideal contingencies for either policy making or the use of military force. The key lessons of 1986 were not so much the failures in U.S. policy, but the failure to define a policy in the proper terms. Further, the failures were not in the U.S. national security system or in its expertise; they were in the Reagan Administration's failure to use that system and expertise.

NOTES

1. For the details, see the author's *The Iran-Iraq War: 1984–1987*, London, Jane's, 1987.

2. *Washington Post*, July 8, 1985, and December 7, 1986; U.S. Senate Select Committee on Intelligence, *Report on Preliminary Inquiry*, January 28, 1987, p. 4; *Tower Commission Report*, New York, Bantam and Times Books, 1987, pp. 19–24.

3. *Washington Post*, December 7, 1986, p. A-25.

4. U.S. Senate Select Committee on Intelligence, *Report on Preliminary Inquiry*, January 29, 1987, pp. 1–2 ; *Tower Commission Report*, New York, Bantam and Times Books, 1987, pp. 23–26.

5. It later became clear that the latter three individuals had been discussing the Iranian initiative since mid-1984.

6. *International Herald Tribune*, January 30–31, 1988, p. 3.

7. U.S. Senate Select Committee on Intelligence, *Report on Preliminary Inquiry*, January 29, 1987, pp. 4–8; *Tower Commission Report*, New York, Bantam and Times Books, 1987, pp. 28–31.

8. Fred Reed was abducted on September 9, 1986, and Joseph James Cicippio on September 12. Edward Austin Tracy was taken on October 21, 1986. In all three cases the U.S. State Department identified pro-Iranian radical groups as being responsible. The net result neatly balanced out the total hostage releases. See the*Washington Post*, December 7, 1986, p. A-25. See the *Tower Commission Report*, New York, Bantam and Times Books, 1987, pp. 153–334, for full details.

9. *Jane's Defense Weekly*, November 29, 1986, pp. 1256–1258; and *Washington Post*, December 7, 1986, p. A-25. France's Prime Minister Jaques Chirac later indicated that Iran had made somewhat similar arms for hostages initiatives, and had rejected them. Most French sales to Iran during this period seem to have consisted of large artillery shipments which were made before France's conservative government came back to power and which may have been made without the full knowledge of the French government. *Washington Post*, December 1, 1986, p. A-7.

10. *Washington Post*, December 7, 1986.

11. U.S. Senate Select Intelligence Committee, *Report on Preliminary Inquiry*, January 29, 1987, pp. 18–20; *Tower Commission Report*, New York, Bantam and Times Books, 1987, pp. 34–49 and 227–334.

12. U.S. Senate Select Intelligence Committee, *Report on Preliminary Inquiry*, January 29, 1987, p. 22.

13. *Washington Times* and *Christian Science Monitor*, February 16, 1986.

14. *Washington Post*, February 11, 1986; *Economist*, February 15, 1986.

15. The Iranians were evidently charged $16,000 for each TOW missile versus a regular price of around $6,000. The surplus funds were diverted to arms dealers and to the Contra rebels. *Washington Post*, December 7, 1986, p. A-25. By this time, the flow of arms from other sources had also broadened significantly. This led to later reports from the Mujahideen that a U.S. arms shipment had landed in Tehran airport on July 4. The Mujahideen reported that the aircraft was a Race Aviation B-707-331C, registration N345FA, and that the flight had been arranged by Rafsanjani and Col. Javidana of the Defense Ministry's logistics department. *Jane's Defense Weekly*, November 22, 1986, p. 1205. For overall reporting on events, see the U.S Senate Select Intelligence Committee, *Report on Preliminary Inquiry*, January 29, 1987, pp. 20–30 and 32–39.

16. *Washington Post*, December 7, 1986, pp. A-30 to A-31. *Al-Shiraa* (Ash-Shira or As Shir a) announced that an American envoy had visited Tehran, and had shipped key tank and radar parts in four C-130 cargo planes from the Philippines. See *Washington Post*, November, 1986, p. A-15; *Newsweek*, November 17, 1986; *Time*, November 17, 1986, pp. 49–52; *Aviation Week*, November 17, 1986, pp. 16–18; U.S. Senate Select Intelligence Committee, *Report on Preliminary Inquiry*, January 29, 1987, p. 38; *Tower Commission Report*, New York, Bantam and Times Books, 1987, pp. 51 and 414–449.

17. The editor of *Al-Shiraa* later declared that the story had been leaked to him by a member of Montazeri's staff. He stated that the story was not Syrian inspired and that this issue of the magazine had been suspended in Syria. *Washington Post*, December 7, 1986; *Baltimore Sun*, December 14, 1986, p. 18A; *Economist*, November 1, 1986, pp. 43–44; *Washington Times*, October 29, 1986, p.

6A; Barry Rubin, "My Friend Satan", *New Republic*, December 15, 1986, pp. 14–15.

18. *Washington Times*, November 25, 1986, p. 5A; and *Washington Post*, December 6, 1986, p. A-13.

19. *Washington Post*, December 7, 1986, p. A-25.

20. These counts are based upon figures provided by USCENTCOM. The Lloyds counts used in most Western reporting are slightly different. Many press counts use Iraqi and Iranian data, which are sometimes wildly inaccurate.

21. Larry Dickerson, "Iranian Power Projection in the Persian Gulf", *World Weapons Review*, August 12, 1987, p. 7.

22. The operational status of these missile is uncertain, and stocks as low as seven missiles have been reported by some sources.

23. Estimates of the number of mines involved are extremely uncertain, and often confuse large holdings of land mines with holdings of naval mines. The core of Iranian naval mine holdings probably consisted of roughly 1,000 moored contact mines made by the USSR in the early 1900s, reconditioned, and then sold to Iran by North Korea.

24. Naval Guards units conducted 17 known attacks on neutral ships between February and late July, 1987.

25. *Jane's Defense Weekly*, July 4, 1987, p. 1417; *Chicago Tribune*, July 12, 1987, p. 1; *Washington Post*, July 26, 1987, p. A-25; *World Weapons Review*, August 12, 1987, p. 7.

26. These assessments are based on various editions of IISS, *Military Balance*; the *Jaffe Center Middle East Military Balance*, and *Jane's Defense Weekly*, July 11, 1987, p. 15.

27. The Silkworm was a Chinese version of the Soviet CSS-N-2 or Styx anti-ship missile, which the Chinese had begun to deliver to Iran in the summer of 1986. The PRC designates this system as the Hai Ying HY-2 or "Silkworm". It can be deployed in both mobile and fixed sites, and the Iranians chose to use a mix of mobile equipment and fixed concrete bunkers and launch rails. In this form of deployment, the launch sites are pre-surveyed to provide precise range and sensor locations, and the missile moves on a wheeled or tracked launcher rail. This allows the missiles and the launch vehicles and support vans to be kept in dispersed locations, where they are safer from attack, and to be moved to the site when needed. Each fire unit has two of four missile launchers.

The Silkworm missile is made by the China Precision Machinery for Import and Export Corporation (CPMIEC). It is not particularly sophisticated by Western standards and has been on Chinese ships for twenty years. It has been exported, however, only since 1984. It is 6.5 meters long, and has a diameter of 75 centimeters. It weights 2,500–3,000 kilograms and has a 500 kilogram or 1,100 pound warhead. This warhead weight is about seven times the weight of that on the Sea Killer and three times the weight of that on the Exocet. It has a maximum range of 95 kilometers, and comes with three different guidance systems: The HY-2 homes in on a target by using the tracking radar at the launch site, the HY-2A homes in using passive IR to defeat electronic counter measures, and the HY-2G uses terminal radar homing plus altimeter.

The missile is most effective at ranges under 40 kilometers, but it has an effective range of 70–80 kilometers if a ship or aircraft can designate the target and allow the Silkworm to reach the point where its on-board guidance can home in on the target. China sold Iran such radar designation capability to use on some of its ships and aircraft.

The missile will reach most targets its range within six minutes of firing. The missile normally climbs to an altitude of 145 meters before dropping to a final approach altitude of about 30 meters. The primary warning a target receives is from radar contact at launch before the missile drops to near sea level and when the missile is completing its final approach and no longer is masked by the curve of the earth. This is at a range of 16–32 kilometers in the case of active radar homing or 10 kilometers in the case of IR—since the IR seeker remains passive until this point. It is large enough to be detected and killed by some surface-to-air missiles, but these require good kill capability against very low-altitude attacks.

28. *Chicago Tribune*, July 8, 1987, p. I-1; *Baltimore Sun*, July 7, 1987; *U.S. News and World Report*, July 6, 1987, p. 41.

29. *New York Times*, January 20, 1987, p. A-1; *Philadelphia Inquirer*, January 21, 1987, p. 5A; *Observer*, January 25, 1987; *Washington Times*, February 5, 1987, p. 7A; *Baltimore Sun*, March 1, 1987, p. 24. Many of these attacks took place outside Iranian waters and off the shore of the UAE. Little effort seems to have been made to identify the targets. The *Wu Jiang*, for example, was a PRC-registered freighter with no cargo of military value.

30. The Sea Killer first became operational in 1984, and was sold to Iran before the fall of the Shah. It is a relatively light missile, about 1.01 meters long and 20.6 centimeters in diameter. It weighs 300 kilograms, and has a small 70 kilogram high-explosive semi-armor piercing warhead. It has a maximum air-to-surface range of 25 kilometers and a 4 range of 6 kilometers. It is a "sea skimmer" which flies at a height of 3 to four meters. It normally rides a radar beam to the target, but it can be radio-directed in a heavy jamming environment.

Iran's Sea Killers could be deployed in both a helicopter-launched and deck-mounted configuration. The deck mount was a five-round trainable launcher which could be integrated with the existing X-band radars, conical scan radars, and shipboard computers on many small combat ships.

The helicopter-fired version could be mounted on any medium-lift helicopter, although it was originally configured for use with the Agusta/Sikorsky SH-3D and AB-212. The heliborne system includes an MM/APQ-706 tracking radar, a missile control console, guidance computer, command link, optical tracker with joystick, pilot display, and Sea Killer/Marte missile.

The heliborne system weighs 865–1,165 kilograms, with 300–600 kilograms for one to two missiles, 400 kilograms for the launch console, 143 kilograms for the radar, and 22 kilograms for the optical sight. Either the missile can be fired directly from a helicopter flying at medium altitude, or the helicopter can acquire the target on its radar, pop down to fly under a ship's radar, and then pop up to fire the missile.

31. *Washington Times*, February 5, 1987, p. 7A.

32. The Armilla Patrol was set up in 1980 with four combat ships, but was reduced to two in 1982, when Britain had to establish a permanent presence in the South Atlantic. A Type 42 destroyer is normally assigned to provide medium-range air defense plus a Type 22 or Leander-class frigate with Sea Wolf to provide an anti-missile capability. Only ships under the British flag, or more than 505 beneficially owned ships, are eligible for British protection, but Britain lets other ships join its convoys. Technically, the British warships accompany cargo ships, rather than escorting them, which implies war-time conditions. *Jane's Defense Weekly*, May 2, 1987, p. 824, *Christian Science Monitor*, January 21, 1987, p. 1; *Washington Post*, January 28, 1987, p. A-1.

33. *Washington Post*, March 24, 1987, p. A-25, and August 23, 1987, p. 12.

34. Free floating mines were becoming a growing problem in the Gulf during this period. The Qatari and other navies were routinely destroying such mines with naval gunfire. *Jane's Defense Weekly*, May 2, 1987, p. 824; *Washington Post*, March 24, 1987, p. A-25.

35. Caspar W. Weinberger, *A Report to Congress on Security Arrangements in the Persian Gulf*, Department of Defense, June 15, 1987, p. 5.

36. Kuwait's conservative production policy has led to a decline in production and to deficit spending. It earned $12.28 billion from oil exports in 1984, $10.46 billion in 1985, and $6.38 billion in 1986. It has the capacity, however, to produce nearly 3 million barrels per day. Its Fund for the Future has $40 billion in holdings and receives 10% of state revenues each year. It holds money in trust for Kuwait until the year 2001, and is one of the largest single foreign investors in the West. It owns such major Western oil firms as Santa Fe International, and has major refining and other downstream investments, as well as a large tanker fleet. Only about 10% of Kuwait's income is now tied to crude in comparison with downstream products. *Washington Post*, July 5, 1987, p. A-1; *Christian Science Monitor*, July 10, 1987, p. 9; *Wall Street Journal*, June 25, 1987, p. 1, July 23, 1987, p. 26;

37. See *War in the Persian Gulf: The U.S. Takes Sides*, Staff Report to the Committee on Foreign Relations, U.S. Senate, November, 1987, for the background to the chronology presented in this chapter.

38. *Washington Post*, June 5, 1987, p. A-1.

39. *Washington Post*, August 28, 1987, p. A-16.

40. Kuwait denies providing any direct military support to Iraq. *Washington Times*, June 22, 1987, p. A-6.

41. These did not include any interagency intelligence assessments of the risks. The were conducted only months later—a fact that several Congressional leaders publicly criticized. *Washington Post*, July 5, 1987, p. A-1; *Philadelphia Inquirer*, June 28, 1987, p. 2E; *Washington Times*, May 8, 1987, p. 8A.

42. This later became public when Secretary of the Navy James H. Webb sent a highly critical memo to Secretary of Defense Weinberger in July, 1987, expressing his concern that the U.S. was being drawn into a morass and building up a massive force without a clear assessment of the risks, limit to the forces involved, or ability to predict the course of the war. *Philadelphia Inquirer*, September 6, 1987, p. 9A.

43. The reflagging effort did, however, have broad political support within

both the State Department and Department of Defense. The NSDDs written in May, 1984, also broadly supported such a policy, and an NSDD signed in June, 1984, stated that it should be "U.S. policy to undertake whatever measures are necessary to keep the Straits of Hormuz open". *Washington Post,* August 28, 1987, p. A-16.

44. The assumption of the USSR's greater reliability was not totally without cause. The Reagan Administration had previously broken several promises to provide Kuwait with improved air defense systems. In November, 1986, it had also been covertly negotiating with Iran to pressure Kuwait to release 17 Shi'ite terrorists as part of its arms for hostages deal. The "Kuwaiti 17" were pro-Iranian Lebanese Shi'ites who had been held since 1984, and who constituted one of the original reasons Iran had supported the hostage-taking effort. Both Colonel Oliver North and his aide, Lt. Colonel Robert L. Earl, made a number of commitments to Iran that virtually pledged the U.S. would act to persuade Iran to free the hostages. *Washington Post,* August 28, 1987, p. A-1.

45. President Reagan made the first statement on February 25, and this was followed by statements by Secretary Weinberger beginning March 22. The U.S. released details of the new Iranian missile build-up during the month of March. *New York Times,* March 21, p. 4, March 24, 1987, p. A-1; *Washington Post,* March 23, 1987, p. A-19; *Economist,* April 18, 1987, p. 39.

46. *New York Times,* June 19, 1987, p. A-1.

47. *Washington Post,* May 5, 1987, p. A-1.

48. Casper W. Weinberger, *A Report to Congress on Security Arrangements in the Persian Gulf,* Department of Defense, June 15, 1987, p. 5.

49. *New York Times,* June 16, 1987, p. A-11.

50. *Washington Times,* April 14, 1987. p. 6A; *Washington Post,* May 5, 1987, p. A-1, and May 14, 1987, p. A-27; *Chicago Tribune,* April 18, 1987, p. Q-8; *Economist,* April 18, 1987, p. 39.

51. *Observer,* January 25, 1987.

52. *Washington Post,* June 5, 1987, p. A-1.

53. *Washington Post,* June 7, 1987, p. A-24; *Washington Times,* June 8, 1987, p. 4D; and *New York Times,* June 19, 1987, p. A-1.

54. The land-based radars for the Silkworm have a maximum range of about 42 kilometers. The longer ranges assume remote designation by a ship or aircraft mounted radar. *Jane's Defense Weekly,* March 28, 1987, p. 532.

55. *Washington Times,* April 24, 1987, p. 12A.

56. *Washington Post,* April 16, 1987, p. A-24.

57. *Washington Times,* April 20, 1987, p. 6A.

58. *Washington Post,* April 21, 1987, p. A-21.

59. *Washington Times,* April 24, 1987, p. 12A.

60. Ibid., April 30, 1987, p. 8A; *Jane's Defense Weekly,* May 23, 1987.

61. *Baltimore Sun,* May 6, 1987, p. 5A.

62. *Washington Post,* May 14, 1987, p. A-33.

63. *New York Times,* May 9, 1987, p. 1; *Washington Post,* May 9, 1987, p. 1; *Washington Times,* May 9, 1987, p. 2A.

64. *Philadelphia Inquirer,* June 6, 1987, p. 8A; *Wall Street Journal,* May 11, 1987, p. 54. The British increase in strength took place during the rotation of the

Glouster and *Hermione*, which were replaced with the destroyer *Cardiff* and two frigates, the *Broadsword* and the *Active*.

65. *Christian Science Monitor*, May 14, 1987, p. 19.

66. *New York Times*, May 12, 1987, p. A-8.

67. Casper W. Weinberger, *A Report to the Congress on Security Arrangements in the Persian Gulf*, June 15, 1987.

68. The outboard motors are made by Outboard Marine in Chicago, and are painted olive drab. Although U.S. regulations prohibit the export of motors over 45 HP as military goods, they are said to range in power from 5 to 235 HP. While some of the small craft are used in the Gulf, they have even more value in the water barriers around Basra. *Philadelphia Inquirer*, May 15, 1987, p. 1E.

69. *Baltimore Sun*, May 14, 1987, p. 17A, *Washington Post*, May 14, 1987, p. A-32; *Jane's Defense Weekly*, May 16, 1987, p. 920.

70. The 272,000 VLCC, *Primrose*, was hit by a mine in a similar position about 34 miles east of Mina al Ahmadi as it left Kuwait on May 26, 1987. The mine was reported to have broken loose from the Shaat al-Arab waterway about 60 miles to the north, and was about 10 miles west of the site where the *Chuykov* was hit. This shows the possible accuracy of the free floating or breakaway mine theory. At the same time, it was all too easy for Iranian naval Guards units to put mines in the ship's path.

71. *Washington Post*, May 16, 1987.

72. The War Powers Act was approved in November, 1973, when the Congress overrode President Nixon's veto, which stated that the act was "unconstitutional and dangerous to the best interest of our nation". The House overrode the veto by a vote of 84 to 125, and the Senate by 75 votes to 18. The law requires that the President notify Congress within 48 hours any time that he sends combat troops into a foreign country or "substantially enlarges" the number of U.S. troops there. He must provide reports to Congress every six months. If troops are sent into hostilities, or situations where "imminent involvement in hostilities is clearly indicated", they must be withdrawn in 60 days unless the Congress authorizes their staying on or declares war. The President can obtain a 30 day extension of the 60 days by claiming "unavoidable military necessity" which requires the troops to stay to ensure a safe withdrawal.

The act had uncertain constitutionality because it mandated withdrawal if the Congress did not act—something the Supreme Court had rejected as violating the separation of powers. President Reagan was careful to comply with the reporting requirements, however, and previous Presidents had complied with the act. President Ford reported on U.S. participation in a relief effort to move refugees out of Da Nang on April 4, 1975; on the use of U.S. troops to evacuate U.S. citizens from Saigon on April 30, 1974; and on the Mayaguez incident on May 15, 1975. President Carter reported on the Iranian rescue attempt on April 26, 1980. President Reagan reported deployment of the MLF to the Sinai on March 19, 1982; on sending 800 Marines to the MLF in Lebanon on August 21, 1982; on deploying 1,200 Marines to help restore Lebanese sovereignty on September 29, 1982; on deploying U.S. air units to help Chad against Libya on August 8, 1983; on the death of Marines in Lebanon on August

30, 1983; on deployment of 1,900 troops to Grenada on October 25, 1983; and on the air strike against Libya on April 16, 1986.

73. *Aviation Week,* May 25, 1987, p. 25.

74. *Washington Post,* June 4, 1987, p. A-3.

75. The technical data in this section are taken from interviews with experts from the Office of the Secretary of Defense and the manufacturers concerned.

76. The missile is produced by Hughes Aircraft (Canoga Park, Calif.), and Raytheon (Bristol, Tenn.)

77. The vendor of the F-15C/D and MSIP package was McDonnell Douglas.

78. *New York Times,* June 12, 1987, p. A-1.

79. *Washington Post,* June 4, 1987, p. A-34, and June 19, 1987, p. A-29.

80. *Philadelphia Inquirer,* May 20, 1987, p. 11A; *Washington Post,* June 5, 1987, p. 24; *New York Times,* June 8, 1987, p. A-9.

81. *Washington Times,* May 19, 1987, p. 8A; *Baltimore Sun,* May 19, 1987, p. 4A; *New York Times,* May 22, 1987, p. A-1.

82. *Christian Science Monitor,* June 1, 1987, p. 1; *Washington Post,* June 1, 1987, p. A-13; *Washington Times,* June 1, 1987, p. 6A.

83. *Economist,* May 30, 1987, p. 29. The 11 tankers composed half of Kuwait's fleet of 22 ships. They were to be reflagged by transfer to a Delaware corporation called Chesapeake Shipping, Inc., after a payment of $7 million. The corporation was formed on May 14, 1987, but the actual reflagging and escort service was delayed while the U.S. Coast Guard inspected the ships and the Administration lobbied Congress to make sure it had suitable political support.

84. The President and the Chairman of the Joint Chiefs disagreed over the risks of war. Reagan stated that," I don't see the risk of a war....I don't see how one could possibly start". Admiral Crowe made it clear that there was a serious risk of military action and of a large-scale conflict.

85. *Washington Post,* June 6, 1986, p. A-1; *New York Times* and *Washington Post,* June 6, 1987.

86. *New York Times,* June 10, 1987; *Boston Globe,* June 12, 1987, p. 10.

87. *New York Times,* June 4, 1987, p. A13; *Washington Times,* June 3, 1987, p. 4D.

88. *Newsweek,* June 2, 1987; *Philadelphia Inquirer,* June 6, 1987, p. 8A.

89. *New York Times,* June 5, 1987, p. A-11; *Philadelphia Inquirer,* June 6, 1987, p. 8A.

90. *Washington Times,* June 2, 1987, p. 10A; *New York Times,* June 7, 1987, p. 1; *Philadelphia Inquirer,* June 6, 1987, p. 8A.

91. *Washington Post,* May 31, 1987, p. 30; *New York Times,* June 2, 1987, p. A-3.

92. *Christian Science Monitor,* June 5, 1987, p. 1; *Washington Post,* June 10, 1987, p. A-23, and June 11, p. A-29; *Washington Times,* June 8, 1987, p. 1A; *New York Times,* June 11, 1987, p. A-8.

93. *Washington Post,* June 7, 1987, p. A-1, June 22, 1987, p. A17; *Washington Times,* June 8, 1987, p. 1A.

94. *Christian Science Monitor,* June 9, 1987, p. 11; *Washington Post,* June 9, 1987, p. A-25.

95. *Washington Post,* June 23, 1987, pp. A-14 to A-15.

96. The UAE also acted at a time of considerable internal instability. The

ruling Sheihk of Sharjah was overthrown by his older brother on June 18 for mismanaging the nation's economy. While Abu Dhabi originally supported the coup, which was backed by most of Sharjah's national guard and business community, it later persuaded Sheik Abdul Aziz to allow his brother to return and become crown prince. *Boston Globe*, June 11, 1987, p. 9; June 12, 1987, p. 10; *Washington Times*, June 16, 1987, p. 6; *Washington Post*, June 19, 1987, p. A-33; *Chicago Tribune*, June 19, 1987, p. I-4.

97. *New York Times*, June 14, 1987, pp. 3 and 6; *Washington Post*, June 13, 1987, p. A-18, June 14, 1987, pp. A-1 and A-30, June 16, 1987, p. A-21.

98. *Washington Post*, June 16, 1987, p. A-6; *Philadelphia Inquirer*, June 17, 1987, p. 1; *Washington Times*, June 16, 1987, p. 6.

99. *New York Times*, June 2, 1987, p. A-3.

100. *Christian Science Monitor*, June 10, 1987, p. 25; *New York Times*, June 10, 1987, pp. A-1 and A-10.

101. *Time*, June 8, 1987, p. 35; *Christian Science Monitor*, June 9, 1987, p. 1; *Washington Post*, June 9, 1987, p. A-1; *USN&WR*, June 15, 1987, p. 26.

102. *Washington Post*, June 2, 1987, p. A-21, June 5, 1987, p. A-23, June 7, 1987, p. A-24; *Washington Times*, June 9, 1987, p. 8B.

103. *Washington Times*, June 24, 25, 27, and 29, 1987; *Washington Post*, June 24, 25, 27, and 29, 1987; *New York Times*, June 24, 25, 27, and 29, 1987; *Washington Times*, July 1, 1987, p. 1.

104. *Washington Post*, June 21, 1987, pp. A-1 and A-27.

11
The West and Naval Conflict in the Gulf: 1987

The previous two chapters have dealt with case studies in Western relations with the Gulf which have fairly clear outcomes. At this writing, however, the outcome of both the Iran-Iraq War and Western naval intervention in the Gulf are far from clear. Even the time when they will be resolved is unpredictable. By and large, however, it seems likely that the West and the Gulf should draw roughly the same lessons from virtually any outcome of the current crisis except the most negative ones: an Iranian victory over Iran and/or Iranian success in forcing U.S. and European forces to leave the Gulf under conditions tantamount to Western defeat.

Both these contingencies now seem unlikely, although scarcely impossible. It is important to note, however, that the U.S. began active military intervention in the Gulf under conditions it can and should avoid in the future.

At the point when the U.S. began its convoy activity in the Gulf, events had created an extraordinarily difficult strategic situation. The pace and nature of the Iran-Iraq War had shifted to the point where the U.S. reflagging effort was far more dangerous than the U.S. had originally calculated. At the same time, the U.S. faced the consequences of a long series of painful mistakes:

- The Reagan Administration was still suffering from the embarrassment of public exposure of its covert arms dealings with Iran.
- Both the Administration and the Congress had created major political and military problems in dealing with the Southern Gulf states.
- The U.S. still could not make major arms sales to friendly Gulf states.
- The U.S. was divided from its European allies.
- The Soviet Union had been able to capitalize on these U.S. mistakes, and improve its relations with Iran.

- U.S. forces were entering a war zone to meet a commitment that had been made without proper analysis of the risks, without a clear definition of U.S. interests and commitments, and without the support of effective interagency organization for what might well be a major military engagement with Iran.

The U.S. scarcely had a monopoly in making mistakes in its relations with the region, but its actions during 1986 and early 1987 had demonstrated the brutal costs of trying to implement unilateral policy based on temporary expediency, and the demands of domestic politics, versus policy based on stable strategic relations with the Gulf states and its Western allies.

U.S. CONVOY ACTIVITY BEGINS

The initial U.S. military performance in the Gulf was scarcely reassuring. The U.S. held a press conference on July 14, 1987, to say that it would begin to escort the reflagged Kuwaiti ships in one week, and that the U.S. *Bridgeton* (*Al Rekkah*) would be one of the first tankers to be escorted by the time the mines were cleared. The U.S. also indicated it had delayed its convoy activity because clearing the Iranian mines from Kuwait's main shipping channel was taking longer than had originally been anticipated, and the other channel was too close to Iranian waters to be safe.[1]

In response to follow-up questions, a U.S. official noted that none of the nine U.S. warships in the Gulf was a minesweeper and that the U.S. had no MH-53 Sea Dragon minesweeping helicopters in the region—although five were ready to deploy.[2] The U.S. stated later that a plan to send a 200 man detachment and the helicopters was deferred because Kuwait planned to lease Dutch minesweepers and obtain Dutch technical assistance. At this time, the mine-clearing effort in Kuwait had a peculiarly international character. Some of the work was done by Soviet and Dutch experts, and other work was done by Saudi minesweepers and an 18 man U.S. Navy Explosive Ordnance Disposal detachment. The mines turned out to be old Soviet MO-8 mines which Iran had bought from North Korea. They could be planted to a depth of 350 feet and had 250 pounds of explosive.[3]

The first U.S. flags went up on two Kuwaiti tankers on July 21, 1987, and the convoy effort—by then codenamed Operation Earnest Will—was under way. The U.S. had had nearly four months to prepare for its first convoy, but the resulting task force still had a complex mix of strengths and weaknesses.

The main strength of the U.S. forces that accompanied the first convoys was that it consisted of a relatively large force to convoy only two tankers. The U.S. had four frigates, three cruisers, and a destroyer in the area around the Gulf and the Straits of Hormuz. These ships had gone through three rehearsals since the last elements of the escort force had arrived in early July, and the escort plan had reached the size of an 80 page document. The U.S. also had a carrier task force, including the carrier *Constellation,* in the Indian Ocean. The battleship *Missouri,* two more cruisers, and a helicopter carrier were in transit to the region.

The convoy plan called for three to four U.S. ships to escort each of the two tanker convoys while the cruisers stood by to provide a defense against air attack. U.S. A-6 and F/A-18 attack aircraft, EA-6B jamming aircraft, and F-14 fighter planes were to provide support from the U.S. carrier *Constellation* in the Arabian sea. The tankers were then to have at least three escorts once they passed beyond the Straits, with USAF E-3A AWACS providing airborne surveillance.

The U.S. planned to provide convoys once every two weeks during July and August. This interval was intended to allow the task force concept to be tested, and to find out Iranian reactions. It also eased the burden of being on constant alert against suicide attacks or other means of irregular warfare. The destroyer *Fox* and cruiser *Kidd* were selected in part because their 76 mm guns provided considerable firepower against sudden raids by the Iranian Navy or shore targets. All of the U.S. combat ships were equipped with long-radar radars, data nets, and Phalanx terminal defense guns.

The U.S. force also, however, had several weaknesses. The U.S. naval forces operating in the Gulf faced the wide array of potential friends and enemies shown in Table 11.1, but they had no allied forces operating in support. They lacked access to most of the local naval and air bases in the Gulf, although they had limited access to facilities in Bahrain, Oman, and Saudi Arabia. They lacked the ability to provide a cross-reinforcing defense of single ships operating outside the convoy to minimize the success of a saturation attack, and they lacked mine warfare defenses. It was the latter weakness that Iran chose to exploit.

The U.S. lack of mine warfare defenses was partly a matter of inadequate planning, and the initial operational plan the Joint Chiefs and USCENTCOM developed for the convoy made no provision for mine forces in spite of warnings from U.S. intelligence officials and the U.S. Navy.[4] The U.S. Navy, however, was not ready to carry out the mine warfare mission. It relied on its European allies for most mine warfare missions since the early 1970s, and the U.S. had made little initial effort to get direct allied military support.

Table 11.1: Naval Forces in the Gulf and Gulf of Oman: July, 1987

Country	Cruisers	Destroyers	Frigates	Corvettes	FAC	Patrol	Amphibious	Mine
U.S.*	2	1	4	–	–	–	–	–
Iran**	–	3	4	2	8	7	8	2
Iraq***	–	2	12	15	11	7	5	
Britain	–	1	1	–	–	–	–	–
France	–	2	–	–	–	–	–	–
USSR	–	–	1	–	–	–	–	2
Kuwait	–	–	–	–	8	50	13	–
Saudi Arabia	–	–	4	4	12	46	15	4
Bahrain	–	–	–	–	4	–	–	–
Qatar	–	–	–	–	3	6	–	–
UAE	–	–	–	–	6	9	–	–
Oman	–	–	–	–	8	9	7	–

*The U.S. had an additional command ship in the Gulf, and a carrier group with the Constellation and its escorts in the Indian Ocean. A battleship was in transit to the area.
**Does not reflect war time damage or losses. Does not include naval Guards units.
***Does not reflect wartime damage or losses and includes ships trapped in port.

Source: Adapted from IISS, *Military Balance, 1987–1988*; and *New York Times*, July 12, 1987, p. E-3.

The U.S. thus faced a critical threat which it was not prepared to deal with. It was unprepared even though it had reports that Iran had had access to a wide range of sources for its mines. While Iran's mine assets were uncertain, there were reports before the U.S. convoys began that it had both surface mines and bottom mines, and at least some timed or interval mines that became active only after a fixed time period or after several ships or minesweepers passed by. Other reports indicated that Iran might have contact, magnetic, acoustic, bow wave, pressure, and temperature mines, and possibly remote-controlled mines as well. Some of these types were large metal mines and could easily be detected by sonar, but many were too small for easy detection and others were non-magnetic. In practice, Iran used only mines whose design dated back to World War I, but this was to prove embarrassing enough.

Another weakness in the U.S. convoy effort consisted of the distance the convoy had to move, and the fact that it was continuously vulnerable from the time it approached the Straits to the moment it entered port in Kuwait. The convey had to follow a route that included a 60 mile voyage from Dibba, outside the Gulf, to the Straits. With a convoy speed of 16 knots, this link took eight hours. For the next 50 miles, the convoy would pass through the Straits, and near Iran's 20 mile exclusion zone and its Silkworm missiles.

It then had to sail for another 90 miles to a point near Abu Nuayr off the coast of Abu Dhabi. This meant passing by Abu Musa and the Tunbs. There was then a 60 mile stretch to the UAE's Zaqqum oil channel, a 60 mile voyage to Qatar's Halul Island, at a 90 mile link to the Shah Allum shoals and a point only a mile from Iranian waters. At this point the convoy still had about 285 miles more of relatively open sailing from a position off the Ras Tanura beacon ship to Kuwait. During all of this time, Iran could choose its point and means of attack.[5]

Finally, the U.S. faced the problem that—in spite of efforts to keep the convoy's composition and schedule and the details of its strengths and weaknesses secret—it received a relentless exposure in the media. Its weaknesses, particularly in regard to mine warfare, were fully communicated to Iran. Since Iran could predict the path of the convoy and its timing, small craft might succeed in hitting a ship even if they did nothing more than drop a few contact mines at night.[6]

THE MINING OF THE *BRIDGETON*

The United States' first convoy sailed on schedule on July 22, 1987. The U.S. sent in four combat ships, including a guided missile cruiser. The reflagged ships included the 414,266 ton supertanker *Bridgeton* and the 48,233 ton gas tanker *Gas Prince*. The convoy transited the Straits safely, amid wide speculation that a week-long lull in the tanker war somehow meant Iran had decided not to attack. In fact, the most Iran did publicly was to send four of its F-4s near the convoy when it entered the Gulf, and to announce new naval maneuvers which were codenamed Operation Martyrdom.[7] The convoy reached the halfway point on July 23, when Iran declared it had a cargo of "prohibited goods".[8]

Unfortunately, the U.S. concentrated far too much on the missile threat near the Straits, and far too little on the risk of other forms and areas of attack. This became brutally clear at about 6:30 A.M. on July 24. The *Bridgeton* struck a mine at a position of 27°58' north and 49°50' east. This was a position roughly 18 miles from the Iranian naval Guards base on Farsi Island. It was about 80 miles southeast of the four mine hits reported between May 16 and June 19 in the approach channel to Mina al-Ahmadi, and about 50 miles north of the Juaymah departure channel.

The *Bridgeton,* which was the only one of the 11 reflagged vessels designed to carry crude oil, was hit in its number-one port cargo tank, and the mine blew a large hole in its hull and flooded four of its 31 compartments. The convoy was forced to slow from a speed of 16 knots to 5, and its warships were forced to follow the damaged *Bridgeton,* which was the only ship large enough to survive another mine hit. The only action the warships could take was to turn on their sonars and put riflemen on the bow to try to shoot a mine if they saw it.

The impact of the Iranian attack was further heightened by the fact that it came shortly after the U.S. government had publicly expressed the feeling that the most threatening part of the passage was over. The day before, Rear Admiral Harold Bernsen, commander of the navy's Middle East Task Force, had declared that "so far it has gone exactly as I thought it would—smoothly, without any confrontation on the part of Iran". The admiral then went on to declare that Iran had been weakened by seven years of war, and that "the Iranian Air Force and Navy are not strong. It would not be in their best interest to utilize their forces in a direct confrontation".[9] It is interesting to note that after the *Bridgeton* was hit, Admiral Bernsen was forced to admit that no one had checked the convoy's route for mines, although intelligence sources had warned him and the convoy commander that mines might be present.[10]

Iran had succeeded in capitalizing on the greatest single vulnerability in the U.S. convoy system, and had done so without leaving a clear chain of hard evidence linking Iran or the Iranian government to the attack. Although it later turned out that Iran had laid 3 different minefields, and at least 60 mines, this activity had not been fully detected and had not been filmed.

This gave Iran a major propaganda victory. Prime Minister Mir

Hossein Mousavi called the attack "an irreparable blow on America's political and military prestige", although he was careful to say that the blow came by "invisible hands". Prime Minister Rafsanjani extended the threat to Kuwait and Saudi Arabia: "From now on, if our wells, installations, and centers are hit, we will make the installations and centers of Iraq's partners the targets our attacks." Rafsanjani also announced a "new policy of retaliation". President Khamenei reiterated this threat on July 27, and called Kuwait "the only country in the region that openly supports Iraq in the war". He specifically threatened Kuwait with the fact that Iranian surface-to-surface missiles could reach any target in Kuwait.[11]

The next day, more mines were sighted south of the area where the *Bridgeton* was hit, and seven more North Korean versions of Soviet mines were discovered on July 27. This made it even more apparent to the world that the U.S. had no immediate contingency plan to deal with the situation. It further strengthened Iran's victory, and those within Iran who favored continuing the war. Khomeini stated on July 28 that there could be no end to the war along as Sadam Hussein was in power.[12]

The most the U.S. could do initially was to airlift eight RH-53Ds to Diego Garcia, and prepare the amphibious landing ship, *U.S.S. Guadalcanal*, to operate four of them in the Gulf.[13] The *Guadalcanal* was operating in the Indian Ocean and could be deployed relatively rapidly, but then had to be equipped and organized for the mine warfare mission. The most the U.S. could do to speed up the deployment of its minesweepers was to load another amphibious ship, the *U.S.S. Raleigh*, with four minesweepers. This still meant a nearly month long voyage to the entrance of the Gulf.[14]

THE INITIAL EUROPEAN REACTION TO THE *BRIDGETON* INCIDENT

The U.S. effort to obtain European support initially had mixed results. The Netherlands government was sympathetic, but faced serious domestic political constraints in actually deploying forces. It preferred to seek some form of WEU action and offer Kuwait training aid and minesweepers for lease or sale.

While Britain had 25 minehunters and 17 minesweepers, Britain had already deployed about 20% of its active combat ships in the Gulf and Indian Ocean. It also had adopted a policy of silently "shadowing" British flag ships through the Gulf and had successfully done this with 160 ships since Iranian attacks had increased in 1986. It was reluctant to join in a highly public effort with the U.S. that might well lead to Iranian attacks on British ships. Further, some British experts felt that Britain would be dragged into an open-ended commitment that was leading to a full-scale naval war with Iran. The British government delayed action on the U.S. request for aid, without immediately making a formal rejection.[15]

France had a long-standing policy of avoiding any military role that put it under the direct command of U.S. forces. France did, however, send a 3,000 man task force with the aircraft carrier *Clemenceau*, two destroyers, and a supply ship to reinforce its units already in the Indian Ocean.[16] France was reacting to a growing confrontation of its own with Iran that had begun when France demanded that the Iranian Embassy in Paris give up Wahid Gordji—a suspected terrorist who served as a translator, but who lacked diplomatic immunity. Gordji was believed to be a member of the Eighth Branch of Iranian intelligence—its overseas action group—and to have been active in several bombing and assassination efforts in France.[17]

Iran flatly refused to turn Gordji over for questioning, and retaliated by accusing a French diplomat with full immunity of espionage, and then demanded that he present himself at Gazvin prison in Tehran. By July 3, the French Embassy in Tehran was virtually under siege, and French police responded by ringing the Iranian Embassy in Paris.

Iran attacked a French freighter on July 13, and issued a "caution" regarding France's treatment of Iranian diplomats. On July 16, Iran threatened to cut relations with France unless the police ringing the embassy were removed in three days, and issued a series of indirect threats against French hostages in Lebanon.

France replied by severing diplomatic relations. This left both embassies ringed by security forces, and French efforts to reach some modus vivendi with Iran and to obtain the release of its hostages in a shambles. France continued to negotiate, but sent two frigates with Exocet missiles into the Gulf. France warned its ships to steer clear of the Gulf on July 20 and, after several more efforts to negotiate the embassy issue with Iran, sent its carrier force towards the Gulf.[18]

SUPPORT FOR THE U.S. FROM THE SOUTHERN GULF STATES

The U.S. faced problems in finding the basing facilities it needed for the helicopters, special forces units, and other elements it wanted to deploy in the Gulf. Although Kuwait had hosted a small U.S. mine-clearing detachment, and had allowed U.S. military aircraft to operate out of Kuwait on 17 occasions in the previous month, it remained reluctant to offer any formal base for a U.S. combat unit for the reasons discussed earlier. Its Crown Prince agreed to discuss the issue, but it was clear that Kuwait wanted to avoid any formal U.S. basing of a force on Kuwaiti soil, and was under considerable pressure from its neighbors to avoid this. Further, the U.S. wanted to avoid any implied commitment to defending Kuwaiti territory.

Bahrain and Saudi Arabia were willing to provide virtually all the support the U.S. wanted, but could not agree to the kind of formal basing arrangements some U.S. officials wanted. Both Gulf nations felt they needed to keep a low military profile at a time when U.S. domestic politics were forcing it to do everything in a spotlight, and the Secretary of Defense was seeking formal basing agreements.[19]

This led to a compromise in which Kuwait agreed to charter three large barges that were to be moored in international waters in the Gulf near Farsi Island, and Bahrain and Saudi Arabia agreed to provide the staging support for them. These barges avoided the problems inherent in any formal U.S. base on Saudi or Bahrainian territory, but gave the U.S. a facility where it could stage attack, reconnaissance, and mine warfare helicopters and state army and navy commando units, and deploy radars, intelligence sensors, and electronic warfare equipment.

The barges were defended by Stinger missiles and Phalanx anti-missile close defense systems, and were heavily sandbagged and compartmented to minimize the damage impact of a single missile or air attack. The U.S. stationed one of these barges, a converted Brown and Root North Sea drilling platform, only about 20 miles from Iran's Farsi Island. The barge, now codenamed "Hercules", paid off immediately in providing the U.S. with a capability to maintain a constant watch over Iranian activity at the island. It was so successful that the U.S. immediately began to plan for another facility near the naval Guards speed boat base at Abu Musa.[20]

Saudi Arabia also agreed to expand the use of its four minesweepers, and these helped discover the presence of additional mines after the *Bridgeton* was hit. It could not, however, fully commit its small force to protecting shipping to Kuwait when it was equally vulnerable. It quietly agreed to provide ship and aircraft fuel, and to provide emergency landing and fueling support for U.S. carrier aircraft. Bahrain agreed to a de facto basing arrangement in which the U.S. leased some extremely large platforms normally used for off-shore oil work and created the equivalent of small bases in the Gulf without formally operating from Gulf territory.

THE RIOT IN MECCA

It soon became apparent that Saudi Arabia and Kuwait had every reason to be concerned with the visibility of their help to the U.S., and with Iran's reactions. On July 31, Iranian "pilgrims" staged a massive riot during the pilgrimage. At the time, there were roughly 157,000 Iranian pilgrims in Iran, the largest group from any nation. Cadres began to mobilize agitators among the 70,000 Iranian pilgrims in Mecca itself at roughly 2:00 P.M. They completed getting ready for a mass meeting at 4:30 P.M. This was typical of similar disturbances Iran had caused in the past, but these had stayed outside the Grand Mosque.[21]

This time, the Iranians slowly began to advance on the Grand Mosque. At about 6:30 they reached the Saudi security line barring the entrance. The Iranians used sticks, knives, and stones, and may have used a limited number of small arms. The Saudi security guards and National Guard replied with tear gas and riot control devices. The incident was over by 7:00 P.M., but the resulting panic and fighting in the crowded pilgrimage area left 402 dead and 649 injured, and its intensity is revealed by the fact that while 303 of those killed were Iranian, 85 were Saudi security officers, 60 were other Saudi citizens, and 201 were of other nationalities.

Iran made claims the next day that the Saudi police had opened fire without warning, and that 650 Iranians were dead or missing and 4,500 were wounded.[22] Iran made these claims in spite of Saudi TV coverage which showed that the riots had been started by organized Iranian mobs throwing rocks at the police and considerable evidence that organized Iranian teams had been instructed to start the riots and infiltrators were equipped with pistols and explosives.[23]

While some Saudi security forces may have fired pistols and even M-16s, any such firing incidents had to have been limited by the armament of the Saudi officers involved. Shuarah II, verse 197, of the Koran, forbids any violence or even verbal abuse in the holy area, and Saudi policemen in the Mosque area are not allowed to carry pistols. In 1979, for example, the Saudi leaders took several days to obtain a *fatwa* to allow the use of force to put down insurgents that seized the Grand Mosque.

The director of the Iranian Red Crescent, Dr. Vahid Dastjerdi, also cast Iran's initial claims into doubt when he made much more modest claims during a press conference on August 16. He talked about pilgrims being stoned and gassed and killed underfoot or from poor medical care. He stated that there were 155,000 Iranians in Mecca and that 332 were killed, and 40 to 50 missing. He claimed 4,000 were hurt, but described them as injured rather than as casualties. He made no effort to substantiate claims about large numbers of deaths from bullets, although he stated that all but 90 of the dead had been returned, and made no effort to substantiate the unrealistically round number of 4,000. It is also important to note that medical representatives from twelve other Islamic countries examined a large number of bodies and signed a statement on August 8, 1987, noting that those hurt in the riots had received good medical care, and that all the deaths occurred from "suffocation, deep wounds, and heavy blows to the head and body using bars and sharp instruments".[24]

Iranian Prime Minister Rafsanjani responded to the events in Mecca at a different level. He stated that "we, as soldiers of God and implementers of divine principles, oblige ourselves to avenge these martyrs by uprooting Saudi rulers from the region".

Carefully coached Iranian mobs in Tehran then sacked four embassies, including the Kuwaiti and Saudi embassies. Iran claimed to have found documents proving that the Kuwaitis were spies and four Saudi diplomats disappeared. One, the Saudi political attache, Mussaid Ghamadi, eventually died from being pushed out of a second story window.

Two bomb attacks followed against Saudi diplomatic facilities in Beirut, and the siege of the Saudi Embassy in Tehran was continued until Saudi Arabia released the supervisor of the Iranian pilgrims in Mecca—Hojatolislam Rezai Karubi—and six confederates and allowed them to fly to Tehran.[25] Khomeini also accused the U.S. of having been responsible, and said that "we hold America responsible for all these crimes. . . . God willing, we shall deal with her, avenging the children of Abraham".[26]

These Iranian actions, however, had the opposite result of what Iran intended. All the Arab states except Libya expressed support for the Saudis, even though the Iranians sent some seven missions to other Islamic states to make the Iranian case. The Saudis did not expel most of the Iranian pilgrims, and only tightened security in the holy places and the oil facilities in the Eastern Province.

The riots failed to reveal any signs of any internal weaknesses in Saudi Arabia, or any significant signs of pro-Khomeini support among the Shi'ites in Saudi Arabia's Eastern Province, although Shi'ites make up about 15-25% of the province's population of 1.5 million and are a substantial part of the labor force in the oil fields.[27] The Saudi royal family had made major efforts to win Shi'ite support after some serious problems in the late 1970s. These efforts included the appointment of more progressive governors—such as Mohammid ibn Fahd—and led to reforms which defused most of the more serious Sunni-Shi'ite tensions in the region.[28]

While some 30 senior Saudis in the royal family did rush home to

Saudi Arabia, this was intended largely to remove potential hostages to terrorism. If anything, Iran's actions catalyzed Saudi Arabia into taking a firm and decisive line in opposition to Iran. Saudi diplomatic officials and members of the royal family made an unprecedented public statement that they would resist any Iranian pressure and reply by force if necessary.

Saudi Arabia also began to provide more direct and comprehensive military cooperation with the U.S.—including supporting the operation of U.S. Sea Stallion minesweeping helicopters. The limits to such Saudi cooperation remained Saudi Arabia's continuing desire to avoid any formal basing arrangement and any public exposure of the depth of U.S. and Saudi cooperation.

Further, the riots allowed those in the Reagan Administration who advocated stronger ties between the U.S. and Saudi Arabia to reopen the issue of selling arms to the Kingdom. In mid-August, sources within the Administration leaked the fact it planned to notify Congress of an arms sale in September or October, and that the sale would include F-15 aircraft to act as a replacement pool for losses by the Saudi Air Force, the upgrading of the F-15s in the Saudi Air Force to a modified MSIP status, the 1,600 Maverick missiles whose sale had been "temporarily" withdrawn on June 11, 1987, M-60 upgrade kits, and the armored artillery support vehicles.[29]

If anyone benefited directly from the events in Mecca, however, it may have been the USSR. During both June and July, it continued to publicly support the Iraqi and Arab position. It openly sold more arms to Iraq and denounced Iran for continuing the fighting. At the same time, it sharply criticized the U.S. for escalating the conflict in the Gulf and indicated that it would not provide a military escort for its reflagged ships. It also rejected the U.S. escort plan, and declared it would leave the Gulf if the U.S. did.

CONTINUING ESCALATION BETWEEN THE U.S. AND IRAN

During the next month, the tensions between Iran and the U.S. steadily escalated. Virtually every week brought a new Iranian effort to strike at the U.S. by indirect means, as Iran exploited every means it could find of attacking shipping in the Gulf that would embarrass the U.S. and potentially force it to withdraw. At the same time, Iran continued to escalate its political and military pressure on Kuwait and Saudi Arabia. It also continued its efforts to limit the effectiveness of the UN cease-fire initiative, although its military actions in the Gulf progressively alienated more European countries and led the PRC to increasingly distance itself from any overt ties to Iran.

The U.S., in turn, steadily improved its military capabilities and showed a growing willingness to confront Iran. The first step in this process of escalation occurred on August 4, when Iran noisily announced a naval exercise in the Gulf called "Operation Martyrdom", which was carefully timed and located to interfere with U.S. escort operations. The exercise also showed off the naval Guards force, a remote-controlled speed boat filled with explosives, and the existence of a small submarine.

On August 6, France announced a ban on Iranian oil imports. This did not affect Iran's revenues because it could find other buyers, but it was politically significant because Iran was France's eighth ranking supplier in 1986, and had been its top ranking supplier in June, 1986—providing 5.03 million barrels out of a total of 35.6 million. It also set a precedent, and encouraged the U.S. to impose a similar embargo. The U.S. had exported only $34 million worth of goods to Iran in 1986, but had imported $612 million worth of goods—of which $460 million worth was oil. U.S. oil imports from Iran were worth $418.5 million in the first five months of 1987.[30]

On August 8, two U.S. F-14s from the carrier *Constellation* encountered an Iranian F-4 which refused to be warned off and appeared to be closing in on a U.S. Navy P-3C patrol plane. The Iranian fighter closed to within threatening range of the P-3C, in spite of the fact that the F-14s were closing directly in on the Iranian fighter, and issued a steady series of warnings.

One F-14A fired an AIM-7F missile whose motor failed to ignite. It then fired an AIM-7M missile which appeared to track the target but which was fired at the limit of its minimum range. This allowed the Iranian pilot to maneuver and dodge the missile, which lacked the rapid turn rate required for dog fight conditions. Once the Iranian pilot turned back, the U.S. fighters did not pursue. The entire exchange took place on radar, since visibility was less than 8 miles, and the closing ranges varied from 20 to 5 miles and ended below 10,000 feet. The E-3A AWACS provided tracking data used to detect the Iranian intercept and to vector the F-14A fighters.[31]

Meanwhile, the *U.S.S. Guadalcanal* entered the Gulf, after stopping at Diego Garcia to pick up the RH-53D Sea Stallion helicopters that had been airlifted to the base. The ship not only gave the U.S. an improved mine warfare capability; it also brought Marine Stinger teams to provide point air defense and Bell AH-1T/UH-1N attack helicopters with TOW missiles. This gave the U.S. an enhanced capability to strike at Iranian small craft and naval Guards operations.

The battleship *U.S.S. Missouri* also arrived in the area, along with another Aegis cruiser, and the *U.S.S. Raleigh,* which was carrying four 65 foot patrol boats operated by SEAL special forces units and armed with 20 mm and 40 mm cannons. Another army surveillance unit with surveillance and counterattack helicopters was also on the way. The U.S. planned to build up to a level of 31 ships and vessels, and more than 25,000 naval personnel, by early September. This compared with a total of 27 vessels during the Iranian hostage crisis.

BRITAIN AND FRANCE ENTER THE GULF

On August 10, a U.S.-operated, Panamanian-registered, 117,200 ton tanker—the *Texaco Caribbean*—was hit by a mine off the coast of Fujayrah, about 80 miles south of the Straits and in the Gulf of Oman. This was the sixth tanker to have hit a mine in the last three months, but the first to be hit outside the Gulf.[32]

At the same time, the second U.S. convoy halted briefly at Bahrain because more mines had been found floating in the area. The U.S. convoy, however, soon sailed on, and reached port without further incident. A Saudi minesweeper preceded the convoy to look for mines, and a rented offshore supply ship equipped with U.S. sonars was used immediately in front of the convoy. It is also interesting to note that the convoy made up only a small fraction of the 68 tankers that entered Kuwaiti waters that month, carrying a total of 14.37 million tons of oil. Out of some 330 ships that had been hit since the beginning of the war, only three of the attacks had been made on the 22 ship Kuwaiti shuttle fleet.[33]

The next day, five more mines were found off Fujayrah. They were all Soviet-made mines that Iran had bought from North Korea and of the type used against the *Bridgeton.* This mining incident had a major international impact because Iran had chosen to escalate its mining operations to cover the "sanctuary area" outside the Gulf where virtually all the tankers entering the area waited before transiting the Straits. While Rafsanjani responded by saying the mines were planted "by the U.S. or its allies, if not Iraq" and announced that the Iranian Navy had been sent to clear the mines, there was no doubt regarding the fact that Iran was responsible.[34]

Whatever Iran's calculations may have been in expanding its mining effort into the Gulf of Oman, it badly underestimated the reaction of the major European states. Iran had gone far beyond low-level attacks on the U.S. convoy. It had dramatically expanded the war zone and had taken actions that threatened every ship moving through the Gulf. Iran may have done so because of the Western European Union's failure to agree on any course of action, and because it mistook calls by the Italian Prime Minister, Giulio Andreotti, for a UN mine warfare force, as symbols of European weakness and paralysis. In fact, the major European states were actively involved in an effort to agree on collective action in order to avoid being identified by Iran as individual political targets.

Once Iran mined the Gulf of Oman, Britain and France felt they could not wait any longer to protect their economic interests. They decided to immediately commit the mine vessels to the Gulf. They did so on a purely national basis, and without any formal notice to the U.S. In fact, Secretary Weinberger was taken somewhat by surprise. He had just called for an international minesweeping force and certainly had no knowledge of what Britain and France would do. Further, both countries had rejected a request from Frank Carlucci, the National Security Advisor to President Reagan, only one week earlier. Britain felt the U.S. would drag it into a high-profile effort that would simply lead to further escalation. France wanted to keep most of its ships outside the Gulf for similar reasons, although it now had a destroyer, three frigates, and a support ship in Gulf waters.

Only eleven days after it had first refused to help the U.S., Britain announced it would send four 615 ton Hunt-class minesweepers to the Gulf to join the destroyer, two frigates, and the supply ship it already had in the Armilla Patrol. Further, the British made it clear they would cooperate with the U.S., and a senior British official described efforts to create some kind of UN or WEU force as "escapism". France announced that it would send three minesweepers and a support ship to protect the Gulf approaches, and would join the carrier *Clemenceau*, two destroyers, and a support ship—which had just reached waters outside the Gulf.

These announcements, however, did not mean immediate military capability. British mine vessels would take 5 weeks to reach the Gulf, and the French vessels 13 days. Still, Iran had succeeded in creating a de facto coalition between the U.S. and Europe. It also did little to intimidate Europe by threatening Britain and France, by alluding to the 1983 truck bomb incident in Lebanon, and by warning that it was "ready to repeat the events in Lebanon, which resulted in their flight".[35] Further, Iran's hard line—and the unilateral actions of Britain and France—led Belgium, Italy, and the Netherlands to realize that they could no longer wait for international action. Each began to actively consider contingency plans to send national mine warfare contingents and military assistance groups.

IRANIAN ATTACKS ON SAUDI ARABIA

The UN peace effort made far less progress than the escalation of the war. On August 11, Iran's Foreign Minister Ali Akbar Velayati delivered a reply to the Secretary General which described the Security Council resolution as "illogical, unbalanced, and impractical", although he did not specifically reject the UN initiative.[36] Iraq kept up its air attacks on Iran's oil facilities, and Iran continued to pound away at Basra. Iran also broadened its "tanker war" to attack targets in the Southern Gulf.

A Saudi Coast Guard vessel was hit by a mine on August 13, and then a small supply boat blew up off the port of Fujayrah on August 15. This second strike in the Gulf of Oman meant that the 60 ships in the region had to be warned to move a distance of at least 10 miles from the port. That same day, two massive explosions rocked an Aramco facility at Juyamah in Saudi Arabia. These explosions were heard as far away as Bahrain, and fires were still burning twelve hours later.

While Saudi Arabia claimed that a "minor fire broke out", this claim was largely an attempt to reduce regional tensions. The actual cause seems to have been sabotage to the propane tanks very similar to a bombing effort in Kuwait on May 22, 1987. A major security cordon was rushed into place around the oil and gas port, and one was put in place at Ras Tanura, some 12 miles to the south. Ironically, the explosion occurred just as Saudi diplomats were returning to the Saudi Embassy in Tehran.[37]

The Reagan Administration responded on August 18, 1987, by announcing that it would send an arms sale request to Congress when it reconvened in the fall. It also announced that the cost of the sale could reach $1 billion. The timing of this announcement was clearly designed to reassure Saudi Arabia, and it was made from the White House, but it was awkwardly timed from a domestic political viewpoint. First, it did not contain any fixed details, largely because the Reagan Administration was still negotiating with Israel, and pro-Israeli lobbyists in the U.S., to try to defuse Congressional opposition to the sale. Second, the announcement was followed by only a token Administration lobbying effort. Third, Saudi Arabia made it clear that it did not want to have the details of its cooperation with the U.S. made public, even if it delayed or defeated the sale.

Once again, the Reagan Administration found itself unable to catalyze an effective political approach to winning Congressional support. It allowed the opponents of any close military relationship between the U.S. and Saudi Arabia to begin lobbying against the sale before it began its own effort, and it failed to inform the Congress of the details of U.S. and Saudi cooperation and the Administration's rationale for the sale until its opponents had already won the support of many key members of Congress.[38]

THE "ESCALATION LADDER" IN LATE AUGUST, 1987

The Reagan Administration faced other problems which went far beyond events in the Gulf, and owed more to the history of the U.S. interventions in Vietnam and Lebanon:

- On the one hand, the domestic political climate in the U.S. was changing. Roughly 100 members of Congress had visited the Gulf by late July. The results were mixed. The mood of Congress shifted towards acceptance of the fact that the President had made a commitment from which the U.S. could not withdraw. At the same time, many Congressmen favored implementing some form of the War Powers Act, or similar legislation, to constrain the President's freedom of action. This was difficult largely because the Senate was far more supportive of the President than the House was, and both houses of Congress had to pass such legislation.
  On the other hand, it was becoming steadily more clear that the U.S. commitment was becoming increasingly open ended. The U.S. was involved in a duel with Iran in which low levels of military force tended to favor Iran. No matter what the U.S. did in reprisal to any given Iran act, its effectiveness had to depend on Iran's fear of the U.S. using its superior strength to destroy a large part of Iran's military forces in the Gulf or escalating to block its oil exports and arms imports.

Limited U.S. strikes on naval Guards units, or the loss of a few Iranian ships and aircraft, could not have any real military effect on Iran, and the resulting "martyrdom" was more likely to encourage Iran's forces than to deter them. Further, the fact that the U.S., Britain, and France had limited their escort activities to ships flying their own flags in carefully designated zones in the Gulf meant Iran could use its peripheral strategy to strike at other ships flying flags of convenience, use indirect tactics like mining, use small Guards units as sacrifice pawns, and strike by using the Guards and pro-Khomeini movements and proxies to commit acts of sabotage, terrorism, and hostage taking.

The same limited encounters forced the U.S. to keep military units in the area costing $1–2 million a month, depending on the definition of the costs to be included.[39] Further, they confronted the Reagan Administration with the fact that even limited U.S. casualties could create enough opposition to the U.S. presence in the Gulf to shift the Senate towards support of the House in limiting the President's freedom of action. Since the Iranians were well aware of these vulnerabilities, a cautious strategy of escalation and steadily growing pressure on Iran tended to favor Khomeini even though Iran lacked the military capability to survive a U.S. attack on its naval and air forces and was extremely vulnerable to any higher level of escalation that threatened its trade in arms and oil.

This situation created the kind of "escalation ladder" that was a recipe for trouble. Each side had strong incentives to test the other to its limits. At the same time, the situation was an invitation to the USSR to attempt to exploit U.S. and Iranian tensions. It left the allies and friends of the U.S. in both Europe and the Gulf with the uneasy prospect that they might back the Reagan Administration only to become associated with U.S. withdrawal and defeat. This, in turn, acted as a further incentive for Iran to refuse any settlement, while it left Iraq with strong incentives to try to create a military situation that would deepen U.S., European, and Southern Gulf state involvement.

Finally, the struggle was becoming more and more a test of the U.S. commitment to maintaining its role as a global superpower, and of Khomeini's historical legitimacy. This test was particularly crucial for Khomeini. While President Reagan had little more to fear than historians criticizing part of his record, Khomeini's status as a revolutionary leader inspired by God was far harder to maintain in the face of major reversals and defeats. His whole heritage was one in which the only "historical" alternative to victory was direct martyrdom, and both his national and religious heritage was filled with examples of leaders whose failure to make good on their claims had virtually destroyed their reputations.

The U.S. had failed to address these risks in the original NSPG assessment of the cost-benefits of reflagging the Kuwaiti tankers. The NSPG had not sought a military risk assessment from the Joint Chiefs, it had not defined clear policy goals, and it had not asked for a National Intelligence Estimate of the probable Iranian, Iraqi, Southern Gulf state reaction to the U.S. commitment. It spent far more time discussing the potential benefits of the reflagging in reestablishing the U.S. role in the Gulf, and U.S. relations with the Arab Gulf states, than it did studying Iran.

The NSPG had allowed itself to be rushed into action by the twin pressures of the Soviet agreement to reflag Kuwait ships and the damage to regional relations being done by the disclosures coming out of the Iran-Contra Hearings. It tended to confuse the pragmatism of the Iranian regime in dealing with politically invisible Western actions, like the quiet British convoying of British flag merchant ships, with the impact of highly visible and political U.S. actions which directly challenged Iran. Further, the American political system had helped lead to a natural conflict between political cultures and methods. The U.S. wanted visibility, and virtually could not avoid it because of the American media.

As for the factors acting to control escalation, the key was the hope that the sheer scale of Iran's military and economic weakness would lead it to accept the UN peace initiative. At the same time, however, the Reagan Administration had reason to fear that escalation would only lead Iran to turn towards the USSR, or create long-term hostility between the U.S. and Iran.

ESCALATION IN THE GULF OF OMAN AND REACTIONS IN THE SOUTHERN GULF

The "escalation ladder" clearly favored continuing conflict, and the next step up the ladder came in the Gulf of Oman. While the Iranian Navy was completing a "minesweeping" effort off Fujayrah that involved six ships and six RH-53 helicopters inherited from the Shah, two Iranian patrol boats attacked a Liberian-registered chemical tanker, the *Osco Sierra,* outside the Straits of Hormuz. This marked the first use of the naval Guards vessels in direct attacks outside the Gulf, and came at a time when the UN Security Council was calling for new pressure on Iran to obey its resolution, and Secretary General Javier Perez de Cuellar was trying to arrange another visit to Iran and Iraq.[40]

The attack also occurred at a time when Iran was making some conciliatory gestures towards the UN and the West. On August 20, the spokesman for the Iranian Supreme Defense Council, Dr. Kamal Kharazi, stated that Iran had no objection in principle to the U.S. escorting Kuwaiti tankers. He also said that Iran was not totally displeased with the UN resolution. He reiterated claims that Iran had not mined the Gulf or Gulf of Oman and was busy sweeping mines. At the same time, however, he made a veiled threat in the form of a discussion of Iran's military options: "In the Persian Gulf, we are powerful and rely on both classical and regular forces, and the non-classical martyrdom forces".[41]

Iran was clearly trying to pursue a two track policy of its own. It did nothing to interfere with the third U.S. convoy moving into the Gulf, although both an Iranian frigate and a Soviet warship shadowed the convoy. The only real problem the convoy encountered was bad weather, which closed visibility down to less than a kilometer and forced the convoy to pause. Iran did, however, step up its search procedures and an Iranian frigate fired on a Yugoslav container ship when it refused to halt.

Meanwhile, the impact of Mecca, and the growing Iranian pressure on traffic to and from the Southern Gulf States, forced Saudi Arabia to take a steadily stronger line. Although it publicly denied it, Saudi Arabia expanded the landing rights it offered U.S. forces and its refueling support for U.S. aircraft and ships.[42] It quietly agreed to allow U.S. fighters and maritime patrol aircraft to land in Saudi air bases for refueling and logistic support under "emergency" and "case of need" conditions. It steadily expanded its cooperation in intelligence and internal security activities with the U.S., Kuwait, Britain, and France, and allowed U.S. Army and SEAL groups to use Saudi facilities for support purposes.

Saudi Arabia also continued to take a strong public line in opposing Iran. On August 23, Saudi Foreign Minister Prince Saud Al-Faisal issued a strong call for an Arab break with Iran. This came one day before an emergency meeting of the foreign ministers of the Arab League in Tunis. The Saudis pressed during the meeting for an Arab embargo in economic relations with Iran, and increased their pressure on Syria, Algeria, and Libya to halt their support of Iran.

This pressure was only partly successful. When the 18 countries in the Arab League met on August 23, they were far too divided to take immediate action, and Syria took the lead in blocking any effort to declare an immediate embargo. The Arab League did, however, condemn Iran for the events in Mecca and delivered an ultimatum indicating that it would meet again in September and consider an embargo if no progress was made towards peace by September 20.[43]

Iran's reply was to accuse the League of trying to impose an "ignominious peace", although Prime Minister Mousavi again stated that Iran was showing an interest in the UN initiative. Even so, the growing Saudi pressure on Syria was significant. Syria's arms race had driven it to the edge of economic collapse. While the Saudis could not force Assad to heal his long-standing conflict with Sadam Hussein, they could exert powerful pressure on Syria to halt any activities that directly aided anti-Iraqi Kurds and political movements.

Saudi action did not stop with the Arab League. On August 25, Interior Minister Prince Naif ibn Abdul Azziz issued a statement that accused Iran of "criminal sedition" in starting the riots in Iran, and said that Khomeini had sought to "shake the security of the Kingdom". He also stated that Saudi Arabia was "capable of defending itself" and would "immediately repel" any attacks. He issued this statement to more than 100 journalists, largely from Islamic countries. He also attacked Iranian claims that Saudi security forces had fired on the Iranian pilgrims, and noted that there were some 70,000 Iranian pilgrims present in the area and only several hundred dead, and that any such firing would have produced casualties in the thousands. Finally, he warned that while Saudi Arabia would continue to permit Iranians to make the pilgrimage, it took its role as the custodian of Mecca and Medina extremely seriously, and would not tolerate further incidents.

The radio and TV media in Saudi Arabia and Iran also engaged in a strident contest in invective. One Saudi broadcast went so far as to state that "Khomeini is Satan". Iranian broadcasts referred to "the new shah Fahd, the corrupting agent", and described the Saudi royal family as "imbeciles" and "infidels". During the week before Ashura—when Shi'ites mourn the death of Hussein—Khomeini stated that "Mecca is the place where all prophets have served since time was created, but unfortunately it is now in the hands of a group of infidels, who are grossly unaware of what they should do. It is a shame for world Moslems to see the sanctity of Mecca, the divine and sacred places broken in such a manner".[44]

At a more material level, Saudi Arabia revived plans to create a massive underground reserve to ensure it could export oil even in the event of attacks. Two Swedish contractors, ABV A.B. and Skanska, were asked to explore the project, which was to be built near the existing Saudi oil port on the Red Sea, and which had a cost of up to $3.9 billion.

The other Gulf states were more divided in their reactions. Bahrain and Kuwait were steadily increasing their support of the U.S. and their European military presence in the Gulf in ways similar to those of Saudi Arabia. Kuwait began to explore having the U.S. reflag two more of its ships, and asking Britain to provide similar protection. Bahrain quietly acted as a communications and logistic base and transit point for the U.S. Middle East Task Force. Oman allowed the U.S. to use Omani facilities and cooperated closely with the British. Qatar largely stood aside, and the UAE split between those Emirates favoring Iraq and those favoring Iran—or simply profiting from the war.

Abu Dhabi, the richest oil state in the UAE, consistently supported Iraq and quietly supported the U.S. reflagging effort. Fujayrah also tended to support Iraq. The tankers waiting off its port of Khor Fakkan had become a major source of revenue that shrank from 70 to 12 vessels when Iran started to mine the Gulf of Oman.

All the states in the UAE, however, had to take account of the fact that they were conducting $1.5 billion in re-export trade—one-third of it with Iran—and that most of their $8 billion in oil export revenue in 1986 had come from very vulnerable offshore oil facilities. None could forget that Iran had already hit the Abu al-Bukhoosh offshore oil platform in November, 1986. With oil revenues estimated at around $11 billion in 1987, Abu Dhabi stood to get $8 billion and Dubai $2.5 billion. Both states had good cause to be careful.

Dubai and Ras al-Khaimah were relatively pro-Iranian. They made substantial profits from Iran and needed to placate large Shi'ite minorities. Some 100,000 Iranians and native Shi'ites lived in the UAE, 30,000 of them illegally, and most were in Dubai and Ras al-Khaimah—with 50,000 in Dubai alone.

Dubai imported about $150 million in fruits and vegetables from Iran and it alone re-exported about 70% of Iran's non-petroleum civil imports—an activity which dominated most of its dhow traffic and which had a major impact on the native labor force. Some 7,000 dhows had left for Iranian harbors in 1986, with a traffic worth well over $390 million, and this traffic was already worth $400 million in August, 1987. Dubai Dockyards (150 ships per year) also competed with Bahrain's Arab Ship Building and Repair Company (50 ships per year) for business in repairing the ships damaged in the "tanker war", although it was not always clear that wartime business offset the loss of regular peacetime service business.[45]

THE TANKER WAR AND ATTACKS ON KUWAIT

Although the war in the Gulf had escalated during the spring and aummer of 1987, most of the tankers and cargo vessels moving through the Gulf in August still did not have any military escorts. Six out of every seven tankers moved on their own, as did dozens of cargo vessels and hundreds of dhows and small coastal vessels. The Gulf "tanker war" was still largely a political and military test of wills, not a major barrier to Gulf shipping. Kuwait was receiving 70 to 80 tankers per month, the vast majority of which traveled without incident. While the number of ships hit in the war had risen to 325, the total casualties were still only about 320, or less than one per ship attack. While the volume of trade was far below its 1984 level, this was due more to the fall in oil prices than to the war.

Tanker freight costs had trebled in the Gulf during the first seven months of 1987, but they still amounted to only about $2 a barrel, and the cost could often be offset by the quiet discounting of oil prices by Gulf oil producers. Insurance also rose from about 2% of the price for a barrel of Saudi crude in 1985 to 12% in 1986, but once again oil prices dropped. In August, insurance cost 0.25% of the value of the cargo, plus a 0.25% supplement for Iraq, and did little to discourage Gulf shipping. Tanker demand, which had shrunk by 36% between peaks in 1979 and 1985, grew by 12% in 1986, and employed tanker capacity rose by 23% in the first seven months of 1987.[46]

Ironically, Iran may have suffered almost as much from its attacks on shipping as did the West and Iraq. Some 35 to 40 ships were calling at ports in the Gulf of Oman and offloading cargo to move by road to the UAE. Although Iran had mined the area, much of this traffic went to Iran. It had been heavily dependent on transshipping through Khor Fakkan since 1985, and the total flow of shipping had increased during 1986. While the number of cargo ships going to Dubai's ports at Port Rashid and Jebel Ali had dropped from 3,229 in 1983 to 2,888 in 1986, the number of dhows and small coastal vessels calling at these ports had risen from 6,366 in 1984 to 10,008 in 1986.[47]

The "tanker" war became more serious, however, during the first part of September. Iraq kept up its air raids on Gulf targets for four straight days from August 29 to September 1, 1987. It hit tankers, oil facilities in the Gulf, and Iranian land-based oil facilities and factories. Iran replied on September 1 by having the naval Guards attack a Kuwaiti freighter with machine gun fire and rockets. Iran was careful, however, to pick a target which was not reflagged and which was then about 350 miles from the nearest U.S. convoy.

Iraq and Iran exchanged another round of threats. Iran stated that it would retaliate on a "blow for blow" basis and charged that Iraq was using its fighters to conduct extensive attacks on cities for the first time since a limited unofficial cease-fire on such attacks in February, 1986. The next day, Iran hit 2 tankers carrying Arab oil. This triggered a round of Iranian and Iraqi attacks that hit 7 ships in one 24 hour period. Iraq had now claimed to have hit 11 naval targets in 5 days, plus 2 power plants and a communications facility. Counting Iranian attacks, 20 ships were hit in the eight days between August 27 and September 3, 1987. This was the most intense series of attacks in the history of the "tanker war", and Lloyds raised its war risk premium by 50%.

Two events then further heightened the tensions in the Gulf. On September 3, Iranian speedboats attacked an Italian container ship, and wounded two Italians in the crew. This led to a meeting of the Italian

Cabinet on September 4, and Italy changed its position from seeking a UN peace force to one of sending Italian warships to join the U.S., British, and French warships in the Gulf. Italy was then receiving 40% of its oil from the Gulf, and was able to obtain a quiet French promise to provide air support in an emergency.

That same day, Iran fired a Silkworm missile at Kuwait from a site on the far southern tip of the Faw Peninsula. The missile fell on an uninhabited strip of coastline, but it was only two miles from one of Kuwait's main oil terminals and two tankers were loading at the time. It was initially uncertain as to whether the missile's terminal guidance system had failed to acquire the ships or Iran had merely intended the firing as a threat. Iran denied responsibility for the firing, but then fired another missile on September 4, and a third on September 5. The second missile hit near Mina Abdullah, 30 miles to the south of Kuwait City, and showed that Iran could bracket the entire country. The third hit near Failaka Island, 13 miles from Kuwait's northern coast.

Kuwait protested the missile firings to the UN, complained to the permanent members of the Security Council, and expelled five of Iran's seven diplomats in Kuwait. This led Iran to charge that the U.S. had pressured Kuwait to expel the diplomats, but that Kuwait would have to take the consequences. It again called for the U.S. to leave the Gulf.[48] The GCC states responded by calling a meeting of their foreign ministers on September 12, 1987, although it was clear that there was little the Southern Gulf states could do.

The U.S. rebuked Iraq for its attacks, but concentrated on attacking the Iranian delay in accepting the UN cease-fire proposal. The UN announced on September 4 that a peace mission was going to the Gulf. This led to a reduction in the intensity of the tanker exchanges between Iran and Iraq. Both Iran and Iraq agreed to a limited cease-fire in the Gulf during September 5 to 7 as a precondition for a peace mission by UN Secretary General Cuellar. Iran obviously hoped to show it was not totally rejecting the UN initiative. Iraq, however, saw the visit as a precondition to any sanctions on Iran.[49]

This cease-fire, however, led to only a momentary lull in the war. Iraq continued to attack land targets and stated it would only promise not to attack Tehran during the Secretary General's visit. The Secretary General's mission reached Tehran on September 11, 1987. When his talks with the Iranians began the next day, both sides were claiming the other had violated the cease-fire. Iran said Iraq had killed 13 civilians. Baghdad said Iran had shelled 8 Iraqi cities and towns in the last 48 hours, and naval Guards attacked a Cypriot supertanker carrying Saudi oil with rockets.

THE GROWING EUROPEAN AND U.S. PRESENCE IN THE GULF

The European role in the Gulf continued to grow throughout the summer of 1987. The French and British decision to send warships to the Gulf on August 11, 1987, was followed by growing British pressure on Belgium, Italy, and the Netherlands to take a more active role. As a result, a special meeting on cooperative action in the Gulf took place in the WEU on August 20. This meeting failed, however, to produce any tangible result other than a call for freedom of navigation. The FRG could not agree on any positive course of action; Italy persisted in pushing for some form of UN initiative; and Britain and France had already moved beyond the point where they felt they could subordinate their freedom of action to a committee.

When Italy unilaterally decided to enter the Gulf on September 3, the Netherlands, which chaired the WEU—and had tried to catalyze a common WEU effort in the Gulf—gave up on collective action and decided to act on its own. It turned to Britain for assistance, and Britain agreed to provide air cover and logistic support for a Dutch minesweeping effort in the Gulf. On September 7, 1987, the Netherlands announced that it would send two minesweepers to the region. Belgium, which had long worked closely with the Netherlands and the U.K. in planning the naval defense of the Gulf, followed the Dutch lead. It announced it would send two of its minesweepers and a support ship to the Gulf. Belgium also agreed to a joint command with the Netherlands, with a Belgian officer in charge of the naval units on the scene and a Dutch officer in command of the overall task force effort.

The Belgian and Dutch ships sailed as a flotilla in late September, but then demonstrated that the U.S. was not the only nation that could run into problems in deploying its mine forces. The Belgian minesweepers were ex-U.S. World War II ships of the Avenger class. One, the *Breydel*, was 30 years old and broke down in the Mediterranean. This meant the task force could not arrive by its scheduled date of November 1. The flotilla also ran into problems because it had no base in the Gulf and no air cover. The best that it could do was to rely on France to provide support from Djibouti and use the Belgian support ship—the *Zinnia*—as a supply shuttle. The British agreed to provide air cover, but only as a third priority after British warships and merchant ships. The Belgian-Dutch force was armed only with 20 mm guns, and this led the Netherlands to reinforce them with Stingers operated by Dutch Marines.[50]

As has been discussed earlier, Italy had also decided to enter the Gulf. The Italian Cabinet announced the dispatch of eight ships, including three Lerci-class minesweepers and one Lupo-class frigate and two Maestrale-class frigates, on September 8, 1987. The ships were put under the command of Rear Admiral Angelo Mariani, Commander of the Second Naval District at Taranto. They set sail on September 15, 1987 (see Table 11.2).[51]

As for the forces already in the Gulf, the French task force now included the *Clemenceau* (a 32,700 ton carrier with 38 aircraft and recently modernized air defenses, including Thomson CSF EDIR defense systems), two guided missile destroyers, a corvette, and three multi-role frigates. There were two replenishment ships and a 2,320 ton repair ship under way. The French force had been reinforced by three minesweepers (two Dompaire and one Tripartite-class vessels) which had been heavily modernized in 1976 and 1979, and which began operations in the Gulf of Oman on September 11.[52]

By late October, the French force had found found nine M-08 mines in the area off the ports of Fujayrah and Khor Fakkan. Two were surfaced by sweeping and seven were detected by sonar at depths of 2 to 7 meters. The French frigate *Dupleix* had detected and destroyed two more mines off Qatar. This brought the total number of mines found since the beginning of the U.S. reflagging effort to eighty.[53]

The British Armilla Patrol included the *HMS Andromeda*, a Batch 3 Leander-class frigate with Sea Wolf short-range air defense systems. It was supported by the destroyer *Edinburgh*, a Type-42 air defense ship, and the *Brazen*, a Type-22 Batch 1 anti-submarine warfare frigate. The British mine force had reached the Gulf of Oman on September 15. It included the 1,375 ton tender *Abdiel*, which carried spare minesweeping gear and cables and acted as a command ship, and three minesweepers from 1 and 4 MCM, which carried lightweight wire sweeps, mine hunting sonars, Barricade anti-missile decoys, and ROVs.[54] Fleet support was provided by the 40,200 ton replenishment tanker *Brambleleaf*, the stores ship *Regent*, and the repair ship *Diligence*—which provided repair, fire fighting, and diving facilities.

As for the U.S., it was now deployed for a major war. The task force led by the *Constellation*, and the forces in the Indian Ocean, included a full carrier wing of 87 aircraft. The *Valley Forge* acted as an air cover control ship, coordination center with the Saudi Air Force, and escort for the *Constellation*. The *Cochrane* provided air defense cover, and two Knox-class frigates—the *Cook* and *Quellet*—provided ASW cover and surface defenses with RGM-84 Harpoon missiles. The 53,600 ton support ship *Camden* provided fuel and munitions, a 16,070 ton Mars-class ship provided combat stores, and the *Niagra Falls* provided spares and aircraft repair parts and equipment. This carrier group was replaced later in the summer by a group led by the carrier *Ranger*.

There also was a U.S. Naval Surface Group Western Pacific, which included the battleship *Missouri*, which had modernized command facilities, short-range air defenses, and anti-ship missiles. These included a battery of 32 Tomahawk cruise missiles with both anti-ship and land attack versions. The *Missouri* was escorted by an upgraded Aegis ship, the *Bunker Hill*, and the cruiser *Long Beach*—the first nuclear powered warship—with Tomahawk and Harpoon missiles. The air defense for the surface group came from the *Hoel*, and the *Curts* provided anti-submarine defense. The supply ship *Kansas City* provided oil, fuel, ammunition, and provisions.

Admiral Bernsen's Middle East Task Force in the Gulf continued to include the command ship *La Salle*. It now, however, had been reinforced by the amphibious assault ship *U.S.S. Raleigh*, which was equipped with a well dock. The *Raleigh* had carried four MSB5-class minesweeping boats, and four SEAL Seafox light special warfare craft, to the Gulf. The cargo ship *St. Louis* had arrived on August 27, with two more MSB5s and two Seafoxes.[55]

The *Guadalcanal* had offloaded its 700-man Marine unit and carried eight RH-53D Sea Stallion minesweeping helicopters. The force also included four AH-1T Sea Cobra attack helicopters, four UH-1N utility helicopters, and two CH-46C Sea Knight medium transport helicopters. There were five missile-armed hydrofoil ships—the *Hercules, Taurus, Aquila, Aries,* and *Gemini* —with 76 mm guns and Harpoon missiles. A U.S. Army Special Force unit provided four MH-6A surveillance and counter-attack helicopters.

The convoys of the reflagged Kuwaiti shuttle tankers were escorted by Destroyer Squadron 14, whose flagship was the guided missile destroyer *Kidd*. The squadron included the guided missile cruiser *Fox*, and the guided missile frigates *Reid, Crommelin, Jarret, and Hawes*. These four frigates had 76 mm guns, and LAMPS 1 SH-2F or LAMPS III SH-60B helicopters. The force was supported by an area defense force including the frigates *Flatley* and *Klakring* and the guided missile cruisers *Worden and Reeves*. There were now 11 U.S. ships, with about 4,500 crew members, actually in Gulf waters. Altogether, these reinforcements raised the Pentagon's estimate of the total cost of the U.S. reflagging exercise to the navy alone to $200 million a year.[56]

By early October, the total number of ships scheduled for operation in the area had risen to 35 European and 35 U.S. ships. Europe was sending a total of 1 aircraft carrier, 3 destroyers, 10 frigates, and 5 large and 5 small support ships. Europe was also sending a substantial minesweeping capability of 14 modern vessels.[57] While Japan did little more than discuss a "general willingness" to help the U.S., the FRG agreed to send its ships south to replace some of the ships that other European nations had moved out of the Mediterranean. The FRG sent the

Table 11.2 Western Ships in the Gulf and Gulf of Oman, or en Route, on September 20, 1987[1,2,3]

U.S. Task Forces[4]		French Forces		British Forces	
Middle East Task Force		Carrier:	Clemenceau (1,338)	Destroyer:	Edinburgh (253)
		Destroyers:	Duquesne (355)	Frigates::	Andromeda (224)
Command ship:	LaSalle (499)		Georges Leygues		Brazen (175)
Cruisers:	William H. Standley (513)		Suffren (355)		Mine Hunters: Bicester
	Reeves (513)	Frigates:	Commandant Bory (167)		Brecon
			Portet (167)		Brocklesby
Destroyer:	Kidd (346)		Victor Schoelcher (167)		Hurworth
Frigates:	Crommelin (200)	Mine hunters:	Cantho		Repair Ship: Diligence
	Rentz (200)		Loire	Mine Support Vessel:	Abdiel
	Flatley (200)		Vinh Long	Supply ship:	Brambleleaf (42)
	Jarret (200)			MPA:	Nimrods
	Klakring (200)	Tendership:	La Garonne		
Transport ship:	Raleigh (429)	Tanker:	La Meuse		
Amphibious assault:	Guadalcanal (754)	Support Ship:	La Marnet		
Mine Support Ship:	Loire				
Other Temporary Assignments					
Italian Forces		Belgian Forces			
Carrier:	Ranger				
Cruiser:	Valley Forge	Frigates:	Scirocco	Minesweepers:	Breydel
Destroyer:	Arcadia		Gregala		F. Bovesse
Frigates:	Cook		Perseo	Support Ship:	Zinnia
	Quellet	Mine Vessel:	Vieste		
Ammunition Ship:	Camden		Milazzo		
Supply Ship:	Niagra Falls		Sapri	Netherlands Forces	
Supply Ship:	Vesuvio				
Battle Group		Salvage Vessel:	Anteo	Minesweepers:	Hellevoetsluls
					Maassluis

Battleship: Missouri
Cruisers: Bunker Hill
Long Beach
Destroyer: Hoel
Frigate: Curts
Support Ship: Kansas City

Minevessels in transit: Enhance
Esteem
Fearless
Inflict
Illusive
Mine tow ships: Barber County
Grapple

---

1. The Soviet Indian Ocean Squadron was commanded by a Vice Admiral, using the *Ladny*, a Krivak 1 class frigate. The Soviets had a total of Sovremenny or Udaloly class destroyer, 2 frigates, 3 Natya-class minesweepers, 1 Urga-class submarine depot ship, an unstated number of submarines, and 3 AGI trawler or intelligence gathering vessels in the region.
2. Figures in parentheses crew numbers and provide a rough indication of ship size.
3. The weapons complements of the ships in the Gulf included 1 ASW helicopter for the William H. Standley; 2 ASW helicopters for each U.S. frigate; 8 MSW helicopters, 4 attack helicopters, 2 transport helicopters, and 4 utility helicopters for the Guadalcanal; 6 MSC, 4 special warfare boats, and 2 FPB for the Raleigh; 20 FBA, 10 AWX, and 10 ASW aircraft for the Clemenceau.
4. U.S. forces were supplemented by E-3A and tanker aircraft in Saudi Arabia, cargo and support ships at Diego Garcia, P-3 aircraft in Masirah in Oman, and 150 support personnel at Jufair in Bahrain.

destroyer *Moelders*, frigate *Niedersachsen*, and support ship *Frieburg*, all of which had arrived in the Mediterranean by mid-October.[58]

IRAN SETS THE PACE IN THE TANKER WAR

Iraq continued to sporadically strike at targets in the Gulf throughout the rest of September, but it was Iran that set the pace. On September 20, an Iranian speed boat attacked a Saudi tanker in the Straits. On September 21, Iran attacked a British flag tanker, the *Gentle Breeze*, setting it afire and killing a crewman. This led the British government to react by expelling all of the personnel in the Iranian Logistic Support Center in London that had been purchasing arms, and to bring its full weight into the effort to halt arms sales to Iran.

Later that day, a more dramatic development occurred. After the *Bridgeton* had hit a mine, the U.S. publicly and privately stated to Iran on several occasions that its rules of engagement would lead it to strike without warning at any Iranian ship that was caught laying mines, or preparing to lay mines, in international waters. At the direction of the Chairman of the Joint Chiefs, Admiral William Crowe, the U.S. also dispatched a U.S. Army special forces unit to the Gulf from Task Force 160—the "wings of Delta Force"—at Fort Campbell, Kentucky.[59]

In early September, Crowe visited Rear Admiral Harold Bernsen on the *U.S.S. La Salle*, and they worked out a plan to track and intercept any Iranian vessels that continued to lay mines. The U.S. used its intelligence satellites, SR-71 reconnaissance aircraft, and the tactical intelligence gathering assets the U.S. had deployed in the Gulf—including E-3As, P-3Cs, and specially equipped helicopters such as silenced OH-6As—to track Iranian minelaying efforts from the loading of the ships in port to the actual minelaying.

The U.S. picked out one of the Iranian vessels—an Iranian Navy LST called the *Iran Ajr*. The U.S. tracked the *Iran Ajr* from the time it left port. It watched while the Iranian ship reached a point about 50 miles northeast of Bahrain and north of Qatar, and laid at least six mines. The Iranian ship seems to have felt it was safe in doing this because it was operating at night. The U.S., however, had deployed the *Guadalcanal* and *Jarret* to the area, and was using OH-6 light attack helicopters from the U.S. Army Special Operations Command which were equipped with night-vision devices including forward looking infra-red (FLIR) equipment. They also had passive light intensification devices, and could track every movement on the Iranian ship.

Two of these helicopters, based on the frigate *Jarret*, flew at 61 meters to a point about 500 meters from the *Iran Ajr*. They watched the ship, and when they visually confirmed that it was still laying mines, they asked for permission to attack. Bernsen approved and the helicopters attacked the Iranian vessel with their 7.62 mm machine guns and 2.75 inch rockets. They immediately disabled the ship, and killed five crewmen. U.S. Navy SEAL teams from the *Guadalcanal* then used small boats to seize the ship and took 26 prisoners—one of whom later died from wounds. A day later, the U.S. blew up and sank the Iranian vessel.

This American strike exposed Iran in a way that publicly confirmed its guilt for all its past minelaying efforts. Iran could no longer make vague references to an "invisible hand" or blame the minelaying on an American plot. Further, the U.S. attack came while Khamenei was at the UN. All he could do was to claim the ship was a "merchant ship" and describe the minelaying charge as "a pack of lies". The U.S., however, found 10 fully armed Soviet-made M-08 mines on the ship and charts showing where the ship had laid mines in the past day, filmed the captured crewmen discussing their minelaying activities, and recovered mines where the chart indicated they had been laid.

In any case, the U.S. attack immediately gained a great deal of U.S. and foreign support. The American people overwhelmingly supported President Reagan, and Congressional attempts to invoke the war powers act proved even more ineffective than in the past. Polls showed that roughly 60% of the American people supported the U.S. deployment to protect Kuwaiti tankers although 55% felt it was likely to involve the U.S. in a war. Some 76% described the Gulf as very important to U.S. strategic interests and 78% had unfavorable feelings towards Iran.

This support enabled Reagan to take a hard line on maintaining the U.S. presence in the Gulf, and he publicly stated that efforts to invoke the War Powers Act were helping Iran. At the same time, the Reagan Administration had shown it could and would hit Iran. This had a powerful impact in correcting the U.S. reputation for military errors and in reassuring both its European and Gulf allies that the U.S. would stay in the Gulf. The incident also seems to have prompted the PRC to openly state in the UN that it would support an arms embargo if Iran did not move towards a cease-fire.

During the next few days, the political impact of the U.S. action was reinforced by the fact that the U.S. found more mines and returned the Iranian sailors it had captured. Further, it was clear that the Iranian actions could not simply be blamed on members of the Guards. A regular Iranian naval vessel was involved, and the documents found on the ship clearly indicated it was acting on the basis of formal orders from a relatively high level.

The incident also pushed the U.S. Congress towards banning all U.S. trade with Iran. On September 29, 1987, the U.S. Senate voted 98 to nothing to ban all imports from Iran. The House passed a similar bill on October 1. This vote reflected broad popular support for an embargo, but its impact was uncertain. Many of the experts in the Reagan Administration felt the "embargo" would be ineffective without broad world support, and that the oil would simply shift to other buyers.

Given the fact that Iran was selling for $17 a barrel at a time when world prices were $21 to $22, this seemed likely to be an accurate assessment. Nevertheless, the U.S. and Iranian balance of trade was politically embarrassing. The U.S. imported $468.3 million worth of oil in 1986, $51.7 million worth of carpets, and $14.1 worth of pistachios. Iran's export trade was also increasing. It rose to $567 million worth of imports in the first six months of 1987. Oil imports from Iran, which averaged only 15,000 BPD in 1986, rose to 45,000 BPD in direct imports and 126,000 BPD shipped through the Virgin Islands. In July, 1987, Iranian shipments rose to 11% of U.S. oil imports, and Iran was seeking some $40 million in oil-field equipment.

U.S. exports to Iran were much smaller, and reached only $33 million in 1986, but they included sensitive items like computers and Scuba gear that turned out to be for the Iranian naval Guards. Further, the issue was politically sensitive because Iran was exporting about 600,000 yards of textiles in the first seven months of 1987, versus 200,000 yards in all of 1986. This was strongly opposed by American cotton growers and textile workers.[60]

The USSR was the only major power which reacted to the U.S. attack by taking a stand favorable to Iran. On September 23, Soviet Foreign Minister Eduard Shevardnadze spoke to the UN, called for an immediate cease-fire, called for the simultaneous appointment of a U.N. commission of inquiry into the causes of the war, and called for the foreign navies in the Gulf to be replaced by an international force. He also made it clear that he would prefer negotiation to any arms embargo. Iran, however, did not get uncritical support from the USSR.

Soviet spokesmen were careful not to imply, however, that Iran had any justification for mining the Gulf. The Soviet Union was still concerned with Arab opinion, and Shevardnadze met with Secretary Shultz the next day and made it clear that while both the U.S. and USSR had agreed to defer an arms embargo and continue diplomacy, even the USSR did not rule out the idea of an embargo in the near future.[61]

IRANIAN AND U.S. ESCALATION REACHES NEW HEIGHTS

At the beginning of October, a pattern had been established whereby the war involved three parallel struggles: The first struggle was the escalating confrontation between Iran and the U.S., Western Europe, and Iran. The second struggle was a continuing political battle over the UN peace initiative, with the U.S. increasingly pressing for an arms embargo, the USSR seeking to strengthen its developing ties to Iran without alienating the Arab world, and Iran in the ambiguous position of either seeking concessions or simply using such demands to delay UN action. The third struggle was the continuing series of exchanges and low-level border conflicts taking place between Iraq and Iran.

The first and second of these struggles were highly interactive. During August 31 to October 1, Iran launched a new set of gunboat attacks. Iranian ships hit three tankers, and often seemed to deliberately pass near Western and Soviet naval forces. Iraq responded in kind, and this raised the number of attacks on Gulf ships to 375.

Iran continued to take the position that any effort to establish war guilt had to begin at the same time as a cease-fire. It claimed that the UN Security Council resolution was being misinterpreted and did not really call for a simultaneous withdrawal to the 1975 borders—a point the text left ambiguous. Further, it raised the issue of which aspect of the Algiers Accords was really in force, since neither Iraq or Iran had ever fully complied with the original settlement between the Shah and Sadam Hussein.

In spite of these problems, the UN kept trying to move the cease-fire effort forward. It adopted a two track approach of continuing to talk to Iran while moving forward towards implementing sanctions. On October 1, the Security Council began to draft two documents: a detailed plan for implementing a cease-fire, and a plan for imposing sanctions on either belligerent. The effort to implement the cease-fire was due to begin right after the Secretary General returned, but was delayed in order to allow the presidency of the Council to rotate to Italy. The presidency had previously been held by Ghana, the most pro-Iranian member, and it was felt that this would block effective action.

The key players, however, still could not agree on a course of action. Secretary Shultz stated that he now saw sanctions against Iran as likely and believed the UN Security Council would have to act, although he said that Iran should be given more time. The USSR, in contrast, pressed for creating a UN force to police the Gulf and delaying any sanction effort. It proposed that a cease-fire and to investigation into war guilt occur at the same time, rather than having the investigation follow a cease-fire and withdrawal to prewar borders. Other nations quietly pressed Iraq to compromise on the sequence of events in the cease-fire and accept some of the Iranian proposals.

This indecision resulted in protests by Iraqi Foreign Minister Tariq Aziz about delays in the UN effort, and growing Iraqi tension with the USSR. The differing positions also blocked effective UN action and pushed the U.S. and Iran towards further confrontation. It was also a confrontation that Iran's leadership clearly understood was coming. On October 2, Rafsanjani stated that, "with great likelihood, we will get involved in a new front in the southern part of the country".

A NEW IRANIAN THRUST AT SAUDI ARABIA

Iran quickly backed its words with actions. On October 1, more mines were spotted off Farsi Island, north of Bahrain, and a British minesweeper was dispatched to sweep the area. On October 2, intelligence sources discovered that Iran was planning a substantial strike on Ras al-Khafji—a large Saudi-Kuwaiti oil field and processing complex with a capacity of 300,000 BPD near the Saudi-Kuwaiti border, and about 110 miles from Kharg. This field was used to produce the oil that Saudi Arabia and Kuwait marketed for Iraq.

After consultation with the U.S., the Saudis put F-15 and Tornado aircraft on alert and began to move ships towards the area. They also warned Tehran they had learned of the plan. Tehran, however, decided to go on with the attack, at least to the point of forcing a Saudi reaction and demonstrating its resolve.

On October 3, 48 to 60 Iranian speed boats assembled near Kharg Island. They then moved towards the Khafji oil field. U.S. E-3A aircraft detected the move, and notified the U.S. commander of the Middle East Task Force and the key command centers Saudi Arabia. The Saudis responded immediately by sending ships and F-15 and Tornado aircraft towards the Iranian force, and the U.S. task force—including the command ship *LaSalle*—sailed in the same direction. When the Iranian force detected these moves, it turned away.[62]

This encounter not only demonstrated how firm the Saudis now were in resisting Iran; it also helped catalyze a shift in Qatar's attitudes toward the war. Qatar had previously tended to take a moderate stand relative to Iran, to distance itself from Saudi Arabia, and to oppose any foreign military presence in the Gulf. The riots in Mecca, however, had led Qatar to begin shifting its position. The new attack on Saudi Arabia led Qatar to sharply reduce its opposition to the U.S. and Western presence in the region. The combination of Iran's attacks on Kuwait,

France's deployments of forces to the Gulf, and this latest Iranian adventure also caused Qatar to cooperate in taking a collectively harder line within the GCC.

On October 5, Iraq launched a new series of long-range air strikes, and struck at tankers loading near Larak, Iran's transloading point for its tanker shuttle. These attacks included a successful hit on the *Seawise Giant*, a 564,739 ton ship which was the largest ship afloat. While reports differed, up to four tankers may have been damaged during this raid. This brought the total number of Iraqi raids on Gulf shipping since late August up to 21, and Iraq stepped up its air attacks to the point where it was flying some 50 sorties a day during October 5–8.[63]

Iran, in turn, launched new Scud strikes on Baghdad for the first time since mid-February, and continued to strike at Gulf shipping. Both the Guards and the regular Iranian Navy were increasingly aggressive. Guard ships often attacked targets near Western warships. For example, they hit a Pakistani oil tanker only three to four miles from the French warship *Georges Leygues*. An Iranian destroyer moved within a mile of the U.S. destroyer *U.S.S. Kidd* and locked its fire control radar on the *Kidd*. The US. ship immediately warned the Iranian ship three times to cease illuminating, and finally warned it to halt immediately or it would fire. The Iranian ship immediately halted and turned away.[64]

Iran hit a number of Japanese ships during this period. Japan again warned its ships not to enter the Gulf, although it was getting more than 50% of its oil from Gulf states. On October 7, 1987, Japan also announced that it would finance a precision navigation system to assist ships of all nations in avoiding mines. Prime Minister Nakasone indicated that Japan might make a major increase in financing its share of the Western military presence in the area. Nevertheless, Japan conspicuously avoided taking sides and making any military commitment, and its actions had little effect on Iran. On October 8, Guard units set afire a 9,400 ton Japanese-owned ship under Panamanian registry, the *Tomoe-8*, about 60 miles east of Jubail. This strike came only one week after Japan had again allowed its ships to enter the Gulf.[65]

MORE U.S. AND IRANIAN CLASHES

On October 8, another direct clash took place between U.S. and Iranian forces. A force of one Iranian corvette and three speedboats moved toward base Hercules, one of the barges that U.S. forces were using as a base near Farsi Island.[66] When the U.S. sent helicopters towards the Iranian force, which was moving directly towards the U.S. base, the Iranian ships fired on the U.S. helicopters.

The U.S. force included three MH-6 night surveillance/attack helicopters, and was about 15 miles southwest of Farsi Island and 3 miles outside Iran's 12 mile limit.[67] The U.S. Army helicopters immediately returned the Iranian's fire. They attacked the Iranian force of four ships, and sank one Boghammer speed boat and damaged two Boston-Whaler type boats. An Iranian corvette either escaped or was not attacked.

Eight Iranians were killed, and six were taken prisoner by SEAL units using Mark III Sea Specter patrol boats. An inspection of the damaged boats revealed that at least one had carried U.S.-made Stinger missiles of the kind being given to the Afghan Freedom fighters. The Afghans later claimed these were part of the six Stingers that Iranians had stolen from a group near the border in April, but U.S. experts felt they might well have been sold. The Iranians claimed they were part of another secret sale by McFarlane in 1986, and that many more had been delivered.

This U.S. strike led to yet another series of Iranian promises to retaliate, and the new engagement between U.S. and Iranian forces was serious enough to reduce U.S. Senate resistance to the House's efforts to invoke the War Powers Act. At the same time, it encouraged Congressional efforts to ban imports from Iran. U.S. Energy Secretary John Herrington, then in the Gulf, described the boycott as "a moral position as much as an economic position".

The Senate also began to discuss imposing user charges on escorted vessels. The U.S. had every reason to be sensitive to this issue. The directly accountable military forces in the Gulf had cost $69 million from the time operations began in mid-July to September 30.[68] Total costs were now about $30 million a month. The costs were so high that they led Secretary Weinberger to cancel plans to send U.S. Coast Guard cutters and aircraft to the Gulf because this might require Congressional debate and approval of the move. Further, the U.S. Navy was instructed not to seek supplemental funding from Congress, and to cut back on readiness elsewhere, although the Gulf operations were raising the navy's operating budget by at least $80 million during the current fiscal year.

The Congressional debate over such costs was made even more intense when reports leaked to the press on October 12 that Admiral Bernsen, commander of the Middle East Task Force, was seeking authority to attack any Iranian vessel attacking a merchant vessel that called for aid, regardless of its flag. The practical military problem was that Iran was now concentrating on targets that did not fly the flags of any of the

Western nations in the Gulf, and was often striking even when Western warships were nearby (see Table 11.3).

The Reagan Administration rejected this request to expand the U.S. role for four reasons. First, it meant a major increase in the U.S. commitment and risk in the region. Second, it meant the U.S. would have to pay to protect ships which did not pay anything in return, including U.S.-owned ships carrying foreign flags. Third, it preferred to emphasize the UN peace effort over escalation. The question the Administration did not answer was how long it could continue to take this position if Iran continued to escalate.

ANOTHER SAUDI ARMS SALE CRISIS

The Administration did, however, provide the U.S. Congress with formal notification that it was sending up an arms sale package for Saudi Arabia. The Reagan Administration had been maneuvering actively to try to bloc Congressional and pro-Israeli opposition to the sale for several months, and had faced severe opposition. In mid-September, Senators Alan Cranston and Robert Packwood had circulated a draft letter from the Congress to the President opposing the sale and claimed to have 64 signatures—enough to bloc Presidential approval—on September 29.

Several of the Senators involved attacked Saudi Arabia for failing to support the U.S. presence in the Gulf and for ailing to make any effort to seek peace with Israel. Israel's Prime Minister, Yitzhak Shamir, sharply attacked the sale in a speech on September 24, 1987, and by the end of September, some 217 members of the House had added their signatures to those of the 64 Senators.[69]

The Reagan Administration faced a difficult political situation because Saudi Arabia made it clear that it did not want the full extent of its cooperation with the U.S. to receive public attention, because it faced a potential Congressional challenge to both the sale and its failure to involve the War Powers Act, and because it had to act during October to give the Congress 50 days to act before it recessed for Christmas.[70] As a result, the Reagan Administration began to work out a compromise whereby it would drop the request for 1,600 AGM-65D Maverick missiles from the sale, and informally agree not to sell Saudi Arabia the F-15E at some later date.

Although the prompt Saudi response to the Iranian patrol boat advance on October 3 helped the Administration convince some Senators that aiding the Saudis was important, Carlucci concluded that there

Table 11.3: Targets in the Tanker War as of October 12, 1987

Year/Country	by Iran	by Iraq	Total
Total Attacks by Source			
1981	0	5	5
1982	0	22	22
1983	0	16	16
1984	53	18	71
1985	33	14	47
1986	66	45	101
1987	62	61	123
Target by National Flag of Ship Involved			
Australia	0	1	1
Bahamas	1	2	3
Belgium	1	0	1
China	1	0	1
Cyprus	9	33	43
FRG	1	4	5
France	5	0	5
Greece	10	22	32
India	4	4	8
Iran	0	48	48
Italy	1	1	2
Japan	9	0	9
Kuwait	11	0	11
Liberia	24	36	60
Malta	1	11	13
Netherlands	0	2	2
North Korea	0	1	1
Norway	4	1	5
Pakistan	2	0	2
Panama	18	28	46
Philippines	3	0	3
Qatar	2	1	3
Saudi Arabia	9	2	11
Singapore	1	5	6
South Korea	3	3	6
Spain	3	0	3
Sri Lanka	1	0	1
Turkey	2	8	10
UAE	1	0	1
U.S.	1	0	1
USSR	2	0	2
Yugoslavia	1	0	1
Unknown	2	42	44

Source: Adapted from the Washington Post, October 13, 1987, p.1 and from information provided by the Center for Defense.

was no way to win Congressional support for the full Reagan arms package—particularly when Senate Majority Leader Robert Byrd warned the President on October 7 not to sell Stingers to Bahrain. Accordingly, President Reagan reached a compromise with the Senate on October 8 that eliminated the Maverick missiles from the Saudi arms sale package. While some Senators still opposed the sale, both Byrd and Senate Minority leader Robert Dole stated that they now felt the package could be approved.[71]

This compromise was better than nothing, but it scarcely sent the kind of unambiguous signs of support the Saudis wanted—particularly since the Reagan Administration waited until Crown Prince Abdullah completed his visit to Washington and did not actually send the compromise arms deal to the Congress until October 29, 1987.

The end result was that the Reagan Administration's need to compromise again raised the political problem for the Kingdom that a U.S. President had been publicly forced to severely cut back on an arms sale because of pressure from Israel and pro-Israeli lobbying. At the same time, the sale was still of considerable importance, and Saudi Arabia needed virtually any sign of U.S. political support.

The Kingdom had just rebuffed an Iranian raid, and Iranian Guards had rocketed a Saudi tanker on October 7. Saudi Arabia was actively trying to build a coalition within the Arab League to impose sanctions on Iran at the League's meeting in Amman on November 8, and already faced considerable opposition from Algeria, Syria, and Libya. Indeed, it later became apparent that Qaddafi—who had sent a mission to Iraq in September saying he was shifting his support to Baghdad—sent another mission to Iran on October 17 and 18 to say he wished "the victory of the Muslim people of Iran against the conspiracy of world arrogance."[72]

Saudi Arabia was under pressure for other reasons. Although its oil production had climbed to over 4 MMBD, roughly 1 MMBD of production was tied to prior barter deals. Another 800,000 BPD went to domestic consumption, and more than 800,000 BPD went to refined product that was sold locally. While prices were higher, the dollar was also weaker. These factors were seriously affecting Saudi Arabia's oil revenues, and led it to shift more and more of its oil sales to the spot market in an effort to cease being OPEC's swing producer—a step that was putting pressure on Iran and the other more radical OPEC states and was leading to further hostility against Saudi Arabia. Further, Saudi Arabia had agreed to expand its military role to provide protection for Bahrain. It had agreed to defend the new 15.6 mile, $3 billion causeway between the two countries.[73]

IRANIAN SILKWORM STRIKES ON KUWAIT

On October 15, Iran again fired a Silkworm missile at Kuwait from Faw, 40 miles to the north. This time, however, it hit a U.S.-owned tanker with Liberian registry. The tanker was 275,932 ton *Sungari*, which was anchored in the Shuaiba Anchorage in Kuwaiti waters. The strike caused serious damage, but no injuries. It also showed that Kuwait had failed to reposition its Hawk defenses to Failaka Island, which was south of Bubiyan and provided an ideal location to defend Kuwait's port. The most Kuwaiti forces on Failaka could do was to fire a few ineffective short-range SA-7 and SA-14 manportable missiles. While they fired these at every Silkworm targeted at Kuwait after the first missile, none of the SA-7s and SA-14s proved effective in head-on attacks against a small cruise missile.

Western intelligence exports estimated that Iran had a total of 50 to 70 Silkworm missiles and 12 launchers. According to some reports, the missile attacks were carried out by the 26th (Salman) Missile Brigade which was located near Salman. The brigade and a newly formed 36th Assef Brigade were stationed in Faw under the command of Derakhshan—an aid of Rafsanjani. These units were affiliated with the Guards Corps 1st Naval Region under the command of an officer named Sotoodeh. Another brigade remained in the Sirrik region, about 50 kilometers south of the town of Minhab in Iran's southern coastal province of Hormozgan, on the eastern side of the Straits of Hormuz. The 36th Brigade had been stationed at Sirrik originally, but had moved through Shiraz by truck.

The U.S. did not react to this missile attack. It stated that the attack was against Kuwait, and not against a U.S.-flagged ship. This again reflected the decision by the Reagan Administration to avoid any commitment to defend Kuwaiti territory. It also reflected a decision not to protect U.S.-owned ships under foreign flags, which was reinforced by the knowledge that such action would meet with intense Congressional opposition. While American ship owners had about 40 supertankers under other flags, and were threatened by the fact that nearly one-third of all Iranian attacks on tankers had now occurred after August 29, they were in a poor position to ask for aid. They were paying less than 50% of the crew wages of U.S.-crewed vessels and no taxes.[74]

Most important, the U.S. did not want to create a further barrier to the UN cease-fire effort, and had previously been told by Kuwait that it would be responsible for any defense of its territory. The UN was then in the midst of trying to reach a compromise between Iran and Iraq over the sequence of events in Resolution 598 leading to a cease-fire. The resulting implementation plan for a cease-fire was now ten pages long, and had reached an almost algebraic complexity, but it did offer Iran some concessions by making the cease-fire concurrent with the "setting into motion" of a commission to decide who was to blame for the war.

Iran, however, proceeded to force the Reagan Administration's hand. On October 16, Iran fired another Silkworm missile into a tanker. This time, however, it hit a U.S.-flagged Kuwaiti tanker, the *Sea Island City*. The missile hit the tanker about seven miles east of the Mina al Ahmadi port and two miles south of the oil terminal at Kuwait's sea island. It blinded the American captain, wounded the American radio officer, and wounded 17 other members of the crew, 8 seriously. To make matters worse, the French minesweeping force found 4 more submerged mines in the lower Gulf that same day.

While it was uncertain that Iran had deliberately chosen a U.S. flag target for its Silkworm strike, the previous pattern of Iranian missile launches did indicate that the Iranians were communicating with spotters in Kuwait, and that the attack was deliberate. In any case, Khomeini sent his personal congratulations to the Guards for their "heroic deed" the next day—a statement that was far less ambiguous than Khamenei's earlier announcement of the strike and statement that "where the missile came from the Almighty knows best."

These Iranian statements came at a time when Secretary of State George Shultz was visiting King Fahd in Saudi Arabia and made it virtually impossible for the U.S. not to respond. After a brief pause, the U.S. stated it would retaliate because it was an attack on a U.S. flag vessel and the U.S. would protect such vessels anywhere in the world. On October 18, President Reagan announced that the U.S. had chosen its option and would act. The U.S. quietly held discussions with Kuwait and Saudi Arabia.

The Kuwaiti cabinet met on the Iranian strike, and officers were sent to Kuwait from USCENTCOM to help Kuwait improve its missile defenses. Anti-aircraft guns and light missiles were deployed to several Saudi oil platforms in the Gulf, and Saudi Arabia deployed attack helicopters at Dhaharan to protect against Iranian counterstrikes. U.S. gunships were put on alert and given surveillance support from three U.S. P-3Cs.[75]

U.S RETALIATION AGAINST IRAN

On October 19, the U.S. retaliated with a carefully limited attack. Its choice of target came only after prolonged debate and a full meeting of the NSPG under Deputy National Security Advisor, Colin L. Powell.

The U.S. had difficulty in attacking the Silkworm sites at Faw for several reasons. The Iranians were well aware of the vulnerability of their missiles. They normally kept them dispersed, set up the missiles at their sites at night, fired them near dawn, and immediately dispersed in camouflaged sites in the swamp terrain. The sites were also so far up the Gulf that trying to hit the missiles while they were at the launch sites presented operational problems for U.S. carrier aircraft. There was a risk that U.S. aircraft might fly more than 600 miles into an area defended by large numbers of land-based air defense weapons and hit nothing but the ramps and bunkers in the built-up site.

Further, the Iranian launch site was at the extreme southern tip of the Faw Peninsula. A U.S. strike on any target in Faw had the disadvantage that it would appear to aid Iraq. As a result, it was likely to do the most damage to the UN peace effort, and further undermine the neutral status of the U.S. It also risked making an open-ended commitment to Kuwait which the U.S. was not ready to make, and which required bases in Kuwait or Saudi Arabia that the U.S. did not have.

These factors help explain why the Reagan Administration decided on a visible show of force elsewhere in the Gulf. Four U.S. warships—the destroyers *Kidd, Leftwich, Young,* and *Hoel*—moved to within visible distance of Iran's Rostam offshore oil platform in the lower Gulf, about 120 miles east of Bahrain. They had air cover from two F-14s and an E-2C and were supported by the guided missile cruiser *Standley* and the frigate *Thatch*. This oil platform had produced about 18,000 barrels a day from the small Rashadat oil field, but oil production had halted about two years earlier. Further, Iraqi aircraft had heavily damaged the platform in November, 1986, and knocked out the bridge between its two main units. It was now being used by Iranian forces and these forces had fired on a U.S. helicopter on October 8.

The U.S. warships radioed to warn the naval Guards on the platform and allowed them to evacuate. They then fired some 1,065 rounds into the platform with their Mark-45 guns.[76] This gunfire set the platform afire, and SEAL teams then seized the two halves of the platform and blew it up. In the process, the SEAL teams saw Iranian troops evacuating from a second platform. They investigated, and found this platform was also being used as a base, and used explosives to largely destroy it. The SEALS who took the platforms found that both were armed and used for weapons storage, and were equipped with radars or other surveillance equipment which the Iranians used to track tankers and other ships in the Gulf.

The entire incident took about 85 minutes and the only Iranian response was to launch an F-4, which turned back the moment its radar acquired a picture of the size of the U.S. forces in the area. The same day, the U.S. started another convoy up the Gulf. It also announced that it had authorized nearly $1 billion in food purchase credits, which was a major rise in aid over the $680 million authorized the previous year, and which made Iraq the largest single recipient of such aid.[77]

While Secretary Weinberger ended his announcement of the U.S. attack by stating that "we now consider this matter closed", the U.S. attack scarcely acted as a definitive deterrent to further Iranian action and it left a number of U.S. planners unsatisfied. In fact, the Chairman of the Joint Chiefs, and other U.S. military planners, had strongly urged an attack on Farsi Island or on an Iranian combat ship. Virtually all U.S. policy makers and planners, however, wanted to avoid a level of escalation that would block any hope that the UN cease-fire proposal would succeed and commit the U.S. to a full-scale war at sea.

The public response to the attack in the U.S. strongly supported President Reagan's actions. Some 63% of those polled felt it was the right use of force, 59% felt the U.S. had the right amount of force in the area, 57% felt it was doing everything needed to avoid war, and 64% continued to favor the U.S. escort effort. Some 57% also, however, felt the response would encourage Iran to do more, 70% felt it would encourage more terrorist attacks, and 74% believed a major military conflict was now at least somewhat likely.

The attitude of the U.S. Congress was also favorable, but President Reagan had now used so much military force that the debate over the War Powers Act took on new meaning.[78] This helped push the Senate towards some form of action. The Senate was already scheduled to vote on a compromise bill the day after the attack, and the Senate voted 54 to 44 to support a bill drafted by Senate Majority Leader Robert C. Byrd and Senator John Warner, a ranking minority member.

This bill, however, scarcely tied the Reagan Administration's hands. It required the President to send a detailed report within 30 days on the range of U.S. military commitments in the Gulf. It then called for the Senate to vote 30 days after the report on an undefined resolution that could be anything from support for the President to opposition. While the resolution put some pressure on the President, it effectively deferred action on the War Powers Act.

The new bill also required support from the House and the President's signature, before it could become law. Congressman Jim Wright, the Speaker of the House, did indicate his support for the bill, but noted that he felt the House should defer action until it got a preliminary reaction from the courts to the suit that a number of Democratic members of the Congress had filed demanding that Reagan invoke the War Powers Act. This was due to be heard in a U.S. District Court on November 10. Wright's position meant a further delay in Congressional action, and the most tangible act the Congress took was for Claiborne Pell, the Chairman of the Senate Foreign Relations Committee, to begin hearings on how to eliminate the potentially unconstitutional aspects of the Act.

As for foreign reactions, the British, French, and West German governments sent messages of support, although the French statement expressed some fear of further escalation and the Belgian, Dutch, and Italian governments were silent. Most of the smaller Gulf states made cautious statements. Saudi Arabia and Jordan expressed their support, but Crown Prince Abdullah ibn Abdul Aziz of Saudi Arabia—who was visiting Washington—made it clear that he wished the U.S. had attacked the Silkworm sites as well. Kuwait issued a statement reiterating that it would not offer the U.S. bases, but that Iran had committed repeated aggressive acts and had been repeatedly warned to stop its attacks before the U.S. reaction.[79]

The USSR, however, continued to distance itself from the U.S. policy. Gorbachev stated that cooperation with the U.S. on the UN peace initiative was uncertain because the U.S. was acting "as it did of old". The Soviets kept their military presence in the Gulf down to one frigate and half a dozen minesweepers and supply ships. They called for restraint on all anti-ship attacks by both Iran and Iraq, and continued to demand the withdrawal of all foreign fleets and reliance on diplomatic solutions. At the same time, the USSR continued to expand relations with Iran and to sell arms to Iran as well as Iraq.

Iran's first response to the U.S. strike was to issue a mix of new threats against the U.S. and to briefly shell northern Kuwait. While the twelfth U.S. convoy up the Gulf left without incident, Iran accused the U.S. of causing $500 million in damage, and cutting Iran's oil production by 25,000 barrels a day. It also accused the U.S. of "launching a full scale war", and promised "decisive retaliatory action" and a "crushing blow". Iran held emergency consultations with Moscow, and the Soviet press denounced the attack as "military adventurism". Iran then shifted its declared policy towards the UN cease-fire proposal. It changed its position to demand that guilt for the war be established before a cease-fire.

Iran had clearly begun to realize that its actions earlier in the year had pointlessly alienated a number of useful foreign governments. It was particularly careful to increase its courting of the USSR. It suddenly dropped all attacks on the USSR from the usual Friday diatribes against foreign nations, and suddenly resumed Aeroflot flights to Tehran. At the same time, Velayati began a visit to Cuba and Nicaragua, and the Iranian Deputy Foreign Minister announced that he was seeking to improve relations with Britain and France.

Iran, however, was still committed to military action. On October 22, Iran fired another Silkworm missile at Kuwait's Sea Island—an oil-loading facility nine miles out in the Gulf from the Ahmadi oil complex. This third Iranian Silkworm strike again raised questions about Iranian targeting, since the Silkworm missile either had to be fired into the harbor area and allowed to home in on the largest target it could find, or had to be retargeted by some observer within line of sight of the target. This made it difficult to be certain whether Iran was choosing specific targets or simply firing into an area where it knew the missile would find a target. The ships normally near the Sea Island had now been moved away, however, and the island's 100 foot height made it by far the largest radar blip in the area. This made it easy for Iran to choose a target outside the direct coverage of the U.S. reflagging agreement.

The strike also had a much more significant effect than the previous strikes against tankers, because the missile fragmented when it hit the Sea Island and did extensive damage to one loading arm. The Sea Island terminal is a 2,500 foot long pier which was the only facility in water deep enough for Kuwait could to load supertankers of up to 500,000 tons. It normally provides some 33% of Kuwait's oil export flow and can handle a peak of 80%. It was then handling roughly 200,000 barrels per day of Kuwait's average production of about 600,000 barrels of crude oil. Kuwait's two other terminals were in shallower channels better suited to tankers of 150,000 tons.

Although the Sea Island terminal was not loading any tankers when it was hit, the oil lines to it were pressurized. The missile started a major oil fire in its overflow tanks and sent up a smoke column over 300 feet in height. The fire was out by early in the afternoon, but enough damage was done to force Kuwait to jury-rig alternative loading facilities and close the island. Kuwait's exports of roughly 600,000 barrels a day of petroleum products were not affected, but the Sea Island had to be closed until late November.[80]

It is unclear how carefully Iran calculated its response, but the U.S. did not respond militarily to the new Iranian attack. It again took the position that it had no obligation to defend Kuwait, and that Kuwait had refused U.S. protection and ship calls—although this U.S. position ignored the fact that a small U.S. mine warfare team and new advisory task forces were then in Kuwait.[81]

The U.S. did, however, quietly help Kuwait plan the redeployment of part of the its 12,000 man army to help secure the island of Bubiyan and prevent any quick Iranian amphibious attack that might seize this strategic location between the Faw Peninsula and the Kuwaiti mainland.

The U.S. speeded up an effort to assist Kuwait in setting up a number of barges with radar reflectors as decoys, in providing other countermeasures to the Silkworm, and in re-siting its Hawk missiles to Failaka to provide coverage against the Silkworm. In fact, the new Iranian strike on Kuwait led to a scramble throughout the rest of the Southern Gulf for air defense systems, as each GCC country sought to try to improve the defense of its own oil facilities. Bahrain asked the U.S. to speed up action on its request for 70 Stingers and 14 launchers, and the UAE quietly revived previous requests for the sale of Stingers and other U.S. air defense equipment.[82]

The U.S. also took added political and economic action against Iran. Secretary Weinberger announced that the U.S. might have to send more combat ships and expand its coverage of commercial traffic. President Reagan announced on October 26, 1987, that he had declared an embargo on all Iranian imports that would continue until Iran agreed to a cease-fire, and on 14 additional "militarily useful categories of U.S. exports—including mobile communications equipment, tractors, boats, diesel engines, electrical generators, hydrofoil vessels, and inboard and outboard motors. Mousavi replied to the U.S. embargo by calling the action, "mere cosmetic measures....in view of the events in the Gulf and the unprecedented nose dive of the stock market." Khamenei went further. He declared that "we have been threatened by an economic blockade by the Western states. We are not afraid of such things".[83]

Finally, the U.S. announced that it had decided to delay all transfer of military technology to the PRC, and was reexamining a $528 million arms sale of radar and navigation equipment for China's F-8 fighters. It sent Undersecretary of State Michael Armacost to Peking to try to persuade the PRC to stop further arms deliveries to Iran. The U.S. was not concerned with the fact that China had already sold at least 30 to 70 Silkworms to Iran and had provided considerable technical assistance to Iran in using the missile. It was reacting to the fact that the PRC had just sold Iran at least 35 to 50 more Silkworm missiles and a new shorter-range anti-ship system called the C-801, which was somewhat similar to the Exocet, and which Iran could launch from its smaller ships. The PRC had been continuing its arms sales both to earn hard currency and to try to counter the growing Soviet influence in Iran, and showed little immediate willingness to accept the U.S. view.

This mix of low-level U.S. actions does seem to have impressed both Iran and the PRC, although Iran took no public notice. In fact, a Chinese foreign ministry spokesman called the U.S. charges "groundless rumors" and blamed the U.S. for trying to shift responsibility for the escalation in the Gulf. Other Chinese spokesmen claimed that China was not making any additional arms sales, and that reports the PRC had sold

Iran anything like $1 billion in arms over the past few years were totally false.[84]

The reaction of the Arab world to the latest Iranian missile strike was as divided as ever. Iraq demanded an "honorable Arab stand" that would help force Iran into a cease-fire. Kuwait, announced the creation of a 300 man civil defense force on October 27, 1987, and that it was registering three of its shuttle tankers under the British flag. Rumors also surfaced that Kuwait was asking for Egyptian military aid, and had already signed a $20 billion contract to obtain Egyptian advisors and 70 pilots. The leading Southern Gulf and moderate Arab states protested to the USSR that it was blocking progress towards a cease-fire. The Foreign Ministers of the six GCC states met in Riyadh on October 25, and warned Iran against further attacks on Kuwait. They said that an attack on Kuwait would be treated as an attack on the other countries. At the same time, they stopped short of any action that would take the form of sanctions.[85]

A SHIFT TO LOWER LEVELS OF CONFLICT

During the rest of October, Iran was careful to avoid direct provocation of the U.S. and the thirteenth U.S. convoy through the Gulf arrived in Kuwait without incident. It seemed to be waiting to see if the U.S. would take the initiative, and suspended most of its naval activity in the Gulf. Iran did, however, claim on October 24 that three of its fighters had flown within 10 miles of a U.S. warship in spite of warnings not to do so (Table 11.4).

Iran continued to use terrorism. A Pan American World Airways office in Kuwait was bombed on October 24. Pro-Iranian groups in Lebanon—such as the Islamic Jihad—revived the threat of terrorism during late October. They exhibited photos of the damage to the U.S. Embassy and Marine Corps barracks in Beirut, showed new films of the hostages they still kept, threatened to take new hostages, and threatened to send suicide volunteers to the Gulf.

Iran also supported at least three successful assassination attempts against anti-Khomeini Iranians living in Britain, and stepped up its efforts to suppress such opposition in other countries. Pro-Iranian elements also exploded a bomb under a police truck near the Kuwaiti Interior Ministry on November 3, although this may have been designed to act as a warning to Kuwait that it was still vulnerable rather than being intended as a serious attack.[86]

Iraq, in turn, continued to fight its tanker and oil wars. It used its air

Table 11.4: The Pattern of Attacks in the Tanker War, 1984-1987

Year	Air Launched Systems Missiles	Rockets	Bombs	Helicopter Launched Missiles	Missiles from Ships	Rockets/ Grenades Gunfire from Ships	Mins	Unknown	Total Attacks
1984									
Iraq	35	–	–	–	–	–	2	16	53
Iran	(18)–	–	–	–	–	–	–	18	
Total	52	–	–	–	–	–	–	16	71
1985									
Iraq	32	–	1	–	–	–	–	–	33
Iran	(10)–	–	–	3	–	–	–	1	14
Total	(42)–	–	–	3	–	–	–	1	47
1986									
Iraq	52	4	1	1	–	–	–	8	66
Iran	(9)	–	–	26	4	1	–	5	45
Total	(65)	–	1	27	4	1	–	13	110
1987 (to October 12, 1987)									
Iraq	57	–	3	–	–	–	–	2	62
Iran	–	–	–	1	14	34	8	5	62
Total	57	–	3	1	14	34	8	7	124
Total: 1984 to 1987									
Iraq	176	4	5	1	–	–	2	26	214
Iran	(37)	–	–	30	18	35	8	11	139
Total	(217)	–	5	31	18	35	10	37	353

Source: Adapted from Bruce McCartan, "The Tanker War", *Armed Forces Journal*, November, 1987, pp. 74–76, and from reporting by Lloyd's and Exxon.

power to demonstrate that it would not accept any form of partial cease-fire that affected only the tanker war. At the same time, it pursued a shift in the air strategy that it had begun in late August. It attempted to hit Iran's refinery system and power plants with sufficient force to make it difficult for Iran to provide fuel and power during the winter. This strategy had the potential advantage that it could cripple Iran's economy more quickly than an embargo on oil exports, and that Iraq could reduce the number of its attacks in the Gulf and the resulting pressure from other states to halt them.

During late October, Iraqi jets hit the Agha Jari oil field in southwest Iran, a refinery in Shiraz, and Iran's shuttle tankers. While Iraq also claimed a hit on a supertanker, it seems to have hit a hulk that was moored near the tankers as a decoy. Iraq claimed that these strikes were in response for the 134 killed and 2,036 wounded that resulted from the four Scud strikes Iran had launched against Baghdad in October.

Iran responded to Iraq's attacks by charging that Iraq was bombing civilian targets, and Iranian radio warned that the rulers of Baghdad should anticipate the "deadly response of the combatants of Islam as long as they continue their wicked acts". It warned all Iraqi civilians living near economic and military targets to abandon their homes, and to seek refuge in the holy cities of Najaf, Karbala, Kadhimain, and Samarra. Iran, however, could carry out only a few symbolic air raids and fire occasional Scud missiles. For example, Iran fired a total of five Scud missiles at Baghdad between October 4 and October 31. While each attack produced casualties, the volume of fire was so sporadic and the targets were so random that they almost unquestionably did more to increase popular hostility to Iran than to deter Iraq's use of its air power.[87]

Iraq had effectiveness problems of its own. It began to make an increasing number of claims to have damaged Iranian ships that later proved false. For example, Iraq claimed to have hit three Iranian tankers on October 28, 1987. In reality, however, no successful Iraqi attack on shipping took place between October 21 and November 4, when Iraqi jets hit the *Taftan*, a 290,000 dwt VLCC which was part of the National Iranian Tanker Company's shuttle fleet and which was loading at Kharg.[88]

Iran did not attack any ship moving in the Gulf between October 21 and November 6, 1987. It then, however, resumed its pattern of choosing the kind of indirect and peripheral targets in the Gulf which could embarrass the West, but which would not provide a major Western response. It started to hit ships that were not flying U.S., British, French, Dutch, Italian, or Belgian flags. Iran seemed to be shifting to a lower-level war of attrition with the West in which it hoped that it could keep up constant low-level pressure on Gulf shipping while forcing the U.S. and its allies to maintain massive military deployments that they would lack a reason to use.

On November 6, 1987, Iran's naval Pasdaran forces carried out a speed boat attack from Abu Musa that fired rocket-propelled grenades at a U.S. operated tanker. This ship was the 105,484 dwt *Grand Wisdom*, which was sailing about 20 miles west of the main UAE port of Jebel Ali.[89] This attack came without warning and was the first attack against a foreign tanker since one of Iran's Silkworm missiles hit the *Sea Island City* on October 16.

The Iranian attack on the *Grand Wisdom* seems to have involved the careful choice of a target which would affect the U.S., but which would avoid the consequences of attacking a ship flying under the U.S. flag. Further, it seems to have been carefully timed to embarrass the U.S. The Iranians hit the *Grand Wisdom* when it was near the guided missile cruiser *U.S.S. Rentz*. The *Rentz* had an attack helicopter on board, but could take no action other than to shadow the crippled *Grand Wisdom* as it sailed back to Jebel Ali.[90]

Both sides kept up their attacks on shipping during the second week in November, and Iraq continued to claim far more hits than it actually scored. Iraq claimed it hit eleven ships during this period, but seems to have hit no more than three.

Iran, in turn, continued its pattern of cautious defiance of the U.S. On November 11, two Iranian gunboats hit a Japanese tanker. It was then sailing within 15 miles of the 17th U.S. convoy through the Gulf, and was near a French warship escorting two French tankers. The gunboats also struck at a point when the *U.S.S. Missouri* and the cruiser *Bunker Hill* had entered the Gulf for the first time in order to escort the 12 ship U.S. convoy as it sailed through the "Silkworm envelope" at the Straits. A Soviet merchant ship reported an unconfirmed sighting of mines in the same general area. The Iranians had again showed that the Western rules of engagement could not protect freedom of navigation in the Gulf, although the U.S. did quietly add Bahrain-flagged tankers to one of its convoys for the first time.[91]

Further, Iraqi and Iranian attacks occurred during November 12–15, 1987, although it is important to note that each side hit less than 1 ship per day. Iraq, for example, claimed to have hit 15 ships between November 9 and November 15, but damaged only 3. Both sides also attacked civilian targets. Iran bombarded Basra and Iraq bombed the district capital of Kamyaran, about 50 miles east of the border. Iraq kept up its attacks on Iran's oil facilities and claimed to have attacked three Iranian oil fields—at Abed al-Khan, Marun, and Kaj Saran—on November 14, 1987.

On November 16, however, Iranian speedboats attacked three tankers in one day. They hit the U.S.-managed Liberian tanker *Lucy* near the Straits of Hormuz. They also hit a U.S.-owned ship under the Bahamanian flag, the *Esso Freeport*. It marked the first time in the war that Iran had hit a ship owned directly by a major U.S. oil company like Exxon. Finally, they hit a small Greek-owned Tanker, the *Filikon L.*

The air and sea wars continued to follow a similar pattern for the rest of the month. Iraq claimed Iran had attacked a hospital on November 19, and two major generating complexes at the Reza Shah and El-Diz dams on November 29. Both Iraq and Iran hit a ship nearly every three to four days, although Iraq claimed a total of 21 successful attacks during the 12 days between November 8 and November 20, but actually scored only 4.[92] U.S. minesweepers found 13 mines in the waters near Farsi island during November 20–25, and U.K. mine forces found 5 more mines northeast of Bahrain—although it was unclear whether these were new or old Iranian mine deployments.

Iran hit the Kuwaiti flag tanker *Umm al Jathatheel* on November 26. The attack was interesting only because Kuwait had painted out the Kuwaiti name on the ship, painted in the name *Dacia*, and was flying the Rumanian flag. Iraq hit only two ships between November 20 and November 31: These were the VLCCs *Stilikon* and *Khark*, both of which were part of the Kharg-Lavan shuttle. As for bombing, Iraq continued to strike at Iran's power plants, and Iran claimed to have flown some attack sorties against Iraqi petrochemical facilities.

Iran's attacks on shipping during November affected only a small proportion of the shipping to and from Kuwait. Nevertheless, they did have some effect. Insurance rates continued to rise, and Kuwait decided on November 26, to keep the now repaired *Bridgeport* out of its tanker shuttle. The 401,382 ton ship was the only one of the eleven reflagged vessels that carried crude oil; the others carried gasoline, LNG, and refined products. This meant that U.S. convoys would escort only smaller tankers carrying product, and that the only ship carrying crude oil was dropped from the convoy system.

Even though Iran did not hit U.S. ships, it succeeded in forcing the U.S. convoy effort to lag far behind schedule. By the end of November, the U.S. had completed only 18 one-way convoys, or less than 5 a month. This compared with an original goal of 10 per month.

The U.S. still faced a mine warfare problem. MSOs deployed from the U.S. joined the MSCs and helicopters already in the Gulf in early November, and became a fully operational mine-sweeping unit by mid-November. British, Dutch, and Italian units also operated independently in the area.[93] Nevertheless, Iran still had the remnants of 3 mine fields with 60 mines in the Gulf. While many of the mines had been swept, and Iranian minelaying activity had halted or been sharply curtailed, there were reports that Iran was building new mines and might be getting more modern-influence mines from Libya.[94]

The only direct military incident between the U.S. and Iran during this period occurred on November 22, when Iran claimed that it had fired on four U.S. helicopters. This latter claim was made by Commodore Mohammed Hussein Malekzadegan, but seems to have been little more than part of a propaganda campaign in which Iran was using the U.S. threat to try to boost internal morale. It occurred a day after Hussein Alaie, a commander of the naval Guards, announced that no political solution to the U.S. presence in the Gulf was possible and that Iran had drawn up plans to destroy the U.S. fleet.[95]

The U.S. did, however, accidently fire at a fishing boat from the UAE. The frigate *U.S.S. Carr* mistook a fishing boat from Sharjah—the *Al Hudei*—that was heading towards the frigate for one of the Pasdaran's Boston Whalers. The *Carr* fired 0.50 caliber machine guns at the fishing boat, damaging it and killing one of the crewmen and wounding others.[96] The *Carr* was new to the Gulf and may have overreacted. Virtually all the U.S., European, and Russian warships in the Gulf, however, had reached a state of alert where they immediately locked their fire control radars, guns, and missiles on any Iranian aircraft or warship that came near. French, Italian, and Russian ships had all been in incidents where they forced Iranian warships to turn away by locking on their radars and threatening to fire.[97]

THE DIPLOMATIC FRONT

Little, if any, progress in reaching a cease-fire took place between October and the end of December, 1987. The major diplomatic developments consisted of a complex series of skirmishes that largely reaffirmed the past positions of the various players and continued to escalate the war.

The first of these skirmishes occurred over the UN peace effort. Russia continued to tilt towards Iran while pursuing a separate peace policy. The USSR announced at the end of October that Soviet officials would visit Baghdad and Tehran in an effort to reach a cease-fire. As for the Soviet naval presence in the Gulf, the USSR continued to keep five ships in the area, but they stuck carefully to self-defense roles. The three Nataya-class Russian minesweepers did not carry out any mining activity between their arrival in August and the beginning of December. The two Russian frigates, and the intelligence gathering ship, the USSR had sent to the Gulf, spent as much time observing the Western forces in the Gulf as those of Iran.

While the USSR continued to "convoy" three tankers chartered by Kuwait, the tankers were long-haul shippers rather than shuttle vessels, and made only a few appearances each month. The only significant military encounter between Iran and the USSR during the fall and summer of 1987 occurred on November 10. An Iranian warship used its radar to illuminate a Soviet warship, and turned away when it was warned not to do so. Iran turned a blind eye to the Soviet freighters moving through the Gulf, including those going to Kuwait, even though the ones carrying arms to Iraq were easy to detect and were all tied up at one special pier.[98]

The second series of skirmishes consisted of diplomatic exchanges between Iran and the U.S. On November 1, Rafsanjani made Iran's harshest comments to date about the U.S. role in the Gulf and the UN peace effort. He accused the UN of "cheating", of having "ill intentions", and of favoring Iraq. He described the U.S. actions in the Gulf as "ugly", and went on to state that "the work on international forums will come after a decisive blow on the battlefield." On November 4, Iran celebrated the seizure of the U.S. Embassy with a series of rallies it declared were in celebration of "death to America" day.

Iran then informed the Secretary General of the UN that its terms for a cease-fire now included the redefinition of the border with Iraq—since the Algiers Accord was no longer valid—and that Iraq would have to pay reparations before any withdrawal. These terms were in addition to Iran's demand that the cease-fire resolution be modified to delay a formal cease-fire until the UN reported on its findings relative to war guilt, although Iran indicated it would accept an informal cease-fire on the day the UN commission began its work.[99]

Iran also rebuffed a new series of U.S. efforts to start face-to-face talks. Although the U.S. had been seeking direct contacts since May, Iran set conditions the U.S. found unacceptable. These included the delivery of arms purchased by the Shah without charging storage costs, and a number of other conditions. Both Khamenei and Deputy Foreign Minister Mohammed Javad Larihani refused to meet with U.S. officials on any basis during their visits to the UN. Iran also refused to discuss the situation directly with the U.S. without prior U.S. concessions after the mining of the *Bridgeton*, the U.S. attack on Iran's minelayers, and the U.S. attack on the Rostam oil platforms.[100]

U.S. EFFORTS TO MOVE TOWARD AN ARMS EMBARGO

This left the U.S. with the alternatives of expanding its military presence, seeking immediate UN Security Council action on sanctions, or waiting and trying to mobilize added support for military and economic sanctions on a bilateral basis. The U.S. chose the latter option, but had little success. The PRC continued to deny it was selling arms to Iran while it continued to accept new orders and made deliveries. Israel made it clear that it would not support the U.S. in its alignment with Iraq. Israeli Defense Minister Yitzhak Rabin warned that the U.S. had allowed itself to be manipulated by Iraq, and that Israel would continue to try to build relations with Iranian moderates.

Further, the Israeli government stated that it was not delivering arms to Iran, but would not interfere which any private Israeli sellers and agents. These declarations occurred at a time when reports surfaced that Israelis had negotiated through a third party in Geneva to sell up to $750 million worth of arms to Iran during the previous summer, including TOW missiles, Gabriel air-to-surface missiles, F-4 and F-5 aircraft engine parts, tanks, and jeeps.[101]

The U.S. had no better luck in its effort to get other nations to reduce their trade. On October 31, Japan told the U.S. it would not support any economic sanctions against Iran, or do anything more than ask its importers to keep their oil imports at current levels—or about 7% of Japan's oil. This concession was meaningless, since Japanese importers were then arguing with Iran over the fact it was not offering the same discounts it had offered Europe, and Iran was shifting oil to the European spot market. Japan's role in the sanction was critical because Iran had been Japan's fifth largest supplier in 1986, had imported $1.1 billion worth of oil, and had supplied $1.4 billion worth of goods including ships and machinery. These exports included many of the 14 categories of civil goods the U.S. was attempting to restrict because of their military value, including mobile communications gear, marine engines, boats, navigation equipment, and submersible items.

The most Japan would do was to repeat its commitment to providing a $10 million navigation system to make it easier for commercial ships in the Gulf to avoid the war zones, and to agree to increase its offset payment for the 50,000 U.S. troops stationed in Japan. Japan's new Prime Minister Noboru Takeshita, who replaced Yasuhiro Nakasone on November 6, also made it clear he would be much more cautious in expanding Japan's national security activity than his predecessor had been.[102] Japan took this position although it obtained 60% of its energy from oil and 55% of that oil from the Gulf. It also normally had about 20

Japanese-owned ships in the Gulf at any given time, and 7 Japanese ships had already been attacked in 1987. [103]

THE ARAB LEAGUE MEETING IN AMMAN

Another set of political skirmishes took place at the Arab League summit meeting on November 8. The meeting took place in Amman, and included all of the key heads of state except Qaddafi and King Fahd. By the time the meeting took place, however, it was already a partial failure. The Saudis had initially sought to pressure Syria into agreeing to, three things: Sanctions against Iran, the return of Egypt to the Arab League and restoration of diplomatic relations. The Saudis hoped to succeed because Syria's economic situation was continuing to degenerate from bad to worse, and Assad desperately needed Saudi and Kuwait economic aid. Saudi aid was particularly important because Kuwait had sharply reduced its aid and the annual Saudi payment of $540 million was critical. Further, Saudi Arabia provided an important wheat subsidy and Syria had little hard currency to make up for a harvest that fell some 1 million tons short of domestic consumption.

In practice, however, the Saudis could only get a Syrian agreement before the meeting to drop a proposed initiative to demand that the U.S. leave the Gulf, to meet with Saddam Hussein, and to condemn Iran for failing to accept UN resolution 598, for its attacks on Kuwait, and for the riots at Mecca. While these agreements were important in political terms, they did not involve any tangible action to end the war. As a result, King Fahd did not attend the meeting, and sent Crown Prince Abdullah—the member of the royal family who had the best personal relations with Assad. This allowed the King to avoid being pressed to make new commitments to provide aid, while it gave Abdullah the opportunity to persuade Assad to compromise.

Abdullah seems to have been usually successful in doing so, at least in part because of the skillful maneuvering of King Hussein of Jordan. The three day summit meeting did not produce major changes in Assad's position, but it did lead to a meeting between Assad and Saddam Hussein that went beyond mere cosmetics. The two leaders met for five hours, and produced at least some signs that the two leaders might reduce the intensity of their long-standing feud. Although both leaders refused to shake hands for pictures after the meeting, they agreed that their foreign ministers would meet in Jordan and negotiate on more specific arrangements to improve their relations.

Syria did not abandon its alliance with Iraq, but it joined the other states in voicing their "indignation at the Iranian regime's intransigence, provocations, and threats to Gulf states". It also supported the condemnation of Iran's "bloody criminal acts" at Mecca in July. Syria was joined by Algeria and the PDRY, and this left Libya as the only state that did not support the demand for a peace based on the original UN peace resolution.

Further, Syria made some partial concessions towards the recognition of Egypt. Syria refused to support such an effort on the grounds that there was no way that the Arab states could"...consider Egypt's readmission to the Arab League when the Israeli flag waves in Cairo from the Israeli embassy...." It did agree, however, not to oppose the resumption of diplomatic relations by individual Arab states, and this led Bahrain, Iraq, Kuwait, Morocco, and the UAE to resume diplomatic relations between November 12 and 14.[104]

The Arab League meeting also eased the problems the Reagan Administration had in gaining full Congressional support for a compromise arms package for Saudi Arabia. It was clear that Saudi Arabia had joined Jordan in leading the effort to restore full Arab recognition of Egypt, and had strongly supported U.S. policy in the Gulf. This helped the Administration when it finally provided formal notification to the Congress on November 22, 1987.

The Arab League meeting did nothing, however, to influence Iranian behavior. On the first day of the meeting, Iran fired a Scud missile at Baghdad. Then, on November 10, Khomeini made a rare appearance to celebrate the birth of Mohammed, and declared that "He made war so we should make war". Mousavi rejected the Arab League's call for a cease-fire more formally on November 12, denounced it as having been dominated by the U.S., and denounced any Arab rapprochement with Egypt. That same day, the Iranian Foreign Ministry stated that "the Arab leaders were in line with the aggressive policy of the United States".[105]

SAUDI PROBLEMS WITH PAKISTAN

Recognition of Egypt was of growing importance for several reasons. Egypt remained the largest military power in the Arab world, and was the only Arab state that could provide major military support, and troops, to Iraq or Kuwait. It also benefited Saudi Arabia, which was encountering serious problems in its relations with Pakistan—which provided a substantial part of the forces in estern Saudi Arabia.

Pakistan had taken a neutral position towards the Iran-Iraq War since its beginning, although it tilted towards Iran. It tended to side with Iran for several reasons: trade, a common border, a common policy towards the Soviet invasion of Afghanistan, and the fear that Khomeini might try to increase the tensions between Pakistan's Shi'ites and Sunnis. Since roughly 10–15% of Pakistan's population was Shi'ite, and much of this population was in the border area and heavily influenced by Islamic fundamentalism, Pakistan had to exercise considerable care.

By late 1987, however, Pakistan's position was becoming considerable more difficult. It faced the potential suspension of U.S. aid because of Pakistan's continuing effort to develop nuclear weapons. This affected some $4 billion in aid over the next six years, and gave Pakistan little reason to support the Western position towards Iran. Pakistan was experiencing a trade gap of nearly $2 billion, and growing internal political and economic problems—including Shi'ite riots. Further, the strain of housing several million Afghan refugees, and Soviet and Afghan government raids on Pakistani territory in retaliation for Pakistan's support of the Mujahideen, had led Pakistan to explore an improvement of its relations with the USSR.[106]

This mix of pressures led to a crisis in Pakistan's relations with Saudi Arabia. Pakistan had kept a 10,000 man force in Saudi Arabia, including a tank brigade at Tabuk, since the early 1980s. This unit served to train Saudi forces and was intended to act as a shield in the event of any Israeli attack—or attack by Syria or Iraq. The Pakistani troops, however, included a strong Shi'ite element, and their position became more uncertain once Iran became a serious threat to Kuwait. Beginning with the Iranian seizure of Faw in 1986, Pakistan increasingly faced the possibility that Saudi Arabia might want to use the Pakistani forces to support Iraq or Kuwait, or that it might face an Iranian attack on Saudi Arabia. This risk grew even more serious with the Iranian attack on Basra. It continued to increase after the riots in Mecca on July 31, and as Iranian and Kuwaiti tensions grew throughout the summer and fall of 1987.

As part of Saudi Arabia's planning for the Arab summit meeting, Saudi Arabia asked the government of Pakistan if it would expand the terms of its contract to include the potential need to intervene in an Iranian attack on Kuwait and Saudi Arabia, and would remove the Shi'ite element in its forces. The Saudi timing was dictated by a mix of the growing tensions in the region, its improving relations with Egypt, and the fact that the contract with Pakistan was due to be renewed in December. Pakistan, however, had just experienced a series of attacks by Iranian covert action forces on Iranian dissidents in Karachi and Quetta, and had good reason to fear major uprisings in Islamabad if it took a position against Iran. Further, the Saudis explored the idea of Pakistan coming directly to the aid of Iraq. This led Pakistan to refuse to change the terms of the contract, and Saudi Arabia replied by cancelling the contract.[107]

This refusal deprived the Saudi Army of some 10,000 men, and of one of its armored brigades in western Saudi Arabia. Further, Saudi Arabia's effort to find similar help from Bangladesh and Morocco quickly revealed that few countries could provide the kind of modern and well-trained troops the Kingdom needed. Given the fact that Saudi Arabia could man only one brigade of its own—which it split between Tabuk and Hafr al-Batin—this seriously cut the number of the combat forces available to the Southern Gulf states in the event that Iraq should lose Basra or Iran directly threatened Kuwait.

It was scarcely coincidental, therefore, that Saudi Foreign Minister Saud bin-Faisal, the Chief of Staff of the UAE Air Force Sheik Mohammad Bin Zayed al Nahyan, Egyptian Foreign Minister Esmat Abdel Meguid, Egyptian Defense Minister Abu Ghazala, and Iraqi Minister of State for Defense Abdeljabber Chenchall met in Cairo on November 11. During the four days that followed, Egypt made it clear that it did not want to send an air mobile brigade or any other combat troops to the Gulf. It did agree, however, to provide additional military advisors and pilots. Agreement was also reached to try to revive the Arab Organization for Industrialization, which had been dormant since Camp David. Iraq expressed a particular interest in buying Alphajet trainers, more Tucano trainers, and Egyptian-built Gazelle anti-tank helicopters.[108]

THE U.S DEBATE OVER STINGER SALES TO BAHRAIN

At the end of November, the U.S. became involved in another arms sale controversy. Iran's Ambassador to the UAE announced on November 22 that Iran had enough Stinger missiles to trigger a major scandal in the U.S.that would be even bigger than the Iran-Contra scandal and which would "put the President's future at stake, and topple heads in the White House". He claimed that Iran had far more Stinger missiles than the 25 it had seized from the Afghan Mujahideen and that Iran would make the details clear at the time when they would be most embarrassing to the U.S.

This Iranian announcement came at a time when pro-Israeli groups were pushing hard for Congressional legislation to block the sale of Stinger missiles to any nations other than formal allies of the U.S. The legislation was sponsored by Mel Levine and the House passed it by 322 votes to 93 on November 18, 1987.

A similar bill, sponsored by Senator Dennis Deconcini, passed in the Senate Foreign Operations Subcommittee on December 3. By this time, the White House had made it clear that Bahrain was fully supporting the U.S. presence in the Gulf with "vital logistic and reconnaissance" facilities. The sale had also been supported by both the chairman and ranking member of the opposition of the Subcommittee, Senate Majority leader Robert Dole, the Defense and State Departments, and the Chairman of the Joint Chiefs.[109]

These actions threatened to block the Reagan Administration's plans to sell 14 Stinger launchers and 60–70 missiles to Bahrain, and any other GCC state, although the $7 million cost of the sale would normally have meant it did not require Congressional approval. It forced the Administration to delay a request for Bahrain until after it was sure it could obtain Congressional approval of the Saudi arms package. Like most such debates in the U.S. Congress, however, the real issue remained the strengthening of U.S. ties with the Arab world.

The Stingers in question were the early version of the Stinger, not the improved version or the far more advanced Stinger POST. By this time, the USSR had sold well over 17,000 manportable or light crewed weapons like the SA-7, SA-13, and SA-14 to the Third World, and nearly 70% had gone to the Middle East and South Asia.

The SA-14 Gremlin was, in fact, a near copy of the same model of the Stinger. The USSR had obtained a complete microfiche of the plans and technical details for the missile from a Greek electronics expert, Michael Megaleconomou, in 1984. He and a Greek naval officer in Athens had also provided enough technical data for the USSR to make major improvements in the countermeasures inits fighters and helicopters used in Afghanistan. Nevertheless, the Senate included a ban on the sale of the Stinger in the foreign aid bill on December 1, 1987.[110]

FRANCE'S HOSTAGE DEAL WITH IRAN

The Western position in the Gulf also suffered in late 1987 because of a French deal with Iran in which France traded an end to the siege of the Iranian Embassy in Paris, and a major loan repayment to Iran, for Iran's freeing of several French hostages.

The first hint that France was secretly dealing with Iran came when two French magazines —*L'Express* and *Le Point*—printed reports on arms sales to Iran under the previous French government and created a major scandal. The magazines leaked portions of a classified report by General Jean-Francois Barba, the controller general of the French Army, which strongly indicated that former French Defense Minister Charles Hernu knew that Luchaire had sold 450,000 105 mm, 155 mm, 175 mm, and 203 mm artillery shells to Iran for $115 million.

Further, the reports indicated that President Mitterand might also have known that the sale involved at least $18 million in kickbacks; that Luchaire's pro-Socialist head—Daniel Dewavrin—might have provided 3–5% of the funds received from the sale to the French Socialist Party; and that Luchaire had sold Iran 1.2 tons of C-4 plastic explosive. This explosive was the same material used by a number of pro-Iranian terrorists, including the groups that had set off explosives in France. While all the Socialists involved denied the reports, the denials were far from convincing.[111] Nevertheless, the conservative Chirac government downplayed the Socialist scandal.

Part of the reason for this surprising reticence became clear on November 29, when France reached an arrangement with Iran in which it agreed to pay $330 million in debt payments to Iran. It also allowed Wahid Gordji to return to Iran after a two hour interrogation, and led to the expulsion of the key representatives of the People's Mujahideen still living in France. In return, Iran agreed to the release of Paul Torri, the French consul in Iran, and two French hostages held in Beirut, but the Hezbollah still retained three French hostages.

The freeing of Gordji involved significant government pressure on the French judiciary. French intelligence groups had found what they regarded as conclusive evidence that Gordji had been responsible for the bombing incidents that had killed 12 and wounded nearly 150. It also led to a considerable amount of hostile public opinion in France, U.S. complaints, and blunt condemnation from Britain's Prime Minister Margret Thatcher.[112]

Nevertheless, France did not back away from its commitment to deploy forces in the Gulf, and it actually increased its coordination with the other European force in the region. Further, France continued to support the U.S. in seeking a UN arms embargo.

In January, 1988, the commander of France's forces in the Gulf went further, and took the hardest line yet towards Iranian attacks on Gulf shipping of any Western power. Rear Admiral Guy Labourie called for all Western ships to come to the aid of any ship that Iran attacked or that was in distress. Labourie took this position after a French frigate, the *Dupleix*, which was escorting a French minesweeper through the Gulf, warned off three Guard gunboats attacking a Liberian tanker. This show of force seems to have been designed as a signal to Iran that the hostage deal would not mean French appeasement.[113]

THE TANKER WAR CONTINUES IN DECEMBER AND EARLY 1988

The military action in December, 1987, concentrated on the tanker war. Iraq hit occasional land targets, including Iran's dams and refineries, and continued to strike at the Iranian tanker shuttle. Iraq hit the tanker *Anax* on December 2, and hit the VLCC *Actina* twice on December 4. Iran did not attack any land targets, and did not hit any ships between November 26 and December 6, but Pasdaran gunboats then hit the Danish tanker *Estelle Maersk*, and crippled the Singapore flag tanker *OBO Norman Atlantic*.

Both sides then continued to step up their attacks on cargo ships for the rest of the month. As a result, roughly 60% of all the attacks on shipping during 1987 took place in the last four months of the year and after the failure of the temporary UN cease-fire on August 29. These attacks reached their peak in December. Twenty-one attacks occurred in November, but 34 occurred in December.[114]

Iran increasingly targeted its strikes against Kuwait and Saudi Arabia. Seventy-three of the 80 attacks Iran conducted on Gulf shipping in 1987 were on ships en route to and from Kuwait and Saudi Arabia. Iran averaged 3–6 attacks on shipping en route to and from Kuwait during each month of the year, and steadily increased its attacks en route to and from Saudi Arabia at the end of the year. The cumulative total of Iranian attacks on shipping to Saudi Arabia was only 1 in February, 3 through March and April, and 7 through June and August. They reached 16 in September, 22 in October, 26 in November, and 38 in December.[115]

These attacks did not affect oil flows from Saudi Arabia, but they did briefly threaten crude oil exports from Kuwait. The U.S. convoy effort provided security for Kuwait's exports of product and liquid natural gas, but most crude oil shipments operated outside the convoys. The missile attack on the *Sea Island City* on October 17 led some tanker companies to be cautious about loading Kuwaiti crude, however, and Iran seems to have tried to capitalize on this situation to try to reduce Kuwait's crude exports.

The attacks also forced tanker captains in the Gulf to try measures like joining the tail end of Western convoys, maintaining radio silence, and even simulating convoys on radar by moving in convoy-like lines at night. These measures may have helped reassure the tanker captains and crews, but they had little real impact on vulnerability because of the constant Iran patrols in the Gulf.

The Iranian attacks peaked shortly before Christmas. While the Guards conducted most attacks, the regular Iranian Navy sent out its frigates, and Iranian forces carried out three major attacks on crude oil tankers en route to Kuwait between December 18 and 23. Iranian frigates shelled a Norwegian tanker on the 18th and a Liberian tanker on the 22nd. The next day, naval Guards forces attacked another Norwegian tanker. These attacks did not injure the crews, but they did make all three ships unfit to carry crude oil. The attacks were serious enough to force Kuwait to renew its use of the *Bridgeton* to ship crude oil, and to include it in the U.S. convoy effort.

The exact sequence of events which halted this escalation is uncertain, but the U.S. seems to have quietly threatened Iran, and Syria seems to have put pressure on Iran to halt its attacks. Iraq also carried out an unusual long-range air strike on the Iranian tanker shuttle near Larak on December 22, and showed it could hit four of the supertankers which had been deployed there as oil storage ships. The 564,739 ton *Seawise Giant*, the world's largest tanker, was one of Iraq's targets.[116]

Regardless of the cause, Iran quickly reduced the frequency and lethality of its attacks on Kuwait's crude oil carriers. Iran then carefully avoided Western-escorted ships throughout the rest of the winter of 1987/1988, and rapidly reduced the intensity of its attacks. As a result, Kuwait was able to increase its crude oil exports from around 970,000 barrels per day during the first two quarters of the year to 1.2 million barrels per day during most of the winter.[117] Further, Lloyd's reduced its war risk premium from 0.75% to 0.45% in February, 1988.[118]

As for the overall pattern of the tanker war, there were a total of 80 Iranian attacks and 83 Iraqi attacks on shipping during 1987. These attacks raised the total number of attacks since 1984 to 180 Iranian attacks and 215 Iraqi attacks. It is important to note, however, that most of these attacks had only limited lethality. At the end of 1987, Iran had destroyed or heavily damaged a total of only 16 ships since the beginning of 1984, and Iraq had destroyed only 49 and heavily damaged 9.

While Iraq increased its total number of attacks on shipping from 65 in 1986 to 83 in 1987, Iranian oil exports through the Gulf were 40% higher in 1987 than in 1986—demonstrating that oil prices and the oil glut were more important than the tanker war. In fact, both Iran and Iraq steadily increased their oil exports throughout the second half of 1987.[119]

DEVELOPMENTS IN THE LAND WAR, AND THE ARMS TRADE TO IRAN IN DECEMBER AND EARLY 1988

The war remained unusually quiet on the land. Iran did continue to launch small offensives after its major attack on Basra. It conducted roughly a dozen small "offensives" with names like Karbala, Nasr, and

Najaf, but these were generally limited night attacks, and many did little more than try to seize a given ridge line or position in the mountains. Iran was most successful in the Kurdish sections in the north. It supported not only Kurdish forces, but also a unit of Iraqi prisoners of war who had agreed to fight Iraq. These forces did not dominate the countryside in the north, but they could occupy many villages at night. They often hit Iraqi convoys and closed roads during the day, and sometimes were able to cut electric power to major cities like Kirkuk.

Iraq was forced to send elements of its fourth division from the 7th Corps in the south into the area, but relied heavily on bombing Kurdish villages and on relocating the people in other villages and demolishing them. This often created more new enemies than it penalized old ones, but Iraq was irritated by the Kurds, not threatened by them.

Iran also carried out its seasonal mobilization in the fall, and built up large land forces in the south opposite Basra. Iranian radio and TV talked of another final offensive, and reports were issued of build-ups of up to 200 battalions of Basij, or 200,000-500,000 men.

In fact, however, the Iranian build-up seems to have ranged from only 60,000 to 100,000 men, allowing for rotations. This was not enough to fill out the force of roughly 20 Iranian combat formations or "divisions" claimed to be in the southern sector. The total number of volunteers seems to have dropped from 80,000 in 1986 to 40,000 in 1987, and forced Iran to extend its conscription period from 24 to 28 months in early January, 1988. Iran also did not deploy all the support forces and supplies it normally deployed before a massive offensive.

Iran was able to use Chinese, Austrian, and North Korean artillery deliveries to virtually double its artillery strength at the front, and reduce its ratio of inferiority from roughly 3:1 to 2:1. It also added some armor, and built new roads to help it rapidly redeploy and reinforce during attacks. Nevertheless, it did not deploy the normal number of tents, trucks, and support equipment.[120]

If anything, Iraq may have done more to prepare for defense than Iran did to attack. Iraq maintained some 900,000 men all along the front versus around 600,000 full-time actives for Iran. It deployed up to 250,000 men in the south around Basra. It completed three rings of defensive positions around the city, and completed a series of parallel north-south defensive lines to provide defense in depth all along the southern and central fronts.

Iraq raised its number of tank transporters from 1,000 to 1,500, and created a new road network to allow units to redeploy from the central to the southern front in 12 to 24 hours. To ensure that Iranian forces could not cut Basra off from Baghdad by thrusting through the north-south roads on the western bank of the Tigris, Iraq completed a new six-lane highway from Safwan on the Kuwaiti border, past Zubair and Nasiriyah, and then north towards Baghdad. Given the fact that Iraq had still another road and a railway along the Euphrates, which were much farther to the west, this mix of north-south lines of communication gave Iraq considerable insurance against an Iranian breakthrough.[121]

While the reasons for Iran's failure to attack are uncertain, they do not seem to have been the result of a shortage of arms, although Iran was having growing problems in getting a Western system. Iran obtained roughly $1.5 billion worth of arms, and got 60–70% of its arms from the PRC and North Korea, 20% more from Eastern Europe, and 20% from the rest of the world. It also doubled its domestic artillery production during 1987, and was fully self-sufficient in small arms and small arms munitions. Its main problems were its dependence on foreign suppliers for fuses and propellants. Further, Iran began to manufacture Phosgene gas, although it lagged badly behind Iraq in producing more advanced chemical weapons.[122]

The PRC supplied some $600 million worth of arms in 1987, largely artillery, ammunition, and missiles. China also sold equipment for the manufacture of arms and missiles. It shipped another $200 million worth of arms in January, 1988, and had agreements to provide $400 million more during the rest of 1988.

North Korea sold another $400 million worth of arms in 1987, including artillery, fast patrol boats, and Soviet-designed Scud surface-to-surface missiles. New North Korean arms shipments, including Scud and Silkworm missiles, arrived in January, 1988. The Warsaw Pact shipped some $350 million worth of arms, including a large number of troop carriers and some self-propelled artillery. None of these countries, however, seem to have delivered jet fighter aircraft, in spite of new rumors of sales by North Korea and the PRC.

Western supplies to Iran included some $150 million worth of ammunition and explosives from Spain and Portugal, at least some of which were actually made by France's Luchaire and the National Power and Explosive Company. They were sold with the same tacit knowledge of senior French defense and intelligence officials under a conservative government that had permitted them under a socialist one. These orders included up to 200,000 shells to be delivered during 1987–1988, 200,000 detonators, 2,500 tons of TNT, and 650 tons of powder.[123]

Japanese firms sold $100 million worth of trucks and spare parts to Iran, and private Swiss and West German firms continued to sell chemical warfare equipment. Iran's major free world supplier, however, was Brazil, which sometimes used Libya to act as a third party. This meant Iran was now experiencing far more serious difficulties in getting

Western arms and parts, particularly critical aircraft parts like those for the F-5.[124]

As for other explanations for the Iranian failure to launch a major offensive, this may partly have been a response to the rains, which were late and allowed Iraqi to use its armor effectively much later in the year. It may have also stemmed from the size of Iranian losses opposite Basra in early 1987, and from the reluctance to take 60,000 more casualties in attacking a front whose defenses had been greatly improved. It may have been affected by Syrian and other pressure to avoid further attacks that would alienate the Arab world, and by the fear such attacks would lead to Soviet and PRC support of a UN arms embargo. It also may have been the result of internal political divisions within the leadership around Khomeini, the desire to avoid any bloody defeat with Iranian elections coming up in the early spring, and Iranian calculations that there was little real chance of victory.

In any case, it is interesting to note that the only significant land action in the south during late 1987 occurred on December 20–21. Two Iranian brigades carried out a limited attack on the northern edge of the Hawizeh marshes in the south-central front. Iran attacked along a river in the Fakkeh border area, near the outpost of Zobeidat and east of the Iranian town of Misan (formerly Amara).

While both sides made the usual conflicting victory claims, Iran seems to have lost several thousand men and Iraq only several hundred. A brigade or division of Iranian troops seems to have tried to find a gap in Iraqi defenses, and to have been caught up in a minefield. Iraqi was then able to use its advantage in artillery to inflict serious casualties. The experience was scarcely one that encouraged an Iranian attack on Basra.[125]

ARMS EMBARGOS AND PEACE EFFORTS AT THE BEGINNING OF 1988

The broader political and economic situation remained confused during early 1988. UN Secretary General Cuellar virtually gave up on his cease-fire negotiations with Iraq and Iran on December 10, 1987, and turned the issue over to the Security Council. Iraq remained unwilling to compromise on the timing of a cease-fire and withdrawal, and Iran insisted that Iraq be identified as the aggressor before the cease-fire.

Getting the Security Council to act, however, presented two problems. It was unclear the PRC would agree to an arms embargo without Soviet support of such an embargo, or that either the PRC or the USSR would honor an embargo if they did agree to it. The USSR continued to be torn between trying to court Iran and seeking some kind of formal Western agreement to a Soviet role in the Gulf as the price of its support of an embargo. It called for a United Nations flagged force in the Gulf as the price of support for an embargo, and the UN force concept was clearly designed to limit Western freedom of action.

By the end of December, the Security Council could only agree on an announcement that it would move towards drafting and adoption of an arms embargo. On December 25, the permanent members of the Council agreed to start drafting a resolution early in 1988, but this agreement ignored the fact that several drafts already existed.

What was more interesting was the fact that an article in the Soviet news agency Tass appeared on December 19, accusing Khomeini and his lieutenants of prolonging the war to maintain power, and to cover up their "total incompetence". The article, by M. Krutikhin, also called Khomeini a "fanatical and stubborn" figure. It was followed by several others, and this was at least an indicator that the USSR might be giving up its attempts to court Iran. Further, a U.S. State Department spokesperson stated on December 28 that the U.S. would be willing to study the UN naval force concept if the Security Council approved an arms embargo, and if it did not affect the ability of the U.S. task force in the Gulf to protect naval shipping.

These events in the UN may have been helped by a Saudi effort to persuade the USSR to support the embargo. The Saudi Foreign Minister, Prince Saud al-Faisal, took the lead in quiet talks with the USSR, which seem to have included discussions of Saudi resumption of formal diplomatic relations with the USSR and the creation of a Soviet embassy in Riyadh. He visited Russia in late January, 1988, and met with President Gromyko and Foreign Minister Shevardnadze. Faisal was accompanied by Prince Bandar, the Saudi Ambassador to Washington and a son of Prince Sultan, the Saudi Minister of Defense.

Soviet attitudes towards the issue seem to have been affected by two other factors. One was a growing reluctance to go on providing Syria military support. Another was the desire to get Western help in withdrawing from Afghanistan. The Soviet ambassador to the UN, Alexander Belonogov, hinted at this in a speech on January 14, and stated that an arms embargo might "become absolutely necessary" if Iran did not respond to further talks.

In mid-February, Secretary of State Shultz was confident enough to direct UN Ambassador Vernon A. Walters to move towards tabling an arms embargo resolution. The U.S. also began to discuss a possible Soviet abstention, since nine of the ten members of the Council could pass the resolution if one abstained, and began to hint at a linkage to the Afghan withdrawal talks then under way between the Soviet Union and Pakistan. Even so, Shultz's meeting with Gorbachev and Shevardnadze in late February did not produce any public Soviet willingness to support the embargo. Eight months after the UN had passed the original cease-fire resolution, there still was no formal agreement over enforcement.[126]

IRAN LOSES THE BATTLE OF OPEC

On the oil front, Iran suffered a more decisive reversal. During much of 1987, Iraq was able to export more oil than Iran. It now had two pipelines through Turkey, with a capacity of 1.5 MMBD. It had another spur line joining the Saudi line to the Red Sea with a capacity of over 500,000 BPD and a Saudi agreement to build an independent line to a loading point near Yanbu with a capacity of 1.1 MMBD. In spite of Iranian talk of pipelines through Turkey, to Jask on the Gulf of Oman, and converting gas lines to ship oil through Russia, Iran was still dependent on Kharg Island and movements through the Straits of Hormuz.[127]

Iran tried to use OPEC to counter this trend. In early December, 1987, Iran attempted to threaten its Gulf neighbors into supporting much tighter OPEC quotas and an effort to raise its official reference price from $18 a barrel to $20. On December 9, it declared it would double its production and start a price war if it did not get suitable support for its position in OPEC. In practice, however, Iran was then accounting for only 11.5% of OPEC production. Saudi Arabia accounted for 24%, Iraq for 13.7%, the UAE for 9.3%, Kuwait for 6.7%, Venezuela for 8.7%, and the other OPEC states for 26.1%.

The Southern Gulf states, particularly Saudi Arabia, felt secure enough to ignore the Iranian threats. In an unusual power play, they confronted the Iranian Oil Minister at the OPEC meeting in Vienna on December 14 and handed him a virtual ultimatum. Although they were producing well beyond their quotas, and actual oil prices were $2–$3 below the $18 reference price, they refused to make any serious effort to restore the quota system. They also refused to insist on any quota for Iraq, and even exempted the up to 300,000 barrels per day of oil that Kuwait and Saudi Arabia were marketing for Iraq from the quota system.

The end result was that the Iranian oil minister had to fly back to Tehran for permission to agree to a pro forma quota system even though Iraq's Oil Minister had pledged that Iraq would not only produce nearly twice its former 1.5 MMBD quota, but produce all the oil it could. The only thing Iran could gain was the lack of a formal OPEC agreement that Iraq should have the same quota as Iran.

Iran, however, could not even meet its 2.4 MMBD quota. It also had to sell at prices as low as $12 to $14 a barrel, including the cost of its oil shuttle and discounts. Saudi Arabia was able to continue producing at 4.2 MMBD, out of a quota of 4.343 MMBD, Iraq to continue producing at around 2.6 MMBD, Kuwait to continue producing at 1.1 MMBD versus a quota of 996 MBD, and the UAE to continue producing at 1.6 MMBD versus a quota of 948 MBD.[128]

THE WAR AGAIN ESCALATES TOWARDS CRISIS

The land and air war remained relatively quiet during the first months of 1988. Iran did claim to have raided Iraq's inactive Al Bakr and Al Amaya oil terminals in the Gulf, to have destroyed missile and radar sites, and to have killed at least 100 Iraqis. These claims, however, seem to have been exaggerated. Further, Iran claimed to have destroyed three Iraqi frigates, none of which were at sea in the Gulf at the time.[129]

Iran did launch a more serious offensive, the "Bait-ol-Moghaddas 2" attack, on January 15, 1988. This attack took place in the Mawat border area at the northern front, east of Sulaimaniya. Iran had been fighting in this area since the spring of 1987, with mixed success. Iran claimed to have taken 42 square miles of new territory, including 11 heights and 29 peaks, to have killed or wounded some 3,500 Iraqi soldiers, and to have taken 750 prisoners.

Iraq denied these reports, but Iran does seem to have scored some gains. While the area involved was sparsely populated, and involved heights of 2,950 to 6,500 feet, it also had some strategic value because it allowed Iran to improve its position in future attacks on the northern front, and again strengthened its ability to supply anti-Iraqi Kurds.[130]

As for the air war, Iraq continued to strike both at ships and at targets like dams, bridges, and refineries throughout December, January, and March. Iran could do little about this, although it did experiment in trying to use its F-4s to fire Maverick missiles at ships in early January. It seems to have concluded that the Mavericks were now so old they had to be used before they became totally inoperable. In practice, however, the Maverick's small warhead and decaying guidance systems made them ineffective.

Iraq hit Kharg Island for the first time since November 4, on February 7, 1988, and constantly probed at Iran's air defenses. This led Iran to commit its F-14s to a rare ambush on February 9, 1988. The F-14s were armed with AIM-9 missiles and were able to close on two Iraqi Mirage F-1s when they turned north into Iranian waters at Farsi Island. At least one Mirage F-1 was shot down. While Iran had lost most of its air power, it still seemed to have about 20 F-4s, 20 F-5s, and 7-9 F-14s operational.[131]

The most dramatic change in the air war, however, began in late February. Iraq carried out a major attack on the refinery in Tehran on February 27, 1988. As was usually the case, Iraq did some initial damage, but failed to follow-up on its attacks or deliver the size of attack to knock out the refinery. Nevertheless, Iraq did serious damage and forced Iran to again start rationing petroleum products.

Iran responded to this Iraqi air raid by renewing the "war of the cities" and firing 3 Scud missiles at Baghdad. This was one of the highest number of missiles it had fired at a single time during the war, and was significant because many experts believed that Iran was down to as few as 20–30 missiles. All 3 missiles hit in largely unpopulated regions south of Baghdad. These strikes were the first Scud attacks on Iraq since November 8, 1987, and raised the total number of missiles targeted on Baghdad to 37.

This Iranian attack led Iraq to a major new escalation of the war. On February 29, Iraq launched five new extended-range variants of the Scud-B. These missiles came as a surprise to both Iran and Western intelligence experts. They seem to have been regular Scud missiles which used a lighter warhead and more of the missile's propellant, although some experts originally claimed they used a Western designed booster that was made in Iraq.[132] In any case, Iraq was able to give the Scud twice its normal 190 mile range, and some missiles flew over 360 miles. This was enough range for Iraq to reach Tehran and Qom from positions south of Baghdad.

By March 6, Iraq had hit Tehran some 33 times and Qom 3 times. The missiles had only a 200–400 kilogram warhead, and were scarcely "city killers", but they killed nearly 60 and wounded 130. Iran, in turn, had fired 16 Scud missiles at Iraq, 12 of which were targeted at Baghdad. Iraq was now firing an average of 3 Scuds a day and Iran was firing 1. By March 10, Iraq claimed to have fired 47 missiles, all but 5 at Tehran, and Iran claimed to have fired 25 missiles at Baghdad and several at other Iranian cities. Both sides had also made use of their air power. Iraq hit as many as 10 Iranian cities a day. Iran could do little more than launch a few token sorties, but it did renew heavy shelling of Basra. The two nations did reach yet another cease-fire on attacks on the cities on March 11, but by this time Iran claimed 165 killed and 440 wounded.[133]

Iraq seems to have launched the missiles to publicize the seriousness of the war and push the UN cease-fire effort forward, and to find a means of striking at Iran which presented fewer political complications than attacks on Gulf shipping. Iraq may also have sought to divide the Soviet Union from Iran.

If so, Iraq had some success. The Iranian government announced a few days later that it had discovered missile fragments showing that the missiles Iraq had launched were Soviet and had been manufactured as recently as 1985 and 1986. Iran publicly blamed the USSR for supplying the missiles Iraq was using. The Iranian government allowed a carefully staged anti-Soviet riot to sack the Soviet embassy in Tehran on March 6, and another riot attacked the consulate in Isfahan that same day.[134]

Iran also countered the Iraqi effort to push towards a cease-fire. It took new steps to try to block any efforts at a UN arms embargo by claiming to accept the UN resolution in a way that was so ambiguous that it did not commit Iran to anything. The U.S., Britain, and France tried on March 5 to get Security Council support for an arms embargo with a 30 to 60 day waiting period, but could not obtain either Soviet or Chinese support. The USSR then proposed a limited cease-fire on missiles, which Iraq angrily rejected. This again left the UN paralyzed and the Secretary General could do little more than invite both sides to send their foreign ministers to New York for intensive consultations.[135]

As for the situation in the Gulf, there were no attacks reported on shipping between February 12 and March 6, one of the longest lulls since the beginning of Western intervention. On March 6, however, Iranian naval Guards units fired on routine U.S. helicopter reconnaissance missions both from boats and from a naval oil platform. On March 8, Iraq hit the first Iranian ship since February 9, 1988. It previously had claimed 23 attacks since the beginning of 1988, but only 9 had been confirmed by shipping companies. It now, however, began to hit Gulf targets regularly and most of its strikes were confirmed by insurance groups or shipping agencies. Iraq also launched a major new raid on Kharg Island on March 19, burning 2 tankers and killing 46 sailors. Iran responded by launching a series of new Pasdaran attacks in the Gulf.

On March 13, the first reports began to surface of an even grimmer battle. Iran began an attack on the northern front in the area of the Iraqi border towns of Halabja, Khurmal, Kholmar, Dojaila, and Darbandkihan, just west of Nowsud in Iran and about 150 miles north of Baghdad. Iran's goal was to take enough territory in Northern Iraq to be able to advance on Sulaimaniyah and to seize control of the Darbandikhan Reservoir. This reservoir was created by one of the largest dams in Iraq and fed hydroelectric power to much of northeast Iraq and Baghdad.

Iran soon claimed to have started major artillery barrages against both towns. It also claimed that a mix of Pasdaran forces, Kurdish rebels, and Iraqi rebels and ex-prisoners of war had captured 7 border villages and 15 square miles of Sulaimaniya Province. Iraq claimed to have repulsed the attack, to have repulsed two attacking brigades, and to have killed 1,000 Iranians.

The truce in the "war of the cities" broke down the same day, and Iraq fired seven missiles at Tehran and used its aircraft to strike at six Iranian cities. It was Iran, however, that won the battle in the border area, and its success was significant.

Iran seems to have succeeded for several reasons. Iraq kept most of its best forces in the south because of its fear of another "final offensive". It had only two divisions in the forward area and these were of mixed quality. Iraq's forces withdrew from many of the heights and more impassable ridges in the area and sent too many of its troops too far forward to keep the towns at the border. Giving up the rough mountain terrain allowed it to be outflanked. Iraq should have kept its troops further in the rear and then directed them at the main Iranian thrust. Iraqi forces may also have had problems in dealing with the Kurds behind their lines of communication.

The end result of these mistakes was that Iraq lost much of its 43rd division, from 1,500 to 4,000 men, and large amounts of tanks, armored vehicles, artillery, support vehicles, and ammunition and spares from its 1st Corps. As in its Faw offensive in 1986, Iran found it could obtain substantial supplies when it overran Iraqi positions.

More happened at Halabja and Dojaila, however, than a border clash. Iran took the city in spite of Iraqi claims, and almost immediately released TV films showing that up to 4,000 civilians had died of mustard gas and other agents, possibly including phosgene, nerve gas, and/or cyanide.[136] It was obvious that Iraq had made extensive use of gas warfare against civilians. At the same time, evidence appeared that Iran had also fired gas shells into the town of Halabja. Both sides had, in effect, created a new level of conflict.[137]

Iraq might well have been trying to signal Iran as to what would happen to Iranian cities if the war continued. If so, it failed to be persuasive. Khomeini made a rare public appearance on March 20, and said he would press on for "final victory". That same day Iran claimed to have fired 13 missiles at Iraq.

Iraq also almost certainly used gas to warn the Kurds because of the growing success of the Kurds in challenging Iraqi control of the northeast. Kurdish forces had scored growing successes in spite of an extremely bloody Iraqi campaign to control and/or eliminate any Kurdish towns that showed any signs of support for the Kurdish rebels. Kurdish forces briefly occupied the Iraqi town of Kanimasi near the

Turkish border in September, and Kurdish and Iranian forces had taken some 100 square miles of Iraqi territory in the area northeast of Kirkuk near Mawat, including some 29 heights and 6 villages on either side of the Little Zab River.

Kurdish forces occupied the border town of Deirlouk in January, 1988, and then conducted a massive raid on the Iraqi resort of Sari Rash, northeast of Arbil. Travel to the city of Rawanduz became unsafe, and far more raids, ambushes, and assassinations occurred throughout northeastern Iraq. Jalal Talabani's Kurdish Patriotic Union of Kurdistan (KPUK) also began to cooperate for the first time with the Barzani clan's Kurdish Democratic Party. This unity seems to have been the result of negotiations which Rafsanjani had conducted in Iran in late 1987.[138] Iraq was still able to secure much of the area with Kurdish security forces that still supported the government, but long before the new Iranian attack, Iraq had had to redeploy some of its troops and paramilitary forces to the north, and had had to put sandbagged fire bases along the major roads in the area and to mine Kurdish rebel infiltration routes.

In spite of Iraq's use of gas against the Kurds, Iran resumed its offensives northeast of Sayyid Sadiq. Iraq had deployed some 7,200 troops (four battalions) from its 7th Army Corps in the South, and had counterattacked, but was repulsed by roughly 30,000 Iranian attackers. Iran then advanced up to 10 miles through the Rishan Mountains and came close to Sayyid Sadiq. The seriousness of the fighting is illustrated by the fact that Iraq claimed it flew some 224 combat sorties per day. Iraq also lost up to 3 planes per day, although Iran may also have lost some of its few remaining F-5s.

This fighting in the north continued during the rest of March and well into April. As the fighting went on, however, Iraq was able to bring more and more of its superiority in airpower and firepower to bear. It reestablished its defensive lines and gradually pushed the Iranian forces back. While it is impossible to determine how much use it made of poison gas, its position in the northern and central fronts seems to have become much more secure by the first week in April. As had been the case in several previous offensives, Iran lacked the technology and firepower to continue its advance once Iraq adjusted for its initial tactical mistakes.[139]

Iraq also retaliated with new missile strikes. It announced on March 22 that it had fired 106 missiles against Iranian cities. Iraq fired 10 missiles a day against Iranian targets. On March 27, 1988, Iran also claimed that Iraq had started to use a new and heavier missile. This started new speculation about possible Iraqi use of the SS-12, although the missile involved may well have been a heavy air-to-surface missile launched from Iraq's Soviet-made Tu-16 bombers. Iraq also launched 3 missiles at Qom, and air raids at 4 Iranian cities. These sorties were still relatively ineffectual on an individual sortie basis, and only about 18% hit a target. Nevertheless, Iraq hit some 37 cities, and did not give Iran time to recover from the Iraqi air attacks on its refineries and petroleum distribution points.[140]

Iraq also began to shift away from the defensive on the land. It initially supported new attacks on Iran by the People's Mujahideen National Liberation Army on a 14 mile front near Shush. At the same time, it began to build up its forces in the south. This build-up initially appeared to be a strengthening of its defensive positions. While Iranian aircraft took the unusual step of beginning to attack Iraqi concentrations south of Basra on March 26, 1988, Iran does not seem to have taken the risk of an Iraqi counterattack seriously. In fact, however, Iraq massively reinforced its 7th Corps and redeployed much of its Republican Guard forces. It built up large stockpiles of artillery and chemical weapons, and on April 17, it launched what it called its "Blessed Ramadan" offensive.

This time it was Iraq that achieved tactical surprise. The Iranian forces in Faw seem to have been under strength, manned with relatively low grade volunteers, and slow to react. Iran quickly made major gains, and took advantage of relatively dry terrain conditions and good weather. Iraq built up for its attack at night, and began to advance at 5:00 A.M. Forces from the 7th Corps, under the command of Major General Maher Rashid, advanced down the bank of the Shatt al-Arab waterway. Republican Guard forces pushed east from positions between Zubair and Umm Qasr, and into the marshy tidal flats, or "Great Salt Lake", west of Faw. At the same time, Iraqi forces launched amphibious attacks on the western side of the Peninsula from Umm Qasr. Iraq claimed that its air force flew 318 fighter and gunship attacks in support of the offensive on the first day.

Iraq scored major advances. The Iranian forces broke, and do not seem to have been able to regroup. Iran's dismay over the attack is indicated by the fact that on April 18, it charged Kuwait with allowing Iraq to launch part of its attack from Bubiyan and claimed that U.S. helicopter gunships had supported Iraq.[141]

The fighting in the Gulf reached a new peak at virtually the same time. Iran began its escalation by supporting pro-Iranian terrorists in seizing a Kuwaiti B-747 airliner on a flight from Bangkok to Kuwait on April 5. The hijackers took advantage of the almost nonexistent security at Bangkok airport and were able to take over 100 hostages. They immediately demanded that Kuwait free the 17 pro-Iranian terrorists it had convicted of acts of sabotage, and began a cycle of killing and intimidation that took the aircraft from Tehran to Larnaca in Cyprus, and finally to Algeria. Iran may well have allowed additional hijackers to board the aircraft while it was in Tehran and seems to have given them grenades and machine guns in addition to the pistols they used to seize the aircraft. The hijackers also showed some indications of advanced training, since they quickly revealed an expert knowledge of both the layout of the aircraft and its technical details.[142]

On Thursday, April 13, 1988, however, Iran took a far more serious step. Small ships of the Iranian Revolutionary Guard laid a new minefield in the Gulf. The next day, the U.S. frigate *Samuel B. Roberts* hit one of the mines as it was returning to Bahrain after completing the 25th convoy of 1988.[143] The *Roberts* was one of 25 U.S. ships still operating in the Gulf, and the mine explosion lifted the ship's stern 15 feet, and blew a massive 22 foot hole. Ten American sailors were wounded, and the engine was flooded.[144]

Like the attack on the *U.S.S. Stark*, the explosion again illustrated the critical importance of modern damage control and fire fighting capabilities. The explosion not only ripped a hole in the ship; it also broke its keel, ripped the ship's engines from their mounts, and sent a fireball 150 feet up its stack. Fuel lines ruptured and major fires broke out. There were no fatalities because the engine room was fully automated and no crew were present.

The crew did, however, face a series of massive problems. It had to lace the ship together with steel wire to keep it from breaking up and had to reinforce the compartments with wood timbers to prevent flooding. Steel plates had to be welded over massive cracks in the deck house. The ship was surrounded by mines and providing assistance was difficult. The *Samuel B. Roberts* did, however, have important advantages over the *Stark*. The crew was able to save the ship because it was fully deployed at battle stations. It had won an award for best performance for advanced damage control training before its deployment that was partly the result of the lessons from the *Stark*. The Gulf was calm when the mine explosion took place, and all firefighting equipment was dispersed and operational.

The U.S. rapidly discovered firm evidence that Iran had deliberately placed the mines in the convoy channel. The U.S. recovered several of the mines, which were Iranian-made copies of the Soviet M-08 mine, and the divers who inspected them found Iranian serial numbers and clear evidence that the mines had been newly planted.[145]

On April 14, the National Security Planning Group (NSPG) met at the White House and considered several options. It eventually decided on a strike at three Iranian oil platforms in the Gulf, all of which were used by the Iranian naval Guards. The NSPG chose this option because it avoided any strike on Iranian land targets, was far from the fighting in the upper Gulf, and demonstrated Iran's acute vulnerability to any interruption to its oil exports. President Reagan approved the drafting of contingency plans the same day, and then the suggested strike plan at a meeting at 8:30 P.M. on April 17. The President briefed the leaders of Congress at 9:00 A.M. on April 18. It is important to note that the U.S. did not know that Iraq planned to attack Faw the same day.

Early in the morning, the U.S. set up an extensive air screen over the Gulf, using aircraft from the carrier *Enterprise*. This air screen included E-2C Hawkeyes in addition to the regular E-3A patrols. It also included F-14A on combat air patrol. The E-3As performed both long-range warning functions, and attack coordination for A-6Es armed with Harpoon missiles, laser-guided weapons, and Mark 20 Rockeye cluster bombs. A-7E light attack aircraft were on standby alert.

At 9:01 A.M. on April 18, the U.S. gave Iran's Saasan oil platform several minutes warning. The 20–40 Guards on the platform abandoned it, leaving four ZSU-23 anti-aircraft guns and several SA-7s. At 9:17 A.M., the U.S. shelled the platform, and then blew it up with explosives. The U.S. used three ships in the attack, the *Merrill,* the *Lynde McCormick,* and the *Trenton*. The *Trenton* was an amphibious ship carrying a 400 man Marine Corps unit which included AH-1T, UH-1N, and CH-46 helicopters. It was part of the Contingency Marine Air Ground Task Force (CM) 2-88. The Marines boarded the platform at 10:39 A.M., and blew up the remaining installations.[146]

At 9:32 A.M., three additional ships, the *Simpson, Bagley,* and *Wainwright,* gave a similar warning to Iran's Nasr oil platform off Sirri Island. Iranian forces started firing ZSU-23-4 anti-aircraft guns at the U.S. ships. The *Wainwright* fired several shells in return and the Iranians abandoned the platform. The U.S. ships then shelled and destroyed the Nasr oil platform by 10:25 A.M. The two oil platforms were roughly 100 miles apart.[147]

The Iranian Navy ship *Joshan,* carrying *Harpoon* and *Sea Cat* missiles, closed within 10 miles of the *U.S.S. Wainwright* and the American ships around the Nasr oil platform at 1:01 P.M. The *Wainwright* warned it to remain clear or come under fire at 1:15 P.M. The *Joshan* then fired what seems to have been a Sea Cat at the *Wainwright.* The *Wainwright* fired chaff to counter the missile, and moments later the U.S. ships responded with four missiles and hit the *Joshan,* which sank less than an hour later. At 1:52 P.M., the *Wainwright* fired two missiles at two Iranian F-4s closing on the ship after they fired a missile, and may have damaged one aircraft. Iranians in a Boghammar speedboat then fired on a helicopter from the *Simpson.*

The only other Iranian aircraft in the area on that day were a P-3 and a C-130, which took no aggressive action.

These attacks led the U.S. to break off the attack on a third oil platform because it was obvious that the fighting might escalate to further naval conflict, and the U.S. did not want to raise the threshold of conflict. Nevertheless, Iran took several other steps to escalate during the morning and early afternoon. At 12:15 P.M., three Iranian speedboats fired on a helicopter from the *U.S.S. Simpson,* and then on the American tug *Willi Tide,* a U.S.-registered supply boat off the coast of the UAE. An Iranian vessel then attacked the 113,000-ton British flag tanker, *York Marine,* and at 2:26 P.M. it attacked the *Scan Bay,* a U.S.-operated oil platform in the Mubarak oil field about 15 miles off the coast of Abu Dhabi.[148]

At 2:23 P.M., U.S. A-6 jets spotted three Iranian Boghammar speedboats attacking the Mubarak oil platform. The resulting command sequence provides an important lesson in the importance of satellite communications. The A-6s radioed the carrier *Enterprise,* which radioed the commander of the Middle East Task Force aboard the Coranado in the Gulf. He radioed the commander of USCENTCOM in Tampa, Florida, who contacted the Secretary of Defense and the Joint Chiefs in the National Military Command Center in the Pentagon. The Chairman of the Joint Chiefs contacted the President's National Security Advisor, who asked the President for permission to attack the ships. The President approved and the A-6s used laser guided bombs to sink one and disable another. The fate of the third was unknown. The entire command sequence from the A-6 to presidential approval was conducted in near real time using global secure communications. The transmittal of the President's approval to the lead A-6 pilot came less than three minutes after his original request.

The U.S. then intercepted communications that two Iranian frigates had been ordered to attack the U.S. force. The Iranian frigates kept closing on the U.S. force, and the U.S. warned them away at 3:59 P.M. At 4:00 P.M., the Iranian frigate *Shahand,* which had long been one of the ships firing at cargo carriers in the Gulf, appeared south of Qeshm Island in the Strait of Hormuz. At 4:21 P.M., the *Shahand* fired on three Navy A-6 attack fighters and the U.S. ships in the area. Between 4:34 and 4:43 P.M., the U.S. aircraft fired laser guided Skipper 2 bombs, and the frigate *U.S.S. Strauss* fired a Harpoon missile.[149] By 5:06 P.M., the *Shahand* was dead in the water and afire. The Iranian frigate sank that night.

At 6:18 P.M., the Iranian frigate *Sabalan* fired a missile at the U.S. frigate *Jack Williams,* and a second missile at U.S. Navy attack jets in the area. Both missiles missed, but a U.S. A-6 attacked the Sabalan with laser guided bombs and seriously damaged it. Although the *Sabalan* was notorious for its strikes on Gulf shipping, Secretary of Defense Carlucci halted further strikes, and at 8:00 P.M., the U.S. allowed the ship to be taken under tow by two Iranian tugs and to limp home to Bandar Abbas. By the time the smoke cleared, the U.S. had scored a major victory, and its only loss was one Marine Cobra attack helicopter from the *Trenton* and its crew of two, which seem to have been lost because of an accident. The next day, the Gulf was relatively calm, although Iran did fire one Silkworm missile into the desert in Kuwait.

U.S. weapons and countermeasures all functioned exceptionally well. There was, however, one major uncertainty as to what happened during the missile exchanges. The *U.S.S. Jack Williams* reported that it had been attacked by five Silkworm missiles, and had turned to fire chaff and optimize its ECM coverage each time.[150] Department of Defense officials stated only that they had no evidence that Silkworm missiles were fired. Witnesses, including members of the press, saw all five missiles, but it was not possible to confirm whether they had been fired from the mobile Silkworm sites near the Straits of Hormuz.

Iran's motive for taking on the U.S. Navy and losing nearly half of its operational major ships can only be guessed at. It is impossible to know if the Iranian government fully approved the new minelaying, although it later became apparent that it was a major Iranian effort. New mines continued to be found, and the U.S. was still finding more as late as April 22.

Iran's mining of the Gulf, and its willingness to commit its ships, may have been related to Iraq's war of the cities, the halt of its offensive in Kurdistan, and Iraq's increasing use of poison gas. These actions caused Iran's leadership to take progressively harder line positions regarding the continuation of the war. Khomeini received a strong majority in the first round of the Iranian election of a new Majlis on April 8, 1988. The elections also bolstered Rajsanjani and Mousavi, and the Majlis gained a number of new "hardliners". For example, Hojatolislam Mehdi Karrubi came in second in the voting. He was a strong advocate of exporting the revolution, and had been the main cause of the previous year's uprising in Mecca.[151]

Iran may also have misunderstood the nature of U.S. naval and air power. Commanders may have executed standard contingency orders without fully reassessing the situation. Iran may have been reacting to its defeat at Faw and to some perception of U.S. and Iraqi collusion. Iran may have felt that such a clash would deter the USSR from agreeing to any arms embargo or build up sympathy in the Third World. Finally, Iran may have hoped that even one successful strike would cause major damage, and that a large-scale clash would force the U.S. to trigger the War Powers Act.[152]

If so, Iran must have been rapidly disillusioned: The Congress gave President Reagan almost universal support. The main Congressional response in terms of the War Powers Act was for several leading Democratic members of the Senate to suggest that it should be modified to remove the automatic withdrawal provisions, and to create an act that would remove any incentive for foreign blackmail or escalation and help create a true consensus behind U.S. action without tying the President's hands.

The Soviet press did attack the U.S. actions in the Gulf as "banditry", but the Western European nations largely supported the U.S. and demanded a halt to Iranian ship laying. Prime Minister Maragret Thatcher called the attack "entirely justified." After consultation and a meeting within the WEU, the Europeans supported the U.S. action and also sent their minesweepers into the area where the mines had been laid and immediately began to provide the U.S. with assistance. The European ships had been withdrawn when the U.S. notified Belgium, Britain, France, Italy, and the Netherlands that it was about to commence combat operations in the area. They found eight newly laid mines the next day.[153]

The regional reaction was more complex. Rafsanjani declared on April 19 that "Time is not on our side anymore. The world—I mean the anti-Islamic powers—has decided to make a serious effort to save Sadama Hussein and tie our hands". The Gulf states also expressed their dismay at the fact that the U.S. had not provided protection for all merchant ships and oil facilities. The U.S. reacted on April 21 by declaring that it would now come to the aid of any ship or target under attack at the time and place of its own choosing, and effectively warned Iran that there were no safe targets in the region.

By April 21, the war in the Gulf was as unpredictable as the war on the land. Western ships had found more than 13 newly laid mines. Iran had hit 31 tankers since the beginning of 1988, versus 20 for Iraq. This compared with 92 Iranian attacks and 87 Iraqi attacks in all of 1987. There were 30 U.S. ships still in the Gulf, including the carrier *Enterprise*.

THE SAUDI LONG-RANGE MISSILE CRISIS

It is perhaps fitting that this analysis should end with the fact that the escalation of the Iran-Iraq War was accompanied by yet another crisis in U.S. and Saudi relations. On March 6, 1987, Saudi Arabia confirmed that it had bought CSS-2, or Dong Feng (East Wind), long-range surface-to-surface missiles from the PRC. These missiles had a range of 1,840 to 2,200 miles and an accuracy of around 1.2 miles—assuming perfect targeting and missile reliability. They could reach any target in the Middle East, including any part of Israel. They were not particularly reliable or accurate, and had a payload of only about 1,650 pounds.[154]

The Saudi government had begun negotiating for the systems in July, 1985, when it first began to believe it could never get full U.S. support for the modernization of its air forces. Ironically, it was the Saudi Ambassador to Washington, Prince Bandar Bin Sultan, who negotiated the deal during a visit to Beijing—at least in part because of his personal frustration over U.S. rejection of the F-15 sale. Two high-level Saudi delegations visted the PRC in July and November, 1987, and the missiles were shipped to Saudi Arabia in late 1987 and January, 1988.

Saudi Arabia evidently bought enough missiles to deploy a force of 20, although some reports indicate that a buy of more than 60 missiles was involved. According to one report, the Saudis disguised their initial deployments from the U.S. They had the PRC ship the missiles to Saudi Arabia along with missiles transiting to Iraq, and the missiles were then trucked into the Rub al-Khali or Empty Quarter. The Saudis told U.S. officials they were building a huge ammunition depot, and the U.S. did not detect that the missiles were staying in Saudi Arabia until some of the missiles were seen moving south from Saudi ports, rather than to Iraq. Other reports indicate that the missiles were to be deployed to fixed and sheltered sites at the new Al-Kharj airbase, about 60 miles south of Riyadh.

The original Saudi motive in buying the missiles may have been a mix of anger over U.S. treatment and a desire to assert Saudi independence. This motive may well have evolved during 1986 and 1987, as the possible threat of Iranian attacks on Saudi Arabia became steadily more real. The missiles offered Saudi Arabia a way to strike back at Iran and any other neighboring states without having to fear the loss of its air force. The missiles also had symbolic value in the region that almost certainly outweighed their limited military effectiveness. Further, regardless of Saudi Arabia's immediate intentions and long-term actions, they raised the constant threat of proliferation.

Saudi Arabia did declare that the missiles did not have nuclear warheads, and that the Kingdom had no intention of procuring them, in a letter from King Fahd to President Reagan on March 12, 1987. The missiles still raised serious concern, however, because even the variants sold to Saudi Arabia had too small a warhead to be effective conventional weapons, given the missiles' poor accuracy. While Saudi Arabia quietly indicated that they were a response to the growth of similar long-range ordnance in the region, and in Iraq as well as Iran, it was unclear to many observers how the missiles could have more than symbolic value, or act as more than an irritant, without chemical, biological, or nuclear warheads.

On March 20, senior Israeli officials expressed their concern by threatening preemptive attacks. These officials made pointed references to Israel's June,1981 attack on Iraq's Osriak reactor. There were reports of Israeli aircraft training for such missions and Israeli fighters pointedly demonstrated Israeli air superiority by overflying Tabuk, and even doing touch and go landings on Tabuk's air strip.

This Israeli pressure led to quiet Saudi efforts to reach a quid pro quo with Israel by offering the U.S. the right to secretly inspect the missiles and assure Israel they were conventional. While the facts involved are secret, the U.S. seems to have accepted this offer.[155]

The end result in the U.S., however, was to reopen the same debate over U.S. sales to Saudi Arabia that has been described in previous chapters. On April 14, 1988, 58 U.S. Senators and 187 members of the House sent a letter to President Reagan demanding that the U.S. halt the sale of a $450 million AWACS support package to Saudi Arabia.

The Congressmen involved had a point. There was no question that the Saudi purchase added yet another dimension of instability to a process of escalation that involved the acquisition of long-range missiles or strike aircraft by virtually every major state in the region; the production or development of gas warfare forces in India, Iran, Iraq, Israel, Pakistan, and Syria; and the development of nuclear weapons in India, Israel, and Pakistan.

At the same time, the Congressional opposition to the AWACS support package owed a great deal to the fact that an election was approaching. Further, it revealed the usual problems in Congressional handling of the situation. The Congress proposed to halt a sale of systems of direct military value to the U.S. forces in the Gulf largely because the supporters of the resolution ignored the fact that the Administration was also requesting a large sale of TOW missiles and Bradley armored personnel carriers.

King Fahd and the Saudi Council of Ministers attempted to ease Congressional concerns on April 25, 1988, by announcing that it would sign the Non-Proliferation Treaty, and agree to inspection of any of its nuclear facilities. It also announced that it was breaking diplomatic relations with Iran.

The U.S. announced the same day that it was delaying the sale of the

$450 million AWACS support package to allow more time for consultation.[156] This U.S. action was virtually designed to do more to provoke Saudi Arabia into similar third country arms purchases, rather than lead it to support U.S. policy. If either the U.S. or Saudi Arabia had learned much from the arms sales crisis over the AWACS and F-15, it was not readily apparent.[157]

THE LESSONS OF THE WESTERN INTERVENTION IN THE GULF

The broader lessons that the West should learn from its experiences in 1987 and 1988 are discussed in the final chapter of this book. Nevertheless, there are several points that emerge from the preceding analysis that are worth emphasizing in spite of the fact that this particular case study is far from over, and it is far too soon to say that the West has been successful:

- The events of 1987 and early 1988 illustrate the validity of many of the conclusions drawn throughout this book. They confirm the strengths and weaknesses of both the West's position and that of the Southern Gulf states. They confirm the need for their cooperation, the pivotal role of the U.S. and Saudi Arabia, the need for effective regional air and naval forces and Western arms transfers, and the need for effective over-the-horizon reinforcement capabilities.
- It is clear that the U.S. in some ways blundered into the Gulf, and that the West will now have to stay in the Gulf until it can collectively blunder out of it. What is far from clear, however, is that the U.S. really had any choice other than to attempt to "muddle through". The U.S. unquestionably could have done many things better, but history is not kind in providing unambiguous needs for action. If history is painful to those who act too quickly, it can be devastating to those who act too slowly. This is particularly true of the defense of long-term strategic interests where the assertion of a strong and continuing regional presence is essential to success.
- The main errors in the initial U.S. entry into the Gulf stemmed from the fact that the entry was rushed for external political reasons, failed to draw properly on the U.S. national security system and its expertise, under-estimated the risks, failed to organize properly for low-level war, failed to fully define the outcome or "end game" the U.S. was seeking, and failed to properly

define the full range of political and military means necessary to achieve that outcome.

- U.S. and Western inability to control events after the U.S. entry into the Gulf in large part stemmed from three factors. First, Iranian values were sufficiently alien that the U.S. could not predict Iranian reactions and responses. Second, the U.S. emphasis on limited escalation made it impossible to confront Iran with cases in which it did not have a wide range of potential responses and could not risk escalating. Third, the war was often driven by Iraq in ways the West could not predict. This acts as a further caution against the thesis that escalation ladders and limited response can be controlled or used with predictable response. In practice, this is virtually never the case.
- What the U.S. cannot be condemned for is treating the containment of both the USSR and a radical hostile Iran as legitimate security objectives. The U.S. can be condemned for not consulting its allies closely enough, but it cannot be condemned for not waiting for their consent. It cannot be condemned for seeking a UN role in reaching peace, while avoiding any reliance on the UN for peacekeeping. No UN force could have engaged Iran at the level of force necessary to have deterred Iran, nor could a UN force have done anything to assert the West's long-term strategic interests.
- Much of the U.S. success during the summer, fall, and winter of 1987 was the result of Iranian extremism. Much also was the result of the growing professionalism of U.S. and European action in the Gulf. The lesson here is simple: The initial wisdom of a policy is often far less important than the skill and consistency with which it is executed, and the ability to capitalize on the mistakes of one's opponent.
- This, however, raises the issue of the War Powers Act. An American administration must be sensitive to the Congress, public opinion, and its allies. The leader of a superpower, however, cannot be hostage to a veto of any or all of these three groups on a day to day basis. Regardless of its constitutionality, the events of 1987 indicate that a U.S. administration should not be hostage to a negative veto like the one inherent in the War Powers Act. It can appear so weak for domestic political reasons that it encourages attack and escalation. It tends to be forced to overreact within the time permitted by the Act, and to be forced into turning ambiguous events and uncertain needs into strategic necessities.[158]
- One also has to be careful about setting impossible standards. The problems in the West's intervention in the Gulf in 1987, have characterized virtually every Western intervention in the Third

> World since the 15th Century, and often have been no barrier to success. These problems are partly the inevitable result of limited means and multiple commitments, and are partly the inevitable impact of the one true lesson of history: While it may be true that those who forget the past are condemned to repeat it, this is equally true of those who remember it.

More generally, the Western intervention in the Gulf in 1987 and 1988 is to some extent a case study in how the West must cope with a continuing set of very uncertain realities. The West has had to react to the international system as it is and will be for the next decade unless the West falters badly, Iran wins the Iran-Iraq War, or one or more of the Southern Gulf regimes collapse into some form of radicalism.

It is always easy to provide a critique of military actions after the fact. It is equally easy to make new and creative suggestions as to substitute or future actions. The difficulty is not one of being creative, it is one of being correct. The West's position in the Gulf is now difficult and uncertain. It is likely, however, to remain equally difficult and uncertain for at least the next-quarter century. The West may be able to vary its military presence in the Gulf, but it can eliminate it only in abject defeat. Success does not consist of leaving the Gulf, but in learning to live with the same basic difficulties and uncertainties the West has had to face in 1987 and 1988.[159]

NOTES

1. Ironically, the reflagging made a significant increase in the U.S. cargo fleet. The U.S. merchant marine was now half the size of the Soviet one. Although the U.S. calculated it would need some 373 vessels in a national emergency, with a deadweight tonnage of 19.4 million, it had fewer than 200 vessels with a capacity of 13.2 dwt and fewer than one-half of the tankers it required. U.S. shipyards were building less than 1% of world shipping, and the number of yards was expected to drop from 116 in 1982 to 65 in 1990. Further, 45 of the tankers that did remain in U.S. service did so only because foreign flags were forbidden to move oil between Alaska and the U.S. This had some interesting implications for any war in the Gulf because U.S. ships had carried 80% of the military cargos in the Korean conflict and 16 million long tons of POL and 65% of other cargo during the Vietnam conflict. Tankers were 30% of the ships Britain deployed to the Falklands.

2. This was not accurate. The force was not ready to deploy and was not airlifted to the Gulf until August 1, 1987.

3. Some of these mines later turned out to date back to 1908. *Washington Times,* July 2, 1987, p. A-1, and July 15, 1987; *Washington Post,* July 2, 1987, p. A-26. *Baltimore Sun,* June 27, 1987, p. 2A, *Christian Science Monitor,* July 2, 1987, p.

11; New York Times, July 2, 1987, p. A-2.

4. The Pentagon released an unclassified version of the plan on June 16, 1987. It makes no reference to planning for the mine warfare threat. *Baltimore Sun,* August 23, 1987, p. 2A.

5. *U.S. News and World Report,* July 13, 1987, p. 39; *Washington Times,* July 2, 1987, p. A-1, and July 15, 1987, p. 20; *Washington Post,* June 27, 1987, p. A-23, 1987, p. A-21, July 21, 1987, P A-1; *Christian Science Monitor,* July 2, 1987, p. 11; *New York Times,* July 19, 1987, p. 12; July 21, 1987, p. A-8, and July 22, 1987, p. A-10; *Baltimore Sun,* July 1, 1987, pp. 1A and 2A, July 24, 1987, p. 2A; *Wall Street Journal,* July 1, 1987, p. 2, July 19, 1987, p. 12.

6. Free floating or moored contact mines can be planted by dropping them over the side of any small ship with a crane. They also are difficult to destroy since a direct hit on the tip of one of the mines' spiked contacts is necessary to make it explode.

7. *Chicago Tribune,* July 24, 1987, p. I-1.

8. *Washington Post,* July 24, 1987, p. A-16.

9. *New York Times,* July 25, 1987, p. 1; *Washington Post,* July 25, 1987, p. A-1.

10. *Washington Post,* July 27, 1987, p. I-1, and August 2, 1987, p. A-23.

11. Kuwait is highly vulnerable to air and missile strikes. It has about 6–8 power and desalinization plants that are critical to economic operations and normal life, and only limited air defense capability. There are three power plants near Doha and Shuwalkh and two more near Shuaiba. The desalinization plants are at Mina Ahmadi, Shuaiba, Mina Abdullah, and Mina Saud. Another power plant is under construction at Failakka Island. Kuwait also tends to layer liquid gas and oil processing and refining facilities, and the propane plant at the Ahmadi refinery is particularly vulnerable. A lucky hit could trigger a series of explosions that could cripple Kuwait's export capability. Iran's missiles, however, are too limited in number and accuracy to be used for surgical strikes, and its air attack capability is so limited it would taker a major risk in committing its remaining active attack aircraft against even a small force like the Kuwaiti Air Force. A successful Iranian strike would require a massive amount of luck.

12. *Economist,* August 1, 1987, p. 38.

13. The *Guadalcanal* is an Iwo-Jima class amphibious assault ship which is designed for amphibious operations. It can operate 4 Sea Stallions simultaneously from its deck, and hold 11 of the helicopters. It is 602 feet long and has a 104 foot flight deck. In addition to the helicopters, it could operate Harrier VSTOL fighters. It has air defense guns and regular guns, but no real missile defenses. Its maximum speed is 23 knots. It carries 47 officers and 562 crew and a landing force of 144 officers and 1,602 enlisted personnel. It was commissioned on July 20, 1963. *New York Times,* July 30, 1987, p. A-1; and *Washington Post,* July 30, 1987, p. A-1.

14. *Aviation Week,* August 3, 1987, pp. 25-26; *Washington Post,* August 4, 1987, p. A-1.

15. *Chicago Tribune,* July 31, 1987, p, I-1; *Washington Post,* August 4, 1987, p. A-8.

16. *New York Times,* July 27, 1987, p. A-1, July 31, 1987, p. A-1.

17. The French had what they regarded as proof that Gordji was guilty. In

early 1987, a Middle Eastern source had provided the French secret service in Tours with strong indications that Iran had organized the bombings in Paris in September, 1986, and had done so in spite of the fact that France had expelled Massoud Rajavi in June, 1986, and had largely settled Iranian financial claims in agreements initialled in July and November. France originally discounted the possibility of Iranian support for the effort because it was slowly freeing French hostages in Lebanon. French intelligence, however, was able to find some of the explosives involved and to trace one of the cars used in the bombing. The car proved to be Gordji's. He was a translator without diplomatic immunity.

18. *Washington Times*, July 12, 1987, p. A-7, and July 15, 1987, p. A-20; *Washington Post*, July 8, 1987, p. A-21, July 9, 1987, p. D-4, July 14, 1987, p. A-10, July 18,1987, p. 1, July 19, 1987, p. A-17, July 21, 1987, p. A-16; *Christian Science Monitor*, July 2, 1987, p. 11; *New York Times*, July 18, 1987, p. 1, July 19, 1987, p. 12, July 21, 1987, p. A-8, and July 22, p. A-10; *Baltimore Sun*, July 12, 1987, p. 7A, July 19, 1987, p. 1A, July 24, 1987, p. 2A; *Wall Street Journal*, July 1, 1987, p. 2, July 19, 1987, p. 12.

19. *Washington Post*, July 21, 1987, p. A-1; *USN&WR*, August 30, 1987, p. 30.

20. *Washington Post*, November 30, 1987, p. A-1, and December 2, 1987, p. A-33.

21. Iran had gone beyond riots. On August 8, 1986, Iranian pilgrims were halted when their suitcases failed to pass through a routine metal check at Jiddah airport. All 95 suitcases on the flight were confiscated, and 51 kilograms of plastic explosive were discovered in secret compartments. The pilgrims stated under questioning that they were asked to bring in the explosive by members of the Revolutionary Guard.

22. These claims were never substantiated. Saudi Arabia returned most of the bodies involved. While some did have what appeared to be bullet wounds, these were only a relatively small number. No indications of large numbers of wounded were made public. Iran claimed at least some Iranians were dead whom Saudi Arabia had actually arrested after the riot.

23. According to one report, the riots were triggered early. They were supposed to be led by Mahdi Karoubi, and to involve a tightly organized effort to use nearly 150,000 Shi'ite pilgrims and take over the mosque. While such reports are uncertain, Khomeini did send a message on July 29, 1987, to Iranian pilgrims in Mecca that stated, "if Moslems cannot denounce the enemies of God in their own home, where can they...? With confidence, I tell you...that Islam will eliminate all the great obstacles, internally as well as externally beyond its frontiers and will conquer the principal bastion in the world". *Washington Post*, August 10, 1987, p. A-l.

24. *Washington Times*, July 2, 1987, p. A-8, July 7, 1987, p. A-1, July 20, 1987, p. A-6; *Washington Post*, July 2, 1987, p. A-23, July 8, 1987, p. A-24, July 14, 1987, p. A-10, July 18, 1987, p. A-20, July 19, 1987, p. A-17, July 21, 1987, p. A-16, July 22, 1987, p. A-14, July 29, 1987, p. A-19, August 10, 1987, p. A-1, August 17, 1987, p. A-16; *Christian Science Monitor*, July 2, 1987, p. 11; *New York Times*, July 18, 1987, p. 1, July 19, 1987, p. 12; July 21, 1987, p. A-8, July 23, 1987, p. A-1; August 17, 1987, p. A-1; *Baltimore Sun*, July 4, 1987, p. A-1, July 19, 1987, p. 1A, July 24, 1987, p. 2A; *Wall Street Journal*, July 9, 1987, p. 24; *Philadelphia Inquirer*, July 3, 1987, p. 10A,

August 11, 1987, p. 4A; *Los Angeles Times*, July 17, 1987, p. I-1; *Insight*, August 3, 1987, p. 30; *USN&WR*, August 17, 1987, p. 23.

25. *Washington Post*, August 2, 1987, p. A-l; August 3, 1987, p. A-1; August 4, 1987, p. A-1; *Christian Science Monitor*, August 18, 1987, p. 2.

26. Ibid.

27. No precise figures are available. Some estimates indicate that Shi'ites make up 40% of the labor force in the oil fields. The total Shi'ite population in Saudi Arabia is estimated at 300,000 to 350,000. Most live in the Eastern Province. While some Shi'ite activists have been arrested, the number has evidently never been high enough to indicate a critical problem.

28. Prince Fahd had completed a political science degree at the University of California at Santa Barbara in the U.S. He did, however, not only maintain the Majlis system, but deliberately used it to treat Sunni and Shi'ites as equals and to break down some of the traditional barriers between them—many of which had been reinforced during the time of local riots by the governor, who had taken a hard line and an anti-Shi'ite stance.

29. *Washington Post*, August 17, 1987, p. A-1, and August 18, p. A-8; *Christian Science Monitor*, August 2, 1987, p. 11; *New York Times*, August 18, 1987, p. A-3; *Baltimore Sun*, August 4, 1987, p. A-1; *Wall Street Journal*, August 18, 1987, p. 58; *USN&WR*, August 24, 1987, p. 26.

30. *Washington Post*, August 7, 1987, p. A-25.

31. The U.S. Navy was having a substantial quality control and reliability problem with virtually all of its radar guided air-to-air missiles. It is not surprising, however, that the AIM-7M failed to achieve a hit. Radar guided air combat general requires large numbers of carefully orchestrated aircraft to achieved high kill rates. *Washington Post*, August 11, 1987, p. A-1, and August 12, p. A-20; *Aviation Week*, August 17, 1987, pp. 22–24; *New York Times*, August 11, 1987, p. 2.

32. *Washington Post*, August 12, 1987, p. A-1; *New York Times*, August 11, 1987, p. A-3.

33. *Washington Post*, August 13, 1987, p. A-3.

34. *Washington Post*, August 14, 1987, p. A-29, September 20, 1987, p. A-25; *New York Times*, August 12 and 14, 1987, p. A-1; *Baltimore Sun*, August 14, 1987, p. 1A; *Economist*, August 22–28, 1987, pp. 42–43; *Wall Street Journal*, August 14, 1987, p. 11.

35. *Washington Post*, August 12, 1987, p. A-1, and August 13, 1987, p. A-29; *New York Times*, August 12, 1987, p. A-1; *Baltimore Sun*, August 13, 1987, p. 1A; *Economist*, August 22–28, 1987, pp. 42–43; *Washington Times*, August 14, 1987, p. A-9.

36. *Washington Post*, August 12, 1987, p. A-19; *New York Times*, August 12, 1987, p. A-1; *Baltimore Sun*, August 13, 1987, p. 1A; *Economist*, August 22–28, 1987, pp. 42-43.

37. *Washington Post*, August 16, 1987, p. A-1; *New York Times*, August 16, 1987, p. A-1; *Baltimore Sun*, August 16, 1987, p. 1A.

38. *Washington Post*, August 17, 1987, p. A-19; *New York Times*, August 18, 1987, p. A-1; *Baltimore Sun*, August 17, 1987, p. 1A.

39. The cost can be calculated either as the total cost of the forces involved, or

as the incremental cost of the operation. On August 21, 1987, the Pentagon announced that roughly 10,000 of the U.S personnel in the Gulf would get danger pay of $110 a month. This alone raised the incremental costs by $1.1 million per month. Ironically, David J. Armor—the acting Assistant Secretary of Defense for Personnel who announced the raise in pay—had told the House Armed Services Committee only three weeks earlier that "the threat to U.S. warships in the Gulf is lower than the threat of terrorism ashore in many Middle Eastern countries. A general threat of terrorism has never been a basis for declaring imminent danger". *Washington Post*, August 27, 1987, p. A-21; *Wall Street Journal*, August 27, 1987, p. 22.

40. *Washington Post*, August 19, 1987, p. A-1, and August 20, 1987, p. A-1; *New York Times*, August 19, 1987, p. A-1.

41. *New York Times*, August 21, 1987, p. A-3 ; *Washington Post*, August 20, 1987, p. A-25.

42. The Saudis denied the cooperation after the details were leaked in the *Washington Post*. These denials were largely pro forma, although the Saudis were careful never to make formal written agreements. *Washington Post*, August 22, 1987, p. A-1, August 23, 1987, p. A-21, August 24, 1987, p. A-1; *Chicago Tribune*, August 23, 1987, p. I-4.

43. *Washington Post*, August 22, 1987, p. A-1; *Economist*, August 29, 1987, p. A-17.

44. *Baltimore Sun*, August 26, 1987, p. 4A; *Washington Post*, August 26, 1987, p. A-1, August 23, 1987, p. A-21, August 24, 1987, p. A-1; *New York Times*, August 25, 1987, p. A-9, August 26, 1987, p. A-1; *Chicago Tribune*, August 26, 1987, p. I-1; *Los Angeles Times*, August 26, 1987, p. I-1; *Christian Science Monitor*, August 19, 1987, p. 11; *Philadelphia Inquirer*, August 26, 1987, p. 1A.

45. *Washington Post*, August 22, 1987, p. A-1, August 23, 1987, p. A-21, August 24, 1987, p. A-1; *Chicago Tribune*, August 23, 1987, p. I-4; *Christian Science Monitor*, August 19, 1987, p. 11.

46. *Washington Post*, August 28, 1987, p. A-16; *Economist*, September 5, 1987, p. 58

47. Ibid.

48. *New York Times*, September 4, 1987, p. A1, September 5, 1987, p. 5, September 6, 1987, p. 3; *Washington Post*, September 4, 1987, p. A-1, September 5, 1987, p. A-1, September 7, 1987, p. A-23; *Washington Times*, September 7, 1987, p. A-2.

49. *Washington Post*, September 3, 1987, p. A-1; *New York Times*, September 3, 1987, p. A-1; *Christian Science Monitor*, September 3, 1987, p. 1.

50. *Jane's Defense Weekly*, October 24, 1987, p. 938.

51. *Jane's Defense Weekly*, September 26, 1987, pp. 671–673.

52. The *Vinh Long* had sailed from Brest on August 17, but had to turn back because of engine problems. It was replaced with the *Orion*.

53. All were old Soviet-made M-08 contact mines which weigh 600 kilograms and have 115 kilograms of explosive. While Iran was reported to have modern influence mines, none were recovered.

54. The Barricade system was fitted to the Hunt-class MCMVs and Abdiel as a result of the attack on the *Stark*. It included chaff, flare, and other

countermeasure rockets. It was fitted in 72 hours. Each Hunt was also armed with two Oerlikon-BMARC 20 mm cannons, a Bofors 40 mm dual purpose gun, and two 7.62 mm machine guns. The mine vessels were given Marconi satellite communications systems, passive "Replica" missile defenses consisting of electronic countermeasures that made the cross-section of the ship seem much larger than it was and created a false apparent location, and "Matilda" missile attack and ESM warning systems. Their sonars were modernized to include the EDO Alconbury 2059 sonar, which linked the standard sonar in the ship to a PAP-104 remotely operated vehicle. This extended the range and pinpointing capability of the ship's sonar system.

55. The Seafoxes are 11 meter light special warfare boats which can go 32 knots and can carry four weapons, including machine guns, mortars, anti-tank weapons, grenade launchers, and 20 mm cannons.

56. *Washington Post*, September 5, 1987, p. A-29.

57. *Washington Post*, August 20, 1987, p. A-25, September 29, 1987, p. A-25; *Economist*, August 22, 1987, pp. 42–43, and September 26, 1987, p. 60; *Journal of Commerce*, September 17, 1987, p. 3B; *Philadelphia Inquirer*, September 16, 1987, p. 10; *Jane's Defense Weekly*, August 29, 1987, p. 361.

58. *Jane's Defense Weekly*, October 24, 1987, p. 941.

59. The group is nicknamed the "deathstalkers" and its motto is "Death waits in the dark".

60. In 1978, the last year the Shah was fully in power, Iran exported $2.9 billion worth of goods and imported $3.9 billion worth. It is important to note that the U.S. oil imports were driven solely by discounting, and that the role of U.S. companies in the world oil market meant that U.S. companies routinely distributed about 25% of Iranian exports because they entered the world market outside the U.S. *Wall Street Journal*, September 2, 1987, p. 18; *Insight*, September 14, 1987, p. 26; *Los Angeles Times*, August 13, 1987, p. I-1, October 1, 1987, p. IV-1; *New York Times*, September 30, 1987, p. A-1; *Washington Post*, September 28, 1987, p. A-1, September 30, 1987, p. A-23.

61. *New York Times*, September 24, 1987, p. A-13, October 1, 1987, p. A-1; *Washington Post*, September 24, 1987, p. A-1.

62. *Washington Post*, October 4, 1987, p. A-1, October 5, 1987, p. A-21; *Philadelphia Inquirer*, October 4, 1987, p. 5A.

63. *Chicago Tribune*, October 6, 1987, p. I-1.

64. *Philadelphia Inquirer*, October 5, 1987, p. 3A; *Chicago Tribune*, October 5, 1987, p. I-5.

65. Ibid.

66. *Washington Post*, November 29, 1987, p. A-41.

67. Guards on an Iranian oil platform in the lower Gulf fired on another U.S. helicopter about 40 minutes later.

68. The navy costs included $20.2 million for fuel, $10.8 million for flying hours, $25 million for transportation of personnel and supplies, $2 million in imminent danger pay, and $1.9 million miscellaneous or a total of $59.8 million. The air force costs were $47.7 million and army costs were $1.5 million. The Coast Guard had spent 1,072 staff days and the U.S. Maritime Administration had spent 68 staff days.

69. *Wall Street Journal*, September 25, 1987, p. 22; *Washington Post*, September 29, 1987, p. A-5, October 9, 1987, p. A-1; *New York Times*, September 29, 1987, p. A-1, October 9, 1987, p. A-1; *Washington Times*, September 29, 1987, p. A-3, October 6, 1987, p. A-12; *Defense News*, October 12, 1987, p. 12.

70. The Congress planned to recess in November. The Administration was required to give the Congress an informal notice of 20 days and 30 days of formal notification during which the vote of two-thirds of the Senate was sufficient to block the sale. This Congressional veto had uncertain constitutionality, but was not something the Administration wanted to challenge. Ironically, the time constraints involved were reinforced by the fact that the Administration had agreed not to send up the sale earlier because this conflicted with the Jewish holidays.

71. *Washington Post*, September 22, 1987, p. A-8; September 30, 1987, p. A-23.

72. The Libyan Foreign Minister, Jadallah Azouz al-Talihi; flew to Iraq in September. The October emissary was Abu Zaid Omar Durda, who met with Khamenei and Mohsen Rafiqdoust. This meeting seems to have led to a new Libyan Scud sale to Iran.

73. *Wall Street Journal*, October 7, 1987, pp. 2 and 52; *Washington Post*, October 7, 1987, p. F-12, October 8, 1987, p. A-34.

74. U.S. companies had 454 ships under the U.S. flag (363 active), and 429 merchant ships registered in other countries, on July 1, 1987; 227 of the foreign flagged ships were registered in Liberia. These U.S.-owned ships flew the flags of 18 different nations. A ship with a foreign crew could cost $700,000 to operate versus $3.4 million for a ship under the U.S. flag which required a 75% U.S. crew. The U.S. flag also meant stringent safety requirements, and that 50% of any repairs done outside U.S. yards had to be paid in duty.

75. *New York Times*, October 19, 1987, p. A-8; *Washington Post*, October 19, 1987, p. A-1.

76. This exercise in "overkill" was supposed to demonstrate that the U.S. could have used far more power if it had wanted to. The guns involved were radar guided guns with track-while scan capabilities coupled to a digital computer. They fire 5" 54 caliber ammunition at a rate of 16 to 20 rounds a minute. Each round cost $1,154 dollars, and the total exercise cost well over $1 million.

77. *Christian Science Monitor*, October 19, 1987, pp. 2 and 16; *New York Times*, October 21, 1987, p. A-10.

78. The President formally complied with the reporting aspects of the War Powers Act and reported the attack on the oil platform.

79. *New York Times*, October 21, 1987, p. A-10; *Washington Post*, October 21, 1987, p. A-27.

80. *New York Times*, October 20, 1987, p. A-1, October 23, 1987, p. A-1, October 24, 1987, p. 3; *Washington Post*, October 20, 1987, pp. A-1 and A-26, October 23, 1987, p. A-1; *Christian Science Monitor*, November 30, 1987, p. 9.

81. The U.S. also announced it would send five dolphins to the Gulf to improve its mine detection capability.

82. Senator Dennis DeConcini and Representative Mel Levine announced they would introduce legislation to block the sale of Stinger for one year because

because of the risk of transfer to Iran that had been illustrated by the fact that Iran had gotten six systems from the Afghan Mujahideen.

83. The general sources for the discussion of the war during the first three Iranian Silkworm strikes on Kuwait include *Economist*, October 12, 1987, p. 41; *Washington Post*, October 2, 1987, p. A-16, October 7, 1987, p. A-23, October 8, 1987, p. A-10, October 13, 1987, p. A-1, October 14, 1987, pp. A-1 and A-22, October 15, 1987, p. A-31, October 16, 1987, p. A-34, October 19, 1987, p. A-15, October 21, 1987, pp. A-1 and A-28, October 22, 1987, p. A-32, October 23, 1987, p. A32; *Christian Science Monitor*, October 2, 1987, p. 11, October 8, 1987, p. 9; *New York Times*, October 2, 1987, p. A-8, October 16, 1987, p. A-8; *Washington Times*, October 8, 1987, p. A-8, October 13, 1987, p. A-11, October 16, 1987, pp. A-2 and A-10; *Time*, October 19, 1987, p. 12; *Wall Street Journal*, October 14, 1987, p. 5, October 15, 1987, p. 2, October 22, 1987, p. 35; *Chicago Tribune*, October 14, 1987, p. I-1; *USA Today*, October 14, 1987, p. 4A; *Newsweek*, October 12, 1987, p. 51; *New York Times*, November 1, 1987.

84. *Washington Times*, November 4, 1987, p. A-8; *New York Times*, November 1, 1987, p. 27; *Los Angeles Times*, November 5, 1987, p. I-7.

85. *Washington Post*, October 26, 1987, pp. A-18 and A-19; October 27, 1987, p. A-23; *Chicago Tribune*, October 26, 1987, p. I-5; *New York Times*, October 27, 1987, p. A-13.

86. *Philadelphia Inquirer*, November 4, 1987, p. 9A.

87. *Washington Post*, October 31, 1987, p. A-2; *Washington Times*, October 8, 1987, p. I-6.

88. The attack on October 21 hit the Khark 4, a 284,000 dwt VLCC.

89. By this time, the oil platform at Rostam had been burning for 17 days.

90. *Washington Post*, November 7, 1987, p. A-16; *Washington Times*, November 7, 1987, p. I-6.

91. *Washington Post*, November 12, 1987, p. A-25.

92. *Washington Post*, November 17, 1987, p. A-22, November 18, 1987, p. A-29; *Chicago Tribune*, November 21, 1985, p. I-5; *Philadelphia Inquirer*, November 30, 1987, p. 13D.

93. The U.S. force ended up consisting of five, rather than six, Aggressive-class MSOs. Three arrived under their own power from the Atlantic fleet. Two more were towed from the west coast of the U.S. by an amphibious landing ship. The third MSO developed engineering problems and had to return to base after service in Pearl Harbor. The MSOs coming from the east coast did not enter the Gulf until late October. The Dutch force consisted of the *Hellevoetsluis* and *Maassluis*. It was operating with the British force in the Gulf of Oman and lower Gulf. It was then committed for 10 weeks. The Netherlands had still not finalized efforts to sell Kuwait two Alkmaar-class minehunters.

94. *Washington Post*, November 28, 1987, p. A-23.

95. *Philadelphia Inquirer*, November 22, 1987, p. 9a; *Chicago Tribune*, November 21, 1987, p. I-5; *Washington Times*, November 23, 1987, p. A-8.

96. *Washington Post*, October 31, 1987, p. A-2; *Washington Times*, October 8, 1987, p. I-6.

97. *Philadelphia Inquirer*, November 25, 1987, p. 5D.

98. *Washington Times*, November 20, 1987, p. A-2.

99. *Washington Post*, November 3, 1987, p. A-32; November 4, 1987, p. A-1; *Washington Times*, November 5, 1987, p. A-9.

100. *Washington Post*, November 2, 1987, p. A-22; *New York Times*, November 2, 1987, p. A-1.

101. *Washington Post*, November 6, 1987, p. A-32; *Philadelphia Inquirer*, November 3, 1987, p. 10A.

102. *New York Times*, November 12, 1987, p. A-9.

103. Nakasone had earlier considered sending mine vessels and then Coast Guard boats to the Gulf, but could not obtain the support of his Cabinet. Japan's politicians generally took the position that Article 9 in the constitution prohibited any military action other than defense of the homeland.

104. *New York Times*, November 1, 1987, p. 3; *Economist*, November 7, 1987, pp. 40–41; *Washington Post*, November 29, 1987, p. A-29; *Washington Times*, November 20, 1987, p. A-8.

105. *New York Times*, November 4, 1987, p. 3, November 11, 1987, p. A-3, November 12, 1987, p. A-1; *Economist*, November 14, 1987, p. 48; *Washington Post*, November 5, 1987, p. A-25, November 11, 1987, p. A-25; *Washington Times*, November 4, 1987, p. A-9, November 12, 1987, pp. A-1, A-8, and F-2; *Chicago Tribune*, November 4, 1987, p. I-4; *Wall Street Journal*, November 11, 1987, p. 23, November 12, 1987, p. 35; *Christian Science Monitor*, November 13, 1987, p. 1.

106. *New York Times*, November 4, 1987, p. 3, November 11, 1987, p. A-3, November 14, 1987, p. 7; *Economist*, November 14, 1987, p. 48; *Washington Post*, November 13, 1987, p. A-29; *Washington Times*, November 4, 1987, p. A-9, November 12, 1987, pp. A-1, A-8, and F-2; *Chicago Tribune*, November 13, 1987, p. I-5; *Wall Street Journal*, November 11, 1987, p. 23, November 12, 1987, p. 35; *Christian Science Monitor*, November 13, 1987, p. 1.

107. *Economist*, November 28, 1987, pp. 33–34; *New York Times*, November 1, 1987, p. 26.

108. *Washington Post*, November 28, 1987, p. 1.

109. *Med News*, Vol. 1, No. 5, November 30, 1987, pp. 1–2.

110. *Washington Post*, December 4, 1987, p. A-24.

111. Iran had obtained 16–31 Stingers. Sources disagreed over whether it had bought 16 from Mujahideen commanders in Western Afghanistan for $1 million had seized 25 when Afghan units accidently crossed the Iran-Pakistani border and skirmished with Iranian units in May, 1987. *Washington Times*, October 29, 1987, p. A-10, November 23, 1987, p. A-8, December 2, 1987, p. A-2; *Washington Post*, November 19, 1987, p. A-10, December 2, 1987, p. A-25, December 8, 1987, p. A-25; *Philadelphia Inquirer*, November 17, 1987, p. 10A; *New York Times*, October 17, 1987, p. 1, October 19, 1987, p. 1A; *Baltimore Sun*, October 19, 1987, p. 1A.

112. Iraq banned all sales to Iraq by Luchaire in early November. It also banned sales by SNPE (Societe Nationale des Poudres et Explosives), although it needed caseless propellant charges for its 76 remaining French-built 155 mm GCT self-propelled howitzers. Iraq had bought these in 1982 as part of a $1.6 billion "Vulcan Project".

113. *New York Times*, November 2, 1987, pp. A-3 and A-6, November 30, 1987, p. A-1, November 11, 1987, p. A-3, November 14, 1987, p. 7, December 3, 1987, p.

A-15; *Economist*, November 14, 1987, pp. 53–51; *Washington Post*, November 17, 1987, p. A-20, November 30, 1987, p. A-1, December 1, 1987, pp. A-1 and A-30, December 2, 1987, p. A-14, December 8, 1987, p. A-24; *Washington Times*, November 3, 1987, p. A-8, November 6, 1987, p. A-9; *Defense News*, November 9, 1987, p. 12; *Baltimore Sun*, December 8, 1987, p. 2A.

114. *Chicago Tribune*, January 22, 1988, p. I-5; *Washington Post*, January 21, 1988, p. A-25.

115. *Baltimore Sun*, January 2, 1988.

116. Based on USCENTCOM briefing data.

117. *Washington Times*, December 23, 1987, p. A-8; *Philadelphia Inquirer*, December 23, 1987, p. 7A.

118. These are Department of Energy working data.

119. *New York Times*, February 4, 1988, p. D-18; *Economist*, January 30, 1988, p. 32.

120 According to State Department sources, Iran received about 400 towed artillery weapons from China during 1987, mostly D-30s. It received about 90 Austrian weapons, and some modern North Korean self-propelled artillery weapons.

121. *Philadelphia Inquirer*, November 30, 1987, p. 13D, December 4, 1987, p. 16C; *New York Times*, December 23, 1987, p. A-3; *Washington Times*, December 7, 1987, p. A-8; *Washington Post*, February 11, 1988, p. A-48; *Jane's Defense Weekly*, November 28, 1987, p. 1247; *Economist*, December 26, 1987, p. 47, and February 13, 1988.

122. Iraq now had massive production facilities for mustard and Tabun gas, and air, artillery, and helicopter delivery systems. It still, however, had poor delivery technology. Many gas weapons did not explode, and Iraq discontinued its use of gas warfare during part of the Iranian offensive on Basra because it could not discriminate its targets well enough to avoid casualties among its own troops.

123. *Philadelphia Inquirer*, December 23, 1987, p. 7A; *New York Times*, December 23, 1987, p. A-3.

124. *Washington Times*, January 25, 1988, p. A-1; *New York Times*, January 25, 1988, p. A-3; *Los Angeles* Times, January 20, 1988, p. I-5.

125. *New York Times*, December 22, 1987, p. A-7; *Washington Times*, December 21, 1987, p. A-8.

126. *Washington Post*, December 10, 1987, p. A-13, December 17, 1988, p. A-46, December 25, 1987, p. A-1, December 26, 1987, p. A-25, December 29, 1987, p. A-12, January 15, 1988, p. A-25, February 10, 1988, p. A-32, February 21, 1988, p. A-1; *Washington Times*, December 8, 1987, p. C-5, January 6, 1988, p. A-8, February 12, 1988, p. A-8; *Philadelphia Inquirer*, December 18, 1987, p. 18C, January 22, 1988, p. 1-A; *New York Times*, December 4, 1987, p. A-13, December 14, 1987, p. D-10, December 17, 1987, p. D-1, January 18, 1988, p. A-6, February 21, 1988, p. E-23; *Los Angeles Times*, January 20, 1988, p. I-5; *Wall Street Journal*, January 29, 1988.

127. *Economist*, January 30, 1988, p. 34; Christian Science Monitor, March 2, 1988, p. C-9.

128. *Washington Times*, December 8, 1987, p. C-5; *New York Times*,

December 10, 1987, p. D-1, December 14, 1987, p. D-10, December 17, 1987, p. D-1; *Wall Street Journal*, December 15, 1987, p. 3, December 17, 1987, p. 3.

129. *New York Times*, January 10, 1988, p. A-3.

130. *Washington Times*, January 19, 1988, p. A-8; *Philadelphia Inquirer*, January 18, 1988, p. 8A.

131. *Philadelphia Inquirer*, February 10, 1988, p. 26.

132. Iraq's Scud holdings, and their source, was uncertain. Iraq was believed to have about 50 Scud missiles before it began this series of attacks, but the number of attacks rapidly showed its holdings were far larger. Rafsanjani claimed Iran had evidence that the missiles were standard Scud Bs, which used reduced warhead weight on March 8, 1988. *Washington Times*, March 1, 1988, p. 3; and *Washington Post*, March 9, 1988, P. A-19.

133. Iraq had announced it had tested a missile with a range of about 400 miles, or 650 kilometers, in August, 1987. These reports were dismissed as propaganda. In practice, it seems likely at this writing that Iraq may have gotten the added range by taking a standard Scud and modifying it by (a) cutting its payload from 400 to 200 kilograms, (b) altering it to burn all the propellant at the cost of reliability, or (c) doing both. The Iraqis regularly moved the missile launch sites during this phase of their attack. *Economist*, March 5, 1988, p. 44; *New York Times*, March 2, 1988, p. A-1, March 4, 1988, p. A-8, March 12, 1988, p. A-3; *Washington Post*, March 2, 1988, p. A-16; *Baltimore Sun*, March 6, 1988, p. 2-A.

134. Soviet officials denied that the Scuds given Iraq could hit Tehran on March 10, 1988.

135. *New York Times*, March 6, 1988, p. 1; *Economist*, March 26, 1988, p. 34..

136. Some reports indicate that Iraqi bombers dropped cyanide gas in 100 liter containers that vaporized on impact. An Iraqi general captured in the fighting, Brigadier Nather Hussein Mustafa, claimed on Iranian TV on March 24, 1988, that he saw clouds of gas arise over Halabja. Two other captured Iraqi officers confirmed this. The casualty figures are very controversial. So are the data on the population of the towns involved. Halabja is said to have had a population of up to 70,000, but nearly half the population seems to have fled after an uprising against the Iraqi military authorities failed in May, 1987. Dojaila has a population of around 20,000.

137. It is easy to underestimate the importance of the moral sanctions against the use of gas. In the 35 years between the end of World War I and the end of the Korean War, there were only two confirmed uses of lethal gas—Italian use of mustard gas against the Ethiopians during 1935–1936 and Japanese use of gas against the Chinese during 1937–1942. *Washington Post*, March 24, 1988, p. A-37; *Economist*, April 2, 1988, pp. 35-37.

138. Some cooperation began in 1986, after the failure of talks between Talabani and the government. The KPUK normally is strongest in the area around Sulaimaniyah and the KDP is strongest to the north and west. The Barzani clan is now headed by Massaoud Barzani, the son of the former leader, Mustafa Barzani. See Patrik E. Tyler's excellent analysis in the *Washington Post*, February 19, 1988, p. A-15.

139. *New York Times*, April 21, 1988, p. 8.

140. *Washington Post*, March 28, 1988, p. A-17, April 15, 1988, p. A-26; *Washington Times*, March 29, 1988, p. A-8.

141. *Washington Post*, April 18, 1988, p. A-17, April 19, 1988, p. A-22.

142. *New York Times*, April 14, 1988, p. 1 and 14.

143. The *Samuel B. Roberts* was an Oliver Hazard Perry-class guided missile frigate. It was 453 feet long, displaced 3,740 tons at full load, and had a crew of 225. It was armed with SAMs, SSMs, a 76 mm gun, torpedoes, and a close in weapons system (CIWS). It has active sonar on the hull and towed array sonars.

144. The *Roberts* was towed to Bahrain by the *U.S.S. Wainwright* and arrived on April 16. Three of the ten wounded were seriously injured.

145. The U.S. had found 44 mines in the Gulf, and 16 of them in 1988. The most recent mine detection was on April 9, 1988. All of the previous mines were encrusted with marine growth and showed signs of having torn away from their moorings. *New York Times*, April 15, 1988, p. 3.

146. The *Merrill* was a destroyer, the *Lynde McCormack* was a destroyer, and the *Trenton* was an amphibious transport. The *Simpson* was a guided missile frigate, the Bagley was a frigate, and the *Wainwright* was a guided missile cruiser. The *Joshan* was a 154-foot Combattante II–class French attack boat with Harpoon missiles and a complement of 21. The *Sahand* and *Sabalan* were Vosper-5 class British frigates commissioned in 1972. They were part of a force of four such vessels with a complement of 125 and a length of 310 feet. The Boghammars were 43 foot long speed boats. The Harpoon missile was introduced in 1977, and upgrade in the early 1980s. Only the U.S. Navy had the improved version. It has a range of over 50 nautical miles.

147. Sources differ as to whether the two platforms were then producing 30,000 or 150,000 BPD.

148. The field was operated by the Crescent Petroleum Company and was owned by Sharjah, one of the most pro-Iranian members of the UAE.

149. The Skipper 2 is a 1,000 pound bomb modified with a Shrike missile motor for greater stand-off capability.

150. Like all the U.S. ships in the Gulf, the *U.S.S. Jack Williamson* was equipped with the AN/SLQ electronic surveillance system, Mark 36 rapid super blooming chaff, and two Phalanx close-in protection systems. These systems seem to have worked well, indicating that the primary problem with the *Stark* was its lack of readiness and other human effort. *Aviation Week and Space Technology*, May 25, 1987, p. 23, and April 25, 1988, pp. 20–22.

151. The war on the cities had done comparatively little physical damage, but many people had fled the cities. Iran forced them back by issuing ration books to those who both returned and voted.

152. At this point in time, Iran still had 3 destroyers, but they were at best capable of limited service, if any. It had 2 frigates, 2 corvettes, 8 fast attack craft, 10–12 hovercraft, 1 minesweeper, 40–50 speed boats and 70 or more small fast motorcraft. The U.S. forces in the area had a total of 29 ships, with 21 in the Gulf and a carrier battle group outside it. The ships in the area included the command ship *Coronado*, the guided missile cruiser *Wainwright*; the guided missile destroyers *McCormack* and *Strauss*; 2 guided missile frigates including the *Simpson*, 2 frigates including the *Bagley*, destroyers including the *Merrit*, the

amphibious landing ship *Trenton*, 6 minesweepers, and 1 support ship. The carrier *Enterprise* was in support outside the Gulf. These ships were due for rotation of the carriers, with the carrier *Forrestal* and its task force replacing the *Enterprise*, and 5 of the frigates being rotated. This gave the U.S. the option of nearly doubling its effective force.

153. *Washington Post*, April 19, 1988, pp. A-1 and A-22-23, April 20, 1988, p. A-28; April 21, 1988, p. A-26, April 22, 1988, pp. A-1 and A-18; *New York Times*, April 19, 1988, p. 1 and 11; April 21, p. 4, April 22, 1988, p. 8; *Baltimore Sun*, April 19, 1988, p. 1, and April 21, 1988, p. 4.

154. The CSS-2 was designated the DF3A by the PRC. It first appeared in 1971, and is 67 feet long. It is a single stage liquid fueled missile, and the PRC initially deployed 50 to fixed sites. The original version had a range of 1,550 miles, but the improved version had a range of 1,840 to 2,200 miles. The PRC deployed 90 of these improved missiles with 2–3 megaton warheads. The Saudis were sold the improved version. The Saudis claim that the PRC modfied the missile to provides an improved warhead weight of 1,650 pounds versus 1,000 pounds for the regular CSS-2. This makes it difficult to estimate the range. Some estimates are as low as 1,500 miles.

To put this system into perspective, the Scuds in the Syrian forces had a range of 180 miles and the FROGs had a range of 70 miles. The extended range Scuds that Iraq used against Iran had ranges in excess of 300 miles. The Israeli Jericho II missile had been tested at ranges of over 500 miles.

The announcement of the sale came one week after Chinese Foreign Minister Wu Xuegian had visited Washington and the Reagan Administration had announced it was resuming the transfer of advanced technology that the U.S. had halted because of PRC Silkworm sales to Iran. Xuegian later defended the sale on the grounds that the Saudi government had made a commitment to only use the missiles for "defensive purposes".

155. *Jane's Defense Weekly*, April 2, 1988, p. 627; *Aviation Week*, March 28, 1988, p. 30; *Economist*, March 26, 1988, pp. 40–41, April 9, 1988, pp. 35–36; *Washington Times*, March 21, 1988, p. A-4; *Christian Science_Monitor*, March 30, 1988, p. C-11; *Washington Post*, March 29, 1987, p. A-1, April 7, 1988, p. A-28.

156. The sale was then already past the 20 day period of informal notice to the Congress and one week into the formal 30 day notice period.

157. *New York Times*, April 15, 1988, p. 11; *Defense News*, April 11, 1988, p. 23; *Washington Post*, April 26, 1988, pp. A-1 and A-9.

158. The War Powers Act was still an issue at the time this analysis concludes. The Senate decided on December 4, 1987, to effectively table the resolution by Senator Brock Adams that would have forced a vote on invoking the War Powers Act on December 20. It left the resolution open, however, and it could be revived and invoked by a majority vote under conditions which made it easy to override a filibuster. The new fighting in the Gulf on April 18, 1988, made it seem likely that the President would report hostilities and trigger the 60 day period in the Act in which both houses of Congress had to approve the Act by a majority vote.

159. This analysis ends as of April 20, 1988. The author is as conscious of his lack of prophetic powers as is the reader.

12
The Future of Western Strategic Relations with the Southern Gulf

The events of the last few years have thrust the West and the Southern Gulf states together. The forces involved, however, are unstable. They can either forge a lasting partnership or end in creating new divisions. For all of the complementary strategic and military interests outlined in this book, the case studies in the last three chapters nations involved still face serious difficulties in creating a stable pattern of strategic relations.

These difficulties are not simply a function of the inability to predict the outcome of the Iran-Iraq War, or a function of the problems inherent in Western naval intervention in the Gulf. Neither the West nor the Gulf states are collective entities. Any future approach the West and the Gulf states take to regional security will be based on a fragile mix of the different interests of the U.S., the individual states in Europe, and the states in the GCC. It will depend upon difficult and unstable compromises among all of the states involved.

The previous analysis has shown that a successful approach to strategic relations must deal with the following basic problems:

- It must recognize the differences in the interests and policies of individual Western and Gulf states. It must recognize the differences between U.S. and European policies and interests and adapt to them. It must recognize that Kuwait, Bahrain, Qatar, Saudi Arabia, the UAE, and Oman have different interests and seek suitable compromises, not forced integration or consensus.
- It must encourage the development of an effective deterrent in the Southern Gulf states that is based on a regional capability to defend against naval and air attacks by either Iran or Iraq, which can check any limited land attack before it seizes Kuwait, and which can help any of the Southern Gulf regimes if they are threatened by any combination of external pressure and internal subversion.

- It must use technology as a substitute for the manpower problems in every Southern Gulf state. The most critical areas concern the creation of effective air and naval units that can shield the Southern Gulf states from attacks by the Northern Gulf states and preserve the flow of shipping through the Gulf. At the same time, there will be a need for land forces that can rapidly reinforce each Southern Gulf state, and for an effective defense of Kuwait from its northern neighbors and Oman and Saudi Arabia from the Yemens and radical Red Sea states.
- It must encourage the kind of military relations between the West and the Southern Gulf states that produce effective military forces, and not simply expensive arms sales. This requires stable military relations between supplier and buyer states, effective military training and assistance efforts, and a real Western effort to help the Southern Gulf states to force integrated and effective forces.
- It must minimize the political and cultural liabilities of a Western military presence in the Gulf by avoiding formal basing arrangements and minimizing Western military deployments when these are not forced by a crisis or actual conflict. Western and Gulf military relations will be strongest and least threatening when they are informal and the West remains largely over the horizon. They will be most dangerous when they have high visibility, involve formal basing arrangements, and require Western forces to be present in the area even when no conflict or crisis exists.
- It must recognize the fundamental difference between U.S. and European power projection capabilities. Europe can continue to play a very important political role, be a major arms seller and advisor, and provide limited military forces in contingencies where U.S. intervention is less politically desirable. Only the U.S., however, will be able to project major land forces, large amounts of air power, and the kind of naval power that can dominate virtually any regional contingency.
- It must seek to end the Iran-Iraq War without Iraq coming under Iranian domination, but in a way that allows the Iranian revolution to work out its own destiny as long as it does not attempt to export its ideology or political authority by force. At the same time, it must recognize that both states may be hostile for decades to come, and that the Iranian revolution may not produce a stable regime for years to come.
- It must offer Iraq a future in which economic and strategic cooperation with the West and Southern Gulf states offers more advantages than competition and hostility.

- It must contain Iran without attempting to isolate it. It must offer Iran an economic and political future in which both the West and the Southern Gulf states are willing to respond to any Iranian efforts to normalize political and economic relations.
- It must recognize the reality that isolating the USSR from the Gulf is no longer feasible or even desirable. The challenge is to create the kind of political and economic relations that will minimize any incentive for Soviet political opportunism or military adventurism. The most practical solution is full diplomatic and economic relations between the USSR and each Southern Gulf state.
- It must recognize that there is little immediate prospect of a general Arab-Israeli peace settlement and that the U.S. divisions with its European allies and friendly Gulf states over American support of Israel are going to continue well into the 1990s, if not much longer.

The basic balance of strengths and weaknesses necessary to create an effective strategic partnership is clear. The Southern Gulf states can and must accept responsibility for their internal security, for dealing with terrorism and low-level threats, and for creating the kind of infrastructure in terms of bases, C^3I/battle management systems, and interoperable equipment and supplies that will make Western power projection effective.

The West must develop effective power projection capabilities which can intervene from "over the horizon" to avoid involvement in local internal political matters. The West must limit its need for regional facilities to prepositioning and contingency bases. It must keep a low profile in the face of the resurgence of Islam, and it must rely on a combination of arms sales and military advisory efforts to build up the strength of friendly states.

The core of such a security system is already in place. The Gulf Cooperation Council may only be a beginning, but it is a beginning. Its members may not have bought wisely, and they may not have the kind of interoperable and integrated forces they need, but they have bought many of the elements of such forces. Saudi Arabia will never be a military "pillar" of the kind the U.S. once sought in Iran, but it can form the core of an effective GCC-wide military effort.

The West may lack many of the power projection capabilities it needs, but the U.S. is developing effective rapid intervention forces that can rapidly deploy to Bahrain, Saudi Arabia, and Oman. Britain and France can provide considerable support against terrorism and low-level violence.

If USCENTCOM and Western naval forces can operate in cooperation with Saudi and other Southern Gulf forces, and *if* Western power projection forces can use bases in Saudi Arabia and Oman, the resulting combination of Western and regional strength should be able to secure the Southern Gulf. It should be able to deal both with the future threat from the Northern Gulf states and with any Soviet-backed threats from outside the region.

The primary challenge is not one of inventing new policies and plans, but rather one of making existing policies effective, and linking them together in a way that will produce effective results. Europe needs to take a more realistic attitude towards making its arms sales militarily effective, and structuring its sales and military assistance to include U.S. over-the-horizon capabilities. Saudi Arabia and the Gulf states need to face their need for mutual interdependence and their ultimate dependence on the U.S. The U.S. needs to show the courage to put Western strategic interests before domestic political advantage.

DEALING WITH THE DIVISIONS BETWEEN WESTERN AND GULF INTERESTS

There is no easy way to bring the Gulf and Western states together. No treaties are possible, and neither the Southern Gulf states nor the key Western states are ever likely to be in full agreement on any given issue or course of action. As the previous chapters have discussed, the Southern Gulf states remain divided in their approach to security policy, the development of their military forces, and their relations with external states—including Iran.

At the same time, the events of 1987 have shown that the Southern Gulf states can cooperate with each other and with the West, and that most of their regimes are remarkably stable under pressure. There also is no requirement to fully unify the Gulf states in a collective security effort. It will be enough to create a loose coalition where their differences remain at acceptable levels. The Southern Gulf has already shown that this is possible, and that it may well be possible to create a more integrated political and military structure over time.

The balance of individual capabilities within the Southern Gulf is now fairly clear. The key military capabilities will be those of Saudi Arabia and Oman. Kuwait will remain a critical political and economic actor, and hopefully will continue to cooperate with Saudi Arabia in building up the kind of contingency capabilities necessary in the upper Gulf. Bahrain can be expected to maintain close relations with the U.S.

and Saudi Arabia, and Qatar can be expected to follow the Saudi lead in most respects. The UAE will be a continuing problem, largely because of the feuding between its smaller or less powerful states and Abu Dhabi. A combination of Abu Dhabi and the federal bureaucracy should, however, be sufficient to keep at least the UAE moving in the right direct.

Given this mix of strengths and divisions, the priority for both regional and Western policy making should be to strengthen Southern Gulf military forces, particularly those of Saudi Arabia, in the areas which will have the most value in deterring regional threats and dealing with low-level wars. At the same time, it will be necessary to establish loose informal links to the power projection capabilities of the U.S. and those European states which retain an active interest in the region.

The West too must learn to live within the limits enforced by its divisions. It is easy to talk about the problems the Gulf states have in achieving any degree of unity, and in establishing some common policy towards the Iran-Iraq War. The West, however, is in a poor position to criticize the Southern Gulf states. The U.S. and Europe obviously lack unity in their approach to the Gulf, and one has only to propose some concept of unified Western action to realize how unlikely it is to ever take form.

The U.S. has taken a much stronger and more interventionist approach to the Iran-Iraq War than has been favored by any European state or Japan. This interventionism, however, has involved rapid and contradictory swings in policy. The U.S. first tried to influence the future course of the Iranian revolution by covert arms transfers. It then tried to secure its position in the Gulf by its decision to reflag the Kuwaiti tankers, and by efforts to force Iran to agree to a cease-fire. In the process, the U.S. thrust itself into a military role in the Gulf which may well bring it into direct conflict with Iran, and which has involved much larger military commitments than its allies would otherwise have made.

As has been suggested earlier, however, the problem is not that the U.S. has chosen to act where much less powerful and much more divided European states have not. The problem is rather that U.S. political appointees and officials have acted without making use of the U.S. national security system, and without doing enough to seek the support of a few key European states like Britain, France, Italy, and Turkey.

The practical challenge the U.S. faces is to both improve the professionalism of its initiatives in the Gulf, and to find better approaches to obtaining the support of those European states that have the strength and will to act. For all its recent mistakes, it is the U.S. which must shape the West's approach to the Gulf for the foreseeable future, and the best the West can hope for is that the kind of ad hoc and informal cooperation that developed between the U.S. and its allies in 1987 can be strengthened and sustained in the future.

If the key states in the West and the Southern Gulf accept the fact that the best that is now obtainable is improved coordination, and some degree of integration and interoperability in key areas like air and naval defense, the differences between individual Southern Gulf and Western states should not be critical. The more radical movements and hostile states in the region are, after all, far more divided and unstable than the moderate and relatively friendly Southern Gulf states.

CREATING AN EFFECTIVE REGIONAL DETERRENT

It is easy to concentrate so much on the political and military weaknesses of the Southern Gulf states that one ignores their potential strengths. Geography, however, helps the Southern Gulf states as well as hurts them. Iran is the most dangerous current threat, but it must defeat Iraq to invade Kuwait, and its current threat to the other Gulf states is limited to small amphibious, air, and missile strikes. For at least the next five years, Iran's air and naval forces will be so limited that the Southern Gulf states should have considerable deterrent and defensive capability. Even when Iran does begin to rebuild its forces, it is unlikely to have anything approaching the surplus wealth commanded by the Shah. It will have effective land forces, but only limited opportunity to use them.

If Iraq is not defeated by Iran, it will emerge from the war as one of the largest military powers in the Third World. It also, however, will constantly have to consider the risk of a new round of fighting with Iran, and it will have little incentive to challenge the Southern Gulf states beyond efforts to seek some improvement in its access to the Gulf. Even if Iraq is tempted towards adventures, it is unlikely to be tempted towards another round of recklessness. It may well be able to conquer Kuwait, and even threaten Saudi Arabia, but it is far from clear that it will do so if this means a major air engagement and facing the risk of unified Kuwaiti and Saudi action backed by military support from the West. The challenge to the Southern Gulf in creating an effective deterrent to Iraq is not to create the forces to defeat it, but rather to have sufficient forces to discourage it. Unless a truly radical regime takes power in Iraq, this should be enough.

Similarly, the potential threats in the Red Sea areas and the West may rush into a power vacuum, and exploit the opportunity to win a low-level war, but they are not strong enough to challenge the kind of forces Oman and Saudi Arabia can create, particularly if they realize that Oman and/or Saudi Arabia will be backed by the West. The example of "Arabia without Sultans" in the Yemens also scarcely seems likely to catalyze internal support in the Southern Gulf states.

The Soviet threat is also far more likely to be one of regional opportunism than deliberate aggression or strategy. It may take the form of backing some new regime in the Yemens or some action in the Red Sea by Ethiopia or some future regime in the Sudan. It may take the form of intervention in some future struggle for power in Iran or even Iraq. It is very unlikely, however, to involve high-level war. If it does, this must inevitably be dealt with by the U.S., and not simply by the regional states.

The internal security problems in the Southern Gulf states also are not inherently serious enough to require outside forces, although they do merit outside expertise and every possible effort to ensure that the Southern Gulf states can reinforce each other in an emergency. No internal security apparatus can ultimately save a regime from its own people, but this is not a serious issue in any Southern Gulf state. There is enough wealth, distributed well enough, so that the peoples of the Southern Gulf states can accept the slower but inevitable expansion of each nation's political base that virtually all of their rulers realize must come in time.

The question, therefore, is how to create the relatively limited mix of air, naval, and land forces that are actually required. This is not a matter of more spending, or even of more manpower. It is a matter of using today's resources more wisely.

Given the stresses within the Southern Gulf and the West, it is pointless to recommend that the Southern Gulf states should pursue the creation of fully integrated military forces, work from common force plans, integrate training and support systems, and carry out the kind of large-scale common exercises that have real military meaning. Even NATO is still unable to fully implement this level of collective security, and if the Gulf states wait until it can be achieved, they will wait forever.

What the Southern Gulf states need to do most is to prioritize, and to focus on actually implementing several of the concepts they have already discussed. To be specific, they need to achieve the maximum possible degree of integration and interoperability in the following areas:

- The key missions for Southern Gulf forces are to establish a land defense perimeter in northern Kuwait and a mix of air and naval forces that can create a defense zone in the middle of the Gulf to deal with threats to Gulf shipping, offshore oil facilities, amphibious attacks, and direct air threats to Southern Gulf territory.

The creation of a perimeter defense of Kuwait cannot guarantee Kuwait against an all-out assault by a victorious Iran, or a new and hostile regime in Iraq, but it can buy time and deter against low-level threats and political and military intimidation. Such a defense will depend largely on Saudi and Kuwaiti cooperation, with possible reinforcement by Egypt or Jordan. It would involve creating a three brigade Saudi-Kuwait force in Hafr al-Batin and Kuwait's military city, creating barrier defenses at the Kuwaiti border, and tailoring Kuwaiti forces for rapid defense of the border area. All of these plans now exist in some form, and the issue is one of making them real, not one of creating new concepts.

Similarly, most of the elements necessary to establish a defense zone in the Gulf also exist. As has been discussed earlier, the GCC has had plans to create such capabilities for several years. Once again, the issue is one of making them real, not one of creating new concepts. There are also areas where high-technology air forces and small ships can perform the mission with relatively limited manpower.

Ideally, Saudi Arabia and Oman should cooperate in creating a similar mix of perimeter defense capabilities, and air-naval defense capabilities to deal with the threats in the Yemens. Both nations have already created most of the defense capabilities they need to deal with the threats against them, however, and a joint approach is one that requires considerable political evolution in their willingness to cooperate. Such cooperation is desirable, but scarcely critical.

Creating limited air mobile forces with attack helicopters, and highly mobile paramilitary forces tailored to deal with terrorist or internal security threats, is also a high-priority mission area. The Southern Gulf states tend to focus far too much on creating regular armored and mechanized forces, and far too little on the critical importance of time, mobility, and the kind of cross-reinforcement that can halt narrowly based coups or attacks. The GCC has talked about a rapid deployment force, but this is the kind of force that is needed just as soon as it can be created.

Once again, the elements are also present. Saudi Arabia and several other Southern Gulf states are already planning to create attack helicopter and expanded heliborne forces. Enough progress has also been made in international security cooperation to overcome at least some of the past rivalries that acted as a barrier to the creation of such forces.

- The highest force improvement priority in the Southern Gulf is not for more military forces, but rather for a GCC-wide communications, air control and warning, and battle management system. This means linking the air battle management and maritime surveillance capabilities of Saudi Arabia's E-3As together with the added airborne early warning and maritime patrol aircraft necessary to provide survivable sensor coverage of the entire Gulf.

  It will be particularly important to set up fully integrated digital data links and common identification of friend or foe (IFF) systems. The Iran-Iraq War has already demonstrated that Gulf forces will not have the warning times to deal with most naval and air threats without highly automated data systems. They also are certain to lack the trained manpower to substitute for such systems for many years to come. A common and technically sophisticated IFF system is critical both for cross-reinforcement purposes, and to allow air, surface-to-air, and naval forces to work together and to substitute for the lack of numbers or "mass" in the local military forces.

- The Southern Gulf is very unlikely to standardize on equipment, or to create common force plans. It can still, however, standardize on some of the elements necessary to create forces which are interoperable with other Southern Gulf forces and with U.S. forces. Aside from the areas of C^3I capability mentioned above, the highest priorities for such standardization lie in ammunition and bulk supplies such as POL, food, and medical supplies, and in creating high-capacity repair and service facilities.

  It is probably too late to standardize on air-launched or naval missiles, but it should be possible to standardize on the avionics and naval electronics, software, and weapons mounting systems to allow a high degree of interoperability. It should be possible to standardize on artillery ammunition and most tank ammunition since GCC forces already have near standardization in these areas, and this standardization includes many high-bulk and hard-to-move items used by the U.S. Army.

  Another important area in which standardization is still possible lies in the procurement of Improved Hawk surface-to-air missiles. The one U.S. competitor to the Hawk, the Patriot, is too sophisticated and too expensive. There is no Western European competitor to such systems, many Gulf states already have them or have them on order, and they provide good area coverage to supplement fighter defenses.

- The other Southern Gulf states should follow the Saudi example of overbuying munitions and high-bulk supplies. Most military contingencies in the Gulf are likely to involve critical reaction times, and the main limits in reinforcement capabilities are often in air lift and rapid sea lift. Major overbuys can sharply reduce the reaction times necessary for the U.S. and Europe to deploy their power projection forces. Further, such overbuys do not require commitments by either individual Southern Gulf or Western states. Overstocking, particularly if quiet attention is paid to key U.S. requirements as well as GCC-wide requirements, can help make deterrence credible with minimal political complications.
- Many of the Southern Gulf states already have excellent military bases. This is particularly true of Kuwait, Saudi Arabia, and Oman. It is critical, however, that the other countries develop the kind of air and naval bases which are well defended and sheltered, and which have sufficient over-capacity to allow rapid reinforcement. At a minimum, Abu Dhabi needs at least one major air base and one major naval base designed for rapid and interoperable reinforcement. Kuwait needs to expand and convert its bases, and provide state of the art sheltering and defense for all military operations. In combination with the existing facilities in Saudi Arabia and Oman, this should provide the equivalent of contingency bases both for the forces of the other Southern Gulf states and for Western reinforcements without any of the political costs of full-time foreign bases.
- The events of 1987 have already shown how important mine warfare and low-level threats can be to the flow of oil through the Gulf. The Southern Gulf states need to create the kind of mine warfare capabilities, small missile patrol boats, naval air attack, and sensor/battle management capabilities to deal with low-level threats in the Gulf. These forces need standardized and interoperable sensors and C^3I systems. Ideally, they need to standardize on armaments. They also need the kind of air/missile defenses and communications links to air units necessary to provide for a reasonable self-defense capability. The Southern Gulf states may be able to leave any major naval threats to the West, but they must develop a regional capability to defend against low-level attacks.
- Passive defense is one of the cheapest forms of defense and deterrence. It involves more, however, than simply sheltering military bases and facilities. As Saudi Arabia has already shown, a pipeline network and strategic reserve can do a great deal to

reduce vulnerability to any given threat. The rest of the Southern Gulf states need to go ahead with plans to create a pipeline link to the Gulf of Oman through Oman or the UAE. All of the Gulf states need to create a matrix of pipelines linking the lines through Saudi Arabia to the Red Sea, through Oman to the Gulf of Oman, through Iraq to Turkey, and through Syria to the Mediterranean.

The Southern Gulf states also need to stockpile emergency off-shore loading points to ensure the flow of crude oil, and to create a self-reinsurance scheme that would ensure that the other members would make surplus export capacity available in the event damage was done to the export facilities of any one member. Much of the vulnerability of the Southern Gulf lies in the ability of some state or radical movement to threaten or attack the export facilities of a given member state. Such attacks would have far less meaning if an attacker faced the problem that the Southern Gulf oil export facilities could not be attacked piecemeal, and sufficient surplus export capacity already exists to implement such a scheme.

At the same time, the Southern Gulf states need to stop creating large over-centralized power plants, desalinization plants, and other critical facilities. The seeming cost-effectiveness of economies of scale vanishes the moment one considers the attractiveness of such facilities as military targets, and the marginal cost of dispersed and redundant plants and facilities is low relative to the cost of added military forces. The same is true of stockpiling long lead time repair parts.

TECHNOLOGY TRANSFER AS A SUBSTITUTE FOR MANPOWER

The earlier chapters of this book have shown why the military forces of the Southern Gulf states must use technology as a substitute for manpower. So has the above list of force improvement priorities. Technology is rarely the solution to Third World military problems, but the Southern Gulf states involve a very different matrix of resources and requirements.

The Southern Gulf states have the money not only to buy technology, but to buy support for it. This is particularly true if they concentrate on their areas of real need, rather than simply on the kind of major weapons systems that provide "glitter" rather than capability. Given the past and current expenditure levels of each state, all of the above force improvements could be fully funded without any increase in defense expenditure. Indeed, a better focus would generally permit a reduction in average defense expenditures over time.

What the Southern Gulf states do not have is manpower, and they will not acquire it for at least the next decade. This is not a function of their failure to create training systems; it is a function of their basic demographics and competing needs for manpower. The Southern Gulf can afford small effective air forces and navies, and some effective army cadres. They cannot man large military forces even if they choose to buy them.

Technology transfer has a powerful deterrent impact when it is part of effective force structures, and is not bought for the sake of technology. Regardless of the theoretical exchange ratios between weapons or force numbers, any threat power is going to hesitate before taking major risks against modern air and naval forces. Similarly, the most precious commodity in most contingencies in the Southern Gulf is speed of reaction. It is infinitely better to meet any attacking air or naval force before it arrives than to wait for it to be effective. It is infinitely preferable to stop any land or internal security threat before it can create facts on the ground or new political realities.

Technology transfer creates an implied linkage to the Western states that sell it, as well as to their power projection forces. No threat can estimate the full military reaction it will encounter when Southern Gulf forces have interoperable high-technology forces, and when these forces are interoperable with the West. This is particularly true if the Southern Gulf states tailor their military technology to this end.

EFFECTIVE WESTERN AND SOUTHERN GULF COOPERATION IN ARMS SALES

Technology cannot be effective, however, without additional attention to choosing the right technology and making it effective. The Gulf states and the West need to pay far more attention both to making their individual arms sales and purchases effective, and to the need to coordinate an overall arms transfer policy between the major Western suppliers to Saudi Arabia and the other Southern Gulf states.

Technology transfer is not a U.S., British, or French problem. It is a problem for every Western state that exports military technology to friendly Gulf nations. Western European states sell roughly the same amount of arms to the region as the U.S. France, in particular, has become a major supplier to Iraq and the Southern Gulf states, and is now the principal arms supplier to all the smaller Gulf states except Oman.

Britain's new arms sales to Saudi Arabia will reassert its role in the region, and the FRG may yet become a major supplier of armor.

There is, therefore, a high priority for finding ways in which Western arms sales can be coordinated to ensure the maximum degree of interoperability and effectiveness. While competition between arms sellers will always be inevitable, it should still be possible for the Western governments involved to work together, and with the Southern Gulf states, to ensure that arms transfers are successful.

The key Western European governments involved have an interest in developing local forces which can fully cooperate on a regional basis, and which will be interoperable with USCENTCOM in an emergency that transcends sheer profit in selling arms. Similarly, the U.S. has an interest in working with its European allies to ensure that its arms sales produce some degree of standardization or interoperability with European reinforcements, and to minimize the support and training costs of the Southern Gulf states.

The U.S., however, needs to work more closely with its allies to determine whether they can combine their arms sales with special reinforcements or high-technology forces for contingency operations. Mine warfare, maritime surveillance, short-range air defense, attack helicopter, anti-terrorist, and anti-armor forces are all examples of areas in which recent history has shown that small elements of Western European military forces could substitute for their lack of ability to deploy large numbers.

Arms sales cooperation may be difficult in a West which is far more oriented towards sales than strategy, but it offers great long-term potential. The West's experience in deploying naval forces to the Gulf in 1987 should also be enough of a warning to satisfy any government that sales are not enough.

THE NEED FOR A BETTER U.S. ARMS TRANSFER POLICY

It is obvious that progress will be extraordinarily difficult without a better U.S. arms transfer policy. Improving the ability of the Reagan Administration and its successors to act as a reliable source of arms and military advice to the Southern Gulf states will scarcely solve all of the complex military problems in the region, but it is an essential catalyst in creating a stable strategic relationship between the West and the Gulf, and between USCENTCOM and friendly Southern Gulf states like Saudi Arabia.

The basic problem is that such a policy requires a fundamental change in the attitudes of the U.S. Congress. As the previous analysis has shown, the U.S. refusal to sell Bahrain, Saudi Arabia, and Kuwait the arms they have requested involves far more than a debate over a single arms sale. It is also symbolic of much broader problems in the entire U.S. arms transfer effort to the Near East and Southwest Asia.

While the Department of Defense and the State Department have made repeated and well structured attempts to use U.S. arms transfers to support U.S. strategic interests in the Gulf and Near East, the Congress has virtually paralyzed U.S. national security policy. However well intentioned, the Congress has created an unhealthy mix of instinctive moral objections to any arms sales—regardless of their impact and necessity—and willingness to put domestic political pressure from various pro-Israeli lobbying groups before U.S. strategic interests.

Such Congressional attitudes have been responsible for forcing Saudi Arabia to turn to France for much of its army and naval equipment. They may well force Saudi Arabia and the other Southern Gulf states to turn to France, Britain, and other European states for all their air force and high-technology military equipment.

Such policies are not realistic in terms of the military trends and threats in the region, and they are not effective in denying the Southern Gulf states access to arms. In fact, the only effect of more Congressional vetos on U.S. arms sales to the Gulf will be to deny U.S. industry the sales, further weaken cooperation between the USCENTCOM and Bahrain, Oman, and Suadi Arabia, and undercut U.S. strategic relations with the Southern Gulf states.

Recent Congressional actions have also undermined the entire U.S. FMS program. Saudi Arabia is unquestionably the key purchaser of U.S. arms in terms of actual cash transfers to the U.S. It signed well over $48 billion worth of FMS agreements and construction agreements during the period 1950–1983, of which about $27 billion worth have been delivered.

Saudi Arabia remained the U.S.'s largest FMS customer in FY1986 and FY1987, in spite of its various arms sale crises with the U.S. It bought $714 million worth of arms in FY1986 and $637 million worth of arms in FY1987. Given Iran's default on roughly $2 billion agreements, and the fact that Egypt and Israel will almost certainly have to seek forgiveness on past FMS loans, this means Saudi Arabia will contribute more in cash payments for U.S. arms sales than all the other Middle Eastern nations combined.

To put the importance of Saudi Arabia in further perspective, its FMS agreements with the U.S. during 1950–1983 totaled roughly one-third of the $64 billion worth of all U.S. military sales agreements with the Middle East, all of Africa, and all the Indian Ocean states. Saudi

Arabia also agreed to $19,873 million in FMS construction agreements during this period, as compared with FMS construction agreements with all other nations totalling only $22 million during this period. These trends are particularly important at a time when major declines are taking place in FMS sales. U.S. sales dropped from $12.5 billion in FY1985, to $7.1 billion in FY1986, and to $6.9 billion in FY1987.

In fact, the whole trend of U.S. arms sales to the Near East has been one in which the U.S. has lost influence to the Soviet bloc. The U.S. has not affected the rate of arms transfers in the region; it has merely diverted them to other exporters. Most of these exporters are far less concerned with Israel's security and with creating effective regional deterrent and defense capabilities. This increases the probability that U.S. forces may have to intervene and decreases the probability that they can be effective.

CREATING AN EFFECTIVE WESTERN CAPABILITY TO REINFORCE FROM OVER THE HORIZON

It is impossible to tell at this point whether the Iran-Iraq War, and Western naval intervention in the Gulf, will force the Southern Gulf states into granting the West *de facto* military bases. It is equally impossible to be certain whether the course of Western intervention in the Gulf will build enough mutual trust and popular support in the Southern Gulf states to allow some elements of Western forces to remain.

The U.S. AWACS detachment in Saudi Arabia, and the deployment of the U.S. Middle East Task Force to Bahrain, has already shown that some Western deployments are possible. Similarly, Oman's granting of contingency basing rights to the U.S.—subject to Omani approval of the use of those bases in any military operation—is an example of how a Southern Gulf state can greatly reduce the risk and burden that power projection imposes on the U.S.

It is doubtful, however, that any series of events will make U.S. or other active Western military bases in the Gulf either desirable or practical. The political situation in the region is simply too delicate and unstable for more than limited deployments of Western forces in times of peace. The moment the immediate external threat disappears, every Southern Gulf state will have to confront the reality that both internal and external politics make Western bases the natural target not only of hostile states, but virtually of all the states surrounding the Southern Gulf. They also are the natural target of virtually every emerging political movement within the Southern Gulf. In the case of the U.S., this is because of its ties to Israel. In the case of Europe, this is because of their former role as colonial powers.

The West is faced with the practical reality that it cannot expect to transform its *de facto* alliance with the Southern Gulf states into a formal treaty relationship, with U.S. or European bases in the Kingdom, or a major military presence. The political forces in the region simply do not permit such actions by even the most friendly local state, except in the face of a clear and present threat. Western military bases or a major Western military presence would probably do more to hurt the Western interests than to aid them. Such bases would be key targets for terrorism, and political attacks from Iran and other radical states, and an ideal point of leverage for external efforts to try to divide the host country involved from Iraq and the other Southern Gulf countries.

The careful distinction that nations like Bahrain, Oman, Saudi Arabia, and Egypt make between reliance on Western military advice and equipment, on Western technicians and support, and U.S. "over-the-horizon" reinforcements—and any formal acceptance of a military treaty or Western military base—may be difficult for many people in the West to understand. It is, however, a vital distinction in a part of the world where most Arab states have had to fight for their independence, where nationalism and anti-colonialism are the preconditions for political legitimacy, and where ties to the U.S. must be kept especially vague and low profile because of U.S. ties to Israel.

Friendly Arab states cannot hope to disguise their informal military ties with the West, but they can hope to clearly assert its sovereignty and independence. They can avoid the stigma of allowing formal bases on its soil or an active military presence. They can demonstrate to other Arab states that it is not a proxy for the U.S. or Europe, and maintain its religious and political legitimacy in the face of charges from hostile or radical political movements and nations.

At the same time, the West should accept the fact that the kind of basing arrangements and formal treaties that Britain and the U.S. used to contain the USSR in the 1950s and 1960s are no longer in the West's interest. Such arrangements have obvious legitimacy and value in the case of NATO, Japan, and Korea, but they present problems for virtually every developing or Third World country that aligns its interests too closely with the West.

The West also has every interest in having the Southern Gulf states assume responsibility for their own deterrence and defense to the maximum extent they can, and in avoiding the need to commit Western forces to military confrontations or low-level conflicts. It has a strong interest in helping the rulers of the Southern Gulf states retain their political legitimacy within the Arab world, and in preserving Saudi

Arabia's religious legitimacy in an area where Islamic fundamentalism remains a serious threat. An "over-the-horizon" reinforcement policy and a suitable arms sale policy offer the West political and strategic advantages that more than offset any military disadvantages they incur from not having forces or formal military bases.

RECOGNIZING THE IMPORTANCE OF U.S. POWER PROJECTION CAPABILITIES

There is a price to be paid, however, for "over-the-horizon" reinforcement, and for the limitations in European power projection capabilities. Both the Southern Gulf states and America's European allies need to fully accept the fact that they are ultimately dependent on the U.S. for protection against mid- and high-level threats. They have to understand that the Southern Gulf cannot hope to develop more than a limited collective deterrent and defense capability before the year 2000; and the present limits on the threat from the USSR, the Northern Gulf, and radical Red Sea states can vanish far more quickly than the facilities can be created that will allow USCENTCOM and U.S. Navy forces to deploy before a conflict begins or escalates.

Saudi Arabia and the other Southern Gulf states already seem to recognize these realities in a crisis. They have privately turned to the U.S. every time Iran has escalated its attacks on Iraq or Gulf shipping. Nevertheless, they have made little real progress in standardization, in ensuring interoperability with USCENTCOM forces, or in creating the kind of C^3I/BM systems that U.S. forces would need. Only Oman and Saudi Arabia have generally accepted the reality of their strategic situation, and they have often been erratic in looking beyond a given procurement or force planning issue. All the Southern Gulf states need is to work harder to transform their forces and the Gulf Cooperation Council into an effective basis for common defense posture, and to create tacit links to USCENTCOM.

At the same time, most of the blame for the failure to build sound Western strategic relations with Saudi Arabia and the other Southern Gulf states clearly lies with the U.S. The key problem is that the U.S. policy has been paralyzed by Congressional fears of a largely imaginary increase in the threat to Israel. This has virtually blocked Gulf standardization on key U.S. arms and C^3I/BM technology, and made it difficult, if not impossible, for Britain, France, and the Gulf states to develop a cohesive regional force improvement strategy that can be linked to reliance on the U.S. and USCENTCOM.

SEEKING AN END TO THE IRAN-IRAQ WAR

The pace of events may have already ended the Iran-Iraq War by the time this book appears in print. Alternatively, they could have changed U.S. policy, have led to a major military confrontation between the U.S. and Iran, or even have led to U.S. and European withdrawal from the Gulf. The fighting described in the previous chapter has shown that the war is becoming steadily more dangerous with time, and that it is broadening to involve the Southern Gulf states as well as Western naval forces.

The "two track" policy the U.S. is now pursuing has important limitations. It depends on Iran coming to accept the UN cease-fire proposal and on Iraq compromising enough to get Iran to accept it. This may take a year or more at the present rate. It also involves very high costs in terms of maintaining Western forces in the Gulf, and the constant risk of full-scale military confrontation with Iran.

The U.S. and its allies may well not be able to avoid this confrontation in trying to bring an end to the war. Western values and those of Khomeini and his revolutionary elite are all too different. Further, it is difficult for a leader who claims to be inspired by God to admit he has been deterred by the U.S. Navy.

Nevertheless, the West needs to take every step it can to bring Iran to a cease-fire without further fighting. Its strategic interests lie in containing the broader regional ambitions of Khomeini and Iran's present leadership, and not in picking a quarrel with Iran's people—who will be around long after Khomeini's revolution has changed in character or faded into obscurity. Further, the West faces at least some risk that a major fight with Iran might force it into a far closer arrangement with the USSR than the present marriage of convenience.

At the risk of trying to predict the unpredictable, the West needs to establish the kind of "carrot and stick" policy that will offer Iran a face-saving road to peace without implying American or European weakness. The U.S. will have to take the lead in such an effort, and the first such "carrot" should be for the President or Secretary of State to define U.S. objectives in the Gulf far more clearly than has been done in the past, and in a way that Iran can understand without being threatened. The U.S. needs to reiterate at the highest levels that it wants an end to the war that does not penalize either side, and will not tolerate any Iranian or Iraqi violation of the territory of a third state or of free passage through international waters.

Another "carrot" should be to clearly reassert what the U.S. means by "neutrality". Senior U.S. officials should state that the U.S. will not let Iraq drag it directly into the land conflict or into directly supporting

Iraq. The U.S. needs to make it unambiguously clear that the U.S. will enter the naval and air conflict only to defend U.S. interests and those of friendly states in the Southern Gulf. It needs to make it clear that the U.S. will not support the hopeless anti-Khomeini movements outside Iran, and wants to preserve strong and viable nationalist regimes in power in both Iraq and Iran as a buffer against Soviet expansion into the Gulf. The U.S. also needs to make it clear that it does not intend to maintain an enhanced military presence in the region once that cease-fire is reached, and wants to see a region capable of managing its own security in peace.

Senior U.S. officials should make it clear that the U.S. continues to condemn aggression in any form and does see Iraq as heavily to blame for starting the war. They should tacitly support the idea that the Southern Gulf states should pay Iran war recovery aid that would be an indirect form of reparation. They should begin to discuss U.S. economic aid and trade options for the post cease-fire period. We and our friends in the Southern Gulf have every reason to reach out to the Iranian people and help them build a sound economy. The only way we will ever find true moderates to deal with in Iran is to give them a base for coming to power.

The key "carrot" will be to make it clear that the U.S. and its Western allies will compromise on the UN cease-fire proposal in any way that will protect Western interests, the Southern Gulf states, and the people of Iraq. The U.S. should at least leak the fact that it can accept Iran's proposal to alter the UN Security Council cease-fire proposal to have the investigation into war guilt begin at the same time as the cease-fire, and allow a reasonable delay in Iran's retreat to its 1975 borders.

Accommodation alone, however, will only be seen as weakness. The West will need to use "sticks" in the form of both covert and public action. The best covert "stick" would be to ask France to quietly give added military help to Iraq in the form most likely to force Iran to agree to a cease-fire. Such help could provide the Iraqi Air Force with real time targeting data and the sophisticated strike planning it needs to make its strikes on Iran's refineries, power plants, and other critical targets more effective. At the same time, it could be used as a lever to pressure Iraq to avoid strikes on population targets, food supplies, and naval targets in the Gulf except for those owned by Iran or under charter to it.

Public "sticks" should include a change in our military posture that will keep Iran from trying to force us out of the Gulf through an endless series of low-level attacks or probes at any target we have not formally agreed to defend. The U.S. should stop limiting its convoy action to reflagged ships and define a clear security zone in the Gulf. It should state that it will protect any international shipping in that zone, including shipping loading and off-loading in port.

If Iran does escalate or challenge the West in response, the West should avoid symbolic "sticks" like firing 1,000 rounds at an evacuated oil platform. The West needs to start hurting Khomeini's ability to go on with the war, not create "martyrs" out of misguided volunteers. It should give Khomeini formal warning and then begin a carefully staged pattern of response.

Ironically, Khomeini may show the West the route it should follow. U.S. forces can use mines to pin down the Iranian Navy and Guards in their bases and ports without any threat to international shipping. If Khomeini does not react to this, the U.S. can mine the waters around the eighteen tankers Iran uses as its oil reserve. Next, it can use mines to deny Iran access to its 250,000 barrels a day of offshore oil production, and it can use them finally to cut off access to the two jetties Iran uses to export from Kharg Island—the source of at least 70% of Iran's oil exports.

Gradually embargoing Iran's oil revenues will force Khomeini to compromise far sooner than an ineffective arms embargo or shooting at targets he not only can afford to lose but wants for propaganda purposes. The selective use of mining will give the U.S. a powerful "escalation ladder" with minimal risk to U.S. servicemen. No Iranians will have to die, and no Iranian civilians will be hurt.

The U.S. should continue to revitalize and sustain "Operation Staunch", the U.S. effort to halt the flow of arms to Iran that began in late 1983. Ideally, the West will also be able to get UN support for an arms embargo that would gradually reduce Iran's ability to invade Iraq. If it does not, an alternative "stick" would be to use the media to openly attack the world arms trade to Iran. Formal official protests will be ineffective, regardless of whether they come from the U.S. or the UN.

The West—in practice, the U.S.—should publicly embarrass every government and dealer sending arms to Iran, including the PRC and the USSR and its East European proxies. It should issue a monthly unclassified white paper with every possible name and detail. It should hold background briefs for the world press whenever its discovers another major arms sale. It should use every aspect of the media it can to expose any Soviet or PRC tilt towards Iran and away from enforcing the Security Council cease-fire resolution.

The final "stick" should be for the U.S. to discuss with Kuwait and Saudi Arabia how we can use their immense surplus production capacity to replace Iran's oil in every Western market. This would reinforce any embargo of Iran, ensure the stability of the world oil trade, and confront

Iran with the fact that its former revenues were pouring into the two nations that have given Iraq the aid it needs to continue the war.

It may be that only Khomeini's death will allow Iran's leaders to negotiate on the basis of intelligent self-interest. The West should not, however, pass up any chance to bring peace without major military action. The proper combination of carrots and sticks will thrust Iran's more fanatic leaders towards a cease-fire, minimize the West's need for direct military clashes with Iran, avoid any unnecessary casualties or damage to Iranian economy, and offer the maximum hope that the West can eventually restore friendly relations with Khomeini's successors.

WESTERN RELATIONS WITH IRAQ

The West has no intrinsic reason to choose either side in the Iran-Iraq War. The fact remains, however that Iraq is now far more friendly to the West and the Southern Gulf states than Iran, and it is Iraq that is threatened by the risk of a successful invasion. Further, no matter how successful the West is in establishing strategic relations with the Gulf, it is not going to create any military pillars that can take the place of the best possible Western relations with all the states in the region.

The key to successful relations will unquestionably be trade and the economic support of Iraq's economic development activities. Iraq's policies towards its neighbors, the West, and Israel moderated in direct proportion to the expansion of its trade and development even before the sobering lessons of the Iran-Iraq War. Revived trade will do far more than diplomacy or arms to ensure that Iraq remains friendly.

It is also important to note that Iraq, and not Iran, is the economic prize in the Northern Gulf. Iraq has extremely large unproven reserves, while Iran has little chance of more major discoveries, and had begun to deplete its reserves before the war began. Iraq has a comparatively small population and oil revenues equivalent to those of Iran. Its economy has suffered far less from the war and revolution, and is now structured towards Western concepts of development. Iran has a very large population which is growing rapidly, and its economy is in many ways a "basket case". It will never be able to afford the kind of sophisticated trade with the West that took place before the fall of the Shah.

At the same time, the present division in the military roles of the

U.S. and Europe should be preserved. U.S. efforts to support Iraq with weapons shipments or other major military assistance are not politically possible or even desirable. This is a role that France and other European nations can perform.

DEALING WITH IRAN

One of the bitter ironies of the Iran-Contra hearings was that they eventually revealed that even those conducting the covert arms sales with Iran realized they were not dealing with moderates, and that they had to deal with "extremists" because they were the only Iranians who could get anything done. It is exceedingly dangerous to ignore the historical reality that the French and Russian revolutions triggered a process of instability and extremism that lasted half a century, and that this has been the rule with mass revolutions that are captured by an ideological elite, and not the exception.

Iran is still at the start of its revolution, and the West may not be able to deal with a stable or friendly Iran for years to come. The most it can do is to establish contacts with every reasonable Iranian faction and to try to use economic ties to restore some basis for correct relations and eventual friendship.

The U.S., its European allies, the Southern Gulf states, and eventually Iraq need to deal with Iran on the basis of constructive non-interference. They need to encourage every moderate action and opening to the West, without becoming involved in the internal affairs of the current regime or the inevitable struggles following Khomeini's death. Once again, the emphasis should be one of supporting trade and development, with the Southern Gulf states playing a role in using their oil wealth to aid Iran's reconstruction once it accepts a cease-fire.

The entire history of covert and overt outside interventions in popular revolutions is a history of disasters and alienation. Iran will have to work out its own political destiny. The most the West and Southern Gulf states can do is to contain it from threatening other states while ensuring it is never isolated from the world economy or the opportunity of friendly relations with other states.

At the same time, the West cannot afford to ignore the various opposition movements in Iran. Iran's current ruling elite may not remain in power, and the West must maintain low-level ties with every major faction, and build a relationship based on economic self-interest, not common political and social goals.

This is a case, however, where the divisions in the West and the Southern Gulf states may actually have advantages. One of the major problems in U.S. policy towards post-revolutionary Iran has been the assumption that the U.S. must play a lead role. In fact, the U.S. may well be the last state that should do so. Western European states, Japan, and the UAE are now far better positioned to try to deal with Iran than the U.S. Given the fact that Iran is not going to be anyone's pillar or prize for the foreseeable future, the best approach may well be to let other nations gradually rebuild their trade and economic ties while reopening U.S. markets the moment a cease-fire is agreed to.

What both the Western and Southern Gulf states must be most careful to avoid is the kind of shortsighted opportunism that gives way to Iranian intimidation. There is no point in more hostage deals, or rushing into economic deals that avoid political capitulation. This will not secure anyone against attack, hostage taking, or terrorism. It will only encourage it.

DEALING WITH THE USSR

The West is going to have to accept the fact that it cannot exclude the USSR from the Southern Gulf, and that the Soviet Union is a major player in the Northern Gulf and the Yemens. The USSR is a major arms supplier to Iraq, and is still rebuilding some of its ties to Iran. The USSR can be expected to continue to play the Iran card, to strengthen its relations with Iraq when this is tactically convenient, seek diplomatic and economic openings to all of the Southern Gulf states, exploit the backlash from the Arab-Israeli conflict, and seek to weaken Western influence in the region.

Glasnost may ease the intensity of Soviet opportunism, but it is scarcely going to end it, and the long-term trend in Soviet strategic interests will inevitably lead it to seek added access to Gulf oil and gas resources and trade.

In practice, this simply increases the urgency of Western action to establish sound strategic ties with the Southern Gulf states, end the Iran-Iraq War, establish the proper relations with Iraq and Iran, and balance relations with Israel and the Arab states. Once again, the issue is not to invent dramatic new policies. No Western action can fully contain the USSR. Rather, it is to improve the implementation of long-standing policies that reduce the opportunities open to the USSR. Ultimately, it is the USSR which invaded Afghanistan, whose Muslim population is suppressed, and whose very opportunism has cost it its relations with Egypt and Indonesia, and much of its position in Iraq.

BALANCING WESTERN TIES TO THE GULF WITH U.S. TIES TO ISRAEL

Strong Western strategic relations with the Southern Gulf states should not mean tilting away from Israel, or favoring the Arabs. The West will gain nothing from "taking sides" in the Arab-Israeli conflict. Its strategic interests depend on ties to both moderate Arab states and Israel. The West can only hope to achieve even a moderate degree of stability in the Near East and Southwest Asia if it preserves a proper balance in its commitments to all its friends and allies.

Relations with Israel, however, have become far more a U.S. problem than a European one—although largely by default. Europe has left the problem of preserving Israel's security, and providing Israel with arms, to the U.S. Europe has concentrated on exploiting the political and economic benefits of friendly relations with moderate Arab states, and has often confused the resulting opportunism with morality. For all its often justified criticisms of American policy, this European opportunism has made it even more difficult for the U.S. to strike a proper balance between the pressures of regional and domestic politics.

The American supporters of Israel and the Arab world have also made things worse. Far too many American Arabs and Jews are more extreme than the Arabs and Jews in the nations in the region. They push the U.S.—and particularly the Congress—to "take sides" in a manner that does nothing to contribute to regional peace and development, and does a great deal to contribute to regional vulnerability and the prospect of future wars.

The resulting lack of balance in U.S. policy, and its failure to meet the military needs of the Southern Gulf states, has been costly and destructive. There are legitimate security issues involved in arms sales to the Gulf, and the technical and military trade-offs in providing Saudi Arabia and the other Southern Gulf states with the military technology they need are complex and require careful judgment.

At the same time, the myth that Saudi Arabia or its smaller neighbors are likely to become a major military threat to Israel serves no one's interest. It ignores the consequences of what happens when the U.S. does not provide the arms Saudi Arabia needs, it ignores the true impact of the moderate and conservative Arab states on the Arab-Israeli balance, and it ignores the importance of Western ties to both the Arab states and Israel in reducing the risks of another Arab-Israeli conflict.

The fundamental reason that Western arms sales to the Southern Gulf state are not a threat to Israel does not lie in the details of the arms they request, or their technology, but rather in the fact that the Southern Gulf states will remain a relatively weak military power that must concentrate on deterrence and defense. They face more direct and much higher-priority threats, and would never take the risk of a major conflict with Israel.

Few senior officials or military officers in the Southern Gulf states have any illusions about the consequences of entering a conflict with Israel, or the overall outcome of any Arab-Israeli conflict well into the 1990s. The Southern Gulf states lack any common border with Israel, and the previous chapters have shown that they lack the military strength and manpower to engage Israel unilaterally at any time in the foreseeable future. They can at most provide the same symbolic forces in support of the Arab cause that several have provided in the past.

Saudi Arabia, for example, did not contribute any significant forces to the 1948 or 1956 Arab-Israeli conflicts. It supplied about 4,500 men, 10 tanks, and 40 aircraft to assist in the defense of Jordan in 1967, out of a total Arab strength of over 250,000 men, 2,000 tanks, and 950 aircraft. Saudi forces, however, arrived only after the lightning pace of the Six-Day War had already ensured a decisive Israeli victory. Saudi Arabia contributed 1,500 men, about 1 tank squad, and some replacement aircraft in the 1973 war, but in a defensive role in support of Jordan and Syria that resulted in only minor combat. Its contribution to the Arab forces arrayed against Israel has otherwise been limited to stationing troops in Jordan and the Golan during periods of peace. Saudi Arabia's forces may be far better developed today than they were in 1973, but Israel's forces have improved at least as quickly as those of Saudi Arabia, and Saudi forces and those of the other Southern Gulf states still lag far behind the current capabilities of Israeli forces.

Even the air forces of the Southern Gulf states will pose only a limited threat. There is a massive practical difference between the ability of Southern forces to engage other Gulf and Red Sea air forces, and their ability to engage the forces of Israel. Although some pilots have logged over 1,000 hours, and a few have even demonstrated excellent flight proficiency in flying against top-grade U.S. "aggressor" squadrons, no Gulf air force is ready to fly air combat or offensive missions against Israel. It would lack the endurance, basing, air control and warning capability, and command structure to avoid extremely high loss-to-kill ratios.

The Southern Gulf states also cannot approach Israel's skill in managing air combat or sophisticated electronic warfare capabilities. Israeli electronic warfare capabilities are far more sophisticated than those of any competing Middle Eastern state, and the advanced state of Israel's defense industry ensures that it not only will lead in hardware, but develop a steadily greater lead in software. Israel not only has proven its technical mastery in the 1982 fighting; it is the only state in the world committed to, and capable of, developing such software in a form tailored to regional combat needs and in a form proven by actual combat.

Perhaps most important of all, the Southern Gulf air forces will lack the aircraft strength and pilot numbers to accept the losses they would suffer in any air combat with Israel without crippling their air capabilities and their ability to defend their oil fields for years to come. The fact that the Kingdom is moving towards a considerable self-defense capability against air threats in the Gulf area does not indicate that it will acquire a significant capability against Israel or that it could afford to send more than token forces against Israel without the risk of catastrophic losses.

The U.S. can also take advantage of its lead in electronic warfare to ensure that Israel has a lead in IFF and ECM protection. Developmental systems of this kind are too complex and highly classified to discuss in any detail, but the U.S. has tremendous technical capability to ensure Israel's security through electronic warfare and intelligence technology without depriving Saudi Arabia of superiority over all radical regional threats.

The U.S. can ensure a lead in Israeli air strength and fighter performance through its support of Israeli Air Force modernization. It can aid Israel in obtaining a suitably sophisticated mix of next-generation AWACS, electronic warfare, and ELINT/ESSM capability to ensure its current edge in battle management well into the year 2000. This Israeli lead in battle management is currently far more important in any war fighting than aircraft numbers or fighter performance.

It is essential, however, that both the U.S. and its allies recognize that the key to balancing the U.S. commitment to Israel against strategic ties to the Gulf states is a strong U.S. aid program for Israel. A weak Israel is simply one unable to make progress towards peace, and one forced to use every possible political means to block arms transfers to every Arab state. It is an Israel thrust towards preemption or reliance on nuclear weapons.

THE IMPACT OF U.S. AID TO ISRAEL

In practice, this means that the U.S. is far better off reaching a *quid pro quo* with Israel that trades technology transfer and higher aid levels for Israeli acceptance of U.S. strategic relations with the Southern Gulf states than it is in trying to fight out the Arab-Israel conflict in the context of domestic American politics.

The U.S. needs to find a stable level of annual military and economic aid that will meet Israel's long-term military needs, provide for a two-way street in military sales, and provide effective technology transfer without becoming trapped in adventures like the Lavi. This means improved mutual force planning, and finding some form of technical cooperation that will benefit both countries, not divide them.

The U.S. also needs to avoid forms of economic aid which hurt Israel and which ultimately increase its long-term aid needs, rather than helping them. This means a mutual effort towards funding economic reform now, before another crisis occurs, rather than waiting for things to go wrong. It also means forgiveness of Israel's FMS and other aid debts as a trade-off for lower overall economic aid. The present aid program tends to give Israel grant aid to pay for its past aid loans, and this simply acts as a break on Israel's economic development.

Finally, Israel may have to be convinced that the U.S. sees it as strategically important enough that it will not compromise Israel's security. The U.S. and Israel have talked a great deal about a strategic relationship based on power projection—but this is a relationship that will never be practical as long as Israel remains at war with most Arab states. The U.S. needs to seriously examine how Israel's air force and navy could play a role in the Mediterranean that could benefit both nations, as well as NATO.

MAINTAINING THE SEARCH FOR PEACE

The key policy issue surrounding the Arab-Israel conflict for both the U.S. and Western Europe is not how to choose sides between the Arabs and Israel, but rather how to gradually build bridges between them. No effort to create stable military relations between the West and the Southern Gulf states can be separated from the search for an Arab-Israeli peace settlement.

This is true even when both Israel and the key Arab actors that must reach a peace settlement are not ready for peace, and no dramatic new peace initiative or solution is imminent. The fact also remains that the search for peace is another area where Western Europe cannot act as a proxy for the U.S. If the West is to defuse the hostility in the Arab world, and give moderate Arab regimes a political basis for strengthening their military ties to the West, both the U.S. and its

European allies must show the same balance and consistency in seeking peace, regardless of the current obstacles, that they show in maintaining military relations with both the Arab states and Israel.

Friendly Arab states can accept the fact that such a search may take years, but they cannot accept U.S. or European abandonment of the peace issue, or attempts to treat it with benign neglect. Most Arab leaders—and certainly all Arab leaders friendly to the U.S.—understand the special nature of the U.S. relationship with Israel. They know that these ties are more than a matter of domestic politics and will be an enduring aspect of American policy. At the same time, they know—as do most Israelis—that only a constant and politically visible U.S.-led effort can hope to bring peace to the region.

The U.S. can keep the long-term support and friendship of states like Egypt, Jordan, and Saudi Arabia only if it presses for a just and balanced peace settlement. Whatever the ironies involved may be, the West can establish a sound strategic position in the Near East only by forging both swords and plowshares.

At the same time, Saudi Arabia and the other Southern Gulf states cannot afford to show too much caution in joining in peace initiatives with the U.S., Egypt, and Jordan. Time is not on the side of territorial settlements on the West Bank, in Gaza, or on the Golan. Saudi Arabia can play a unique role in creating a long-term bridge between Syria and the peace process. While the U.S. may have been too demanding in seeking a linkage between Saudi Arabia and Camp David, and then Saudi Arabia's agreement to King Hussein's peace initiatives, there is a time for courage. To put it bluntly, the opportunism the Southern Gulf states and other Arab states have shown in backing hardline Arab positions simply to deflect political attacks at the cost of blocking any progress towards peace has done as much to hurt the ultimate success of every peace effort as all the mistakes of the West and Israel combined.

There will never be an ideal conjunction of U.S., Israeli, and Arab willingness to move towards peace, but both Western and moderate Arab interests in the Gulf require that the search for a just and balanced peace settlement should be pursued on terms that are realistic and offer both Israel and its neighbors full security.

Abbreviations and Acronyms

AA	Anti-Aircraft
ABV A.B.	Swedish corporation
AC&W	Air Control and Warning
ADV	Air Defense Variant
AEW	Airborne Early Warning
AG	Air-to-Ground
ALINDIEN	French Indian Ocean Command
AMRAAM	Advanced Medium Range Air to Air Missile
AMX	French armored weapon prefix
AOC	Air Operations Center
ARM	Anti-Radiation Missile
ASW	Anti-Submarine Warfare
ATGM	Anti-Tank Guided Missile
AWACS	Airborne Warning and Air Control System
BAPCO	Bahrain Arab Petroleum Company
BDM	Braddock, Dunn, and McDonald
BLU	Cluster bomb
BM	Battle Management
BPD	Barrels Per Day
BVR	Beyond Visual Range
CAP	Combat Air Patrol
CDIP	Continuously Displayed ImpAct Point
CFM	Jet engine manufacturer
CIA	Central Intelligence Agency
COC	Command Operations Center
CPMIEC	Chinese Precision Machinery Import Export Corporation
CSF	French company
C^3I	Command, Control, Communications, and Intelligence
DCS	Defense Communications System
DMS	Defense Marketing Services
DSAA	Defense Security Assistance Agency

DWT	Dead Weight Tons
ECCM	Electronic Counter-Countermeasures
ECM	Electronic Countermeasures
EIU	Economist Intelligence Unit
ELINT	Electronic Intelligence
EO	Electro-Optical
EOIR	Electro-Optical Infra-Red
ESM	Electronic Support Measures
ESSM	Electronic Systems Support Measures
EXPEC	Saudi computer facility
EW	Early Warning
EWWS	Electronic Warfare System
FAR	Forces d'Action Rapide
FAST	Conformal fuel tank for F-15
FEBA	Forward Edge of Battle Area
FET	Foreign Economic Trade
FLIR	Forward Looking Infra-Red
FMS	Foreign Military Sales
FRG	Federal Republic of Germany
G	Gravities
GCC	Gulf Cooperation Council
GCT	French self-propelled weapon
GLCM	Ground Launched Cruise Missile
G.m.b.H.	Firm incorporated in West Germany
GNP	Gross National Product
GOC	General Operating Center
GOSP	Gas-Oil Separator
HP	Horsepower
IDS	Intercepter Day Strike
IFF	Identification of Friend or Foe
IISS	International Institute for Strategic Studies (London)
IMF	International Monetary Fund
INS	Inertial Navigation System
IR	Infra-Red
ISF	Intermediate Staging Facility
KOTC	Kuwaiti Oil and Tanker Company
LANTIRN	Low Altitude Navigational and Targeting Infra-Red System for Night
LCAC	Landingcraft Air Cushion
LNG	Liquid Natural Gas
LOA	Letter of Approval
LOC	Line of Communication
LOTS	Landing Over the Shore

LPD	Landing Platform Dock
LPG	Liquefied Propane Gas
LPH	Amphibious Platform Helicopter
LSL	Landing Ship Logistic
LST	Landing Ship Tank
MAF	Marine Amphibious Force
MAP	Military Assistance Program
MAU	Marine Amphibious Unit
MBD	Thousands of Barrels Per Day
MCM	Minecraft Medium
MCMV	Minecraft Multiple Version
MD	Military District
MEF	Marine Expeditionary Force
MER	Multiple Ejection Racks
MEU	Marine Expeditionary Unit
MLF	Multi-Lateral Force
MMBD	Millions of Barrels Per Day
MOD	Ministry of Defense
MSC	Mineship–Coastal
MSIP	Multi-Stage Improvement Program
MSO	Mineship–Oceangoing
NATO	North Atlantic Treaty Organization
NGL	Natural Gas Liquid
NM	Nautical Mile
NORAD	North American Air Defense
NSC	National Security Council
NSDD	National Security Decision Directive
NSPG	National Security Planning Group
OAPEC	Organization of Arab Petroleum Exporting Countries
OECD	Organization for Economic Cooperation and Development
OPEC	Organization of Petroleum Exporting Countries
PACS	Programmable Armament Control Set
PAVETACK	Radar
PCG	Patrol Craft–Guided Missile
PDRY	People's Democratic Republic of Yemen
PFLOAG	Party for the Liberation of the Arab Gulf
POL	Petroleum, Oil, and Lubricants
POST	Passive Optical Seeker Technology
PRC	People's Republic of China
PTG	Chinese missile designation
PUK	Popular Union of Kurdestan
QRA	Quick Reaction Alert

RAF	Royal Air Force
R&D	Research and Development
RO-RO	Roll On–Roll Off
ROVs	Remote Ocean Vehicles
RWR	Radar Warning Receiver
SACEUR	Supreme Allied Commander Europe
SACLANT	Supreme Allied Commander Atlantic
SAR	Sea-Air Rescue, or Search and Reconnaissance
SAS	Special Air Service (British commandos)
SCADA	Supervisory Control and Data Acquisition
SCC	Special Coordinating Center
SDR	Special Drawing Right
SEAL	U.S. Navy commandos
SHORADS	Short Range Air Defense Systems
SIPRI	Stockholm International Peace Research Institute
SLAR	Side-Looking Radar
SLEP	Service Life Extension Program
SNIE	Special National Intelligence Estimate
SNEP	Saudi Naval Expansion Program
SPM	Single Point Moorings
STUFT	Ships Taken Up From Trade
TEWS	Tactical Electronic Warfare System
TOW	U.S. crew-served guided anti-tank weapon
TRF	Theater Reserve Facility
UAE	United Arab Emirates
UHF	Ultra High Frequency
UKNL	U.K.-Netherlands
USAF	U.S. Air Force
USCENTCOM	U.S. Central Command
USGS	U.S. Geological Survey
USN&WR	U.S. News & World Report
VAB	French armored vehicle
VLCC	Very Large Cargo Carrier
VSTOL	Vertical & Short Take Off and Landing
WEU	Western European Union
YAR	Yemen Arab Republic

Bibliography

Abdel, Majid Farid, ed., *The Red Sea: Prospects for Stability*, London, Croom Helm, 1984, pp. 84–94.

Abdulghani, Jasim M., *Iraq and Iran: The Years of Crisis*, Baltimore, Johns Hopkins University, 1984.

Abir, Mordechai, *Oil, Power, and Politics: Conflict in Arabia, the Red Sea and the Gulf*, London, Frank Cass, 1974.

———, "Saudi Security and Military Endeavor", *The Jerusalem Quarterly*, 33 (Fall 1984), pp. 79–94.

Al-Ebraheem, Hassan Ali, *Kuwait and the Gulf: Small States and the International System*, Washington, Georgetown University, 1984.

al-Farsy, Foud, *Saudi Arabia: A Case Study in Development*, London, Stacey International, 1978.

Albrecht, Gerhard, *Weyer's Warships of the World 1984/85*, 57th ed., Anapolis, Md., Nautical & Aviation Publishing Co.

Akins, James E., et al., *Oil and Security in the Arabian Gulf*, New York, St. Martins, 1981.

Allen, Calvin, *Oman: The Modernization of the Sultanate*, Boulder, Westview Press, 1986.

Allen, Robert C. "Regional Security in the Persian Gulf," *Military Review*, 63, 12 (December 1983), pp. 17–29.

Al-Yassini, *Relgion and State in the Kingdom of Saudi Arabia*, Boulder, Westview Press, 1985.

Ali Sheikh Rustum, *Saudi Arabia and Oil Diplomacy*, New York, Praeger Publishers, 1976.

Aliboni, Roberto, *The Red Sea Region*, Syracuse, Syracuse University Press, 1985.

Amirsadeghi, Hossein, ed., *The Security of the Persian Gulf*, New York, St. Martin's Press, 1981.

Amuzegar, Jahangir, "Oil Wealth: A Very Mixed Blessing", *Foreign Affairs*, 60 (Spring 1982).

Anthony, John Duke, "The Gulf Cooperation Council", *Journal of South Asian and Middle Eastern Studies*, 5 (Summer 1982).

———, Goals in the Gulf: *America's Interests and the Gulf Cooperation Council*, Washington, National Council on U.S.-Arab Relations, 1985.

ARAMCO, *Yearbook* and *Facts and Figures* various editions.

ARCO, Series of Illustrated Guides, New York: Salamander Books, ARCO.

———, *Weapons of the Modern Soviet Ground Forces*.

———, *The Modern U.S. Air Force.*
———, *The Modern Soviet Air Force.*
———, *Military Helicopters.*
———, *The Israeli Air Force.*
———, *The Modern Soviet Navy.*
———, *The Modern U.S. Navy.*

Arlinghaus, Bruce, *Arms for Africa,* Lexington, Mass., Lexington Books, 1983.

Armed Forces Journal International, various editions.

Army, Department of, *1985 Weapon Systems,* Washington, D.C., Government Printing Office.

———, *Saudi Arabia, A Country Study,* DA Pam 550-51, Washington, D.C., 1985, pp. 32–322.

———, *Soviet Army Operations,* IAG-13-U-78, April 1978.

Army Armor Center, Threat Branch, *Organization and Equipment of the Soviet Army,* Fort Knox, Kentucky, January 1981.

Arnold, Anthony, *Afghanistan: The Soviet Invasion in Perspective,* Stanford, Calif., Hoover Institution, 1981.

Auer, Peter, ed., *Energy and the Developing Nations,* New York, Pergamon, 1981.

Aviation Week and Space Technology, "F-15C/D Display Nears Test Stage", February 20, 1985, pp. 77–81.

———, "Integrated Systems Evaluated on F-15", April 11, 1985, pp. 47–57.

———, "Mirage 2000 Fighter", June 24, 1985, pp. 38–39.

———, "Saudis, British Define Terms of Tornado Sale", September 30, 1985, p. 29.

Axelgard, Frederick W., "The Tanker War in the Gulf: Background and Repercussions", *Middle East Insight,* 3, 6 (November-December 1984), pp. 26–33.

———, *Iraq in Transition: A Political, Economic, and Strategic Perspective,* Boulder, Colo., Westview Press, 1986.

Ayoob, Mohammad, ed., *The Middle East in World Politics,* London, Croom Helm, 1981.

Azharly-EL, M.S., ed., *The Iran-Iraq War: A Historical, Economic, and Political Analysis,* New York, St. Martin's, 1984.

Aziz, Tareq, *Iraq-Iran Conflict,* London, Third World Center, 1981.

Bakhash, Shaul, "The Politics of Oil and Revolution in Iran", Staff paper Washington, D.C., Brookings Institution, 1982.

———, *The Reign of the Ayatollahs: Iran and Islamic Revolution,* New York, Basic Books, 1984.

Banks, Ferdinand, *The Political Economy of Oil,* Lexington, Mass., Lexington Books, 1980.

Barker, A.J., *Arab-Israeli Wars,* New York, Hippocrene, 1980.

Barker, Paul, *Saudi Arabia: The Development Dilemma,* Special Report No. 116, London, Economist Intelligence Unit, 1982.

Bass, Gail, and Bonnie Jean Cordes, *Actions Against Non-Nuclear Energy Facilities: September 1981–September 1982,* Santa Monica, Calif., Rand Corporation, April 1983.

Batatu, Hanna, "Iraq's Underground Shi'a Movements: Characteristics, Causes and Prospects", *Middle East Journal,* 35, 4 (Autumn 1981), pp. 578–594.

Baylis, John, and Segal, Gerald, eds, *Soviet Strategy,* Totowa, N.J., Allanheld, Osmun & Co., 1981.

Be'eri, Eliezer, *Army Officers in Arab Politics and Society,* New York, Praeger Publishers, 1970.

Beling, Willard A., ed., *King Faisal and the Modernization of Saudi Arabia,* Boulder, Colo., Westview Press, 1980.

Ben Horin, Yoav, and Barry Posen, *Israel's Strategic Doctrine,* Santa Monica, Calif., Rand Corporation, September 1981.

Benton, Graham M., and George H. Wittman, *Saudi Arabia and OPEC: An Operational Analysis,* Information Series No. 138, Fairfax, Va., National Institute for Public Policy, March 1983.

Bernstam, Mikhail S., "Soviet Oil Woes", *Wall Street Journal,* January 10, 1986.

Bernard, Cheyrl, and Zalmay Khalilzad, *The Government of God: Iran's Islamic Republic,* New York, Columbia University, 1984.

Bertram, Cristoph, ed., *Third World Conflict and International Security,* London, Macmillan, 1982.

Betts, Richard K., *Surprise Attack,* Washington, D.C., Brookings Institution, 1982.

Bishara, Ghassan, "The Political Repercussions of the Israeli Raid on the Iraqi Nuclear Reactor", *Journal of Palestine Studies,* Spring 1982, pp. 58–76.

Blake, G. H., and Lawless, R. E., *The Changing Middle Eastern City,* New York, Barnes and Noble, 1980.

Blechman, Barry M., and Stephan S. Kaplan, *Force Without War,* Washington, D.C. Brookings Institution, 1978.

Bligh, A., and S. Plant, "Saudi Modernization in Oil and Foreign Policies in the Post-AWACS Sale Period", *Middle East Review,* 14 (Spring-Summer 1982).

Bloomfield, Lincoln, "Saudi Arabia Faces the 1980s: Saudi Security Problems and American Interests", *Fletcher Forum,* 5, No. 2 (1981).

Borowiec, Andrew, "Turks Seek Aid to Upgrade Army", *Washington Times,* May 16, 1986, p. 7.

Bradley, C. Paul, *Recent United States Policy in the Persian Gulf,* Hamden, Conn., Shoe String Press, 1982.

Braibarti, Raoph, and Abdul-Salam, Al-Farsy, "Saudi Arabia: A Develop-mental Perspective", *Journal of South Asian and Middle Eastern Studies,* Fall 1977, p. 1.

Brassey's Defense Yearbook (later *RUSI* and *Brassey's Defense Yearbook*), London, various years.

Brodman, John R., and Hamilton, Richard E., *A Comparison of Energy Projections to 1985,* International Energy Agency Monograph Series, Paris, OECD, January 1979.

Brossard, E.B., *Petroleum, Politics, and Power,* Boston, Allyn and Bacon, 1974.

Brown, Professor Neville, "An Out of Area Strategy?" *Navy International,* October, 1982, pp. 1371–1373.

Brown, William, *Can OPEC Survive the Glut?* Croton-on-Hudson, N.Y., Hudson Institute, 1981.

Bussert, Jim, "Can the USSR Build and Support High Technology Fighters?" *Defense Electronics,* April, 1985, pp. 121–130.

Campbell, John C., "The Middle East: House of Containment Built on Shifting Sands", *Foreign Affairs,* 1981, pp. 593–628.

Carlsen, Robin Woodsworth, *The Imam and His Islamic Revolution,* New York, Snow Man Press, 1982.

Carroll, Jane, *Kuwait,* 1980, London, MEED, 1980.

Carver, Michael, *War Since 1945,* London, Weidenfeld and Nicholson, 1980.

Center for Strategic and International Studies, "The Economic and Fiscal Strategy of Saudi Arabia", Georgetown University, Middle East Conference, March 20–21, 1985.

Chalian, Gerald, *Guerrilla Strategies,* Berkeley, University of California Press, 1982.

Chicago Tribune, various editions.

Choucri, Nazli, *International Politics of Energy Interdependence,* Lexington, Mass., Lexington Books, 1976.

Christian Science Monitor, various editions.

Chubin, Shahram, "Gains for Soviet Policy in the Middle East", *International Security,* Spring 1982, pp. 122–173.

———, *Security in the Persian Gulf: The Role of Outside Powers,* London, International Institute for Strategic Studies, 1981.

Chubin, Shahram, ed., *Security in the Persian Gulf: Domestic Political Factors,* London, International Institute for Strategic Studies, 1980.

Cittadino, John, and McLeskey, Frank, "C^3I for the RDJTF", *Signal,* September 1981.

Clark, Wilson, and Page, Jake, *Energy, Vulnerability, and War,* New York, W. W. Norton, 1981.

Clarke, John I., and Bowen-Jones, Howard, *Change and Development in the Middle East,* New York, Methuen, 1981.

Clemens, Walter C., Jr., *The U.S.S.R. and Global Interdependence,* Washington, D.C., American Enterprise Institute Studies in Foreign Policy, 1978.

Cleron, Jean Paul, *Saudi Arabia 2000,* London, Croom Helm, 1978.

Collins, John M., and Mark, Clyde R., *Petroleum Imports from the Persian Gulf: Use of U.S. Armed Force to Ensure Supplies,* Issue Brief IB 79046, Washington, D.C., Library of Congress, Congressional Research Service, 1979.

Collins, Michael, "Riyadh: The Saud Balance", *Washington Quarterly,* Winter 1981.

Combat Fleets of the World 1986/87, Their Ships, Aircraft, and Armament, A.D. Baker III, ed., Anapolis, Md., Naval Institute Press, 1986.

Commerce, Department of, "Saudi Arabia", *Foreign Economic Trends and Their Implications for the United States,* FET 84-80, July, 1984.

———, "Saudi Arabia", *Foreign Economic Trends and Their Implications for the United States,* FET 85-79, September, 1985.

Conant, Melvin A., and Fern Racine Gold, *Access to Oil: The U.S. Relationship with Saudi Arabia and Iran,* Washington, D.C., Government Printing Office, 1977.

———, *The Oil Factor in U.S. Foreign Policy, 1980–1990,* Lexington, Mass., Lexington Books, 1982.

Congressional Budget Office, *Cost of Modernizing and Expanding the Navy's Carrier-Based Air Forces,* Washington, D.C., Congressional Budget Office, May 1982.

———, *Rapid Deployment Forces: Policy and Budgetary Implications,* Washington, D.C., Government Printing Office, 1981.

Congressional Presentation for Security Assistance Programs, Vols. 1 and 2, Fiscal Year 1987.

Congressional Research Service, Library of Congress, *Soviet Policy and the United States Response in the Third World,* Washington, D.C., Government Printing Office, 1981.

Conine, Ernest, "Soviets Sit on Oil's Power Keg", *Los AngelesTimes*, February 17, 1986.

CONOCO, *World Energy Outlook Through 2000*, April, 1985.

Conway's All The World's Fighting Ships 1947–1982, London, Conway Maritime Press, 1983.

Cordesman, Anthony H., "After AWACS: Establishing Western Security Throughout Southwest Asia", *Armed Forces Journal,* December 1981, pp. 64–68.

———, *American Strategic Forces and Extended Deterrence,* Adelphi Paper No. 175, London, International Institute for Strategic Studies, 1982.

———, "The Crisis in the Gulf: A Military Analysis", *American-Arab Affairs*, 9 (Summer 1984), pp. 8–15.

———, "Defense Planning in Saudi Arabia", *Defense Planning in Less-Industrialized States*, edited by Stephanie Neuman, Lexington, Mass., Lexington Books, 1984.

———, "The Falklands Crisis: Emerging Lessons for Power Projection and Force Planning", *Armed Forces Journal* , September 1982, pp. 29–46.

———, *Jordan and the Middle East Balance*, Washington, D.C., Middle East Institute, 1978.

———. "Lessons of–the Iran-Iraq War", *Armed Forces Journal,* April-June 1982, pp. 32–47, 68-85.

———, "Oman: The Guardian of the Eastern Gulf", *Armed Forces Journal International*, June 1983.

———, "The 'Oil Glut' and the Strategic Importance of the Gulf States", *Armed Forces Journal International*, October, 1983.

———, "Saudi Arabia, AWACS and America's Search for Strategic Stability", International Security Studies Program, Working Paper No. 26A, Washington, D.C., Wilson Center, 1981.

———, "The Saudi Arms Sale: The True Risks, Benefits, and Costs", *Middle East Insight*, Vol. 4, Nos 4 and 5, pp. 40–54.

———, "U.S. Middle East Aid: Some Questions", *Defense and Foreign Affairs*, June, 1986, pp. 15–18.

———, *The Iran-Iraq War and U.S.-Iraq Relations: An Iraqi Perspective*, Washington, D.C., National Council on U.S.-Arab Relations, 1984.

———, *Western Strategic Relations With Saudi Arabia*, London, Croom Helm, 1986.

———, *The Iran-Iraq War and Western Security, 1984–1987*, Jane's, London, 1987.

———, *The Arab-Israeli Military Balance and the Art of Operations*, Washington, D.C., University Press of America–AEI, 1987.

Cottrell, Alvin J., and Robert J. Hanks, "The Strait of Hormuz: Strategic Chokepoint", in Cottrell and Hanks *Sea Power and Strategy in the Indian Ocean*, Beverly Hills, Calif., Sage Publications, 1981.

Cottrell, Alvin J., and Michael L. Moodie, *The United States and the Persian Gulf: Past Mistakes, Present Needs*, New York, National Strategy Information Center for Scholars, 1981.

Croan, Melvin, "A New Afrika Korps", *Washington Quarterly*, No. 3 (Winter 1980), pp. 21-37.

Cummings, J. H., H. Askari, and M. Skinner, "Military Expenditures and Manpower Requirements in the Arabian Peninsula", *Arab Studies Quarterly, 2* (1980).

Danziger, Dr. Raphael, "The Persian Gulf Tanker War", *Proceedings of the Naval Institute*, May, 1985, pp. 160–176.

Darius, Robert G., John W. Amos II, and Ralph H. Magnus, *Gulf Security into the 1980s: Perceptual and Strategic Dimensions*, Stanford, Hoover Institution Press, 1984.

Davis, Jacquelyn K., and Pfaltzgraff, Robert L., *Power Projection and the Long Range Combat Aircraft*, Cambridge, Mass., Institute for Foreign Policy Analysis, June, 1981.

Dawisha, Adeed I., *Saudi Arabia's Search for Security*, Adelphi Paper No. 158 London, International Institute for Strategic Studies, Winter 1979–1980.

———,"Iraq: The West's Opportunity", *Foreign Policy*, No. 41 (Winter 1980–1981) 134_154.

———, "Iraq and the Arab World: The Gulf War and After", *The World Today*, March, 1981.

de Briganti, Giovanni, "Forces d'Action Rapide", *Armed Forces Journal*, October, 1984, pp. 46–47.

Deese, David A., and Joseph Nye, eds., *Energy and Security*, Cambridge, Mass., Ballinger, 1981.

Defense and Foreign Affairs, various editions,.

———, "France's Special Operations Forces", June 1985, pp. 32–33.

Defense News, various editions.

Defense Update, "Helicopter Special", No. 60, March, 1985.

de Galard, Jean, "French Overseas Action: Supplementary Budget", *Jane's Defense Weekly*, December, 14 1985, p. 1281.

De Gaury, Gerald, *Faisal: King of Saudi Arabia*, New York, Praeger Publishers, 1966.

Dougherty, James E., *The Horn of Africa: A Map of Political-Strategic Conflict*, Cambridge, Mass., Institute for Foreign Policy Analysis, 1982.

Dunn, Keith A., "Constraints on the U.S.S.R. in Southwest Asia: A Military Analysis", *Orbis*, 25, No. 3 (Fall 1981), pp. 607–629.

Dunn, Michael C., "Gulf Security: The States Look After Themselves", *Defense & Foreign Affairs*, June, 1982.

Dupuy, Trevor N., *Elusive Victory: The Arab-Israeli Wars, 1947–1974*, New York, Harper & Row, 1978 .

Economist, various editions.

Economist Publications, London and New York.

———, "Growing Pains, The Gulf Cooperation Countries, A Survey", *The Economist*, February 8, 1986.

———, "Oil Turns Manic Depressive", *The Economist*, February 15, 1986, pp. 61–62.

Economist Intelligence Unit, *The Gulf War; A Survey of Political Issues and Economic Consequences*, London, Economist Publications, 1984.

———, *EIU Regional Review: The Middle East and North Africa,1985*, Economist Publications, London, 1985.

———, *EIU Regional Review: The Middle East and North Africa,1986*, London, Economic Publications, 1986.

El Mallakh, Ragaei, *Qatar: Energy and Development*, London, Croom Helm, 1985.

Epstein, Joshua M., "Soviet Vulnerabilities in Iran and the RDF Deterrent", *International Security*, Vol. 6, No. 2 (Fall 1981), pp. 126–180.

———, *Strategy and Force Planning: The Case of the Persian Gulf*, Washington, Brookings Institution, 1987.

Eshel, David, *Born in Battle*, Series Nos. 1, 3, 12, and 16, Tel Aviv, Eshel-Dramit, 1978.

———, *The Israeli Air Force*, Tel Aviv, Eshel-Dramit, 1980.

———, *Peace for Galilee*, Special edition of the Born in Battle Series, Tel Aviv, Eshel-Dramit, 1982.

———, *The U.S. Rapid Deployment Forces*, New York, Arco Publishing, Inc., 1985.

Evron, Yair, *An American-Israel Defense Treaty*, No. 14, Tel Aviv, Center for Strategic Studies, Tel Aviv University, December 1981.

Farad, Abd al-Majid, ed., *Oil and Security in the Arabian Gulf*, London, Croom Helm, 1981.

Fairlamb, David, "Why the Saudis Are Switching Investments", *Dunn's Business Month*, May, 1985.

Faquih, Osama, "Similarities in Economic Outlook Between the U.S. and Saudi Arabia", February 22, 1985.

Feldman, Shai, "A Nuclear Middle East", *Survival*, 23, No. 3 (May-June 1981), pp. 107–116.

——— *Israeli Nuclear Deterrence, A Strategy for the 1980s*, New York, Columbia University Press, 1982.

Fesharaki, Feridun, and David T. Isaak, *OPEC, the Gulf, and the World Petroleum Market*, Boulder, Colo., Westview Press, 1983.

Feuchtwanger, E. J., and Nailor, Peter, *The Soviet Union and the Third World*, London, Macmillan, 1981.

Fiecke, D., B. Kroqully, and D. Reich, "The Tornado Weapons System and Its Contemporaries", *International Defense Review*, No. 2, 1977.

Financial Times, London and Frankfurt.

Fischer, Michael M. J., *Iran: From Religious Dispute to Revolution*, Cambridge, Mass., Harvard University Press, 1980.

Flavin, Christopher, *World Oil: Coping with the Dangers of Success*, Worldwatch Paper 66, Washington, D.C., Worldwatch Institute, 1985.

Forbis, William H., *The Fall of the Peacock Throne*, New York, McGraw-Hill, 1981.

Fricaud-Chagnaud, General, "La Force d'Action Rapide", July 2, 1986.

Fukuyama, Frances, *The Soviet Union and Iraq Since 1968*, Santa Monica, Calif., RAND, N-1524, AF., 1980.

Furling, R.D.M., "Israel Lashes Out", *International Defense Review* (Geneva) 15, No. 8, 1982, pp. 1001–1003.

———, "Operational Aspects of the F–15 Eagle", *International Defense Review*, 3/1975, pp. 129-139.

Gail, Bridget, "The West's Jugular Vein: Arab Oil", *Armed Forces Journal*

International, 1978, p. 18.

Ghassan, Salameh, "Saudi Arabia: Development and Dependence", *Jerusalem Quarterly*, No. 16, Summer 1980, pp. 137–144.

Golan, Galia, *The Soviet Union and the Israeli War in Lebanon*, Research Paper 46, Jerusalem, Soviet and East European Research Center, 1982.

Golub, David B., *When Oil and Politics Mix: Saudi Oil Policy, 1933–1985*, Cambridge, Harvard University, 1985.

Goldberg, Jacob, "How Stable Is Saudi Arabia?" *Washington Quarterly*, Spring 1982.

Grayson, Benson Lee, *Saudi-American Relations*, Washington, D.C., University Press of America, 1982.

Grayson, Leslie E., *National Oil Companies*, New York, John Wiley, 1981.

Green, Richard, ed., *Middle East Review, 1986*, London, Middle East Review Company, 1986.

Griffith, William E., *The Middle East 1982: Politics, Revolutionary Islam, and American Policy*, Cambridge, Mass., M.I.T. Press, 1982.

———, "The Revival of Islamic Fundamentalism: The Case of Iran", *International Security*, 5, No. 4, Spring 1981, pp. 49–73.

Grimmett, Richard F., *Trends in Conventional Arms Transfers to the Third World by Major Supplier, 1978–1985*, Washington, CRS Report 86-99F, May 9, 1986.

Grummon, Stephen R., *The Iran-Iraq War: Islam Embattled*, Washington Paper 92, Center for Strategic and International Studies. New York: Praeger Publishers, 1982.

Gulf Cooperation Council, *Cooperation Council for the Arab States of the Gulf, Information Handbook*, Riyadh, Bahr Al-Olum Press, 1982.

Gunston, Bill, *Modern Airborne Missiles*, ARCO, New York, 1983.

———, *Modern Soviet Air Force*, New York, ARCO, 1982.

———, Martin Streetly, "Su-24 Fencer C; Major Equipment Change", *Jane's Defense Weekly*, June 22, 1985, pp. 1226–1227.

Haffa, Robert P., Jr., *The Half War: Planning U.S. Deployment Forces to Meet a Limited Contingency, 1960–1983*, Boulder, Colo., Westview Press, 1984.

Halliday, Fred, *Arabia Without Sultans*, London, Pelican, 1975.

———, "Yemen's Unfinished Revolution: Socialism in the South", MERIP *Reports*, various editions.

Halloran, Richard, "Poised for the Persian Gulf", *The New York Times Magazine*, April 1, 1984, pp. 38–40, 61.

Hameed, Mazher, *An American Imperative: The Defense of Saudi Arabia*, Washington, D.C., Middle East Assessments Group, 1981.

———, *Arabia Imperilled: The Security Imperatives of the Arab Gulf States*, Washington, Middle East Assessments Group, 1986.

Hanks, Robert, *The U.S. Military Presence in the Middle East: Problems and Prospects*, Cambridge, Mass., Institute for Foreign Policy Analysis, 1982.

Hardt, John P., "Soviet Energy: Production and Exports", Issue Brief No. 12B75059, Library of Congress, Congressional Research Service, Washington, D.C., 1979.

Hargraves, D., and S. Fromson, *World Index of Strategic Minerals*, New York, Facts on File, 1983.

Harkabi, Yehoshafat, "Reflections on National Defence Policy", *Jerusalem*

Quarterly, No. 18, Winter 1981, pp. 121–140.

Hartley, Keith, "Can Britain Afford a Rapid Deployment Force?", *RUSI Journal*, Vol. 127, No. 1, March, 1982, pp. 18–22 .

Hawdon, David, ed., *Changing Structure of the World Oil Industry*, London, Croom Helm, 1985.

Hawley, Donald, *Oman and Its Renaissance*, London, Atacey International, 1980.

Hedley, Don, *World Energy: The Facts and the Future*, London: Euromonitor, 1981.

Heikal, Mohammed, *Iran: The Untold Story*, New York, Pantheon, 1982 (also published as *The Return of the Ayatollah*, London, Andre Deutsch, 1981).

Heller, Mark, Dov Tamari, and Zeev Eytan, *The Middle East Military Balance*, Jaffe Center for Strategic Studies, Tel Aviv University, 1985.

Helms, Christian Moss, *The Cohesion of Saudi Arabia*, Baltimore, John Hopkins University Press, 1981.

———, *Iraq: Eastern Flank of the Arab World*, Washington, Brookings Institution, 1984.

Henze, Paul B., "Arming the Horn", Working Paper No. 43, Washington, D.C., International Studies Program, Wilson Center, July 28, 1983.

Herzog, Chaim, *The Arab-Israeli Wars*, New York, Random House, 1982.

Hetherton, Norris, S., "Industrialization and Revolution in Iran: Force Progress or Unmet Expectation", *Middle East Journal*, 36, No. 3, Summer 1982, pp. 362–373.

Hickman, William F, *Ravaged and Reborn: The Iranian Army*, 1982 Staff Paper, Washington, D.C., Brookings Institution, 1982.

Hiro, Dilip, *Iran Under the Ayatollahs*, London, Routledge and Paul, 1985.

Holden, David, and Richard Johns, *The House of Saud*, London, Sidgwick and Jackson, 1981.

Horwich, George, and Edward Mitchell, eds., *Policies for Coping with Oil Supply Disruptions*, Washington, D.C., American Enterprise Institute, 1982.

Hottinger, Arnold, "Arab Communism at Low Ebb", *Problems of Communism*, July-August 1981, pp. 17–32.

———, "Does Saudi Arabia Face Revolution?" *New York Review of Books*, June 28, 1979.

Howarth, H. M. F., "The Impact of the Iran-Iraq War on Military Requirements in the Gulf States", *International Defense Review*, 16, No. 10, 1983.

Howlett, Lt. General Sir Geoffrey, "NATO European Interests Worldwide—Britain's Military Contribution", *RUSI Journal*, Vol. 130, No. 3, September, 1985, pp. 3–10.

Hunter, Shireen, ed., *Political and Economic Trends in the Middle East*, The Center for Strategic and International Studies, Boulder, Colo., Westview Press, 1985.

———, *The Gulf Cooperation Council: Problems and Prospects*, Washington, CSIS, 1984.

Hurewitz, J. C., *Middle East Politics: The Military Dimension*, New York, Praeger Publishers, 1969.

Hyman, Anthony, *Afghanistan Under Soviet Domination, 1964–81*, London, Macmillan, 1982.

Ibrahim, Saad Eddin, *The New Arab Social Order: A Study of the Social Impact of Oil Wealth*, Boulder, Colo., Westview Press, 1982.

International Defense Review, Switzerland, Geneva, various editions.

International Defense Review, Special Series, various editions.

International Energy Statistical Review, Washington, D.C., National Foreign Energy Assessment Center, CIA, various editions.

International Institute for Strategic Studies, *The Middle East and the International System*, Parts I and II, Adelphi Papers No. 114 and 115, London, 1975.

———, *The Military Balance*, London, various years.

International Journal of Middle East Studies, New York.

International Monetary Fund, *Direction of Trade Statistics*, various editions.

———, *Direction of Trade Yearbook*, Washington D.C., various years.

Isby, David C., "Afghanistan: The Unending Struggle", *Military Annual*, London, Jane's, 1982, pp. 28–45.

———, "Afghanistan: 1982: The War Continues", *International Defense Review*, 11 (1982), pp. 1523–1528.

———, *Weapons and Tactics of the Soviet Army*, New York, Jane's, 1981.

Ismael, Tareq Y., *The Iran-Iraq Conflict*, Toronto, Canadian Institute of International Affairs, 1981.

———, *Iraq and Iran: Roots of Conflict*, Syracuse, N.Y., Syracuse University Press, 1982.

Ispahana, Mahnaz Zehra, "Alone Together: Regional Security Arrangements in Southern Africa and the Arabian Gulf", *International Security*, 8, 4, Spring 1984, pp. 152–175.

Iungerich, Ralph, "U.S. Rapid Deployment Force—USCENTCOM—What Is It? Can It Do the Job?" *Armed Forces Journal International*, 122, 3, October 1984.

Jacobs, G., "Afghanistan Forces: How Many Soviets Are There?" *Jane's Defense Weekly*, June 22, 1985, pp. 1228–1233.

Jaffe Center for Strategic Studies, *The Middle East Military Balance*, Tel Aviv, Tel Aviv University, various years.

Jane's, *All the World's Aircraft*, London, various years.

———, *Armour and Artillery*, London, various years.

———, *Aviation Annual*, London, various years.

———, *Combat Support Equipment*, London, various years.

———, *Defense Review*, London, various years.

———, *Fighting Ships*, London, various years.

———, *Infantry Weapons*, London, various years.

———, *Military Annual*, London, various years.

———, *Military Communications*, London, various years.

———, *Naval Annual*, London, various years.

———, *Naval Review*, London, various years.

———, *Weapons Systems*, London, various years.

Jenkins, Brian Michael, et al., "Nuclear Terrorism and Its Consequences", *Society*, 17, No. 5, July-August 1980, pp. 5–25.

Johany, Ali D., *The Myth of the OPEC Cartel: The Role of Saudi Arabia*, New York, John Wiley, 1982.

Johnson, Major Maxwell Orme, U.S.M.C., *The Military as an Instrument of U.S. Policy in Southwest Asia: The Rapid Deployment Joint Task Force, 1979–1982*,

Boulder, Colo., Westview Press, 1983.

———, "U.S. Strategic Operations in the Persian Gulf", *Proceedings of the Naval Institute,* February 1981.

Jones, Rodney W., *Nuclear Proliferation: Islam, the Bomb and South Asia,* Washington Paper No. 82, Center for Strategic and International Studies. Beverly Hills, Calif., Sage Publications, 1981.

———, ed., *Small Nuclear Forces and U.S. Security Policy,* Lexington, Mass., Lexington Books, 1984.

Jordan, Amos, "Saudi Arabia: The Next Iran", *Parameters: The Journal of the Army War College,* 9 (March).

Jordan, John, *Modern Naval Aviation and Aircraft Carriers,* New York, ARCO, 1983.

Joyner, Christopher C., and Shah, Shahqat Ali, "The Reagan Policy of 'Strategic Consensus' in the Middle East", *Strategic Review,* Fall 1981, pp. 15–24.

Judge, John F., "Harpoon Missile Targets Ships and Cost", *Defense Electronics,* April, 1985, pp. 92–98.

Jureidini, Paul, and R.D. McLaurin, *Beyond Camp David,* Syracuse, N.Y., Syracuse University Press, 1981.

Kanovsky, Eliyahu, "Saudi Arabia in the Red", *Jerusalem Quarterly,* No. 16, Summer 1980, pp. 137–144.

Kaplan, Stephen S., *Diplomacy of Power,* Washington, D.C., Brookings Institution, 1981.

Karsh, Efraim, *The Cautious Bear,* Boulder, Colo., Westview Press, 1985.

———, *Soviet Arms Transfers to the Middle East in the 1970s,* Tel Aviv, Tel Aviv University, 1983.

———, *The Iran-Iraq War: The Military Implications,* London, IISS Adelphi Paper No. 220, 1987.

Katz, Mark, *Russia and Arabia: Soviet Foreign Policy and the Arabian Peninsula,* Baltimore, Johns Hopkins University, 1985.

Kazemi, Farhad, *Poverty and Revolution in Iran,* New York University Press, 1980.

Keegan, John, *World Armies,* New York, Facts on File, 1979.

———, *World Armies,* 2nd ed., London, Macmillan, 1983.

Kelly, J. B., *Arabia, the Gulf and the West: A Critical View of the Arabs and Their Oil Policy,* New York, Basic Books, 1980.

Kerr, Malcolm, and El Sayed Yassin, eds., *Rich and Poor States in the Middle East,* Boulder, Colo., Westview Press, 1982.

King, Ralph, *The Iran-Iraq War: The Political Implications,* London, IISS Adelphi Paper 219, 1987.

Klare, Michael T., *American Arms Supermarket,* Austin, Texas, University of Texas Press, 1984.

Korb, Edward L., ed., *The World's Missile Systems,* 7th ed., Pamona, Calif., General Dynamics, Pamona Division, 1982.

Kraft, Joseph, "Letter from Saudi Arabia", *New Yorker,* July 4, 1983.

Krapels, Edward, N., ed., "International Oil Supplies and Stockpiling", Proceedings of a conference held in Hamburg, September 17 and 18, 1981, London, Economist Intelligence Unit, 1982.

Kuniholm, Bruce, "What the Saudis Really Want: A Primer for the Reagan

Administration", *Orbis,* 25, Spring 1981.

Kurian, George, *Atlas of the Third World*, New York, Facts on File, 1983.

Kuwait, *Annual Statistical Abstract,* Kuwait City, Ministry of Planning, Central Statistical office, various editions.

Kuwaiti News Service, *The Gulf Cooperation Council,* Digest No. 9, KUNA, Kuwait, 9th issue, 3rd ed, Kuwait, Kuwaiti Universal News Agency, December, 1982.

Lacey, Robert, *The Kingdom*, London, Hutchinson & Co., 1981.

Lachamade, Pierre, "The French Navy in the Year 2000", *Jane's Naval Review*, London, 1985, Jane's, pp. 79–90.

Lackner, Helen, *A House Built on Sand: The Political Economy of Saudi Arabia*, London, Ithaca, 1979.

Laffin, John L., *The Dagger of Islam,* London, Sphere, 1979.

Leites, Nathan, *Soviet Style in War,* New York, Crane, Russak & Co., 1982.

Leltenberg, Milton, and Gabriel Sheffer, eds. *Great Power Intervention in the Middle East*, New York, Pergamon Press, 1979.

Lenczowski, George, "The Soviet Union and the Persian Gulf: An Encircling Strategy", *International Journal*, 37, No. 2, 1982.

Library of Congress, "The Persian Gulf: Are We Committed?" Washington, D.C., 1981.

Lieber, Robert J., *The Oil Decade: Conflict and Cooperation in the West*, New York, University Press of America, 1986.

Liebov, Robert J, "Energy, Economics and Security in Alliance Perspective", *International Security,* Spring 1980, pp. 139–163.

Limbert, John W., *Iran: At War With History*, Boulder, Colo., Westview Press, 1987.

Litwak, Robert, ed., *Security in the Persian Gulf: Sources of Inter-State Conflict*, London, International Institute for Strategic Studies, 1981.

Long, David E. "Saudi Oil Policy", *Wilson Quarterly*, 1979.

———, "U.S.-Saudi Relations: A Foundation of Mutual Need", *American-Arab Affairs,* No. 4, Spring 1983, pp. 12–22.

———, *The United States and Saudi Arabia: Ambivilant Allies*, Boulder, Colo., Westview Press, 1985.

Long, David E., and John A. Shah, *Saudi Arabian Modernization: The Impact of Change on Stability*, Washington, CSIS, 1982.

Looney, Robert E., *Saudi Arabia's Development Potential*, Lexington, Mass., Lexington Books, 1982.

Los Angeles Times, various editions.

Lottam, Emanuel, "Arab Aid to Less Developed Countries", *Middle East Review*, 1979–1980, pp. 30–39.

Luckner, Helen A., *A House Built on Sand: A Political Economy of Saudi Arabia*, London, Ithaca Press, 1978.

MacDonald, Charles G., "The U.S. and Gulf Conflict Scenarios", *Middle East Insight*, 3, No. 1, May-July 1983, pp. 23–27.

Maddy-Weitzman, Bruce, "Islam and Arabism: The Iraq-Iran War", *Washington Quarterly*, Autumn 1982.

Mansur, Abdul Karim (pseud.), "The Military Balance in the Persian Gulf: Who Will Guard the Gulf States from Their Guardians?" *Armed Forces Journal International*, November 1980.

Martin, Lenore G., *The Unstable Gulf: Threats from Within*, Lexington, Mass., D.C. Heath, 1984.

Macksey, Kenneth, *Tank Facts and Feats*, New York, Two Continents Publishing Group, 1974.

Makinda, Samuel, "Shifting Alliances in the Horn of Africa", *Survival*, January-February, 1985, pp. 11_19.

Male, Beverly, *Revolutionary Afghanistan*, London, Croom Helm, 1982.

Mallakh, Ragaei El, *OPEC: Twenty years and Beyond*, Boulder, Colo., Westview Press, 1982.

———, *Saudi Arabia: Rush to Development*, London, Croom Helm, 1982.

Mallakh, Ragaei El, and Dorothea H. El Mallakh, *Saudi Arabia: Energy, Development Planning, and Industrialization*, Lexington, Mass., Lexington Books, 1982.

Mansur, Abdul Kasim (pseud.), "The American Threat to Saudi Arabia", *Armed Forces Journal International*, September 1980, pp. 47–60.

Marcus, Jonathan, and Bruce George, "French Rapid Deployment Force", *Jane's Defense Weekly*, April 28, 1984, pp. 649–650.

Masters, Charles D., "World Petroleum Resources—A Perspective", USGS Open File Report, pp. 85–248.

Masters, Charles D., David H. Root, and William D. Dietzman, "Distribution and Quantitative Assessment of World Crude-Oil Reserves and Resources", Washington, USGS, unpublished, 1983.

McDonald, John, and Clyde Burleson, *Flight from Dhahran*, Englewood Cliffs, N.J., Prentice-Hall, 1981.

McLaurin, R.D., "U.S. Strategy in the Middle East and the Arab Reaction", *Journal of East and West Studies*, 11, 2, Fall-Winter 1982.

———, McLaurin, R.D., and Lewis W. Snider, *Saudi Arabia's Air Defense Requirements in the 1980s: A Threat Analysis*, Alexandria, Va., Abbott Associates, 1979.

McNaugher, Thomas L., "Arms and Allies on the Arabian Peninsula, *Orbis*, Vol. 28, No. 3, Fall 1984, pp. 486–526.

———, *Arms and Oil: U.S. Military Security Policy Toward the Persian Gulf*, Washington, D.C., Brookings Institution, 1985.

———, Shireen Hunter, ed., *Gulf Cooperation Council: Problems and Prospects*, CSIS Significant Issues Series, 6, 15 (1984), pp. 6–9.

———, "Rapid Deployment and Basing in Southwest Asia", in *Strategic Survey* (London, International Institute for Strategic Studies, April 1983), pp. 133–137.

———,"The Soviet Military Threat to the Gulf: The Operational Dimension", Working paper, Washington, D.C., Brookings Institution, 1982.

Meir, Shemuel, *Strategic Implications of the New Oil Reality*, Boulder, Colo., Westview Press, 1986, p. 55.

MERIP Reports, "The Arabian Peninsula Opposition Movements", February, 1985, pp. 13–19.

Mets, Lt. Colonel David R., "The Dilemmas of the Horn", *Proceedings of the Naval Institute*, April, 1985, pp. 49–57.

Middle East, "Guarding Turkey's Eastern Flank", April, 1986, pp. 9–10.

Middle East Economic Digest, London.

———, *Oman: A Practical Guide*, London, 1981.
———, *Saudi Arabia: A Practical Guide*, London, 1981.
———, *UAE: A Practical Guide*, London, 1981.
Middle East Economic Digest, Special Report Series, *Bahrain*, London, September 1981 and September 1982.
———, *France and the Middle East*, May 1982.
———, *Oman*, November 1982.
———, *Qatar*, August 1981 and August 1982.
———, *UAE: Tenth Anniversary*, November 1981.
———, *UK and the Gulf*, December 1981.
Middle East Insight, various editions.
Middle East Journal, Washington, D.C., Middle East Institute, various editions.
Middle East Review, 1985, World of Information, Saffron Walden, England, 1985.
Middle East Review, 1986, World of Information, Saffron Walden, England, 1986.
Miller, Aaron David, *Search for Security: Saudi Arabian Oil and American Foreign Policy*, 1939–1949, Chapel Hill, University of North Carolina Press, 1980.
Moyston, Trevor, *Saudi Arabia*, London, MEED, 1981.
———, *UAE*, London, MEED, 1982.
Mottahedeh, Roy Parviz, "Iran's Foreign Devils", *Foreign Policy*, No. 38, Spring 1980, pp. 19–34.
Naff, Thomas, ed., *Gulf Security and the Iran Iraq War*, Washington, D.C., National Defense University Press, 1985.
National Foreign Assessment Center, *International Energy Statistical Review*, Washington, D.C., Photoduplication Service, Library of Congress, 1978–1986.
Nakhleh, Emile A., *The Gulf Cooperation Council: Policies, Problems, and Prospects*, New York, Praeger Publishers, 1986.
Natkiel, Richard, *Atlas of the 20th Century*, New York, Facts on File, 1982.
Navy, Department of, Office of the Chief of Naval Operations, *Understanding Soviet Naval Developments*, Washington D.C., Government Printing Office, April 1985.
Nearby Observer, "The Afghan-Soviet War: Stalemate or Solution?", *Middle East Journal*, Spring 1982, pp. 151–164.
Neuman, Stephanie, *Defense Planning in Less-Industrialized States*, Lexington, Mass., Lexington Books, 1984.
Neumann, Robert G., and Shireen T. Hunter, "The Crisis in the Gulf: Reasons for Concern but not Panic", *American-Arab Affairs*, 9, Summer 1984, pp. 16–21.
Nevo, Joseph, "The Saudi Royal Family: The Third Generation", *Jerusalem Quarterly*, No. 31, Spring 1984.
Newhouse, John, "The Diplomatic Round, Politics and Weapons Sales", *New Yorker*, June 9, 1986, pp. 46–69 .
New York Times, various editions.
Newell, Nancy Peabody, and Newell Richard S., *The Struggle for Afghanistan*, Ithaca, N.Y., Cornell University Press, 1981.
Niblock, Tim, ed., *State, Society, and the Economy in Saudi Arabia*, London, Croom Helm, 1982.
———, *Social and Economic Development in Arab Gulf States*, London, Croom Helm, 1980.

Nimatallah, Yusuf, "Arab Banking and Investment in the U.S.", *IMF*, February 22, 1985.

Nimir, S.A., and M. Palmer, "Bureaucracy and Development in Saudi Arabia: A Behavioral Analysis", *Public Administration and Development*, April–June 1982.

Nonneman, Gerd, Iraq, *The Gulf States, and the War: A Changing Relationship*, London, Ithaca Press, 1986.

Novik, Nimrod, *Encounter with Reality; Reagan and the Middle East*, Boulder, Colo., Westview Press, 1985.

Noyes, James H., *The Clouded Lens*, Stanford, Calif., Hoover Institution, 1982.

Nugent, Jeffery B., and Theodore Thomas, ed., *Bahrain and the Gulf: Past Perspectives and Alternative Futures*, New York, St. Martin's, 1985.

O'Ballance, Edgar, "The Iran-Iraq War", *Marine Corps Gazette*, February 1982, pp. 44–49.

Ochsenwald, William, "Saudi Arabia and the Islamic Revival," *International Journal of Middle East Studies*, 13, No. 3, August 1981, pp. 271–286.

Odell, Peter R., and Kenneth E. Rosing, *The Future of Oil: A Simulation Study*, London, Nichols, 1980.

O'Dwyer-Russel, Simon, "Beyond the Falklands—The Role of Britain's –ut of Area Joint Forces", *Jane's Defense Weekly*, January 11, 1986, pp. 26-27.

OECD/IEA, *Oil and Gas Statistics, 1985*, No. 4, Paris, 1986.

Oil and Gas Journal, "Worldwide Report", December 31, 1984.

Olson, William J., "The Iran-Iraq War and the Future of the Persian Gulf", *Military Review*, 64, 2, March 1984, pp. 17–29.

———, ed., *U.S. Strategic Interests in the Gulf Region*, Boulder, Colo., Westview Press, 1987.

Oman and Its Renaissance, London, Stacey International, 1980.

Oman: A Practical Guide, London, MEED, 1982.

Oman, Sultanate of, *Oman in Ten Years*, Muscat, Ministry of Information, 1980.

———, *Second Five Year Plan*, 1981–85, Muscat Development Council, 1981.

Organization of Petroleum Exporting Countries, *Annual Report*, Vienna, various years.

Osama, Abdul Rhman, *The Dilemma of Development in the Arabian Peninsula*, London, Croom Helm, 1987.

Osbourne, Christine, *The Gulf States and Oman*, London, Croom Helm, 1977.

Pahlavi, Mohammed Reza, *The Shah's Story*, London, Michael Joseph, 1980.

Paul, Jim, "Insurrection at Mecca", *MERIP Reports*, No., 86, October 1980.

Peck, Malcom C., *The United Arab Emirates: A Venture in Unity*, Boulder, Colo., Westview Press, 1986.

Perlmutter, Amos, Michael Handel, and Uri Bar-Joseph, *Two Minutes Over Baghdad*, London, Corgi, 1982.

Perry, Charles, *The West, Japan, and Cape Route Imports: The Oil and Non-Fuel Mineral Trades*, Cambridge, Mass., Institute for Foreign Policy Analysis, 1982.

Peterson, J. E., "Defending Arabia: Evolution of Responsibility", *Orbis*, 28, 2, Fall 1984, pp. 465–488 .

———, *Oman in the Twentieth Century: Political Foundations of an Emerging State*, London, Croom Helm, 1978.

———, *Yemen: The Search for a Modern State*, Baltimore, John Hopkins

University Press, 1982.
———, *Defending Arabia*, New York, St. Martin's, 1986.
Petroleum Intelligence Weekly, New York.
Pierre, Andrew J., "Beyond the 'Plane Package': Arms and Politics in the Middle East", *International Security*, 3, No. 1, Summer 1978, pp. 148–161.
———, *The Global Politics of Arms Sales*, Princeton, N.J., Princeton University Press, 1982.
Pipes, Daniel, "Increasing Security in the Persian Gulf", *Orbis*, 26, Spring 1982.
Plascov, Avi, *Security in the Persian Gulf: Modernization, Political Development and Stability*, Aldershot, Gower, 1982.
Platt's Oil Price Handbook, New York.
Poullada, Leon B., "Afghanistan and the United States, The Crucial Years", *Middle East Journal*, Spring 1981, pp. 178–190.
Pradas, Col. Alfred B., *Trilateral Military Aid in the Middle East: The Yemen Program*, Washington, D.C., National Defense University, 1979.
Pridham, B.R., ed., *The Arab Gulf and the West*, New York, St Martin's Press, 1985.
———, *Oman: Economic, Social, and Strategic Developments*, London, Croom Helm, 1987.
Pry, Peter, *Israel's Nuclear Arsenal*, Boulder, Colo., Westview Press, 1984.
Quandt, William B., "The Crisis in the Gulf: Policy Options and Regional Implications", *American-Arab Affairs*, 9, Summer 1984, pp. 1–7.
———, "Riyadh Between the Superpowers", *Foreign Policy*, No. 44, Fall 1981.
———, *Saudi Arabia in the 1980s: Foreign Policy, Security and Oil*, Washington, D.C., Brookings Institution, 1982.
———, *Saudi Arabia's Oil Policy: A Staff Paper*, Washington, D.C., Brookings Institution, 1982.
Ra'anan, Uri, *The USSR Arms the Third World*, Cambridge, Mass., M.I.T. Press, 1969.
Randol, William L., "Petroleum Monitor", First Boston Corporation.
Ransom, David M., Lt. Colonel Lawrence J. MacDonald, and W. Nathaniel Howell, "Atlantic Cooperation for Persian Gulf Security", *Essays on Strategy*, Washington, D.C., National Defense University, 1986.
Ramazani, R.K., *Revolutionary Iran: Challenge and Response in the Middle East*, Baltimore, Johns Hopkins University, 198 .
Record, Jeffrey, *The Rapid Deployment Force*, Cambridge, Mass., Institute for Foreign Policy Analysis, 1981.
Risso, Patricia, *Oman and Muscat*, New York, St. Martin's, 1986.
Roberts, Hugh, *An Urban Profile of the Middle East*, London, Croom Helm, 1979.
Rosen, Barry, ed., *Iran Since the Revolution: Internal Dynamics, Regional Conflict, and the Superpowers*, New York, Columbia University, 1985.
Ross, Dennis, "Considering Soviet Threats to the Persian Gulf", *International Security*, 6, No. 2, Fall 1981.
———, Soviet Views Toward the Gulf War", *Orbis*, 18, 3, Fall 1984, pp. 437–446.
Rouleau, Eric, "Khomeini's Iran", *Foreign Affairs*, Fall 1980, pp. 1–20.
———, "The War and the Struggle for the State", *MERIP Reports*, No. 98, July-August 1981, pp. 3–8.
Rowen, Hobart, "Reassessing Saudi Arabia's Economic Viability", *Washington*

Post, July 20, 1986.

Royal United Services Institute/Brassey's, *International Weapons Development*, 4th ed., London, Brassey's, 1981.

Rubin, Barry, *Paved with Good Intentions*, New York, Oxford University Press, 1980.

Rubinstein, Alven Z., "Afghanistan: Embraced By the Bear", *Orbis* 26, No. 1, Spring 1982, pp. 135–153.

———, *The Great Game: Rivalry in the Persian Gulf and South Asia*, New York, Praeger Publishers, 1983.

———, "The Last Years of Peaceful Co-Existence: Soviet-Afghan Relations 1963–1978", *Middle East Journal*, 36, No. 2, Spring 1982, pp. 165–183.

———, *Soviet Policy Towards Turkey, Iran, and Afghanistan: The Dynamics of Influence*, New York, Praeger Publishers, 1982.

Rumaihi, Muhammed, *Beyond Oil: Unity and Development in the Gulf*, London, Al Saqi Books, 1986.

Russi, Pierre, *Iraq, the Land of the New River*, Paris, Les Editions, J.A., 1980.

Rustow, Dankwart, *Oil and Turmoil: America Faces OPEC and the Middle East*, New York, Norton, 1982.

Ruszkiewicz, Lt. Col. John J., "A Case Study in the Yemen Arab Republic", *Armed Forces Journal*, September 1980, pp. 62–72.

Sabah-Al, Y.S.F., *The Oil Economy of Kuwait*, Boston, Keegan Paul, 1980.

Sabini, John, *Armies in the Sand: The Struggle for Mecca and Medina*, New York, W. W. Norton, 1981.

Safran, Nadav, *Saudi Arabia: The Ceaseless Quest for Security*, Cambridge, Mass., and London, Belknap Press of Harvard University Press, 1985.

Saikal, Amin, *The Rise and Fall of the Shah*, Princeton, N.J., Princeton University Press, 1980.

Salameh, Ghassane, "Checkmate in the Gulf War", *MERIP Reports*, 14, 6-7, July–September 1984, pp. 15–21.

al-Salem, Faisal, "The United States and the Gulf: What Do the Arabs Want?" *Journal of South Asian and Middle Eastern Studies*, 6, Fall 1982.

Sandwick, John A., ed., *The Gulf Cooperation Council: Moderation and Stability in an Interdpendent World*, Washington, Westview Press, 1987.

Saudi Arabia, Kingdom of, *Annual Report of the Saudi Fund for Development, 1984–1985*, Saudi Arabia, 1985·

———, *The Kingdom of Saudi Arabia: Relief Efforts*, Ministry of Finance and National Economy, Saudi Arabia, 1985.

———, Ministry of Finance and National Economy, Central Department of Statistics, *Population Census*, 14 Vols., Dammam, 1977.

———, Ministry of Finance and National Economy, *Statistical Yearbook*, annual, Jidda, various years.

———, Ministry of Planning, *Second Development Plan*, 1975–1980, Springfield, Va., U.S. Department of Commerce, Bureau of International Commerce, 1975.

———, *Saudi Arabia, Foreign Trade Statistics, 1984 AD.*

———, Saudi Arabian Monetary Agency, Research and Statistics Department, *Statistical Summary*, Riyadh, various years.

———, "Soviet Air Force in Afghanistan", *Jane's Defense Weekly*, July 7, 1984, pp.

1104–1105.

———, *Third Development Plan*, 1980–85, Riyadh, Ministry of Planning Press, 1980.

Saivetz, Carol R., *The Soviet Union and the Gulf in the 1980s*, Boulder, Colo., Westview Press, 1986.

Sciolino, Paulo, "Iran's Durable Revolution", *Foreign Affairs*, Spring 1983, pp. 893–920.

Schmid, Alex P., *Soviet Military Interventions Since 1945*, New Brunswick, N.J., Transaction, Inc., 1985.

Schmitt, Richard B., "U.S. Dependence on Oil, Gas Imports May Grow", *Wall Street Journal*, April 23, 1985.

Schrage, Daniel P., "Air Warfare: Helicopters and the Battlefield", *Journal of Defense and Diplomacy*, Vol. 3, No. 5, pp. 17–20.

Schultz, James B., "New Strategies and Soviet Threats Spark EW Responses", *Defense Electronics*, February, 1985, pp. 17–21.

Sella, Amon, *Soviet Political and Military Conduct in the Middle East*, London, Macmillan, 1981.

Senger, F.M. von, and Etterlin, *Tanks of the World 1983*, Anapolis, Md., Nautical & Aviation Publishing Co., 1983.

Shamir, Yitzhak, "Israel's Role in a Changing Middle East", *Foreign Affairs*, Spring 1982, pp. 789-802.

Shaw, John, "Saudi Arabia Comes of Age", *Washington Quarterly*, Spring 1982.

Shaw, John A., and David E. Long, *Saudi Arabian Modernization*, Washington Papers 89, New York, Praeger Puglishers, 1982.

Sick, Gary G., *All Fall Down: America's Tragic Encounter with Iran*, New York, Random House, 1985.

Sick, Gary G., and Alvin A. Rubinstein, ed., *The Great Game: Rivalry in the Persian Gulf and South Asia*, New York, Praeger Publishers, 1983.

SIPRI, *World Armaments and Disarmaments;:SIPRI Yearbook 1985*, London, Taylor & Francis, 1985.

Snyder, Jed C., and Samuel F. Wells, Jr.,. eds., *Limiting Nuclear Proliferation*, Cambridge, Mass., Ballinger Publishing Co., 1985.

———, *Defending the Fringe: NATO, the Mediterranean, and the Persian Gulf*, Boulder, Colo., Westview Press, 1987.

"Special Report, Middle East Aerospace: Saudi Arabia", *Aviation Week and Space Technology*, May 23, 1983.

Staudenmaier, William O., "Military Policy and Strategy in the Gulf War", *Parameters: The Journal of the Army War College*, 12 (June 1982).

Stauffer, Thomas R., *U.S. Aid to Israel: The Vital Link*, Middle East Problem Paper No. 24, Washington, D.C., Middle East Institute, 1983.

Stempel, John D., *Inside the Iranian Revolution*, Bloomington, Indiana University Press, 1981.

Stewert, Richard A. "Soviet Military Intervention in Iran, 1920–46", *Parameters, Journal of the U.S. Army War College*, 11, No. 4 (1981), pp. 24–34.

Stobach, Robert, and Daniel Yergin, eds., *Energy Future*, New York, Random House, 1979.

Stockholm International Peace Research Institute, *Tactical Nuclear Weapons:*

European Perspectives, New York, Crane, Russak & Co., 1978.
———, *World Armaments and Disarmament: SIPRI Yearbook*, various years (computer print out for 1982), London, Taylor E. Francis, Ltd.
Stookey, Robert W, *The Arabian Peninsula: Zone of Ferment*, Stanford, Hoover Institution, 1984.
———, *South Yemen: A Marxist Republic in Arabia*, Boulder, Colo., Westview Press, 1982.
Sullivan, William H., "Iran: The Road Not Taken", *Foreign Policy*, No. 40 (Fall 1980), pp. 175–187.
———, *Mission to Iran*, London, W. W. Norton, 1981.
———, "A Survey of Saudi Arabia", *Economist*, February 13, 1982.
Sweetman, Bill, "New Soviet Combat Aircraft", *International Defense Review*, 1, 1984, pp. 35–38.
Szuprowicz, Bohdan O., *How to Avoid Strategic Materials Shortages*, New York, John Wiley, 1981.
Tahir-Kheli, Sharin, and Shaheen Ayubi, *The Iran-Iraq War: New Weapons, Old Conflicts*, New York, Praeger Publishers, 1983.
Tahir-Kheli, Sharin, and William O. Staudenmaier, "The Saudi-Pakistani Military Relationship: Implications for U.S. Policy", *Orbis*, Spring 1982, pp. 155–171.
Taylor, Alan, *The Arab Balance of Power*, Syracuse, N.Y., Syracuse University Press, 1982.
Thompson, W. Scott, "The Persian Gulf and the Correlation of Forces", *International Security*, Summer 1982, pp. 157–180.
Tillman, Seth, *The United States in the Middle East*, Bloomington, Indiana University Press, 1982.
Truver, Dr. Scott C., "Mines of August: An International Whodunit", *Proceedings of the U.S. Naval Institute*, May 1985, Volume 3, No. 5, 1987, pp. 94–118.
Turner, Louis, and James M. Bedore, *Middle East Industrialization: A Study of Saudi and Iranian Downstream Investments*, London, Saxon House, 1979.
United Arab Emirates, Ministry of Information and Culture, *A Record of Achievement, 1979–1981*, Abu Dhabi, 1981.
U.S. Arms Control and Disarmament Agency, *World Military Expenditures and Arms Transfers*, various editions, Washington, D.C., 1980.
U.S. Central Intelligence Agency, *Economic and Energy Indicators*, DOI, GIEEI, Wahington, D.C., Government Printing Office, various years.
———, *Handbook of Economic Statistics*, various editions.
———, *International Energy Situation: Outlook to 1985*, 041-015-00084-5, Washington, D.C., Government Printing Office, 1977.
———, *International Energy Statistical Review*, NFAC, GI-IESR, Washington, D.C., Government Printing Office, various years.
———, *USSR Energy Atlas*, Washington, CIA, 1985.
———, *World Factbook*, Washington, D.C., Government Printing Office, various years.
U.S. Congress, House of Representatives, Committee on Appropriations, *Foreign Assistance and Related Programs Appropriations for 1982. Part 7: Proposed Airborne Warning and Control Systems (AWACS), F-15 Enhancement Equipment, and Sidewinder AIM 9L Missiles Sales to Saudi Arabia*, 97th Cong.,

1st Session, Hearings.
———, Committee on Foreign Affairs, *Activities of the U.S. Corps of Engineers in Saudi Arabia*, 96th Cong., 1st Sess., 1979.
———, *Proposed Arms Sales for Countries in the Middle East*, 96th Cong., 1st Sess., 1979.
———, *Proposed Arms Transfers to the Yemen Arab Republic*, 96th Cong., 1st Sess., 1979.
———, *Saudi Arabia and the United States*, Congressional Research Service Report, 97th Cong., 1st sess., 1981.
———, *Saudi Arabia and the United States: The New Context in an Evolving 'Special Relationship'*, No. 81-494 0, Washington, D.C., Government Printing Office, 1981.
———, *U.S. Interests in, and Policies Toward, the Persian Gulf, 1980*, No. 68-1840, Washington, D.C., Congressional Printing Office, 1980.
———, *U.S. Security Interests in the Persian Gulf*, No. 73-354-0, Washington, D.C., Government Printing Office, 1981.
———, Committee on Foreign Affairs and Joint Economic Committee, *U.S. Policy Toward the Persian Gulf*, 97th Cong., 1st sess., 1975.
———, Committee on Armed Services, *U.S. Military Forces to Protect "Reflagged" Kuwaiti Oil Tankers*, Hearings, June 5, 11, 16, 1987.
U.S. Congress, Senate, Committee on Armed Services, *Military and Technical Implications of the Proposed Sale of Air Defense Enhancements to Saudi Arabia. Report of the Hearings on the Military and Technical Implications of the Proposed Sale of Air Defense Enhancements to Saudi Arabia, Based Upon Hearings Held Before the Committee in Accordance with Its Responsibilities Under Rule XXV (C) of the Standing Rules of the Senate*, 97th Cong., 1st Session.
———, Committee on Energy and Natural Resources, *Geopolitics of Oil*, No. 96-119, Washington, D.C., Government Printing Office, 1980.
———, Committee on Foreign Relations, *Arms Sales Package to Saudi Arabia—Part 2*, 97th Cong., 1st Sess., 1981.
———, *Fiscal Year 1980 International Security Assistance Authorization: State Department Briefing on the Situation in Yemen*, 96th Cong., 1st Sess., 1979.
———, *The Future of Saudi Arabian Oil Production*, Staff Report, 96th Cong., 1st Sess., 1979.
———, *Persian Gulf Situation*, 97th Cong., 1st Sess., 1981.
———, *The Proposed AWACS/F-15 Enhancement Sale to Saudi Arabia*, 97th Cong., 1st Sess., Staff Report, 1981.
———, *Saudi Arabia*, A report by Senator Mike Mansfield, 94th Congress, 1st Sess., October 1975.
———, *U.S. Arms Sales Policy*, 94th Cong., 2nd Sess., 1976.
———, *War in the Gulf*, 98th Cong., 2nd Sess., Staff Report, 1984.
———, *War in the Persian Gulf, The U.S. Takes Sides*, Staff Report, November, 1987.
U.S. Defense Security Assistance Agency, *Foreign Military Sales, Foreign Military Construction Sales and Military Assistance Facts*, Washington, D.C., Government Printing Office, various years.

U.S. Department of Defense, *Soviet Military Power*, Washington, D.C., Government Printing Office, various years.

———, *Foreign Military Sales, Foreign Military Construction Sales and Military Assistance Facts*, September 1984 .

———, *Saudi Arms Sale Questions and Answers*, February 24, 1986.

U.S. Department of Energy, *Secretary of Energy Annual Report to the Congress*, DOE-S-0010(84), September, 1984.

———, *Energy Projections to the Year 2000*, DOE/PE-0029/2, October, 1983.

———, *Annual Reports to Congress*, Washington, D.C., Government Printing Office, various editions.

———, *Petroleum Supply Monthly*, various editions.

———, *World Energy Outlook Through 2000*, April, 1985.

———, International Affairs, *International Energy Indicators*, DoE/IA-0010, Washington, D.C., Government Printing Office, various years.

U.S. Department of State, Bureau of Public Affairs, *Afghanistan: Three Years of Occupation*, Special Report no. 106, Washington, D.C., December 1982.

U.S. Energy Information Administration, *International Energy Annual* Washington, D.C., DOE/EIA-02 (84).

———, *Impacts of World Oil Market Shocks on the U.S. Economy*, DOE/EIA-0411, July, 1983.

———, *Monthly Energy Review*, Washington, D.C., Government Printing Office, various editions.

———, *International Energy Annual*, Washington, D.C., Government Printing Office, various editions.

U.S. General Accounting Office, *Perspectives on Military Sales to Saudi Arabia*, Report to Congress, October 2, 1977.

U.S. Library of Congress, Congressional Research Service, Foreign Affairs and National Defense Division, *Saudi Arabia and the United States: The New Context in an Evolving "Special Relationship"*, Report prepared for the Subcommittee on Europe and the Middle East, Committee on Foreign Affairs, U.S. House of Representatives, 1981.

———, *Western Vulnerability to a Disruption of Persian Gulf Oil Supplies: U.S. Interests and Options*, 1983.

U.S. News and World Report, various editions.

Van Creveld, Martin, *Military Lessons of the Yom Kippur War: Historical Perspectives*, Washington Paper No. 24, Beverly Hills, Calif., Sage Publications, 1975.

Van Dam, Nikolaos, *The Struggle for Power in Syria*, London, Croom Helm, 1981.

Van Hollen, Christopher, "Don't Engulf the Gulf", *Foreign Affairs*, Summer 1981, pp. 1064–1078-

Volman, Daniel, "Commanding the Center", *MERIP Reports*, 14, 6/7 (July–September 1984), pp. 49–50.

von Pikva, Otto, *Armies of the Middle East*, New York, Mayflower Books, 1979.

Yegnes, Tamar, "Saudi Arabia and the Peace Process", *Jerusalem Quarterly*, No. 18 (Winter 1981).

Vertzburger, Yaacov, "Afghanistan in China's Policy", *Problems of Communism*, May-June 1982, pp. 1–23.

Wall Street Journal, various editions.

War Data, Special editions of the "Born in Battle" series, Jerusalem, Eshel-Dramit.

Washington Post, various editions.

Washington Times, various editions.

Weinbaum, Marvin G., *Food Development and Politics in the Middle East*, Boulder, Colo., Westview Press, 1982.

Weissman, Steve, and Herbert Krosney, *The Islamic Bomb*, New York, Times Books, 1981.

Wells, Donald A., *Saudi Arabian Development Strategy*, Washington, D.C., American Enterprise Institute, 1976.

Wenger, Martha, "The Central Command: Getting to the War on Time", *MERIP Reports*, 14, 9 (Fall 1984), pp. 456–464.

Whelan, John, ed., *Saudi Arabia*, London, MEED, 1981.

White, B.T., *Wheeled Armoured Fighting Vehicles In Service*, Poole, Dorset, Blandford Press, 1983.

Wiley, Marshall W., "American Security Concerns in the Gulf," *Orbis*, 28, 3 (Fall 1984), pp. 456–464.

Wittam, George H., "Political and Military Background to France's Intervention Capability", National Institute for Public Policy, McLean, Va., June 1982.

Witton, Peter, *UAE—10th Anniversary*, London, MEED, 1981.

Wohlstetter, Albert, "Meeting the Threat in the Persian Gulf", *Survey*, 25, 2 (Spring 1980), pp. 128–188.

Wolfe, Ronald G., ed., *The United States, Arabia, and the Gulf*, Washington, D.C., Georgetown University Center for Contemporary Arab Studies, 1980.

World Industry Information Service, *Energy Decade: A Statistical and Graphic Chronicle*, San Diego, Calif., 1982.

World of Information, *Middle East Review*, London, various years.

Yodfat, Aryeh Y., *The Soviet Union and the Arabian Peninsula: Soviet Policy Towards the Persian Gulf and Arabia*, New York, St. Martin's, 1983.

Zahlan, Rosemarie Said, *The Creation of Qatar*, London, Croom Helm, 1979.

Zelniker, Shimshon, *The Superpowers and the Horn of Africa*, Center for Strategic Studies, Tel Aviv University, Paper No. 18, September 1982.

Zuhair, Ahmed Hafi, *Economic and Social Developments in Qatar*, London, F. Pinter, 1983.

Index